CAMBRIDGE LIBRARY COLLECTION

Books of enduring scholarly value

Life Sciences

Until the nineteenth century, the various subjects now known as the life sciences were regarded either as arcane studies which had little impact on ordinary daily life, or as a genteel hobby for the leisured classes. The increasing academic rigour and systematisation brought to the study of botany, zoology and other disciplines, and their adoption in university curricula, are reflected in the books reissued in this series.

The Naturalisation of Animals and Plants in New Zealand

George Thomson (1848–1933) was born in Calcutta, grew up in Scotland and emigrated to New Zealand at 20. He settled there, working as a teacher and analytical chemist, and was eventually elected to the House of Representatives in 1908. Thomson had an interest in natural history, but he was especially fascinated by the biological battles between native species of plants and animals and more recent arrivals. Realising New Zealand's unique advantage in having written records about the introduction of new species from the period of Captain Cook's second voyage in 1773 onwards, Thomson was able to trace the origins and spread of many plants and animals. This study, published in 1922, notes their locations and dates, and includes lists of foreign species officially designated as pests. It is a comprehensive guide to the non-native flora and fauna of New Zealand and provides valuable information about the country's ecological history.

Cambridge University Press has long been a pioneer in the reissuing of out-of-print titles from its own backlist, producing digital reprints of books that are still sought after by scholars and students but could not be reprinted economically using traditional technology. The Cambridge Library Collection extends this activity to a wider range of books which are still of importance to researchers and professionals, either for the source material they contain, or as landmarks in the history of their academic discipline.

Drawing from the world-renowned collections in the Cambridge University Library, and guided by the advice of experts in each subject area, Cambridge University Press is using state-of-the-art scanning machines in its own Printing House to capture the content of each book selected for inclusion. The files are processed to give a consistently clear, crisp image, and the books finished to the high quality standard for which the Press is recognised around the world. The latest print-on-demand technology ensures that the books will remain available indefinitely, and that orders for single or multiple copies can quickly be supplied.

The Cambridge Library Collection will bring back to life books of enduring scholarly value (including out-of-copyright works originally issued by other publishers) across a wide range of disciplines in the humanities and social sciences and in science and technology.

The Naturalisation of Animals and Plants in New Zealand

GEORGE M. THOMSON

CAMBRIDGE UNIVERSITY PRESS

Cambridge, New York, Melbourne, Madrid, Cape Town, Singapore,
São Paolo, Delhi, Dubai, Tokyo, Mexico City

Published in the United States of America by Cambridge University Press, New York

www.cambridge.org
Information on this title: www.cambridge.org/9781108108317

This edition first published 1922
This digitally printed version 2011

ISBN 978-1-108-10831-7 Paperback

THE NATURALISATION OF ANIMALS & PLANTS IN NEW ZEALAND

THE NATURALISATION OF ANIMALS & PLANTS IN NEW ZEALAND

BY

HON. GEO. M. THOMSON
M.L.C., F.L.S., F.N.Z.Inst.

CAMBRIDGE
AT THE UNIVERSITY PRESS
1922

IN MEMORY

OF

MY YOUNGEST SON

JOHN HENRY THOMSON

WHO GAVE HIS LIFE IN THE SERVICE OF THE EMPIRE, AND AFTER THREE AND A HALF YEARS OF ACTIVE SERVICE IN GALLIPOLI AND FRANCE, DIED ON 5 APRIL, 1918

⊕

HIS BODY WAS LAID NEAR DOULLENS IN FRANCE

HIS SOUL IS WITH HIS GOD

PREFACE

A HISTORICAL account of the introduced animals and plants of New Zealand has long been a felt want in this country. Changes had been going on for the last century and a half, but records and references to these changes were much scattered, and it was very difficult for many persons interested in the natural history of the country to acquire any exact knowledge of the subject. This has been one of the reasons which induced me to accumulate the facts recorded here. The work has led me into a very large correspondence, but I have been gratified by the interest manifested by those appealed to, and by their readiness to assist me. The whole question of naturalisation appeals to most intelligent persons, and my efforts to elicit information have been most pleasantly received, and readily seconded on all sides.

To secure accuracy as far as possible, especially in connection with those groups of animals and plants with which my acquaintance was very imperfect, I sought and most ungrudgingly received the cooperation of local specialists, and I desire here to acknowledge my deep debt of gratitude to these gentlemen, who have checked my lists and supplied me with many of the facts recorded. They include the late Major Broun of Auckland who went over the Coleoptera; Messrs G. V. Hudson of Wellington, A. Philpott of Invercargill, G. Howes of Dunedin, and D. Miller, Government Entomologist, who dealt with Insecta generally, and the last-named especially with the Diptera; Mr G. Brittin, late of Christchurch, the Coccidæ; Dr Reakes, Director of Agriculture, the Trematode, Cestode and Nematode parasites of our imported animals; and Professor Benham, F.R.S., of Otago University, the Oligochætes. These gentlemen have also given me much valuable general information.

Invaluable assistance has been afforded me in regard both to introduced animals and plants by Mr T. F. Cheeseman of Auckland; by Mr W. W. Smith of New Plymouth, whose experience as a field naturalist is second to none in the Dominion; by Mr B. C. Aston, chemist of the Agricultural Department, who also is a most observant naturalist; by Dr F. Hilgendorf, of Lincoln Agricultural College; by Dr C. Chilton, Rector of Canterbury College, Christchurch; and by Mr A. Cockayne, Biologist of the Agricultural Department. My old Otago friends and fellow-workers, Dr D. Petrie, now of Auckland,

and Dr L. Cockayne, F.R.S., now of Wellington, have contributed much valuable information in regard to plant life. Mr F. L. Ayson, Chief Inspector of Fisheries, has assisted me very materially in bringing the knowledge of introduced fishes up to date.

In addition to all these I take this opportunity of expressing my indebtedness for facts and suggestions to Messrs Edgar F. Stead, Elsdon Best, Chas. Hedley (of the Australian Museum, Sydney), James Drummond, T. W. Kirk, the late Henry Suter, my sons Dr W. M. Thomson, Dr J. Allan Thomson and Mr G. Stuart Thomson, and to a large number of valued correspondents whose names are recorded in the following pages.

This work has given me a great amount of pleasure in the preparation, and I trust it will prove both interesting and useful to its readers.

G. M. T.

DUNEDIN,
August, 1921.

CONTENTS

PART I

INTRODUCTION AND HISTORICAL RECORDS

PART II

NATURALISATION OF ANIMALS

PART III

NATURALISATION OF PLANTS

PART IV

Part I

INTRODUCTION AND HISTORICAL RECORD

Chapter I

INTRODUCTION

THE naturalisation of animals and plants in any country is a most interesting and fascinating subject, as well as being one of very great and far-reaching importance. In the present work I have endeavoured to state what is known of the subject, as far as it relates to New Zealand. I have stated the facts regarding the first introduction of every species into the country, as far as these can be ascertained, and its subsequent success or failure in establishing itself.

In gathering the information required and working out the material, it was soon evident to me that the subject was unique. It had never been attempted before—as far as I am aware—for any country. Indeed it was seen that New Zealand was the only country in which such a bit of history could be attempted with any prospect of success. The islands forming the group lie isolated at a great distance (over a thousand miles) from any other extensive land area. We possess a fairly accurate record of what was here when Europeans first visited these shores, and we have been able to follow the later introductions of new species with a certain measure of success. The missing records and the blank pages are very numerous, but they do not vitiate the general accuracy of these statements.

I first approached this subject from the point of view of natural selection and (in Chap. XIII) have given an outline of the reasons which led me to investigate this question. But while the biological question of the origin of species was the *raison d'être* of this work, there are other aspects of the study which are of importance.

Thus the generation of people now growing up in this country is living under conditions which are largely different from those which prevailed when the first settlers colonised the islands. The surroundings at present are partly determined by the primitive conditions, and partly by the introduction of many new animals and plants. Both the face of the country and its inhabitants have been largely changed, but hitherto no connected account has been available of the agencies which have brought about these profound changes. It is important then that such an account should be prepared, because every year as it passes makes it increasingly difficult to gather the materials. Then the educational value of the knowledge is considerable. The first generations of settlers have already passed away, leaving only isolated records behind them. The generation now passing witnessed the great outburst of acclimatisation zeal in the sixties, but it also failed to keep good records. The acclimatisation societies themselves were very careless in the matter. The Auckland Society has a lapse apparently of some 20 years in its history; the record is somewhere, but it is not available. Nelson has entirely lost its early records; it was one of the earliest societies to enter on the work of introducing new forms of animal life, yet no one seems to have thought it worth while to preserve a complete report of its doings. If such exists it has not been forthcoming. Otago has kept a complete record, but neither the society itself, nor any of its members can show a full set, and some annual reports are missing. And so on with many other societies. The information, therefore, which has been accumulated in this work has been gathered piecemeal. But by so putting it together, it will be possible to make a fresh start in regard to the present position, and any further additions to the fauna or flora can be noted and added to the lists now prepared.

An important consideration is the practical value of such a statement as is presented in this work, in shaping the future policy of acclimatisation. It has hitherto been carried on in the most haphazard and irresponsible manner, districts, societies and individuals acting quite independently of, and often in direct opposition to, one another. One district protects hawks because they destroy rabbits and small birds; another destroys them because they attack game. One district imported stoats and weasels in order to cope with the rabbit pest; another destroyed them wherever found because they threatened the total destruction of the native bird life. There has been no settled policy. This has largely been due to the total failure of the community to grasp the scientific aspect of the question, or even to realise that it has a scientific side. This consistently British attitude towards things scientific (which it is to be hoped the war will largely modify, and in part dispel) has led to neglect of ordinary precautions in nearly all past acclimatisation experiments. Even as late as 1916 several of the societies were contemplating the contribution of a jointly raised sum for the purpose of introducing Australian swallows into the country, presumably to cope with some aspect of the insect trouble. Apparently no biologist was consulted in connection with the proposal. No one seemed to think it worth while to ascertain what was known as to the life-histories of the Australian swallows, for instance as to what insects they fed upon, or whether the birds were migratory and would stay in the country, if introduced. No particular species was pointed out as the desirable one, indeed it is doubtful whether any one of those who were responsible for recommending the step knew one species from another. Further, no one seemed to know that specimens of at least two species of Australian swallows (the Australian Tree Swallow (*Pterochelidon nigricans*) and the Australian Swift (*Cypselus pacificus*)) visit our shores nearly every summer, and that natural agencies have been trying to achieve on a very large scale what some of our acclimatisation experts proposed to do on a small scale with very little prospect of success.

Still more recently (1916–17) an animated discussion has been going on in Auckland as to the desirability of introducing the "stubble quail or partridge" (*Coturnix pectoralis*), as a sporting bird, some persons being keenly in favour of, others just as keenly opposed to, the step, on account of the harm the bird might do to the farmers. Apparently the species has been already introduced three times into the country, nearly fifty years ago, at Christchurch, Auckland and Hokianga, but it did not become established.

The whole history of acclimatisation efforts in New Zealand abounds in similar bungles and blunders, and while a certain measure of good has been achieved—notably in stocking our nearly empty rivers and lakes with fine food- and sport-fishes, yet the record of harm done is enormously greater. So-called acclimatisation societies to-day are only angling and sporting clubs, and it is a question whether the whole control should not be taken up by the Government. At any rate the public wants education on the question, and this work is a contribution towards this aspect of it.

On entering on this task I did not realise how vast it was, and how fragmentary was the sum of the existing knowledge, but having commenced it, I had no thought of turning back, or of abandoning the project. Even if the record be imperfect, it will be of some use to future workers to have pieced together the available material.

In writing some account of the introduced animals I at first thought of confining my attention to mammals, birds and fishes, but this seemed so inadequate that I went on from group to group until I found that my list included over 600 species, commencing with the Marsupials and ending with the Medicinal Leech. The line had, however, to be drawn somewhere, so I have left the microscopic forms for some specialist to deal with. Having launched out on the subject, it seemed inadvisable to stop at the animals, and therefore, having some bowing acquaintance with the floras of Britain, North America and Australia, in addition to that of New Zealand, in due course I added the introduced plants to my previous lists. The two groups can hardly be separated in this connection, and on account of their inter-relations it is best to study them together.

This work does not purport to be merely a list of naturalised animals and plants. I have recorded the introduction of a great number of species which have not succeeded in establishing themselves, though in some cases repeated attempts were made to naturalise them. The reasons for these failures are often so obscure that no plausible explanation has yet been given. For example the greenfinch and the chaffinch have thriven remarkably, the allied linnet has quite failed. Among fishes, the Pacific-coast Salmon (*Onchorhynchus Quinnat*) has become strongly established on the east coast of the South Island; while all attempts to naturalise the Atlantic Salmon (*Salmo salar*), though carried on unceasingly for half a century and in half a hundred different streams, have absolutely failed. The different attempts made are recorded under the various species, and such reasons as can be suggested for failure are also recorded. It seems to me that the failure of a species to become established in a new country into which it has been introduced, under what appear to be most favourable conditions, is as important a biological problem as the success of another species, and that the causes of the failure are worthy of examination.

In order that the various species referred to in this work might be recognised with a minimum possibility of mistake, I found it necessary to adopt some authoritative and readily-accessible scheme of classification and nomenclature. It was impossible to go into all the niceties (or obscurities) of zoological and botanical nomenclature; all that appeared to be essential was that the species referred to should be readily recognisable. Accordingly for the introduced animals I adopted, as far as possible, the schemes used by the various authors of the *Cambridge Natural History* (Macmillan & Co., London, 1895–1909); and for the plants the *Manual of the New Zealand Flora* by Mr T. F. Cheeseman (Wellington, 1906).

A considerable, indeed the major portion of this work is necessarily a compilation, but the information has been secured only by a laborious examination of all the available literature on the subject, and by very extensive correspondence. There is no doubt a great deal of information buried in the columns of the daily press of old days, which I have not been able to consult except in isolated instances. An immense amount of sifting of the wheat from the chaff has also been necessary, for a vast deal of the information communicated to me in all good faith was manifestly unreliable and had to be received with caution. I have endeavoured to secure scientific accuracy, so that the record may be of use to succeeding naturalists; at the same time I trust it may not be of the dry-as-dust type. The work has been a labour of love, and will, I hope, be found of use and interest to many who do not profess to be naturalists, but who are interested in natural phenomena.

An important aspect of the question is the legal one. A study of all the legislation which has been passed, first by the various provincial legislatures, and later by the Government and Parliament of New Zealand is extremely interesting from many points of view, and I have added this at the end of this work.

Chapter II

HISTORICAL RECORD

THE history of the naturalisation of animals and plants in large island areas has never, to my knowledge, been fully studied anywhere. Isolated introductions have frequently been dealt with, especially in recent cases, but apparently no one has sought to work out the history of the whole of the introduced fauna and flora of any country. The reason almost certainly is that, with one notable exception, the beginnings of the introductions could never be ascertained. The one exception is New Zealand. Here we have an area of land of very considerable extent lying far away from any other large areas, in which the first introduction of a majority of the species which now occur and are not indigenous to the country, can be traced. We can tell when and how many of the species which are now so abundantly represented first came into the country. We can learn of numerous attempts to introduce species which have, however, failed to establish themselves.

On the other hand we find that a vast number of species, both of animals and plants, have found their way into the country, as it were, by chance. We do not always know with certainty where they came from, though we have a knowledge of their geographical distribution which enables us to form a fairly correct impression. We often cannot tell the time of their introduction, nor the means by which this was accomplished. The most we can do—and even this is not always possible—is to record the first notice of their appearance in the country and their subsequent history.

The first date which we can fix upon as that at which a definite introduction of new species commenced is that of the arrival of Captain Cook in New Zealand on his second voyage, in 1773, when he landed at Dusky Sound, and later at Queen Charlotte Sound. On these occasions besides leaving various animals, he sowed several kinds of European seeds, mostly garden vegetables. Some of these are known to have survived.

Previous to that date the native inhabitants had brought with them from Polynesia, and perhaps from Melanesia, certain species of plants which they cultivated, and apparently also they had carried with them a species or rather a variety of dog. Unintentionally also they probably introduced the Polynesian rat (*Mus exulans*), as well as at least one species of flea—probably *Pulex irritans* (some think two species). Mr Best considers the Europeans are responsible for the introduction of the fleas. According to Maori tradition two species of louse (*Pediculus*) were also introduced by Polynesian immigrants. Mr Cheeseman has pointed out that the Polynesians were great cultivators, and carried their cultivated plants from one part of the Pacific Ocean to another. He considers that they knew of the existence of New Zealand, of the occurrence of greenstone and of the moa, and that their migrations were not accidental, but were conducted on definite principles.

While it is not possible to fix even approximately the date of introduction of any of the species of animals and plants which occurred wild in New Zealand in 1773, and which were common to this country, and any other land areas, it is advisable to take a brief survey of these common species and see from what region the most recent introductions before that date appear to have come.

To begin with, it must be borne in mind that the introduction of living organisms has been going on continuously throughout all the ages during which New Zealand has existed as a distinct land-area, and that the process still continues naturally. It is impossible to arrive at any accurate testimony of the results of this process, but certain considerations point to its existence.

Of the two bats which occur in New Zealand the Long-tailed Bat (*Chalinolobus morio*) is also found in South-eastern Australia; the other belongs to an endemic genus, *Mystacops*.

The bird-fauna contains a number of endemic genera and species, the affinities of many being obscure. Of those which belong to readily recognised types of land birds, the majority have affinities with the Australian avifauna, but as Hutton has pointed out, only with that section of it which is allied to that of Malaysia.

The lizards do not help us here, for, excluding the Tuatara, which is a survival from archaic times, they belong to genera of very wide distribution, and are probably of very considerable antiquity. As regards the relationships of the land and fresh-water mollusca, Hutton, as far back as 1883, stated that "our closest connection appears to be with North Australia, but there is a considerable generic affinity with the faunas of New Caledonia, Polynesia and South America."

Taking Suter's *Manual* as our guide, we find that there are 34 genera of land and fresh-water mollusca in New Zealand. Of these 13 are confined to these islands; three range into Tasmania and Australia, but no further; 13 are found in Australasia, but are more or less widely distributed outside the region; while five range into the Pacific, but are not Australian. Closer analysis bears out the general accuracy of Hutton's generalisation.

These 34 genera are represented by 236 species, all but one of them being endemic, viz., *Ophicardelus australis*, which is also found in Tasmania, Australia and New Caledonia. *Planorbis corinna* which is world-wide in its distribution is precinctive to New Zealand, but the genus *Planorbis* has a universal distribution.

Mr Suter informs me that "the genera of land mollusca which we have in common with Tasmania and Australia are far better represented in the former country, but disappear gradually as the north-east is reached. The affinities of our land and fresh-water molluscs are strongly marked on the line extending over Lord Howe Island to New Caledonia."

The relationship of New Zealand insects to those of other regions is dealt with in a number of papers scattered through many publications, but the knowledge of the subject is still very fragmentary. Meyrick, in his papers on Lepidoptera, favours the theory of introduction of several groups (e.g. Caradrinina) from South America *via* Antarctica. But leaving the general question and confining myself to species derived from the nearest present land surface, the following summary of the distribution of the Lepidoptera, for which I am indebted to Mr A. Philpott, is of interest:

Total number of species hitherto recorded in New Zealand ...		1040
Common to Australia and New Zealand		63
Cosmopolitan species?	24	
Introduced from Australia to N.Z., by shipping (say) ...	6	
Introduced to Australia from N.Z., by shipping (say) ...	3	
	—	33

Leaving for the question of origin, only 30 species, or say, 3 per cent. common to both countries.

These figures are not very conclusive one way or another.

The nearest land-surface of any extent is the continent of Australia and, as might be expected, immigrants from thence are by no means uncommon. Within the last score or so of years a great many species of Australian birds have been recorded as occasional visitants to New Zealand. The same remark applies to some of the stronger flying insects. This shows that though the fauna recognised as indigenous has originally been introduced from several directions in former ages, there has been and still is a constant stream of immigrants from Eastern Australia into these islands. The remarkable thing then is that there should be so comparatively little direct connection between the two countries so far as the fauna is concerned. The fact is that it is very difficult for a species of animal to establish itself in a new country, even assuming that many individuals arrive at the same time. The immigrants on arrival are certainly in an exhausted state and physically incapable of defending themselves from the assaults of enemies. The shores of the new land are patrolled by great numbers of gulls and similar predaceous birds, which would make short work of any travel-worn immigrants that landed and did not immediately find cover. The chances of getting food are also problematical. But even assuming that the individuals survived and throve, the chances of their finding mates are very remote; so that altogether the probabilities are against the establishment of the species. As a matter of fact they do not succeed. The only bird which appears to have come into New Zealand since the days of European settlement and to have established itself, is the Wax-eye or Blight-bird (*Zosterops cœrulescens*).

In taking a survey of the existing Flora of New Zealand in connection with its relationship to other plant-associations, and taking Cheeseman's *Manual* as my authority for the following figures, I desire to state at the outset that I do not attach too much importance to numerical comparisons, because I realise the enormously different values attached by systematists to different species. These values depend largely upon the personal equation, and further on the amount of detailed study given to any specified groups of organisms. There are certain genera of New Zealand plants which are apparently in a state of flux even at the present time. These have been submitted to close examination, a vast amount of material has been gone through, and in consequence innumerable differences have been recognised, and a large number of species defined. Such, for example, are *Ranunculus* with 37 New Zealand species, *Epilobium* 28, *Coprosma* 39, *Olearia* 35, *Celmisia* 43, *Senecio* 30, *Veronica* 84, and *Carex* 53. Many of these are sharply defined, easily recognised species, but for others the specific diagnosis is only the central rallying point for a large group of individuals showing considerable divergencies in many directions. I am safe in asserting that if similar detail were gone into with all the plants grouped under such common names as, for example, *Acæna microphylla*, *Gaultheria antipoda*, or *Pimelea lævigata*, and many others which might be named, it would be found that each deserves to be separated into a group of distinct species. Keeping this reservation in mind, we can still form an approximate estimate of the relationships shown by any given aggregation of species. Thus of the total number of 1396 species of New Zealand flowering plants recorded by Cheeseman, no fewer than 263 (or almost 19 per cent.) are also found to occur in Australia. Of these 134 occur both in Australia and Tasmania (eight in Tasmania alone), while the remaining 129 have a wider range, some being common tropical or sub-tropical weeds, while others are found throughout the temperate zone in both hemispheres. The endemic species, which do not range outside of New Zealand, number no less than 1069, or 76·6 per cent. of the whole (viz. 860 dicotyledons and 209 monocotyledons). This brings out the affinities of the remaining elements more strongly than ever, for it shows that of 327 species which are common to New Zealand and other countries, no less than 80 per cent. are also found in Australia. The remaining elements—Antarctic and Polynesian—are few as compared with the Australian.

It would appear from the above analysis that immigration of flowering plants from Australia into New Zealand has been going on steadily, and an examination of many of the individual species leads to the conclusion that much of it is quite recent. Thus of pappus-bearing composites, ten species are confined to New Zealand and Australia[1]; six more are found in New Zealand and Australia, but have a wider range outside[2].

No plant of South American, Polynesian or Antarctic affinity is furnished with a pappus. The list of Australian plants includes four species of *Epilobium*, furnished with pappus-like hairs on the seed; and 14 species of Orchids (out of a total of 53 species, the remainder being endemic) furnished with very minute light seeds which are easily carried by wind. These facts tend to show that species whose seeds can be distributed by wind are fairly abundant among those plants which are common to New Zealand and Australia, and the probability is that many were thus introduced into these islands[3].

I regret that I cannot give the date for the following interesting occurrence (I think it was about 1877), but it was so striking a phenomenon that it fixed itself in my memory at the time. It occurred in Dunedin in the autumn (February or March). One bright forenoon the sky became strangely overcast from the west, and the sun at midday assumed a coppery appearance. Some persons attributed the phenomenon to bush fires in the western districts, but no such fires were recorded anywhere in New Zealand. Others more accurately thought it was due to a smoke-cloud from Australia. This proved to be the case. Vessels voyaging between

[1] *Celmisia longifolia*, *Vittadinia australis*, *Gnaphalium traversii*, *G. collinum*, *Craspedia uniflora*, *Erechtites prenanthoides*, *E. arguta*, *E. quadridentata*, *Senecio lautus* and *Microseris forsteri*.

[2] *Gnaphalium japonicum*, *G. luteo-album*, *Picris hieracioides*, *Taraxacum officinale*, *Sonchus asper* and *S. oleraceus*.

[3] Linnean Soc. 30th Nov. 1916 (London). Using a wind-dispersal apparatus Mr Jas. Small, M.Sc., found that in a light air the fruit of *Senecio vulgaris* travelled at the rate of 1·6 miles per hour through the air, and of *Taraxacum officinale* 1·5 miles per hour.

Australia and New Zealand, and others passing up the east coast of Australia at the time, reported dense smoke-clouds from Gippsland and North-west Victoria, and also the falling of considerable quantities of ash and charred vegetable matter. The westerly winds drove the smoke right across the Tasman Sea, and at a distance of about 1200 miles it still exerted such an influence on the upper atmosphere as to make the whole sky lurid for a period of three or four hours. A wind which could carry such a body of smoke such a distance could probably easily transport seeds and spores, and though the usual course of the wind-currents is not so directly from west to east, yet such high winds apparently do occur, and that not unfrequently.

Another agency by which seeds are carried to oceanic islands is by means of birds, which bear them attached to their feet or plumage, and in some cases carry them in their crops. Darwin, Wallace and others have given numerous instances of this fact in plant distribution[1]. Apart from regular migrants which come to New Zealand every year from Australia, Polynesia and the Northern Hemisphere, a considerable number of stragglers are blown or stray over from Australia each year. The wonder, therefore, is not that Australian species of plants are met with in considerable numbers in New Zealand, but rather that they are not more common than is found to be the case.

As far as all truly indigenous species of animals and plants are concerned it is quite impossible to give dates for any which may have been introduced in long past ages, as for example those which are common, say, to New Zealand and Australia. But when we come down to recent times and reach the period of human immigration, it becomes possible to give some approximation to definite dates.

According to Maori tradition, New Zealand was discovered by two Polynesian voyagers named Kupe and Ngahue, but authorities are not yet agreed as to the period of this discovery.

The first Polynesian settlement in the time of Toi took place 30 to 32 generations, that is approximately 800 years, ago. On the arrival of these immigrants, they found the east coast, north and Taranaki districts occupied by the Mouriuri, Moriori or Maruiwi folk in considerable numbers, descendants of crews of three drift canoes, which had apparently come from the north-west. Whether these people had brought any animals or plants with them it is now impossible to say. According to east coast traditions, the Toi tribes had the Hue Gourd (*Lagenaria vulgaris*) in cultivation at an early

[1] Darwin in a letter to Dr J. D. Hooker in January, 1860, says: "Birds do not migrate from Australia to New Zealand," a curious error for such a good observer to make, and showing the danger of generalising from imperfect data. Many species regularly cross, notably the Shining Cuckoo and the Dotterel.

period, so that this plant was probably introduced from 24 to 30 generations ago, that is between 1150 and 1300 A.D. Communication was kept up with Polynesia for about 200 years more, new settlers coming over from time to time. The last batch of vessels, including the Arawa, Tainui and other canoes, arrived about 20 generations or 500 years ago, say about 1400 A.D. Reference has already been made to the introduction by some of these early voyagers, of the dog, the native rat, one or more species of flea, and two species of lice.

The Kumara (*Ipomœa batatas*) appears to have been introduced first about 1300 A.D., tradition saying that certain voyagers left Whakatane for Polynesia about that time, for the express purpose of bringing over that plant. Subsequent immigrants by the Aotea, Arawa, Tainui, and other canoes, also brought the plant. Indeed it is probable that it was continuously introduced by many of the new arrivals.

Mr Cheeseman in the *Manual* (p. 100) states in regard to *Pomaderris apetala*: "The Maoris assert that it sprang from the rollers or skids that were brought in the canoe 'Tainui' when they first colonised New Zealand." Mr Elsdon Best, to whom I referred this point, tells me that about 1879 he saw a grove of these trees

on a terrace near the mouth of the Mohakatina river. Local natives told him that the tree was called *Te Neke o Tainui* (the skid of Tainui), and that the grove had originated from the skids of the canoe Tainui, used in hauling the vessel ashore on her arrival here twenty generations ago, the skids having been brought from oversea. On my return to New Plymouth I met Mr Wilson Hursthouse, who, I found, was acquainted with the Maori name of the tree and the myth connected with it.

Pomaderris apetala is an Australian as well as a New Zealand species, but is not found in any part of Polynesia. It is difficult to conjecture, therefore, how such a myth could have arisen.

Perhaps about the same time, that is about 1400 A.D., the introduction of the Taro (*Calocasia antiquorum*) and the Ti (*Cordyline terminalis*) took place. One tradition says that they arrived in the Nukutere canoe, brought by one Roua, that is about 500 years ago. The same tradition narrates that the Karaka (*Corynocarpus lævigata*) was introduced at the same time, by the same individual. If so, it may have been brought from Western Polynesia by way of the Kermadecs, where it is a common tree. At the same time the genus is quite peculiar, and is endemic to New Zealand. If it did not originate in this country, then the home whence it came has lost it, for its botanical position and relationships are by no means clear.

After the arrival of the main migrations about 20 generations ago, there are no definite traditions of further Polynesian immigration, but voyagers left the shores of New Zealand for Polynesia as late as ten generations or 250 years ago, and presumably others arrived from time to time.

With the arrival of Captain James Cook in New Zealand we can begin to assign definite dates to many of the introductions.

In October, 1769, Captain Cook landed at Poverty Bay, and later at Anaura Bay, and at both places Messrs Banks and Solander made collections of native plants. He next stayed a week at Tolaga Bay, and 11 days at Mercury Bay. On 21st November a landing was made some miles up the Thames River, and then six days were spent at the Bay of Islands. On 16th January, 1770, he anchored in Queen Charlotte Sound, and made a stay of three weeks. Again on 27th March he was four days in Admiralty Bay to the west of Queen Charlotte Sound. There is no word in all these landings of his introducing any animals or any seeds, yet it is more than probable that Black Rats (*Mus rattus*), the common ship's rat, were on board the 'Endeavour,' and that some got ashore. It is also possible that some European seeds may have been accidentally introduced. The voyage was one for exploration only, as far as New Zealand was concerned, and the ships were quite differently equipped on later visits.

In December, 1769, only two months after Cook's arrival, De Surville spent three weeks in the 'Saint Jean Baptiste' in Mongonui Harbour.

In 1772 the French expedition under Marion du Fresne which had such a fatal ending as far as New Zealand was concerned, spent over two months in the Bay of Islands; and it is stated by both Taylor and Polack, I do not know on what authority, that Crow Garlic (*Allium vineale*), which is so abundant in that district, was introduced by him. No collections of plants were made during either of these French expeditions, but it is quite possible that some animals or plants found their way into the country.

Crozet, who took up the command of the expedition on Captain Marion's death, writes (in 1772):

> I formed a garden on Moutouaro Island, in which I sowed the seed of all sorts of vegetables, stones and the pips of our fruits, wheat, millet, maize, and in fact every variety of grain which I had brought from the Cape of Good Hope; everything succeeded admirably, several of the grains sprouted and appeared above ground, and the wheat especially grew with surprising vigour. The garden on Moutouaro Island alone was not sufficient to satisfy my desires. I planted stones and pips wherever I went, in the plains, in the glens, on the slopes, and even on the mountains; I also sowed everywhere a few of the different varieties of grain, and most of the officers did the same.

Captain Cook in his second voyage in the 'Resolution,' spent five weeks—from the 26th March to 1st May, 1773—in Dusky Sound, and while there cleared a piece of ground of about an acre in extent to make a garden, and sowed "a quantity of European seeds of the best kinds." No list of these seeds is given, though cabbages, onions, and leeks are mentioned, but they were in all probability the same sorts as were sown later at Queen Charlotte Sound. Apparently not one of them was able to establish itself in the moist climate of the Sound, and as predicted by George Forster in his *Journal*, the native vegetation quickly re-asserted itself, and obliterated all trace of the introduced plants[1].

That Cook hoped to introduce useful plants and animals into a country which he knew by his previous experience did not furnish much food for voyagers, is shown by his leaving geese at Dusky Sound, and *these were the first animals which were introduced of set purpose*. He had five geese on board his ship, and these were liberated at a spot which he called Goose Cove. This first experiment in acclimatisation, like hundreds of others made in later years, was quite unsuccessful, and nothing was ever seen or heard of the birds again.

Lieut. Menzies, the botanist of Vancouver's expedition in 1791, says:

> As Captain Cook had left five geese in this cove, we were in hopes of meeting with some of their offspring, and thereby partaking of the fruits of his benevolence, but as they were left in the autumn, I am apprehensive they did not survive the first winter, for not the least traces of any could be seen at this time about the cove, and though there was a scarcity of other birds on account of this being the season of incubation, yet it appears to be the most eligible place in the whole Sound for Game at a proper time of the year.

Meanwhile his colleague Captain Furneaux, in the 'Adventure,' had put into Queen Charlotte Sound on 7th April, 1773, and was joined there by Cook on 18th May. They stayed till 7th June, and then went southward in search of an antarctic continent. At the Sound, Cook liberated a ram and ewe he had brought with him from the Cape of Good Hope, but they were in a very bad state of health, and died very shortly after being landed. They were supposed to have eaten some poisonous plant.

Captain Furneaux landed a boar and two breeding sows, and turned them into the woods. They were not to be seen, nor were there any traces of them found the following year, but the members of the expedition thought that the animals had taken themselves off into the denser forest. When Cook came back in 1777 he could learn

[1] In a *Journal* of the voyage of the 'Endeavour' printed anonymously in 1771, it is stated at p. 58: "At Otaheite we had likewise planted many European seeds, of which none, except mustard, cresses and melons were found to vegetate."

nothing about them, so he gave the natives another boar and sow, with instructions not to kill them. It is probable that these original pigs were the ancestors of the long-nosed wild pigs which afterwards became so common in the South Island.

Cook also landed two goats, a male and a female, on the east side of the Sound, but there is reason to believe that the natives killed them. He gave them another pair in 1777, and it is popularly believed that most of the wild goats found in the South Island in the early days of settlement are descended from these.

In West Bay, Cook liberated some fowls, and though he could not find any trace of them when he visited the spot in October, 1774, yet in his later visit in February, 1777, he stated that "all the natives whom I conversed with agreed that poultry are now to be met with wild in the woods behind Ship Cove; and I was afterwards informed by the two youths who went with us, that Tuitou, a popular chief amongst them, had a great many cocks and hens in his separate possession."

During this stay of two months, ground was cleared at more than one spot, and numerous kinds of vegetable seeds were sown, including turnip, cabbage, white mustard, radish, purslane, peas, beans, kidney beans, parsley, carrot, parsnip, onion and leek: potatoes also were planted. Of these, cabbage, and apparently also turnip, onion and leek succeeded in establishing themselves; radishes seeded freely, but the peas, beans and kidney beans were eaten by rats. It is more than probable that some European weeds of cultivation were introduced at the same time.

On 2nd November when near Cape Kidnappers, Cook gave some pigs and fowls to natives who came off in a canoe, the first introduction of these two kinds to the North Island. On the following day he once more entered Queen Charlotte Sound, and waited till the 25th for his consort, but as she had not arrived by that time, he left for a cruise in the Antarctic Ocean. The 'Adventure' arrived in the Sound five days later, and remained over three weeks, during which time the unfortunate massacre of ten of her crew took place. After a long cruise in the Antarctic and Pacific Oceans, Cook returned to Queen Charlotte Sound on 19th October, 1774, and finally left for England on 10th November.

The important thing about this voyage, from our present point of view, is that Cook brought with him various animals and plants for the express purpose of introducing them, having experienced on his first voyage the lack of fresh food in the country. beyond that which the natives were able to supply them with. To this voyage we can assign the introduction and subsequent naturalisation of the pig and the goat and perhaps of fowls; and among plants, of the cabbage, turnip and potato. Other attempts to naturalise plants mostly failed.

Cook visited Queen Charlotte Sound again on his third and last voyage to the Pacific, entering it on 12th February, 1777, and leaving it on the 25th. There is no record of any attempts to introduce further species, except the pigs and goats previously referred to.

In 1791, Vancouver visited Dusky Sound, and Lieut. Menzies reported that in the garden (made by Cook eight years previously) there had grown up a dense covering of brushwood and fern, which completely obliterated all sign of the old clearing, and only the fact that its position was recorded and described enabled the spot to be identified.

In view of the struggle between indigenous and introduced plants which exercised the minds of many eminent naturalists, and to which reference is made in the writings of Hooker, Darwin, Wallace and others, the record of further visits to Dusky Sound is interesting.

The value of the seal and whale fisheries of Southern New Zealand soon drew enterprising sailors to these waters, and a wholesale destruction of these animals took place. Dusky Sound had been charted by Cook, its harbour was not only safe, but it provided abundance of fish, wood and water, hence it made a good rendezvous, and the base of a good hunting ground.

On 3rd November, 1792, the 'Britannia' from Sydney anchored in Facile Harbour, Dusky Sound, and landed a party of twelve sealers, with store of provisions, etc. These men were not relieved till September, 1793, when the 'Britannia' revisited the spot, and took them off. During the early part of the same year the Sound was visited by the Spanish corvettes 'Descuvierta' (commanded by Don Alexandro Malaspina), and 'Atrevida' (Don Jose Bustamente). I do not know how long they stayed.

Captain Raven of the 'Britannia' in reporting from Norfolk Island on 2nd November, 1793, says: "The animals I left had fed themselves on what they found in the woods, and were exceedingly fat *and prolific*." It would be interesting to know what animals these were, and whether any had gone wild, or had been left, or if they were all carried away again. Unfortunately we have no information on the subject.

On 19th September, 1795, the 'Endeavour,' Captain Bampton, of 800 tons, and the brig 'Fancy' of 150 tons, sailed from Sydney for India, and called in at Dusky Sound—perhaps to load some spars. They had no less than 244 people on board the two ships. There they found a small vessel, which the twelve men left by the 'Britannia' had built during their ten months' stay in the Sound, but which they had not taken off the stocks. Captain Bampton completed this little vessel, and called it the 'Providence.' On 19th January, 1796, the 'Fancy' and the 'Providence' arrived at Norfolk Island, and reported that the 'Endeavour' had been wrecked at Dusky Sound. She had been found utterly unseaworthy, and had been emptied, abandoned and sunk there. An enormous amount of stuff must have been carried ashore. Owing to the small size of the two remaining vessels, no less than 35 men had to be left behind, no doubt with abundance of stores. These derelicts were not rescued till May, 1797, when the 'Mercury' left Sydney for Dusky, picked them up, and landed them at Norfolk Island, after twenty months' detention in the Sound.

Sealing and whaling vessels continued to visit the Sound at intervals, and parties of men were certainly there in 1803, 1804 and 1805. I have myself gone down in much more recent years with sealing parties to the south, and have some notion of the equipment they used to carry. In addition to bags of flour, meal, sugar, etc., they nearly always carried considerable quantities of potatoes. During these fifteen or sixteen years referred to (between 1791 and 1805) many men lived on shore, often for lengthened periods, and almost certainly took with them large quantities of stores, which must have frequently contained seeds of many European weeds of cultivation. An example of this is shown in the case of four men (members of a sealing party) who were left on the Solanders for four and a half years, and were rescued in 1813. They had attempted to raise potatoes and cabbages, of which plants one of them happened to have some seed when they were unhappily driven upon the island, but the sea-spray rendered cultivation impracticable. In the same year ten men were rescued from Secretary Island, in Thompson Sound, who had been left there in 1809.

Yet it is an interesting fact that in the West Coast Sounds region practically no European plants are to be found, except on the Milford track, which has been much frequented by tourists in recent years.

A Sydney paper of 4th September, 1813, reports an interview with Captain Williams, who stated that "the natives of the coast of Foveaux Strait attend to the cultivation of the potato with as much diligence as he ever witnessed. He saw one field of considerably more than one hundred acres, which presented the appearance of one well cultivated bed." In 1824, De Blosseville of the 'Coquille,' writing from Captain Edwardson's report says: "Potatoes, cabbages and other vegetables introduced by the Europeans are grown." These southern natives had not seen pigs up to the time of Edwardson's visit; so he gave them some.

In 1826 the schooner 'Sally,' with a large number of immigrants, together with many cattle, sheep and other stock from England, called in at Stewart Island—presumably at Port Pegasus—and stayed for a period of three weeks. Apparently the 'Rosanna' also called with immigrants. She then went on to Hokianga, where a settlement was made, but Captain Herd and most of the settlers took fright and sailed for Sydney, only four men remaining.

In 1820 Major Cruise, who spent ten months in the north of New Zealand, says: "The excellent plants left by Captain Cook" (in Queen Charlotte Sound?) "viz., Cabbages, turnips, parsnips, carrots, etc., etc., are still numerous, but very much degenerated; and a great part of the country is over-run with cow-itch which the natives gave Marion the credit of having left among them." (I do not know what plant he refers to here.) "Water melons and peas were raised while we were in the country, with great success, and the people promised to save the seeds and sow them again. The missionaries have got some peach trees that bear very well, and an acorn and a seed of an orange were sown by a gentleman of the ship near Pomarrees village, and the place rigidly tabooed by the inhabitants." Cruise also reports that the natives (at Wy-ow Bay) brought a cat for them to cook and eat, which he remarks must have come from the shipping at the Bay of Islands or from the Coromandel.

In 1832, d'Urville—who spent four months on the coast of the South Island—found a gang of six men—sealers—working at Mason Bay, Stewart Island. In his visit in 1840, he entered Port Pegasus and learned that 20 English sailors had settled on the shores of Foveaux Strait, where they had married native women. They grew potatoes and various other vegetables, and reared fowls. They told d'Urville that as many as 20 vessels anchored in Port Pegasus annually.

In this same year (1840) Major Bunbury in his report on the proclamation of Stewart Island as Her Majesty's possession, says of Paterson Inlet, "the Europeans there employ themselves at boat building and in the culture of wheat and potatoes, with which they supply the whalers, as also with pigs and poultry."

Previous to this, Waikouaiti was one of the best known whaling stations on the Otago coast. In 1838 this was purchased from its Sydney owners by Mr John Jones, who two years later sent down several families to engage in farming and cattle raising, and at the end of 1840 the population of the settlement numbered about 100 persons. They had enclosed some 6000 acres of land, and had about 100 acres in crop; while the live stock numbered about 100 horses, 200 cattle, and 2000 sheep.

In 1840 also a small settlement was made where Christchurch

now stands, for the cultivation of wheat for certain Sydney mills. About 30 acres were grown, but the place was abandoned soon after on account of rats, difficulties of shipment, and fires.

In 1842 Captain Wm Mein Smith, chief surveyor of the Wellington Land Company, visited the south-east of Otago, and writes of one settlement there as follows:

> At Tautuku Bay (30 miles from Molyneux River) is a good deal of land cultivated by a number of industrious men who are, through the winter, engaged in the whale fishery. In the summer they are occupied in their gardens. They produce abundance of fine potatoes, and as much wheat and barley as they can consume. They have many pigs, goats, and a rapidly increasing stock of poultry.

It is quite probable that several of the European weeds of cultivation which are now so common in the south end of New Zealand were introduced in these days of early and casual settlement. But few animals would be thus brought in, except perhaps certain flies and other domestic insects, and perhaps some worms, wood-lice and such familiar accompaniments of human settlement.

Turning to the north of New Zealand, though the visitation was greater, the record has not been worked out so thoroughly as for the south. But from the end of the 18th century greater numbers of vessels visited northern ports for the whale fishery. Captain King, Governor of New South Wales, had landed in the Bay of Islands in 1793, and gave the natives some pigs, as well as wheat, maize, and no doubt other things not mentioned. The Rev. Samuel Marsden sent them wheat in 1810, and a further lot in 1811. When he visited the island in 1814, he brought with him the mission party, which was established at Kerikeri and Waimate near the Bay of Islands, and the live stock accompanying the party included one entire horse, two mares, one bull and two cows, with a few sheep and poultry. From this date onwards there is no doubt numerous introductions of plants and animals were made. In 1822 the Wesleyan Mission station at Kaeo-Wangaroa was established, but the party were driven out of there and shifted their ground to Hokianga. The occurrence of exotic historic trees of great size at the present day in these regions testifies to the activity of the missionaries as pioneers in this work of introducing new forms of life in the country. Then too, the quantity of flax, potatoes and other produce, exported from New Zealand and supplied to ships in these pre-settlement days, was very great, and this shows that there must have been much trade and intercommunication between the natives and the Europeans. Numbers of weeds and of animals must have been introduced into the north in this way. About 1826 the 'Rosanna' (already mentioned) with some

60 settlers on board, came into Hokianga with the intention of founding a settlement, but as a tribal war was being waged among the natives at the time, the party did not remain, but went off and landed at Sydney.

As I am not writing a history of New Zealand except in so far as it relates to the facilities which existed for the introduction of new forms of animal and plant life into the country, I must hurriedly pass over these pre-settlement days, merely pointing out that a great deal of communication must have been going on with outside ports from many parts of the country. The township of Russell or Kororareka in the Bay of Islands, was founded in 1830 by Benjamin Turner, an ex-Sydney convict, who built the first grog-shop there. Two years later the population numbered about 100, and in 1838 about 1000. "As many as thirty-six whalers were anchored there at one time, and in one year 120 vessels sailed in and out."

The first regular settlement scheme commenced in 1839 when the 'Tory' with Captain Wakefield, Dr Dieffenbach, and others, arrived in Port Nicolson, and after trying a site for a town near Petone, founded what is now Wellington. Early in the following year the immigrant ships began to arrive, and by the end of 1840 the population of Wellington numbered about 1100 persons.

The first official capital of New Zealand was Kororareka or Russell, but the seat of government was shifted to the Waitemata, and Captain Hobson selected the site of the future town there, which he called "Auckland," in September, 1840. The same year saw the commencement of the Taranaki settlement, and by the end of 1841, the population of New Plymouth numbered some 500 persons. In 1841 Nelson was founded, and in January and October of 1842, four vessels with some 850 passengers arrived in Nelson harbour. In 1848 the Otago settlement was founded and 278 immigrants were landed on the site of Dunedin. In 1843 the Deans brothers settled near the present site of Christchurch, but it was not till the close of 1850 that the pioneers of the Canterbury settlement, numbering 800 souls, landed in Port Cooper.

In the First Annual Report (for 1843) of the Agricultural and Horticultural Society of Auckland it is stated that the following trees were then in cultivation: peaches, nectarines, apricots, almonds, figs, lemons, oranges, olives, vines, plums, cherries, mulberries, pears, apples, quinces, walnuts, filberts, loquats, gooseberries, red and black currants, raspberries and strawberries; the Cape gooseberry (*Physalis edulis*) is said to be "almost indigenous; it grows wild in every part of the country."

In those early days of settlement voyages between Great Britain and the colony were long, extending from three to five months, and it must have been difficult to convey many animals on board the small ships which were the only carriers. But the immigrants occasionally brought out pets, especially cats and dogs, with probably fowls, pigeons, rabbits, canaries and other song birds. Certainly also they introduced most of the common weeds, such as chickweeds, thistles, groundsel, and others. I have more than once observed the plants which have grown up round a heap of ashes and rubbish where immigrants' old bedding and refuse were burned, and only regret now that I did not keep a record of the species at the time.

For several years the settlers were too busy founding homes and bringing their land into cultivation to attend much to any but the most essential things; but after about a score of years had passed, and there was time for leisure and reminiscence, new ideas came to them, or perhaps it is more correct to say, original ideas re-asserted themselves as they seemed to be capable of realisation.

The beginning of the rush of immigration dates from between 1840 and 1850, and the process has been continued with more or less intermission ever since. But in a general sketch of the subject of animal and plant introduction, we need not concern ourselves further as to dates; these will be given as far as possible in the case of each individual species. Here we are concerned only with the general result and its causes.

The early settlers of New Zealand found themselves in a land which, as far as regards climate and natural conditions, seemed to them to reproduce many of the best features of the homeland from which they came. They thought with affection and with the glamour of youthful remembrance of the lakes and rivers, the woods and the fields, the hills and the dells of that homeland. They recalled the sport which was forbidden to all but a favoured few, but which they had often longed to share in—the game preserves, the deer on the mountains or in the parks, the grouse on the heather-clad hills, the pheasants in the copses and plantations, the hares and partridges in the stubbles and turnip fields, the rabbits in the hedgerows and sandy warrens, and the salmon of forbidden price in their rivers—and there rose up before their vision a land where all these desirable things might be found and enjoyed. Their thoughts went back to the days when they guddled the spotted trout from under the stones of the burns and brooks, to the song birds which charmed their youthful ears, to the flowers and trees which delighted the eye. They recalled the pleasant memories of hours passed on the hills and in the woods of their beloved native land. Here, in a land of plenty, with few wild animals, few flowers apparently, and no associations, with streams almost destitute of fish, with shy song birds and few game birds, and certainly no quadrupeds but lizards, it seemed to them that it only wanted the best of the plants and animals associated with these earlier memories to make it a terrestrial paradise. So with zeal unfettered by scientific knowledge, they proceeded to endeavour to reproduce—as far as possible—the best-remembered and most cherished features of the country from which they came. No doubt some utilitarian ideas were mingled with those of romance and early associations, but the latter were in the ascendant. They recked not of new conditions, they knew nothing of the possibilities of development possessed by species of plants and animals which, in the severe struggle for existence of their northern home had reached a more or less stable position.

This wonderful wave of sentiment manifested itself especially in the sixties. From Auckland to the Bluff the people founded acclimatisation societies for the purpose of introducing what seemed to them desirable animals, and they allowed their fancy free play. In their private capacities they got their friends at home to send them seeds of the wild flowers they had loved, and they sowed these in all sorts of localities, wherever it seemed to them that they would grow. No biological considerations ever disturbed their dreams, nor indeed did they ever enter into their calculations. I have been on the council of an acclimatisation society, and I know the enthusiasm, unalloyed by scientific considerations, which animates the members. As far as flowering plants were concerned disappointment followed many of their efforts; the primroses and bluebells, the heather and the wood violets, refused to grow either in the bush or in the open country, and the sowers were frankly disappointed. Even when the seed was sown in the garden or the greenhouse and the plants were put out in the open, they would not reproduce their kind. Most of these early colonists recked not of such things as cross- and self-fertilisation, and those who did know were not prepared to recommend an insect invasion to secure the fertilisation of their favourite wild flowers.

In time some of the plants and animals which had been introduced not only established themselves securely, but increased at a rate which upset all calculations. Conditions were produced which had never been anticipated and the introductions became dangerous and expensive pests. Then public measures had to be taken to check the newcomers, and in some cases their natural enemies had to be introduced. This has led to further complication and unexpected results. These natural enemies, like the things they were meant to check, did not always do what was expected of them; they frequently failed to achieve the purpose for which they were introduced, and took to destroying things which it was desirable should be preserved. Legislation had to be resorted to in order to destroy some introduced things and to protect others. Noxious Weeds Act, Animal Protection Acts, Injurious Birds Acts, and so on, have been passed into law, together with countless Regulations and Orders in Council dealing with the same subject in its multifarious aspects. By way of commentary and satire on the whole business, the Government in many cases is itself the chief offender against the laws of its own making.

At the close of nearly 150 years since Cook first visited these shores, the country has not yet realised the necessity of a scientific treatment of the whole question of naturalisation. Species are still being introduced. In nearly every case now it is claimed that this is done for beneficent purposes, but the same argument justified the early settlers who introduced insectivorous birds to eat up the caterpillars which were destroying their grain crops, no doubt also the sheep farmers who helped to bring in stoats and weasels to enable them to grow wool and mutton, instead of rabbits. There is still no general principle underlying the work, and not sufficient knowledge of the possibilities of each problem.

Part II

NATURALISATION OF ANIMALS

In Chapters III to VI species which have become thoroughly established are distinguished by an asterisk.

Chapter III

MAMMALIA

Of the 48 species of Mammalia which have been introduced into New Zealand, 44 have been brought in purposely by human agency, and four accidentally. The latter are the mouse and three species of rats, but one of the latter, the Maori rat (*Mus exulans*), has been exterminated since European settlement began.

The following 25 species are truly feral at the present time in certain districts, some in limited areas, others very widely distributed: wallaby, common opossum, sooty opossum, pig, horse, red deer, fallow deer, Sambur deer, wapiti, white-tailed deer, moose, cattle, sheep, goats, chamois, cat, ferret, stoat, weasel, black rat, brown rat, mouse, rabbit, hare and hedgehog. The following three have been somewhat recently introduced, but cannot be said to have been naturalised yet: Japanese deer, black-tailed deer and thar.

The classification adopted in the succeeding list is that used by Frank E. Beddard in the *Cambridge Natural History*, 1902.

Order MARSUPIALIA

Family Macropodidæ

Apparently about 12 species of marsupials have been introduced into New Zealand at various times, but only three species have established themselves and become feral. These are a wallaby and two species of phalangers, which are popularly known as opossums. Those who introduced them knew little or nothing about the exact relationships or the systematic position of these animals and no one seemed to have thought it worth while to identify them. The information about them is, and always has been, very vague; they were introduced by acclimatisation societies, private individuals and dealers, under various popular names, as kangaroos, bush kangaroos, wallabies, rock wallabies, etc., but the importance of knowing and recognising their specific distinctness with all that this involves in difference of habits, never troubled the introducers.

*Common Scrub or Black-tailed Wallaby (*Macropus ualabatus*)

In 1867 the Auckland Society had three wallabies in their gardens, and a fourth was added in 1874; but there is no possibility of identifying the species, and there is no record of what came of them.

In the same year A. M. Johnson brought over some from Tasmania for the Canterbury Society. A Christchurch newspaper dated April, 1870, says:

> The merit of the introduction into Canterbury province of the brush-kangaroo of Tasmania is due to Captain Thomson, and from the thriving condition of those in the Society's gardens, their adaptability to the province has been proved, whilst their increase has been such as to now render their liberation desirable in suitable localities.

I cannot help thinking that this is the species which Mr Michael Studholme either imported direct from Tasmania, or bought from the Canterbury Society, and liberated at Waimate, South Canterbury. There they have increased to an extraordinary extent. Mr E. C. Studholme writing to me in February, 1916, says:

> I can just remember seeing them turned loose here, two does and one buck being the number liberated. For a week or two they hung about the homestead, after which they were not seen for about two years, when some one sighted them on the hill near Waimate Gorge. They gradually spread along the adjoining hills, and are now to be found as far north as Bluecliffs. It is very hard to estimate the number there are at the present time, but it is quite safe to say there are thousands of them. Parties which go out shooting have killed as many as seventy in a day or two. They live chiefly in the bush, scrub, and fern about the gullies and gorges, coming out in the evenings to feed in the open ground. Their food chiefly consists of grass, but they are very hard on certain trees, barking many of them, particularly the Ohaus or five-leaf (*Panax arboreum*). There are well-defined tracks through all the bushes and scrub they frequent, much on the lines of pig tracks. I understand they are quite easy to snare, a good many being caught in that manner. If not kept in check they would, no doubt, become a great nuisance to farmers. Some years ago I sent the late F. C. Tabart of Christchurch (who was a Tasmanian) one for eating, and he wrote me saying it was a delicacy. Personally I have never eaten the meat, but the tails make very good soup. The skins of those taken in Winter make splendid rugs, being very heavy in fur, and they are much sought after. I believe they are not a wallaby, but scrub-kangaroo, as they are quite large, some of the old bucks weighing over 60 lbs.

About 1870, Sir George Grey introduced a number of species of marsupials into the island of Kawau, and among these was a wallaby (there is no record of where it came from) which increased in an almost incredible manner. Colonel Boscawen informs me that these animals have all been killed off, "except the small brown rock-wallaby, of which very few are now left." This latter species (*Macropus ualabatus*) was also imported to Auckland by Mr John Reed, who liberated them on Motutapu Island, where they are still common. They also crossed the narrow neck of land to Rangitoto Island, where they found a haven of rest, and where they are now abundant. Colonel Boscawen says: "The Wallaby furnishes great sport in shooting, and it is harder to hit than a rabbit, as when driven the animal does not hop, but goes on all fours and dodges from side to side, running at a great rate." Mr Cheeseman tells me that when the Island of Kawau was sold, the new owners encouraged shooting parties to go down—indeed contracts were let to kill the marsupials off the island—and the slaughter was great. One informant, whose name I have lost, told me that even in Sir George Grey's time, as many as two hundred wallabies would be killed in a battue. This gentleman considered them to be useless creatures, fit neither for food nor fur. The consensus of opinion is that the flesh is not particularly attractive, but that the tails make excellent soup. This same informant told me that at Kawau they ate out most of the vegetation, and starved out most of the other animals, being assisted in this by the hordes of opossums. They came out at nights in the fields, grazing like sheep, and in the summer went into the garden, stripping it of fruit and vegetables.

There are still a few left about Kawau, not more than a dozen or two, according to Colonel Boscawen.

Pademelon Wallaby (*Macropus thetidis*)

The Auckland Society had some specimens of this species in 1869, but the number is not specified, nor what came of them.

Kangaroo (*Macropus species*)

Under this name various animals were introduced and liberated, but it is quite impossible to identify the species.

Note. du Petit-Thouars, who visited New Zealand in 1838, says in the account of his voyage (p. 115): "Kangaroos have multiplied very well, but it is much to be regretted that there, as in New Holland, the colonists have not taken the trouble to look after them and increase their numbers, instead of leaving them to perish." I have no idea what animals he is referring to.

The Canterbury Society received a pair of kangaroos from the Rev. R. R. Bradley in 1866, and in 1868 a single large specimen from Sir George Grey. The Society's *Report* for 1872 states that there were "about 15" in the gardens, but no further information is vouchsafed.

The Otago Society introduced one specimen in 1867, and apparently others were privately introduced but not recorded, for the late Mr F. Deans (Curator of the Society) wrote me in 1890:

I do not know when these were liberated, but in 1869 I saw one on several occasions where the Northern Cemetery (Dunedin) now is; he went bounding out of that gully while I was passing down to my work. I heard of one or two having been killed by dogs in the gully above the rifle range.

In 1868 Mr Christopher Basstian liberated three specimens on the Dunrobin Station, but nothing was heard of them afterwards. In the same year the captain of a vessel brought three kangaroos to the Bluff, one male and two females. These were purchased by the Southland Acclimatisation Society and liberated on the range of hills there. Nothing further was ever heard of them.

Wallaroo or Euro (*Macropus robustus*)

Some of these kangaroos were introduced into Kawau by Sir George Grey in the sixties, but there is no subsequent record of their occurrence there.

Rock Wallaby (*Petrogale xanthopus?*)

In 1873 the Auckland Society received a rock wallaby from Sir James Fergusson, which was quite distinct from any previously recorded, and which it is surmised belonged to the above species. There is no later report of it.

Kangaroo Rat (*Potorous tridactylus*)

The Auckland Society introduced this species in 1867, but no later report of the Society mentions them.

Family PHALANGERIDÆ

*Common Opossum, Grey Opossum, Brush-tailed Opossum, or Vulpine Phalanger (*Trichosurus vulpecula; Phalangista vulpina*)

*Sooty Opossum (*Trichosurus fuliginosus*)

The Australian and Tasmanian phalangers, or, as they are popularly called "opossums," which are now so common in many forest-covered parts of New Zealand were first introduced into Southland by private individuals, and a few later on into other districts by some of the societies. Details of these early introductions are somewhat inexact and difficult to obtain. One report (Wellington Acclimatisation Society, 1892) says:

These animals were first liberated in the bush behind South Riverton in 1858 by Mr Basstian. Some years after, one or two opossums (presumably Australian Grey Opossums) escaped from confinement in the same neighbourhood. In 1889 they were found to have increased enormously.

Mr T. D. Pearce of Invercargill writes (2nd August, 1915): "The opossums in Southland owe their origin, not to the Council, but to private enterprise. They were liberated between 1865 and 1868 in the Longwoods by Mr Christopher Basstian, who brought them from Victoria or Tasmania." Mr J. L. Watson of Invercargill writes (October, 1890): "One pair were liberated by the late Captain Hankinson at Waldeck, Riverton, in 1875 or 1876. They have increased marvellously and are plentiful in the South Longwoods." This, no doubt, refers to a later introduction, and Mr C. Basstian was evidently the first person to liberate them in New Zealand. The Auckland Society imported some (number not stated) in 1869; five more in 1874–75; and four more in 1876. There is no record as to where they were liberated. Most of these came through Sir George Grey; who liberated several grey opossums on Kawau.

In 1892 the Wellington Society obtained 19 black Tasmanian opossums (*T. fuliginosus*), and liberated them on the ranges behind Paraparaumu.

In 1895 the Otago Society obtained 12 silver-grey opossums (*T. vulpecula*) from Gippsland, and liberated them in the Catlins district.

This appears to complete the record of introductions.

By 1890 these animals had increased to a great extent in the forest covering the Southern Longwood Range, and the Southland Society caught and distributed in that year some 236 to "the Auckland Islands, Stewart Island, various districts of Otago (including the Te Anau and West Coast Sounds region), North and South Canterbury, West Coast of South Island, Nelson, Wellington and Gisborne." In succeeding years more were obtained and distributed throughout other parts of New Zealand, e.g. to Kapiti and Wainui-o-mata in 1893, and to Taranaki in 1896. For the last 30 or 40 years grey opossums have been very abundant on Kawau. In 1893, Captain Bollons, in N.Z.G.S. 'Hinemoa,' liberated 72 opossums in the West Coast Sounds. They are now abundant from far north of Auckland to Stewart Island. In all localities they appear to have increased to a great extent, becoming so abundant in some parts that people began to destroy them for their skins, while others—especially the Acclimatisation Societies—claimed protection for them and demanded the introduction of restrictive legislation. Some idea of their increase may be gathered from the statement made by the President of the Otago Society that in 1912 no less than 60,000 skins were taken in the Catlins district alone. Mr R. S. Black of Dunedin, a well-known exporter of rabbit and other skins, tells me this number is not an over-estimate.

W. W. Smith (31st July, 1918) reports opossums as common about New Plymouth. They feed on the leaves of the hou-hou (*Panax arboreum*) and come to the shed where horsefeed is kept, and help themselves to the oats.

Protection and Legislation. In 1891, protection of the opossums was urged on the Government by an Invercargill merchant who stated (in *Southland Times* of 20th January) that some New Zealand skins were worth 10*s*. each, and he noted that the supply of skins from Australia and Tasmania was diminishing.

At the same time complaints began to be made by settlers in bush districts that the opossums were robbing their fields and orchards, and destroying plantations—apparently an attempt to justify the destruction of the animals which was then commencing. Such a charge was not supported by evidence. On the other hand Mr T. C. Plante of Melbourne, writing to the Premier of New Zealand on the subject (in 1891) says:

> Tasmania is the orchard of Australia, yet so little harm is caused by this animal and so well is its commercial value appreciated, that a close season is prescribed for it, and indeed for all marsupials. Although the species of Victoria yield a fur of little value, except such as live in the cold and mountainous parts, the case is different with the Tasmanian species, which are of much greater value; the animal is larger, producing fur denser and of much better quality, and the colour is black or reddish-black. Now this is the kind that has been introduced into New Zealand, and from specimens caught in Riverton bush that have been shown to me, I can vouch that in New Zealand they grow even larger and produce fur of better quality. At the October (1890) fur sales Tasmanian skins realised up to 8*s*. 6*d*. each.

Mr Plante recommended trapping from June to September when the fur is fully grown, with a close season during the summer months.

Owing to the increasing destruction which went on in succeeding years in all districts where opossums were found, the societies interested brought pressure to bear on the Government, and in 1911 an Order in Council was issued (*Gazette*, 16th November), declaring these animals to be "Imported Game within the meaning of the Animals Protection Act, 1908." Thus it became illegal to catch or destroy them. By this time, however, the settlers in bush districts—at least in Otago—had found the trapping of opossums a very profitable business, and though they do not appear in published returns of exports, the probability is that their skins were classed and counted with rabbit-skins. Accordingly they set to work through their representatives in Parliament and got the restrictions removed. By *Gazette* notices of 22nd August, 1912, it was stated that "opossums of every variety shall cease to be deemed to be imported game," and "all protection of opossums has consequently been removed, and they may be taken or killed without restriction, and their skins sold." This see-sawing legislation immediately produced an outcry from all the societies in the country, and so much feeling was expressed that the Government reconsidered their decision and another Order was issued on 7th August, 1913, declaring opossums to be absolutely protected in certain specified counties—practically in all the bush-covered districts in New Zealand. A further warrant was issued in 1916 absolutely protecting opossums in the Wellington Acclimatisation District.

The position therefore in 1919[1] briefly was as follows:

> Opossums have ceased to be imported game and they have been absolutely protected in certain areas. There is therefore no existing law in force giving power to declare an open season for these animals unless they were again declared to be either imported or native game, and this is not practicable as they would then automatically be protected in parts of the Dominion where protection is not desired; there being no existing power to enable them to be declared imported game in part only of the Dominion.

In spite of these regulations it is the opinion of some at least of the societies that the law is habitually broken and that the protection is very imperfect, and the Wellington Society in its report for 1915 says that "opossums are being slaughtered wholesale." I learn also from the Comptroller of Customs that the number and value of opossum skins exported during the year ended 31st December, 1915, was as follows:

Wellington	...	...	173 skins valued at		£43
Nelson	...	...	191 „	„	48
Dunedin	...	...	2115 „	„	361

It is known that thousands more go out of the country, nominally as rabbit-skins.

Food, Habits, etc. Mr F. Hart of Round Hill, who had a long experience in catching opossums for the Southland Society, wrote a report on the habits of these animals to Mr Eustace Russell of Invercargill, from which I extract the following. The technical names of the plants given are added by myself:

> The food the opossum lives on is chiefly seeds of Broadleaf (*Griselinia*), Kamai (*Weinmannia*), Broad-gum (*Panax*), Maple (*Pittosporum*), Rata blossoms (*Metrosideros*), Supplejack berries (*Rhipogonum*), Fuchsia, Mako-mako (*Aristotelia*), and practically all the seeds and blossoms that grow in this part of the bush. The opossum is not a grass-eating animal. They will eat white, or red clover, sweetbriar shoots, and seeds, but if an opossum is caged up and fed on grass, he will die of starvation. Also, if he were fed on turnips, it would take as much to feed one sheep, in quantity, as would feed twelve opossums. When I and my brother were catching

[1] For recent regulations (May, 1921) see Appendix A, p. 556.

opossums for the Society, we fed them on carrots, boiled wheat, bread, boiled tea-leaves with sugar, and anything sweet.

The damage the opossums would do running at large would be very little, seeing that they never come on to open country. The animal is blamed for barking apple-trees, but the opossum does not bark a tree. He might scratch the bark with his teeth, but he does not strip it off. The opossum has one young one once a year. The young one is from five to six months old before it leaves its mother, and is very nearly half-grown. The first four months it is carried in its mother's pouch, and after it leaves the pouch it rides on the mother's back, until it is able to look after itself. The proper season for catching opossums would be from April to the end of July; that would save destroying so many young ones.

Mr Hurrell of Ararata (Hawera) tells me they are destructive to fruit trees in his district, eating the shoots in spring-time and the fruit in autumn. This applies to apples and plums. At Kawau, they were reported as very destructive to the shoots of young plants, and to fruit.

Colonel Boscawen of Auckland, who is a most reliable authority, states that as long as there is plenty of green stuff available, opossums do not interfere with fruit, but that the damage they are often credited with is the work of rats.

On Kapiti Island they feed extensively on Kohekohe (*Dysoxylum spectabile*), Mahoe (*Melicytus ramiflorus*), *Passiflora tetrandra* and *Fuchsia excorticata*, trees of the latter species being sometimes completely destroyed by them.

In "Nature Notes" in the *Lyttelton Times* of 19th October, 1912, by Mr Jas. Drummond this passage occurs:

Mr A. J. Blakiston, Manager Orari Gorge Estate, South Canterbury, where opossums are very plentiful, says: "My experience here is that they do very little damage. The garden adjoins the native bush and in the fruit season they eat and knock down some fruit, but do us no great harm."

Mr Dudley le Souef, Director of the Zoological Gardens, Melbourne, writes:

Opossums are protected in Tasmania for half the year, in Victoria all the year round, and in South Australia and New South Wales for half the year during the breeding season. We find them only occasionally troublesome in apple, pear and peach orchards; but as they are easily snared and shot, one seldom hears of any complaints even from the large orchard districts[1].

[1] All orchardists are not of this opinion as the following extract from an Auckland letter shows:

"If you want to see how opossums and fruit trees thrive together, take a run down to Motutapu. Opossums you will see, but it will need a guide to show you where the fruit trees were planted. I have several acres in orchard, which today is free from Opossums, and needs only the regular care to combat moth, scale, scab, mildew, blight, dieback, fungus, leech, collar rot, birds, rabbits, picnickers

Professor Flynn of Hobart states that even with the protection given to the opossum in Tasmania their position in the State is seriously endangered. It is estimated that 100,000 were killed in 1911 for their skins.

Ring-tailed Opossum (*Pseudochirus peregrinus*)

The Canterbury Society introduced two of these animals in 1867, but do not seem to have liberated them.

Family DASYURIDÆ

Australian Native Cat (*Dasyurus viverrimus*)

In 1868 the Canterbury Society received two of these animals from a Captain Thomson. Presumably they were not liberated, as there is no further record of them. The introduction of hurtful carnivorous animals, except under Government sanction, has always been forbidden in New Zealand.

Family DIDELPHYIDÆ

Bandicoot (*Perameles obesula*)

The Auckland Society obtained some bandicoots, how many is not specified, from a Mr E. Perkins in 1873, but there is no record as to what was done with them. These were probably the short-nosed bandicoot (*Perameles obesula*) which is common in Australia and Tasmania.

Order UNGULATA

Family SUIDÆ

*Pigs; Wild Boar (*Sus scrofa*)

The first pigs landed in New Zealand were two little ones which De Surville presented to the Chief of the natives at Doubtless Bay in December, 1769. It is not known what happened to these early juvenile immigrants.

Captain Cook introduced pigs on his second voyage to New Zealand as he states that while in Queen Charlotte Sound in June, 1773, "Captain Furneaux put on shore, in Cannibal Cove, a boar and two breeding sows, so that we have reason to hope this country will, in time, be stocked with these animals, if they are not destroyed by the natives before they become wild, for, afterwards, they will be in no danger."

and small boys. These I can manage to fix during the daylight, but cannot see why a set of cranks, who have nothing of their own to destroy, should compel me to sit up at night to shoot further vermin. If I am counted out in the assumption, will some 'boobs' join me in bringing pressure on the Government, for the introduction of rattlesnakes, tigers and other interesting subjects, because the rattle and claws are beautiful, and the meat would compete with local grown bully?"

Forster, in his *Journal* (vol. I, p. 221), says "they were turned into the woods to range at their own pleasure." In the following year, October, 1774, he says (vol. II, p. 467):

We took the opportunity to visit the innermost recesses of West Bay, in order to be convinced, if possible, whether there was any probability that the hogs, brought thither about a year before, would ever stock those wild woods with numerous breeds. We came to the spot where we had left them, but saw not the least vestiges of their having been on the beach, nor did it appear that any of the natives had visited this remote place; from whence we had room to hope, that the animals had retreated into the thickest part of the woods.

On 2nd November when off Cape Kidnappers Cook gave some pigs to natives who came in their canoe.

On his third voyage, he gave a boar and sow to a native chief (?) in February, 1777, and they made him a promise not to kill them. He adds: "The animals which Captain Furneaux sent on shore here, and which soon after fell into the hands of the natives, I was now told were all dead." I think, however, that this refers chiefly to the goats, for he says: "I was afterwards informed by the two youths who went away with us, that Tiratou, a popular chief amongst them, had one of the sows in his possession." There is little doubt that the wild pigs of the South Island, "Captain Cooks" as they came to be called, were the progeny of those originally left at Cannibal Cove, though Cook himself says in 1777: "I could get no intelligence about the fate of those I had left in West Bay and in Cannibal Cove, when I was here in the course of my last voyage." They have in later years had their numbers added to, and their breed modified by pigs which escaped from settlers, but the type remained dominant, and is still found in most wild parts of the country in great abundance.

Dieffenbach (in 1839) states that "the natives had come from Cannibal Cove to catch pigs, which overrun the island" of Motuaru.

The North Island wild pigs, which are also abundant in nearly all wild country from Cook Strait to North Cape, are largely the progeny of animals given to the natives in later years. Governor King (of New South Wales), during his visit to New Zealand, in 1793, gave the natives at the Bay of Islands, ten young sows and two boars. Dieffenbach states that these animals were mistaken by them for horses, they having some vague recollection of those which they had seen on board Captain Cook's vessels. They forthwith rode two of them to death; and a third was killed for having entered a burying-ground. A very old man, who had known Captain King, related this singular story to me. The introduction of the pigs may have been correctly reported, only Dieffenbach is not very trustworthy, and his credulity seems to have been played on as regards the horses; it is most improbable that any horses were on board either the 'Resolution' or the 'Adventure.'

There is no doubt that the abundance of wild pigs in the country was of great value to explorers, particularly to prospectors, and also to shepherds, miners and back-block settlers. Dieffenbach says: "the natives have great quantities of pigs, which have run wild, but are easily caught by dogs" (this was in the Piako).

Dr Monro, who accompanied Mr Tuckett on his trip through Otago in 1844, speaking of the hill country south-west of Saddle Hill, says: "There is a famous cover for pigs, too, between the upper part of the Teiari (Taieri) Valley and the sea....The whalers come up the river in their boats and kill great numbers of pigs here; as the Maoris told us."

As to the breeds of these wild pigs, it is evident that they were quite distinct in the two islands, due, of course, to their different origin. Mr Robert Scott, M.P. for Central Otago, writing me in January, 1916, says:

> They were originally a variety of the Tamworth breed, long snout, razor-backed, built for speed rather than for fattening, quick and agile in movement, as I have often seen when watching two boars fighting, and as many a dog found to his cost. The predominating colour was red, or sandy red, with some black, and a few black and white, but these may have come from an occasional tame boar which strayed and became wild. At the time when they were most numerous (in Otago) they were decidedly gregarious, usually three or four generations running together in mobs numbering from half a dozen up to forty or even fifty. When attacked by dogs, if cover, such as flax, scrub or high grass was handy, they made for it and would form a circle, with the older pigs on the outside ring, and the younger ones in the centre for greater protection. The boars, particularly old ones, lived alone and roamed far and wide. The habits of the wild pig were clean, and in the case of those tamed exceptionally so.

Angas in *Savage Life and Scenes in Australia and New Zealand*, vol. II, p. 37 (published in 1847), says: "The New Zealand pigs are generally black; and on the approach of a European they erect their bristles, and, grunting, gallop off like wild boars."

Dieffenbach says:

> Pigs have only of late been generally introduced into many parts of the country, and in some places where tribes have been broken up they are found wild in large numbers....The New Zealand pigs are a peculiar breed, with short heads and legs and compact bodies.

They were evidently quite distinct from the "Captain Cooks" of the South Island.

The Rev. Richard Taylor (in *The Past and Present of New Zealand*, in 1868) states that:

there are three kinds of pigs which have been naturalised, whether the produce of the original pair left by Captain Cook, or from later importations it is impossible to say. The ordinary one, which has stocked the forest, is black, with a very long snout, almost resembling that of a *Tapir*; this pig was probably the original one. The next is a grey one, commonly known by the name of *Tonga tapu*, and may therefore be supposed to have been thence derived. The third variety is generally of a reddish brown, marked with lateral black or dark stripes, running the whole length of the body.

Taylor was neither a good observer, nor much of a naturalist, and accepted a great deal of information without sifting its accuracy; still the above may be quite correct for the part of New Zealand which he knew.

Mr Robert Gillies, writing in 1877, says: "In 1848 (the year of the settlement of Otago) wild pigs were very common on the site of Dunedin." In 1854, he and a party killed 70 pigs at the back of Flagstaff in two days.

The long-pointed snout, long legs and non-descript colours of the true wild pig showed them to be quite a different breed from the settlers' imported pig. Their flesh tasted quite different from pork, being more like venison than anything else.

Mr Jas. D. Drummond quotes Mr E. Hardcastle of Christchurch on this subject:

In most parts of the Dominion black is the commonest colour of wild pigs, and he believes that the Berkshire probably was the dominant type in the pigs of our early days. The red coat of the Tamworth type has defied time. It has lost most of its lustre but he thinks that nobody can doubt that the sandy, long-snouted wild pig has Tamworth blood in its veins.

Black pigs with a white stripe over the back or the shoulder were plentiful in Canterbury. The markings still may be found in that province and in other parts of the Dominion.

"These pigs," Mr Hardcastle writes, "were ascribed to an original cross between black pigs and white pigs, but there are in England at least two breeds with those markings, and probably some of these were introduced with other ancestors of our wild pigs. The Hampshire has a white belt round its body, including the shoulder and the front-legs; the saddle-back, or white-shouldered pig, which now is being brought under notice in England, does not seem to have as much white as the Hampshire has."

Mr E. C. d'Auvergne, formerly of Rangiora, and now of Waihoa Forks, Waimate, South Canterbury, states that the late Captain Forster, of Oxford, imported some white-shouldered pigs from England many years ago and gave one to Mr d'Auvergne's father. From these facts, Mr Hardcastle builds up the theory that the white-shouldered pig amongst the wild New Zealand pigs is the descendant of a distinct breed.

"Perhaps the most interesting specimen of the wild pig in this Dominion," he adds, "is the blue pig found in the Mount Grey and Karetu districts, North Canterbury. The blue colour is produced by a blend of apparently equal numbers of white and black hairs. So fixed is the type that blue pigs may be found in a litter with blacks or black and whites. The blue pig, evidently, is the result of a cross between a black pig and a white pig, and the progeny crossed and inbred until the two breeds are absolutely blended as far as colour is concerned."

Mr J. Drummond (1907) says:

They multiplied astonishingly, and enormous numbers assembled in uninhabited valleys far from the settlements. At Wangapeka Valley, in the Nelson Province, Dr Hochstetter in 1860, saw several miles ploughed up by pigs. Their extermination was sometimes contracted for by experienced hunters, and Dr Hochstetter states that three men in twenty months, on an area of 250,000 acres, killed no fewer than 25,000 pigs, and pledged themselves to kill 15,000 more.

Even much earlier they must have been very abundant, both tame and wild, for nearly every sealing and whaling vessel which visited these islands between 1800 and 1830 took away quantities of pork as part of the cargo to Sydney.

Aston (1916) speaks of the wild pigs in the high country of Marlborough as being remarkably tame, apparently from never seeing human beings.

Two sows, in response to our grunts, came out of the bush on to the ridge, and as we remained perfectly still, they came up close and smelt us. My companion made a grab at one leg, and pig and man went rolling down the hill together.

At the present time they are still common in nearly all bush country which is not too near settlement, and to those who like the element of danger in their hunting they afford good sport. They are usually pursued by dogs, often especially trained for the purpose, which after a time succeed in bailing up their quarry. They prefer to take their stand in the hollow of a tree or some such locality, and an old boar will often do considerable damage to the dogs before he is despatched. The orthodox manner of attack is to run in and stab them, but a man without a gun has little chance if he ventures to close quarters with a bailed-up boar.

As regards the Southern Islands, pigs were landed on the Auckland Islands in 1807 by Captain A. Bristow, and were reported as numerous by Hooker in 1840, and by Enderby in 1850. Captain Musgrave, who was wrecked on the Auckland Islands in 1865, found no traces of wild pigs, however. More recently Captain Bollons of the 'Hinemoa,' and others have landed and liberated pigs. Hooker reported them as feeding chiefly on *Pleurophyllum criniferum*, while McCormick, who was surgeon on the 'Erebus,' states they fed on *Stilbocarpa polaris*. Waite, writing in 1909, says:

There can be small doubt that the introduction of pigs to the Auckland Islands has already resulted in considerable havoc among the ground-nesting birds, by destroying both eggs and young. Traces of pigs were very plentiful, not only their spoor but their rootings also being abundantly apparent. Native plants are also suffering, for we found whole patches turned over, *Bulbinella* and *Pleurophyllum* evidently being favourites. On several occasions we came across the pigs themselves, but they were very wild and were approached with difficulty. Of four seen on one occasion, one was black, two white, and one pied. One of them was shot, and proved to be a lean, long-legged, and long-snouted animal, apparently reverting to the characteristics of a wild type.

In 1865 Captain Norman liberated three pigs on Campbell Island, but they appear to have died off.

Dr Cockayne informs me that in the Chatham Islands, the magnificent forget-me-not, known as the Chatham Island lily (*Myosotidium nobile*), formerly grew commonly as a coastal plant, forming a fringe of vegetation round the islands, but that it has been nearly exterminated by wild pigs—aided in part by wild cattle—so that it is now found only in inaccessible spots. They have also helped to reduce the number of plants and nearly exterminate *Aciphylla Traversii*, one of the most characteristic plants of the Chatham Islands, by digging it up and eating the succulent tap-root. Formerly two species of spear-grass—*Aciphylla squarrosa* and *A. Colensoi*—were extremely abundant, especially in the South Island. Vast quantities of these plants were grubbed out by the wild pigs, which are particularly fond of their succulent and aromatic root-stocks and roots.

Aston states that they eat down *Gaya Lyallii*. They root up the ground wherever the bracken fern (*Pteris aquilina*, var. *esculenta*) is found, the starchy rhizomes furnishing abundant food. They are also especially fond of the thick root-stocks of spear-grasses (*Aciphylla*) and other umbelliferous plants, such as *Ligusticum* and *Angelica*.

In some parts of New Zealand wild pigs are destructive to sheep. I am informed that in North Canterbury an old boar has been seen to come down from his hill fastness into a paddock in which were a number of ewes, charge into the midst of them, and kill two of them, "seemingly," said my informant, "more out of mischief than for want of food." Mr W. R. Bullen of Kaikoura writes me in August, 1916:

> It is a well-known fact that wild pigs are very destructive among newly born lambs. I myself have watched a wild boar working a lamb like a dog so as to get straight above him on a hillside, and catch the lamb with a downhill rush, as the latter was too nimble for the boar to catch him otherwise. I think, however, that an old sow with a litter of young ones does more damage, as they follow up the ewes when lambing. We always endeavour to reduce their number before lambing commences both by hunting and laying poison. Phosphorus is usually employed in the latter case.

Mr Kennedy of Greentown, Kaikoura, supplements this information, and informs me that when boars once begin to eat lambs, they will travel long distances to get them; fortunately the habit is not common. He thinks the habit is learned by their finding hoggetts which have got caught and hung up in lawyers (*Rubus*), and dying there. Sows that have a litter of young ones also attack and destroy lambs, but they do not travel any distance to do so. He adds that pigs are very destructive to rabbits, eating the young ones when they take refuge in shallow burrows; and states that where pigs are abundant, very few rabbits are to be found.

The following species of native plants, in addition to those named, are eaten by wild pigs: *Gastrodia Cunnunghamii* and *G. sesamoides* and *Marattia fraxinea*.

Family CAMELIDÆ

Alpaca (*Lama huanacos*)

Two of these animals were imported by the Otago Society in 1878, and were liberated on the property of Mr John Reid of Elderslie, Oamaru. They never increased.

Family EQUIDÆ

*Horse (*Equus caballus*)

It seems rather strange that in such a small country as New Zealand there should be any wild horses, but there are several areas very inaccessible and rarely visited, where escapes appear to have congregated and bred. The natives frequently have very imperfect fences, and stallions have from time to time got away and run free. Mr E. Phillips Turner of the Forestry Department and in charge of Scenic Reserves, informed me (January, 1916) that "wild horses occur on Mt Tarawera, round the base of Ruapehu, and in many places on the volcanic plateau." Mr Yarborough of Kohu Kohu states that at one time wild horses were numerous in the bush country of Hokianga and the west coast of the Auckland peninsula. The natives used to snare them, but they were mostly so inbred as to be valueless for any purpose. They are now (1916) very scarce.

Horses were first imported into the Chatham Islands in the forties, and were commonly wild in the unsettled districts in 1868. There are probably still a few of them on the table land.

Zebra (*Equus zebra*)

Sir Geo. Grey, among his numerous other introductions, imported a pair of zebras into Kawau about 1870, apparently in the hope that they would breed. But one got killed, and the other had to be shot.

Family Cervidæ

The desire to stock the mountain country of New Zealand with large game, so that the Briton's delight in going out and killing something might be satisfied, has led to the introduction of no fewer than ten kinds of deer, in addition to other large animals. Of these, four species—red deer, fallow deer, white-tailed deer and Sambur deer—have established themselves in different parts of the country and are included among the animals for which licences to shoot are now issued. By law they are strictly preserved, but much poaching has always been and still is done. At the same time it must be remembered that the poaching is chiefly done by two classes of people, viz. residents in the neighbourhood of the districts where the game abound, and mere pot-hunters. For the first class it may be said that many farmers, who take no interest in acclimatisation work or in so-called sport, and who were not consulted in any way on the subject, object to the incursions of animals which ignore or break down their fences, harass their stock, and eat their hay and turnips. Therefore some of this destruction of imported game takes the form of reprisals for injury done to crops, fences and stock. There is practically no poaching on the property of private individuals such as is characterised by the name in the mother country, and consequently destruction of game in New Zealand is not looked upon as a heinous offence, as are breaches of the iniquitous game-laws of Britain. The game in New Zealand are either the property of the State or of the acclimatisation societies, and public opinion on the subject of their destruction is lax in comparison with what it is in countries where game is looked upon as something reserved for and sacred to the sporting instincts of a small class. Still a very fair measure of protection is ensured to the animals, and they have increased in most districts where they have been liberated. It is recognised, too, that a wealthy class of tourists can be induced to visit the country, if, in addition to scenic attractions, there can be added those things which appeal to the sporting instincts of humanity. This has led the Government of the Dominion in recent years to devote some attention to the subject of introducing various additional kinds of big game to those already brought in by the acclimatisation societies. Several experiments have been made in this direction, and most of them seem likely to be successful.

*Red Deer (*Cervus elaphus*)

(*a*) According to Mr Huddleston, whose father was curator of the Nelson Acclimatisation Society, a red deer stag and doe were landed in that district in the fifties. The doe was killed, but the stag remained near Motueka, and ultimately joined those which were introduced in 1861. In Judge Broad's account of Nelson, he states that Felix Wakefield landed one stag in 1851. He further states that in September, 1854, the first stag was turned out on the hills near the mouth of the Waimea, brought in the ship 'Eagle.' Two hinds were sent for to England.

(*b*) The next importation of red deer into New Zealand was in February, 1861, when a stag and two hinds, presented by Lord Petre from his park in Essex, England, were landed in Nelson. The progeny of these animals increased and rapidly spread themselves over a great part of the high country in the provincial districts of Nelson and Marlborough—of late years they have further spread into North Canterbury, and over towards the west coast. Mr Hardcastle, who in 1906 wrote a report on the red deer herds in the country, says:

> The heads obtained in Nelson are of a good dark colour and fairly massive, but compared with those of Wairarapa and Hawea, they have not the same average of span or spread....Lord Petre's herd had had no new blood introduced into it for many years, so that a particular type of antler had been fixed from which there is no throwing back.

According to Mr Hardcastle the type of head of the first imported stag continues to persist, and dominates all the deer of the Nelson herd. In 1900 a herd, descended from Nelson deer, was started in the Lillburn Valley, west of the Waiau River, in Southland.

(*c*) In 1862 a stag and two hinds presented by the Prince Consort to Governor Weld were handed over by him to Dr Featherstone, then Superintendent of Wellington Province. The deer (six in number) were captured in Windsor Park, and housed there for some time as preparation for their long sea voyage. One stag and two hinds were shipped by the 'Triton,' for Wellington, and after a passage of 127 days, during which one hind succumbed, arrived on 6th June, 1862.

About the same time the remaining three were shipped for Canterbury, but as one only arrived it was forwarded to Wellington to join the other two. For some months these animals were kept in a stable close to "Noah's Ark," Lambton quay, and subsequently Mr C. R. Carter (then M.P. for Wairarapa) arranged to have them conveyed to Wairarapa. Owing to there being no trains in those days, the animals were placed in the crate in which they came from England, and were carted over the Rimutaka Ranges to the Taratahi Plains, where they were handed to Mr J. Robieson. This gentleman, being an Englishman, took a special interest in the animals, and kept them for some considerable time. Early in the year 1863 he liberated the deer on the Taratahi Plains, and for some time they were constant visitors to the farmers, accepting all kinds of food. Later, however, they crossed the Ruamahanga River, and took up their abode on the Maungaraki Ranges, where they rapidly increased. Mr Hardcastle reported in 1906:

The Wairarapa Forest is "probably the best stocked red deer ground on the globe. On Te Awaite run bordering on the East Coast, the deer may now be seen in bunches of up to a hundred head. At the beginning of last year it was estimated that there were fully 10,000 head on the station. According to information given in *The Field* of September 15th, 1906, the Windsor Park herd" (from which the original stock came), "has been replenished from English, Scottish, German and probably Danish stock. The result has produced in the Wairarapa herd, stags that are remarkable for their massive antlers, some of which are of the German type, and others again more resembling the Scottish form. The antlers do not grow to great length, but some are very wide in spread, and there is a great proportion of Imperials, the most number of points recorded being 22. The stags mature their antlers early.... A number of heads have been shot on Te Awaite station, showing the abnormal development of the back tines on one antler, such as is seen to be the case of the great Warnham Park stags in England, and is probably due to the highly favourable conditions of climate, food and shelter."

From these ranges some of the finest heads in New Zealand have been secured. There is no doubt whatever that the exceptionally rich limestone formation and the English grasses were responsible for the large growth of horn.

(*d*) In 1871 the Otago Society imported 15 red deer, some of which were sent to the care of Mr Rich of Bushy Park, Palmerston, while seven were liberated on the Morven Hills run east of Lake Hawea. Those at Bushy Park spread over into the Horse Range, but they did not succeed, and no definite explanation of the failure has been given. Probably the country was not high and wild enough; on one side they were encroaching all the time on well-stocked sheep country, and on the other on old-settled farm land, besides which there were many old diggers still about the neighbourhood. From one cause or another they did not succeed well. Mr Hardcastle states that they are still to be met with on the Horse Range, but they have always been heavily shot by settlers.

The seven which were liberated on the Morven Hills were part of a shipment received from the estates of the Earl of Dalhousie in Forfarshire, Scotland. They are the only lot of pure Scottish red deer in the country. They multiplied at a great rate and have spread over the country between Lakes Wanaka, Hawea and Ohau. They have worked their way up the Hunter and Makarora rivers, across the Haast Pass into South Westland, and right up to the neighbourhood of Mount Cook. Most of this country runs from 3000 to 7000 feet in height, and much of it is very steep and rugged. But it contains much bush in the valleys and gullies, and the open country is well grassed in summer. Hardcastle says:

> The North Otago Stags maintain the true Scottish type of antler, but they grow to much greater length than the antlers of any stags that have been shot in the British Isles. The antlers are also remarkable for their symmetry and perfection in the development of the tines, and particularly the lower tines....Some magnificent heads have been got, including a 17- and 18-pointer, and two Royals each 46 inches in length of antlers. The coats of the stags are generally shaggy, owing, no doubt, to the severe climate in winter.

Recently (1918) Mr Hardcastle informs me that the record length for an Otago red deer head is 49 inches (J. Forbes, Christchurch); record spread 50½ inches (J. Faulks, Makarora); and record points 20 (J. Fraser, Mount Aspiring); "and I think a 20-pointer was got in the Makarora." In 1895 the Otago Society obtained two fine stags from the Hunt Club, Melbourne, to add to the North Otago herd. I do not know what special strain these belonged to. Again in 1913 the Society imported a stag and six hinds from Warnham Park, England, the object being to introduce new blood into the herds.

(*e*) One stag was brought over from Hobart to Christchurch in 1867 by Mr A. M. Johnson, and was kept in the Acclimatisation Gardens for a time. In 1897 the Canterbury Society imported nine red deer, but it is not recorded from whence, and liberated them in the gorge of the Rakaia River. They have increased rapidly since, herds of 40 and more having been seen from time to time. Some of the heaviest heads secured in New Zealand have been got from this herd. According to Mr Hardcastle the record length of a head from the Rakaia Gorge is 48½ inches (Williams, England); the record spread 46¾ inches (G. Sutherland, Christchurch), and the record points 24, from the same head. But in 1918–19 Mr Barrer of Wellington secured one with a length of 50 inches.

More recent importations have been as follows:

In 1900 the Southland Society received one stag and two hinds from Sir Rupert Clark, Victoria; and in 1901 two stags and eight hinds from the same source. From these, herds were started at Lake Manapouri, and at the Hump, to the west of the Waiau.

In 1903 either seven or eight fawns were obtained from Victoria, presented by Miss Audrey Chirnside of Werribee Park, and were liberated at Mount Tuhua in Westland. In 1906 four more from the same source were added to this herd, and eight were liberated at Lake Kanieri.

In 1903 the Tourist Department obtained eight deer from the Duke of Bedford, and liberated them at Lake Wakatipu. In 1908 four were obtained from Warnham Park, Sussex, England, and were liberated at Paraparaumu. In 1909 three were liberated at Dusky Sound.

The five original importations of red deer account for the vast numbers of these animals which are now to be met with in so many mountainous parts of both islands, for many of the societies as well as the Tourist Department have obtained deer from one or other of the original herds and have started new herds in other districts, e.g. the country round Taupo and Rotorua, the West Coast Sounds of the South Island, and Stewart Island, and these are all increasing. In regard to the last named locality, six fawns taken from the Wairarapa herd were liberated on the banks of the Fresh Water River at the head of Paterson Inlet in April, 1901. In the following year twelve more from the Werribee Park herd were liberated in the same locality. From a report made for the Southland Acclimatisation Society by Mr Moorhouse, who inspected the Stewart Island herds in 1918, it is evident that the deer are now very numerous in the wooded northern and western parts of the island. Stewart Island is a sanctuary for native birds, and this stocking of the island with deer means the opening up of it to stalkers. Mr Woodhouse says:

> Indications to be seen in this big belt of bush clearly go to prove that the deer must be very numerous. Well-beaten tracks lead from the bush to the various clearings, on which grow flax and a coarse tussock. In the bush can be seen their various camping grounds, and the trees and shrubs show where they have been feeding on the barks and leaves.

To the naturalist it is deplorable that an animal should have been introduced into this sanctuary, which compels men with guns—and probably with dogs—to go in, in order to keep them in check to some extent. Writing to me in August, 1918, Mr Hardcastle says:

> Deer increase more rapidly in New Zealand than in the northern hemisphere. Whether there is a larger percentage of calves born, I cannot say, probably there is, considering the conditions here. But the large

increase is mainly due to the hinds calving a year earlier. In Europe hinds do not calve until they are three years old; here they calve at two years. I am speaking of Otago, and the conditions there are not as favourable as further north. At sixteen months (April 1st) a young hind is as large as her mother; and these young animals can only be distinguished by their rounder and neater bodies, and the darker rufous colour of their hair. They are quite big enough therefore to be served by a stag. Further there are not more of these young hinds in a herd than would represent the female progeny of one year. If they did not take the stag till they were twenty-eight months old there would be so many more of them.

The great want of a deer herd is either proper culling by human agency or the presence of carnivora to weed out the old and weakly, but above all to break up the family life and prevent inbreeding. Left alone, deer adopt the family life, and where a hind has once bred she will stay, unless forced away by one means or another. The pioneers of the herd in search of new ground, where there is scope as in Otago, are the big stags, after they have reached their third or fourth year, and are living for ten months away from the hinds. They are followed by young hinds. An old hind on the outskirts of the herd in the line of migration is a rarity. A young stag at twenty-eight months will get a few hinds if he can; a forty months stag will frequently have a good herd, and so will a fifty-two months (four years old) stag. A strong three-year old, that is likely to grow into a good shootable head, will, say in 1918, serve a number of hinds; in 1920 when he is five, he will be serving his own daughters (a stag always makes back to his previous year's rutting ground, if he is not driven off it). In 1922, when he is seven and a quarter years of age, he will be serving his own daughters and grand-daughters! As he only got his royal head at six years, and it may take a few more years to grow it to its maximum weight, he has escaped the stalker until he has done a considerable amount of inbreeding. The opinion of those who have had much experience in Otago is that most of the big heads are of deer that are between eight and twelve years of age. Many of them show signs of their teeth going, and as stags are said to live well over twenty years, one would not expect to see the teeth much worn in the first half of its life. Of course, only a certain percentage of stags get good heads, and, of course, the inferior are left.

In the case of the largest herds attempts are continually being made to thin out the weeds and deer with malformed antlers. Some malforms arise from injury to the horns during the velvet stage of growth, but this injury is often due to the fact of the deer being a weedy specimen in the first instance and in poor condition. Polled stags, that is those without antlers at all, are occasionally met with, but these have apparently suffered from lack of food in the early stages of their growth, for there is no doubt that in some parts the country is already greatly overstocked and severe winters reduce the deer to a poor condition. Mr Hardcastle says:

the great majority of malforms are malformed in the skull itself, and not merely in the bones. A common form is for the pedicle to be misplaced, nearly always being in front of its proper position, and sometimes as low as just above the brow. The horn of the misplaced pedicle is sometimes a switch, and grows either up or down over the face. Sometimes the pedicle is bent outwards,—I have never seen it bent in or back. Sometimes malforms have three distinct pedicles and horns, and four have been found; unicorns are also not uncommon. In nearly all these cases there is no apparent sign of injury, nor would it be possible to misplace the pedicle without killing the animal. The country where malforms appear most in Otago, is open tussock and open birch (*Nothofagus*) bush. Malforms, except an odd one or two that have probably migrated from more overstocked country, are not to be found in the rugged gorges of the Hunter or the Makarora. They did not appear in the rough and dense bush country in the Wairarapa until a few years ago, when the bush was more cleared, and sheep competed more strongly with the deer for food. We do not know whether the defects of the skull are hereditary or not, but from the fact that there are so many to be found in different types, one would think they are. Another question is whether the calf is born with the defect, or at what time it begins to manifest itself. Want of nourishment either in quantity or quality of food rapidly leads to degeneration in stags' heads, and in deer generally, but why it should affect the bone of the head in the way it does is remarkable.

It is clear that there are several distinct strains of red deer in the country, recognised chiefly by the form and growth of the antlers, which are chiefly what sportsmen look to. This mixing of breeds probably tends to the production of a strong race, and the efforts of the main societies are directed, often, it must be admitted, rather blindly, to the elimination of defective deer. In the 1918–19 season the Otago Society had 1000 head shot in the Hunter Valley—Makarora herd—and 667 head in 1919–20. The problem is an interesting one from the eugenic standpoint.

The vast number of red deer found in New Zealand enables the various leading societies to offer shooting privileges to sportsmen, who come from all parts to enjoy this form of sport. The attraction of red deer shooting is now to be reckoned as one of the assets of the country from a tourist's point of view.

Effect of deer on the native vegetation. In the North Island it is stated that *Fuchsia* is the principal food of the deer in spring and summer, but that in winter they take to Koromiko (*Veronica salicifolia*) and other shrubs. Probably they eat the majority of the native shrubs in the bush, but how far they destroy the vegetation of the higher country is not recorded. They are reported as not caring very much for grass. In the North Otago forest the following are mainly eaten: broadleaf (*Griselinia*), native gum (species of *Panax*), ribbonwood (*Gaya Lyallii*), various species of *Coprosma*, pepper tree (*Drimys colorata*), milk tree (*Paratrophis heterophylla*) and Tutu (*Coriaria*).

But when these are scarce they will eat almost any shrub. They will not eat birch or beech (*Nothofagus* sp.) nor celery-leaved pine (*Phyllocladus*), till other food is exhausted. In the thickly-stocked districts all the undergrowth of the bush, as high as the deer can reach, is eaten out by them, and this is mostly done in the winter, when the high open country is covered with snow and they take to the forest for food and shelter. For the rest of the year the grass country is in the undisturbed possession of the deer, as they have no sheep to compete with them for the food. Mr B. C. Aston in an account of the crossing of the Ruahine Range in January, 1914, says:

After getting up about 3200 feet in *Fagus fusca* and *Fagus cliffortioides* forest, where there was a sprinkling of *Phyllocladus alpinus* saplings, we found many with the bark rubbed off, which R. A. Wilson (an experienced deer-stalker) informed me was done by the deer, which always select this tree to rub their horns on. Mr Wilson was surprised that the deer in this district had left bunches of *Loranthus flavidus* and *L. tetrapetalus* hanging within reach, whereas in the South Island they are so fond of *Loranthus*, that they are frequently found hanging by the feet, caught in the *Fagus* trees in an endeavour to jump higher....On an open clearing at a height of 4000 feet, where there was an abundance of *Aciphylla squarrosa* and *Hierochloe redolens* growing together, we found the deer had eaten the grass back into the *Aciphylla*, until the spinous leaves of the latter had pricked their noses.

Mr Hansen, lighthouse keeper at Cape Palliser, reports (April, 1911) on the Waitutumai Creek, the gully of which is here eight miles long, three miles broad, and surrounded by hills from 2000 to 3000 feet high and completely bush-clad:

The terraces have been made passable by the Red Deer, which have eaten away all the lower branches and foliage....There are no pines, ratas, fuchsias, native currants or other berry-bearing trees, on which many native birds make a living. There are no native birds seen, except a few bush-wrens, and one tui was heard. Silence reigned. The deer, mostly stags, come out of the forest from the middle of September to the end of January, when they are in the 'velvet,' and are very tame.

W. G. Morrison of Hamner Springs in giving evidence before the Royal Commission on Forestry in 1913, said the red deer were very destructive to forests both of indigenous and planted trees. They were particularly fond of *Nothopanax Colensoi*, and stripped them to a height of 9 ft.—the trees mostly dying. He had counted as many as 15 trees damaged in a space of 20 yards square. They also destroyed larch and *Pinus laricio*.

*Fallow Deer (*Cervus dama*)

Hon. S. Thorne George, who lived on Kawau from 1869 to 1884, says that the first fallow deer in the colony were introduced there by his uncle Sir Geo. Grey; but he cannot give the exact date of their introduction. In 1864 the Nelson Society received three fallow deer from England, and from these there has descended a well-known herd, but I cannot find any record of the increase and disposal of the original importation. In 1867 the Otago Society introduced two deer, in 1869 twelve, and in 1871 one. All these were liberated on the Blue Mountains, Tapanui, where they have increased to a vast extent, and now form one of the most important herds in New Zealand. Licences to shoot them have been issued for over 25 years. The most recent report (1921) from this district is that so many deer are being shot by the settlers that the herd is threatened with extinction. In 1871 the Canterbury Society had four fallow deer in their gardens, but there is no record now obtainable as to where they came from, nor definitely as to what was done with them. But in later years some were running on the Culverden Estate and two more deer—obtained from Tasmania—were added to them. This herd did not increase, and apparently has been gradually destroyed since. In 1876 the Auckland Society received 28 deer (out of 33 shipped from London), and liberated 18 on the Maungakawa Range, Waikato; while 10 were sent down to Wanganui. The former herd has increased very largely, and is noted for the fine heads of the stags, due, no doubt, to the abundance of food and the favourable climatic conditions. The Wanganui herd is now also a large one. On Motutapu in the Hauraki Gulf, there is a very large herd numbering a thousand or more, and these were probably obtained in the first instance from the Waikato herd. Smaller, more recently established herds occur near Timaru, Hokitika and Lake Wakatipu. It is thus seen that the species is widely spread. Mr Hardcastle informs me that the rutting season for fallow deer is about April 13th to 15th, depending upon the weather. Frosty nights and clear days bring both fallow deer and red deer into season a little earlier, while warm weather delays the rut.

Axis Deer or "Chital" (*Cervus axis*)

In 1867 the Otago Society imported seven of these deer, which were liberated in the Goodwood Bush near Palmerston S. In 1871 another stag was landed and added to the herd, which at that time numbered about 30. In 1881 the inspector reported that he had seen over 40. Then complaints began to come in from the settlers that the deer were a nuisance, and their numbers gradually diminished. Gradually they disappeared, apparently destroyed by the settlers in the district, and none has been seen for the last 20 years. In 1898 the Wellington Society received a pair from the Zoological Society of Calcutta, and placed them on Kapiti Island in Cook Strait. They had not increased by 1902. In 1907 the Tourist Department liberated five deer at Mount Tongariro in the North Island; and in 1909 five at Dusky Sound in the South Island. No reports have as yet been received regarding either of these experiments.

*Sambur Deer or Sambar (*Cervus unicolor*)

In 1875 the Auckland Society received a buck from a Mr Larkworthy, and in the following year a doe. There is no further record of these deer in the Society's reports[1]. But in the annual report of the Wellington Society for 1894 it is stated:

> The Ceylon Elk (Sambur Deer) imported into the Carnarvon district, Manawatu, by Mr Larkworthy, have been brought under the provisions of the Animals Protection Act, and are at present under the control of the Society. It has been reported that the herd now numbers about thirty.

There is no word of these deer in any previous report of the Wellington Society. Then in 1900 the herd is reported to number about 100, "but there is good reason to think that they are really more numerous. ...A pair of antlers were found on the hills near Cambridge, and two deer were shot there," some 200 miles from Carnarvon.

In 1906 the Wellington Society (Marton Branch) reported that "Stag-shooting (*Sambur*) was opened for the first time this season in this district,...but we fear that numbers of stags have been shot by persons unauthorised to do so." This poaching has gone on regularly for many years past, and though the herd seems now a fairly large one, the local rangers complain of indiscriminate destruction in season and out of season. In 1907 the Tourist Department imported two deer (from Noumea) and liberated them in the Rotorua district, adding to them some others secured in the Manawatu, so as to form the nucleus of a new herd.

*Wapiti or Elk (*Cervus canadensis*)

Sir George Grey introduced a pair of these deer into Kawau Island some time in the seventies. The doe died, and the buck had

[1] The difficulty of getting accurate and authoritative information on this subject is characteristic of the manner in which many of the reports of the acclimatisation societies have been kept. The governing bodies of these societies frequently included enthusiasts who took an interest in the work of introducing what they considered desirable forms of animals; but the secretaries in many cases were selected for their capacity in keeping the business of the society in order and in conducting correspondence. The secretaries and the personnel of the committees were also frequently changed. The result has been a great want of continuity in many cases, so that there is now no consecutive record of the work done. Thus in the case of the Sambur deer referred to, the only record of introduction is that of the two specimens received by the Auckland Society in 1875–76; yet it is almost certain there were others. If not, then all the Sambur in New Zealand up to 1907 were the progeny of one pair, and of course are very closely interbred.

to be shot as he became dangerous. In 1905 the Tourist Department obtained 10 of these deer, three bucks and seven does (presented by President Roosevelt), and liberated them at the head of George Sound on the S.W. coast of the South Island. The country, which is eminently suitable for all kinds of deer, is very seldom visited. But Mr Moorhouse, Conservator of Fish and Game (Rotorua), who was sent down by the Government in February, 1921, reports that these deer are now well established in the neighbourhood of the Sound. In April, 1921, they are reported to have crossed over into the Lake Te Anau district.

Japanese Deer (*Cervus nika*)

In 1885 the Otago Society received three of these deer from Mr J Bathgate, and they were liberated on the Otekaike estate near Oamaru Five years later they were reported as "doing well and growing into a nice little herd." In the report for 1892 it is stated that "little or nothing has been heard about these deer on the Otekaike estate." Apparently they have all been destroyed as there is no further record of them. In 1905 Government obtained six Japanese deer and liberated them on the Kaimanawa Ranges, near Taupo.

Black-tailed Deer; Mule Deer (*Odocoileus hemionus*)

In 1905, five of these deer, purchased in America, were imported by the Tourist Department and liberated at Tarawera, Hawke's Bay. The Hawke's Bay Society reports them as increasing in March, 1915.

*Virginian Deer; White-tailed Deer (*Cariacus virginianus; Odocoileus virginianus*)

The Tourist Department imported 18 of these deer in 1905. Of these nine (two stags and seven hinds) were sent to Port Pegasus on Stewart Island, and nine to the Rees Valley, Lake Wakatipu. There is no further record regarding those in the last named locality; but the herd in Stewart Island has increased greatly, and threatens to destroy much of the vegetation, besides opening up the country—which is a reserve for native birds—to deer-stalkers.

South American Deer (*Cariacus chilensis*)

In 1870 the Auckland Society received three South American deer, probably of this species, from Mr W. A. Hunt. Beyond the mention of their receipt, there is no further record of them.

*Moose; Elk (*Alches machlis*)

The first attempt to introduce these animals was made by the Government in 1900, when 14 young ones were shipped on board the 'Aorangi' at Vancouver. Owing, however, to the rough voyage only four—two bulls and two cows, nine months old—arrived in New Zealand. They were liberated in 1901 near Hokitika, but appear soon to have separated, as in 1903, one cow was in one district, another at the gorge of the Hokitika River, while nothing was known of the bulls. In 1913 the cow at the junction of the Hokitika and Trews rivers was "in splendid condition, and as tame as a kitten." The others seem to have disappeared. In 1910 the Government obtained ten more, and these were liberated on the shores of Dusky Sound. Mr Moorhouse found (in February, 1921) that these deer were in considerable numbers round the Sound. Their food seems to consist of certain mosses, and the tops and ends of punga ferns (*Cyathea dealbata*).

Family Bovidæ

Gnu (*Connochætes gnu*)

Sir George Grey introduced one or more of these quaint animals into Kawau about 1870, but there is apparently no record of what happened to them.

*Ox (*Bos taurus*)

From the earliest days of settlement, cattle were run in large numbers on the open country, seldom seeing men, and running practically wild. They were gathered together by stockmen at certain times of the year in order to brand the calves, castrate the young bulls, and separate marketable animals. Otherwise they ran wild, each herd or mob occupying its own particular area of country, and this they kept to, except in winter, when they roamed into the forest and fed on *Panax*, *Melicytus*, and other trees, of which they are very fond. It was inevitable that numbers of them should become truly wild, escaping altogether from the musterers, and getting right away into the back country. Consequently wild cattle have been very abundant in all the back country for the last seventy years.

Apparently the first recorded introduction of cattle into New Zealand took place at the Bay of Islands, for the Rev. R. Taylor says that Marsden brought them over from New South Wales. This must have been in the twenties of last century. Dr McNab states that on 30th March, 1833, John Bell set out from Sydney for Mana Island with ten head of cattle. He adds:

> With the exception of the domestic animals which accompanied the expeditions of Cook and Vancouver, this is the first record of any such having been taken to New Zealand, though it is incredible that sheep, cattle, goats and rabbits were unknown at the shore whaling stations of Preservation, Otago, Cloudy Bay, Queen Charlotte Sound and Kapiti.

E. J. Wakefield saw wild cattle in 1839 on the hills at the entrance of Pelorus Sound. In 1840 he states that they were abundant on Kapiti, and that they were the descendants of some given to the natives there in exchange for flax. I recall in 1868 how they used to come out of the Southland bushes during the winter season to feed on the paddocks of English grass. They raided these during the night, and when disturbed in the morning used to jump the fences and ditches just like deer. The Hon. S. Thorne George, M.L.C., writes (February, 1916): "When I first went to Kawau (1869) there was a large number of wild cattle. The island was originally occupied as a cattle station, but owing to the rough country and heavy bush, very many were lost and became quite wild." Mr A. C. Yarborough of Kohu Kohu informs me (August, 1916) that 40 years ago wild cattle were very numerous in all the bush country, and in those days Hokianga and the West Coast were nearly all covered with bush. The natives used to kill them in large quantities for the sake of their hides, which were valued at from 6*s.* to 12*s.* each. In later years these wild cattle have been driven further and further back, until they are now found only in the ranges distant from settlement. These cattle are merely the descendants of tame ones which have wandered—the Maoris' fences being usually of a defective character—and are not of any distinct character. Wild cattle are found in the high country between Lake Wakatipu and the West Coast. Their tracks were numerous in the Valley of the Rockburn.

Cattle were first introduced into Chatham Island in 1841, and soon became wild; and they used to be trapped by the natives in the early sixties. Wild cattle are now very numerous on the table land.

In regard to the Southern Islands, cattle were landed on the Auckland Islands in 1850 by Captain Enderby, but they were all killed off by sealers. In 1894, cattle were landed from the 'Hinemoa' on Enderby Island and Rose Island, where (according to Cockayne) these were about 10 and 15 head respectively in 1903. Aston says that on Enderby Island they have exterminated the tussocks of *Poa littorosa.* Cattle were landed on Antipodes Island at various times between 1886 and 1900, but they either died or were killed off by castaways. Three more were landed in 1903; these have disappeared also.

Effect of Cattle on Native Vegetation. Aston, who was over the country in 1914 and 1915, says:

> Wild cattle are abundant in unfrequented valleys and gorges of the Tararua Range. They are apparently Hereford cattle gone wild. They eat out many species of native plants, and have destroyed great numbers of *Ligusticum dissectum*, which is one of the most abundant and characteristic plants of the higher ground.

He adds:

> Cattle are particularly fond of certain native trees and shrubs, e.g., Tahoe or Hina hina, *Melicytus ramiflorus*; Karamu, *Coprosma grandifolia* and *C. tenuifolia*; Broadleaf, *Griselinia littoralis*, Mangrove, *Avicennia officinalis*; Tawa, *Beilschmiedia tawa*; and Karaka, *Corynocarpus lævigata*.

According to Mr Maxwell, caretaker of the Waipoua Kauri Forest Reserve, cattle eat out the following plants from the undergrowth of the forest: *Melicytus ramiflorus*, *Pittosporum tenuifolium*, *Hoheria populnea*, *Coriaria ruscifolia*, *Corynocarpus lævigata*, *Panax* (*Nothopanax*) *arboreum*, *Schefflera digitata*, *Coprosma robusta*, *Myrsine* (*Rapanea*) *Urvillei*, *Olea lanceolata*, *Geniostoma ligustrifolia*, *Solanum aviculare*, *Veronica salicifolia*, *Vitex lucens*, *Freycinetia Banksii*, and *Cyathea medullaris*. In addition to these, cattle chew the leaves of bracken fern (*Pteris aquilina*), of the flax (*Phormium tenax*) and cabbage tree (*Cordyline australis*); and occasionally eat Ngaio (*Myoporum lætum*) and anise (*Angelica gingidium*).

*Common Sheep (*Ovis* sp.)

The first attempt to introduce sheep into New Zealand was made by Captain Cook on his second voyage, and was unsuccessful. He brought away two rams and four ewes from the Cape of Good Hope, but by the time the 'Resolution' entered Dusky Sound in March, 1773, only a ram and an ewe survived, and they were in such a bad state, "suffering from an inveterate sea-scurvy," that their teeth were loose, and they could not eat the green food which was given to them. Forster in his *Journal* states that they "were in so wretched a condition, that their further preservation was very doubtful." However, they must have improved, for considering the country about Dusky Sound too rough and forest-clad for them, Cook took them on to Queen Charlotte Sound, which was entered on 18th May. In his *Journal* he says:

> On the 22nd in the morning, the ewe and ram, I had with so much care and trouble brought to this place, were both found dead, occasioned, as was supposed, by eating some poisonous plant. Thus my hopes of stocking this country with a breed of sheep were blasted in a moment.

According to the Rev. R. Taylor, Marsden brought over a merino ram and four ewes from Sydney in the twenties. These animals, which were a present from the King, were the originals of the first flock of sheep in New Zealand. I cannot find when sheep were next brought into the Colony, but as soon as settlement began they were imported freely from New South Wales. In those early days fences were very rough, and little or no attempt was made to keep the sheep within enclosures. They were therefore allowed to roam freely over the open country, and were only mustered at rare intervals for shearing, tailing the lambs, etc. It was inevitable, therefore, that numbers escaped the musterers, especially on high and inaccessible country, and that thus wild sheep became very common, especially in the mountain districts of the South Island.

Twenty or thirty years ago when the minds of naturalists were saturated with Darwinian views, it was somewhat confidently anticipated that isolation would lead to the rapid development of new varieties and species, and that such changes might well be looked for in New Zealand. At the meeting of the Australasian Association in Christchurch in 1891, I read a paper "on some Aspects of Acclimatisation in New Zealand" from which I take the following extract:

In the district of Strath-Taieri, in Otago, some years ago, certain sheep on one of the runs—probably the progeny of a single ram—were found to be evidently short-winded. Apparently the action of the heart was defective, for, when these sheep were driven, they would run with the rest of the flock for a short distance, and then lie down panting. The result of this peculiar affection was that, at nearly every mustering, these short-winded sheep used to be left behind, being unable to be driven with the rest. Sometimes they were brought on more slowly afterwards; but, if it happened to be shearing-time, they were simply caught and shorn where they lay. As a result of this peculiar condition, a form of artificial selection was set up, the vigorous, active sheep being constantly drafted away for sale, etc., while this defective strain increased with great rapidity throughout the district; for, whenever the mobs were mustered for the market, shearing, or drafting, these "cranky" sheep (as they came to be called) were left behind. This defective character appeared in every succeeding generation, and seemed to increase in force, reminding one of the Ancon sheep referred to by Darwin. At first, of course, the character was not recognized as hereditary; but, as the numbers of this "cranky" breed increased to a very serious extent and spread over the district, it came at last to be recognized as a local variety. When the runs on which these sheep were abundant were cut up and sold, or re-leased in smaller areas, the purchasers found it necessary, for the protection of their own interests, to exterminate the variety, of which hundreds were found straggling over the country. This was easily and effectually done in the following manner. As soon as a sheep was observed it was pursued; but, after running for a couple of hundred yards at a great rate of speed, it would drop down panting behind a big stone or other shelter, and seemed incapable for a time of rising and renewing its flight. It was immediately destroyed; and, in this manner a useless—but, to the naturalist, a very interesting—variety was eliminated.

Wild sheep are still abundant in some of the wilder parts of the country, and are especially numerous in the high limestone country of Marlborough. Mr Aston says:

On the North-west side of Isolated Hill is a gently-sloping tussock land stretching down towards the Ure river, on which are hundreds of wild sheep in small flocks of about half-a-dozen in each. All,—rams, ewes, and particularly the lambs, are, as far as we could see, in excellent condition. Some were curiously marked and coloured. One had a brown body, black legs and face, and white forehead. The rams had large horns, and all were tamer than ordinary domestic sheep. Their food appears to consist of the Silver Tussock, *Poa cæspitosa*, which was well eaten down; a Poa like *P. colensoi*; the Spear Grass, *Aciphylla Colensoi*; and several other native plants and shrubs.

In another place he says: "these sheep destroy the Gaya trees" (the mountain ribbon-wood, *Gaya Lyallii*), "by eating the bark, which we watched one stripping off in large sheets."

Sheep have been liberated on the Auckland Islands at various times since 1890, and on the Antipodes between 1886 and 1900, but they either died off or were killed by castaways. They were also liberated on Campbell Island between 1888 and 1890. In 1896 the island was taken up as a sheep run (a piece of vandalism on the part of the man who did it, and the Government which granted it), and in 1903 there were about 4500 sheep on it. The changes produced in the vegetation have been described and discussed at length by Dr Cockayne. In 1907, according to Laing, there were some 8000 sheep on the island, and the transformation and destruction of the native flora was going on at a great rate.

They were introduced into Chatham Island in the early forties. but as late as 1855 there were only about 200 of them. When sheep stations were organised in 1866 there were about 2000 on the island, and by 1900 they had increased to about 60,000, and a number of them were wild. They have profoundly altered the native vegetation by eating out many species, such as *Myosotidium nobile*, *Aciphylla Traversii*, *Veronica Dieffenbachii* and allied species, all of which they eat greedily.

At the present time (1919) several hundred wild sheep are running on the island of Kapiti which is now a plant and animal sanctuary. Steps are being taken to destroy these animals. Nearly all of them carry long, filthy dags; very many of them have the wool torn more or less completely from the back by the bushes. Not only do they prevent to a very large extent the growth of young trees, but they open up the forest to the sweep of the wind. They prepare it for invasion by grass, tauhinu (*Pomaderris phylicæfolia*), manuka (*Leptospermum scoparium*), and other hardy plants. Although the manuka is one of the least objectionable of these invaders, yet in dry situations, such as some of the spurs, where it harbours no moss or liverworts, very little humus is formed, and that little is quickly washed away by rain. On some spurs—for example, on one just south of Waterfall —where manuka has replaced the forest, much soil has been removed, and in no great time the manuka itself will be unable to retain its footing. In such cases the manuka marks a phase in the passage to utter barrenness.

I quote this from the report on Kapiti Island recently made by Professor H. B. Kirk and Mr W. E. Bendall, as showing the far-reaching effects of introduced animal life on the physiography of the country.

In connection with the introduction of sheep into New Zealand, it is of interest to note the remarkable development of the carnivorous habit in the kea or mountain parrot (*Nestor*), a bird which originally fed, chiefly, if not exclusively, on a vegetable diet. In 1867 it was observed in regard to certain sheep in the Wanaka district of Otago that they were wounded or badly scarred on the loins. It was found that this was done by keas, which lighted on the backs of the sheep, and attacked them with their powerful beaks. Many shepherds in the district saw the birds attack the sheep, especially when the latter were in snow or were in poor condition. The keas lighted down on the wool, and bit into the loin generally above the kidneys. Numbers of sheep succumbed to the injuries received, the loss in the Lake Hawea region being estimated at 5 per cent. annually over the whole of the flocks. In the Amuri highlands in North Canterbury the annual loss of 7½ to 8 per cent. was estimated to have risen to 15 per cent. in 1906. All keas do not attack sheep. The habit was originally acquired in the Wanaka district, and spread from there; but it has now been recorded from the Takitimos in the south to Amuri in the north. The origin of the habit is not very clear, but it is probable that it was first learned by keas picking the fat off sheep-skins which were hung on stockyards or on the wire fences, that then they attacked dead sheep—which are common enough in the high country, especially, after heavy snowfalls—and that from these they learned to attack living sheep. Keas shot on mountain country have often been found to have a good deal of both flesh and wool in their stomachs, but it is quite possible that this has been taken from carcasses lying in the snow drifts, where they are often preserved for a long time.

Bharal; Himalayan Bhurrel Sheep; Blue Sheep (*Ovis nayaur*)

In 1909 the Tourist Department liberated three of these animals in the Mount Cook district. Mr J. R. Murrell, guide at the Hermitage, writing in October, 1915, says: "Three were liberated, one of which was in poor health. Another was caught disturbing ewes on a neighbouring station, and was *perhaps* destroyed. The third has not been seen since being liberated."

*Goat (*Capra ægagrus*)

The introduction of goats dates from Captain Cook's second voyage. He says in his *Journal*:

> On 2nd June, 1773, I sent on shore, on the East side of the Sound, (Queen Charlotte), two goats male and female. The former was something more than a year old, but the latter was much older. She had two fine kids some time before we arrived in Dusky Bay, which were killed by cold.

Forster in his *Journal* says they were left by Captain Furneaux in an unfrequented part of East Bay, "this place being fixed on in hopes that they would there remain unmolested by the natives, who indeed were the only enemies they had to fear."

On the third voyage, the 'Resolution' was in Queen Charlotte Sound from the 12th to 25th February, 1777, and Captain Cook says:

> I gave Matahouah two goats, a male and a female with kid, (and to Tomatongeauooranuc two pigs, a boar and a sow). They made me a promise not to kill them; though I must own I put no great faith in this. The animals which Captain Furneaux sent on shore here, and which soon after fell into the hands of the natives, I was now told were all dead.

It is popularly believed that all the wild goats of New Zealand are descended from those introduced by Captain Cook, but while this may be partly true of those in the South Island, especially at its northern end, it can hardly explain those found in the North Island. It is more likely that they are descended from escaped animals; they are now abundant in many parts of New Zealand. Mr F. G. Gibbs tells me that goats were imported into Nelson some time in the forties. "In the fifties a large number were kept tethered on some hills in the Maitai Valley, still called the Goat Hills. Some of these goats escaped into the back country, and were the progenitors of the wild goats."

In the high country of Marlborough they are mainly of three colours, black—which is perhaps the commonest—khaki and white. In a trip through the cañon of the Ure River, Mr B. C. Aston says: "the fusillades of stones showered down on us by the goats which we had disturbed were a source of ever present danger."

Great numbers of them are to be met with in the rocky and precipitous country at Palliser Bay, near Wellington. Except when they move they are difficult to see, as their colours blend almost undistinguishably with that of their natural surroundings. They are abundant on Kapiti Island and unfortunately are also common in the Mt Egmont reserve, where they are doing much damage.

They also occur, though not commonly, on the sparsely scrub-clad faces of the west coast, north and south of Hokianga, as well as on the outskirts of bush land. They are not therefore considered to be of any commercial value.

Writing of those in the Lake Wakatipu district, Mr L. Hotop of Queenstown says (April, 1916):

> There is an immense number spread all over the Lakes district,—a moderate estimate gives them as many as 30,000....They are principally at Moonlight, Skippers, Sandhills, and at the lower end of the Lake, seriously interfering with the pasturage in these localities; one runholder has paid year after year for as many as a thousand during the season. At Moonlight, a digger, during the past nine months, has shot 550. My informant tells me he was offered 2*s*. 3*d*. a skin for as many as he could send.

Mr W. H. Gates of Skippers writes (April, 1916):

> there are a lot of wild goats here, almost within rifle-range of my cabin.... One sheep-farmer gave a shilling per pair of ears, and a shilling for each pelt. The male is a rough-looking customer; some have horns 15 inches in length, and $2\frac{1}{2}$ inches by $1\frac{3}{4}$ inches at the root; and they grow in a slightly spiral form....I think there is a strain of many breeds running through them all. Some have long hair, but are not the Angora breed. Some are almost white, but the chief colours are black and white, or black and tan. I have noticed here (and also on the West Coast) that the female has her young in winter, when food is not plentiful. Why this is I never could understand.

Goats are still found wild on the Galloway Station, Central Otago, though not so abundant as in former years. They live in the high country, and do not come down to the settlements. Mr A. Gunn, who managed this large run for many years, tells me:

> they are of great use to sheep farmers, as they keep down the "lawyers" (*Rubus australis*), and thus save the sheep from being entangled. In shooting them, if the wind is coming from them, you can smell them before you see them; and while they are feeding a billy-goat is always standing on guard. While they are of all colours, black and white is the commonest, though brownish-red, grey and even occasionally a white one are found. They live in the roughest places they can find.

They are also found in considerable numbers round the south-west corner of the South Island, but whether they have escaped from the settlements about Preservation Inlet, or have worked overland from Southland it is not possible to say with certainty. Probably the former is the explanation of their occurrence from Puysegur Point inland.

Mr W. R. Bullen of Kaikoura informs me (August, 1916) that they are numerous on his run, but while they eat very much the same food as the sheep do, they keep the scrub and bush open, so that the sheep can move through it.

The attempts made from time to time to acclimatise goats on the out-lying Southern Islands are of interest. Captain Enderby landed some on Enderby Island in 1850, and Captain Norman landed them both on the Auckland and Enderby Islands in 1865, but none appears to have survived. Cockayne says: "Two or three were landed on Ewing Island in 1895, but none have been seen recently. On Ocean Island, a very small island in the Auckland Group, goats are numerous at the present time, but I have no details as to how they got there." Captain Bollons writing me in February, 1916, speaks also of the last-named island, and adds:

Goats have been sent down from time to time to the Auckland Islands since 1890, most of which have either died or been killed off for food by castaways. At the Snares they were liberated about 1889, but soon died off. At Campbell Island some were landed in 1888 and 1890, and several were alive when the main island was taken up for a sheep run in 1896. At the Antipodes several were liberated between 1886 and 1900, but were either used for food by the castaways or died off.

Special breeds of Goats. In 1867 the Canterbury Society introduced three Cashmere goats, but it is not stated what was done with them. In the same year they introduced a pair of Angora goats, and these commenced to breed at once. From a newspaper cutting dated 1876, I find that "a flock of 120 Angora Goats on the Port Hills (Lyttelton), chiefly descended from two pairs introduced into New Zealand by the Melbourne Acclimatisation Society, has recently been dispersed and sold." The Otago Society imported four in 1867, and liberated them, but it is not stated where. The Auckland Society in 1869 also imported a number, and sold them to a Mr Howick. In addition to these, Angora goats were frequently introduced by private individuals, and in some cases became wild. Mr Aston writes (1916): "I hear that the Angora is hybridizing with the common goat in some parts of Marlborough." In the report of the Agricultural Department for 1903 it is stated:

The original flock (of Angora Goats) imported from Victoria and South Australia has now assumed considerable proportions, partly through the natural increase and the purchase of nineteen grade nannies from Mr Taylor White, of Wimbledon, Hawke's Bay.... The mob has been running at the Weraroa Experimental Station up till now.... The usual wire fence will not keep them in, consequently wire netting must be resorted to.... A few of the ordinary goats, with a pure Angora billy, have been sent to the natives in the Urewera Country, Bay of Plenty. *If not allowed to run wild* they should in a few years become of some commercial value.

No native fence will keep a goat in. The Angora goat is now being bred in fairly large numbers especially in the Auckland province in order to keep the blackberry pest in check. They are usually tethered close to the bushes, and shifted frequently as they eat them down. The total number of these animals registered in the Dominion in 1917 was only 6836.

*Thar; Himalayan Goat (*Capra jemlaica*)

In 1904 six of these animals were received from the Duke of Bedford, and were liberated near the Hermitage, Mt Cook. In 1913 three more were liberated near the Franz Joseph Glacier. Mr J. R. Murrell, guide at the Hermitage, writing in October, 1915, says:

> Other guides and I saw a few days ago a nice "mob" of 13 Thar on the Sealey Range; these were in the pink of condition and doubtless will become plentiful. Previously a much larger number were seen, but doubtless there are a number of mobs on this range.

By the end of 1920 these herds had increased very considerably.

*Chamois (*Rupicapra rupicapra*)

In 1888 enquiries were set on foot by the late Sir Julius von Haast and the author with the object of obtaining chamois for New Zealand. Dr von Hochstetter of Vienna who was communicated with was hopeful of obtaining some partially-tame animals from the King of Bavaria's park near Munich, and arrangements were made with Hagenbeck of Hamburg for their transmission to the Colony. To meet the expenses of shipment a vote of £150 was placed on the estimates by the Government, and the passage of this vote through the House of Representatives led to a scene of historic interest, and one of the most amusing incidents in the history of the House. The vote was objected to by Mr Kerr, member for Motueka, a goldfield's representative more remarkable for his vigour than for his knowledge or the accuracy of his information. The following is an extract from the *New Zealand Times* of 28th June, 1889:

Mr Kerr on the Chamois

> The vote of £150 appearing on the estimates for the importation of Chamois afforded Mr Kerr an opportunity last night of protesting vigorously against the introduction of more pests into the Colony. Amidst considerable merriment the honourable member said he was reliably informed that this animal was a cross between a pig and a sheep, and that it bred scab; and, in case it might be a goat he reminded the Government that there were already plenty of these animals running wild. The climax was reached when Mr Kerr unsuspectingly quoted from the book handed to him by Mr Turnbull (and which proved to be Mark Twain's *Tramp Abroad*)—a remarkable history of the habits of "small deer," under which name the celebrated American humourist concealed the identity of the flea.—"Was it reasonable," Mr Kerr asked, "to spend money on the importation of

animals no bigger than a mustard seed?" The House, however, was quite resolved, and deliberately passed the vote, in spite of the earnest protests of the member for Motueka[1].

I cannot recall now, nor find any record, as to why the introduction of chamois was not carried out in 1889, but I think the cause was that the animals could not be procured. No further attempt was made till recently. In 1907, the Government received *eight* chamois, a present from the Emperor of Austria, and these were liberated on Mt Cook. In 1913, *two* more, from the same source, were received and were set free in the same locality. Unfortunately one of the latter, a buck, attacked a party of tourists near the Mueller Hut, and was killed by the guide. It is most unusual for chamois to attack persons, but this particular animal is believed to have been in captivity for some years prior to importation. By latest reports (August, 1920) the flock is increasing fast and the animals are in very fine condition, herds of 30, 40 and 70 being noticed at one time. There is, therefore, no doubt that this species will be shortly strongly established in the Southern Alps.

Order CARNIVORA

Family FELIDÆ

*Cat (*Felis catus*)

Wild cats have been found in New Zealand from the early days of settlement, though for long they never strayed very far from the abodes of men. But after rabbits began to increase in many parts at such a rate as to reduce the sheep-carrying capacity of the country, sheep farmers began to purchase cats in the towns. These were taken

[1] The passage which Mr Kerr quoted, in which he spoke of the animals he objected to as "shammies," is as follows:

"Within a day or two I made another discovery. This was that the lauded chamois is not a wild goat; that it is not a horned animal; that it is not shy; that it does not avoid human society; and that there is no peril in hunting it. The chamois is a black or brown creature no bigger than a mustard seed; you do not have to go after it, it comes after you; it arrives in vast herds and skips and scampers all over your body inside your clothes; thus it is not shy; but extremely sociable; it is not afraid of man, on the contrary, it will attack him; its bite is not dangerous, but neither is it pleasant; its activity has not been overstated,—if you try to put your finger on it, it will skip a thousand times its own length at one jump, and no eye is sharp enough to see where it lights. A great deal of romantic nonsense has been written about the Swiss chamois and the perils of hunting it, whereas the truth is that even women and children hunt it, and fearlessly; indeed, everybody hunts it; the hunting is going on all the time, day and night, in bed and out of it. It is poetic foolishness to hunt it with a gun; very few people do that; there is not one man in a million can hit it with a gun. It is much easier to catch it than it is to shoot it, etc., etc., etc."

The same gentleman is credited with another amusing "acclimatisation" blunder. When the Nelson Borough Council proposed to import half a dozen Venetian Gondolas to be placed on the lake in the Public Gardens, he protested against such extravagance—"Why not import a pair, and then let Nature take its course?"

out to the back country, turned out and fed for a time, till they were established. No doubt some died, but most became more or less wild and learned to subsist on the smaller animals of the neighbourhood. They certainly destroyed many young rabbits. They cleared off the rats which were formerly so common, they also largely exterminated native lizards, and did much to destroy many native and introduced birds. Mr Chas. J. Peters, of Mount Somers, considers that wild cats are far more effective in keeping down rabbits than stoats and weasels, and estimates that a cat will kill more rabbits in a month than one of the others will in six months. Dieffenbach, writing of the Piako district (Auckland) in 1839, says: "the cats which, on becoming wild, have assumed the streaky grey colour of the original animal while in a state of nature, form a great obstacle to the propagation of any new kinds of birds, and also tend to the destruction of many indigenous species." This statement about the colour of wild cats has been made much of. It is only true to a limited extent, and I have always felt that such statements coming from a traveller who had only limited means of observing the facts, and who apparently founded his conclusions on a few isolated observations of the settlers, are not always safe to generalise from. In this instance they led Darwin (in *The Variation of Plants and Animals under Domestication*) to quote him, and to use the statement as a proof of the strong tendency to reversion shown by the cat when it escaped from domestication. At the time Dieffenbach wrote, settlement was quite in its infancy, and cats had not long been introduced. It is probable, therefore, that his statement, whether the result of his own or other people's observations, referred to cats which were themselves the progeny of grey animals. It certainly is the case that in Central Otago, where cats were freely liberated to cope with the rabbit pest, animals of many colours are now found wild[1].

Mr Robert Scott, M.P. for Otago Central, who had exceptional opportunities for observing the facts, has recently given me most interesting information regarding this question. He says:

the wild cat was no doubt the descendant of the shepherds' and miners' tame cat. The predominating colour was grey-striped, or tiger-striped—as some people called them,—occasionally yellow, and rarely black or black and white. The time I write of was the seventies, say from 1870 on to the time when poisoning the rabbits with phosphorized grain came in. The cats, though not numerous, were fairly common especially in districts where cover, such as fern and scrub, was plentiful. They grew to an

[1] In a paper entitled "Red Cats and Disease" (*Trans. N.Z. Inst.* XXXI, p. 680), Mr Richard Henry refers to the occurrence of distemper among wild cats at Manapouri Station in 1881, and states that red cats—which were always males—seemed to survive, when those of other colours succumbed to the disease. He also states that cats which live wholly on rabbits are very liable to disease.

immense size and were game to the last if attacked; in fact no dog would tackle one single-handed. They were always in the pink of condition, which may be accounted for by the abundance of feed available in the shape of wekas, ducks and rats, with perhaps a dead sheep or bullock occasionally. When the rabbit poisoning came in that class or variety of cat disappeared along with the wild pig and the weka. The reason for the extermination of the cat is because it prefers the entrails to the flesh. Since that time up to the present cats have been turned out in considerable numbers, but the rabbit-trapping has effectually prevented their increase, and the survivors still retain their original colours, that is black, black and white, grey, grey and white, etc., but they are much smaller than the wild cat of forty years ago. My opinion is that had the original cat survived till to-day the colour would have invariably been grey, or rather grey-striped.

Mr H. C. Weir of Ida Valley Station, Otago, states that on high country where rabbit-traps are seldom if ever used, they grow to a very considerable size, and are most commonly of a grey colour, but yellow, grey and white, and black are also to be met with. He adds: "I cannot say I ever saw any approaching the tiger-like stripe of the home country Wild Cat, and I have seen a good few of them in the wilds of Sutherlandshire, Scotland."

Some people consider that wild cats are responsible for much of the failure which has followed the constantly-renewed attempts to naturalise game birds. At the annual meeting of the Wellington Society in 1898, a member said: "cats are more destructive to game than all the hawks, weasels and stoats in the colony. Most of the bush coverts are full of these cats, a fact which he himself proved near Fielding where, with the assistance of traps baited with smoked fish, he caught many." I think they may have contributed to some extent to this failure, but only in a few parts of the country, and then chiefly in the neighbourhood of settlements. I do not think wild cats have had much to do with the extermination of game.

Mr B. C. Aston, in a paper on the Kaikoura Mountains, speaks of the half-wild cats which are found about deserted fencers' and musterers' camps, as retaining

> all their love for man's comradeship if encouraged, but they invariably refuse to eat anything that they have not killed themselves. They probably exist on rabbits, birds and mice. As a result of their hunting habits their chest and foreleg muscles are largely developed, and they have a different look to the ordinary domestic cat, being leaner, and quicker in action.

When the Russian Commander Bellingshausen visited the Macquaries in 1820 he found numbers of wild cats, which hid among the foliage. There were at the time, however, two parties of traders (seal hunters?) on the island, one of 13 and the other of 27 men, and these probably accounted for the cats.

Captain Musgrave, who was a castaway from the schooner 'Grafton' when she was wrecked on the Auckland Islands in 1864, found a cat in a trap, more than a year after the date of the wreck. "She soon cleared the hut of mice, which were dreadfully common."

In 1868, H. H. Travers in his account of a visit to the Chatham Islands states that wild cats were very abundant, and that they had destroyed a great number of the indigenous birds. Mr F. A. D. Cox, writing to Mr Jas. Drummond in 1911, from the Chatham Islands, reports that on Mangare, a small island of the group, there is a colony of tortoise-shell cats; the progeny of some liberated on the island in order to destroy the rabbits which were present in large numbers. He adds: "I do not know whether they have succeeded in killing out the rabbits, but they certainly have exterminated the small native birds." Presumably the Chatham Island Fern-bird (*Sphenœacus rufescens*), which was only found on Mangare, has now ceased to exist. Mr J. Grant of Wanganui informs me that cats frequently catch eels; he has four or five direct observations of the fact (1918). Cats are also responsible for the destruction of birds and tuataras on Stephen's Island in Cook Strait, where they have exterminated the little wren *Traversia Lyalli*, peculiar to this island.

*Dog (*Canis familiaris*)

When Captain Cook arrived in New Zealand in 1769, he found that the dog and a species of rat were the only mammals in these islands. The dog had been brought with them by the Maoris, and was similar to the form which was common in Polynesia. Most of the histories of the migrations of the Maori refer to the fact of their bringing dogs with them, so that they had probably been in the country for some centuries before the advent of Europeans.

Crozet saw them in 1772 and described them as follows:

> The dogs are a sort of domesticated fox, quite black or white, very low on the legs, straight ears, thick tail, long body, full jaws, but more pointed than that of the fox, and uttering the same cry; they do not bark like our dogs. These animals are only fed on fish, and it appears that the savages only raise them for food. Some were taken on board our vessels; but it was impossible to domesticate them like our dogs; they were always treacherous, and bit us frequently. They would have been dangerous to keep where poultry was raised or had to be protected; they would destroy them just like true foxes.

Forster, in his account of the second voyage, 1773, writing of Queen Charlotte Sound natives, says:

> "A good many dogs were observed in their canoes, which they seemed very fond of, and kept tied with a string round their middle; they were of a rough, long-haired sort, with pricked ears, and much resembled the

Common Shepherds' Cur, or Count Buffon's *chien de berger*. They were of different colours, some quite black, and others perfectly white. The food which these dogs receive is fish, or the same as their masters live on, who afterwards eat their flesh and employ the fur in various ornaments and dresses." Later on in the same journal he says: "The officers had ordered their black dog to be killed, and sent to the captain one half of it; this day (9th June) therefore we dined for the first time on a leg of it roasted, which tasted so exactly like mutton that it was absolutely undistinguishable....In New Zealand, and in the tropical isles of the South Sea, the dogs are the most stupid, dull animals imaginable, and do not seem to have the least advantage in point of sagacity over our sheep. In the former country they are fed upon fish, in the latter on vegetables."

Bellingshausen, who visited New Zealand in 1820, says: "We saw no quadrupeds except dogs of a small species. Captain Lazarew bought a couple. They are rather small, have a woolly tail, erect ears, a large mouth and short legs."

Dieffenbach, writing nearly seventy years after Cook's visit, remarks that:

the native dog was formerly considered a dainty, and great numbers of them were eaten; but the breed having undergone an almost complete mixture with the European, their use as an article of food has been discontinued, as the European dogs are said by the natives to be perfectly unpalatable. The New Zealand dog is different from the Australian dingo; the latter resembles in size and shape the wolf while the former rather resembles the jackal; its colour is reddish-brown, its ears long and straight.

The Rev. R. Taylor says: "The New Zealand dog was small and long-haired, of a dirty white or yellow colour, with a bushy tail; this the natives state they brought with them when they first came to these islands." Then he adds: "it is not improbable, however, that they found another kind already in the country, brought by the older Melanesian race, with long white hair and black tail; it is said to have been very quiet and docile."

S. Percy Smith saw several Maori dogs in a native village at Warea, near Cape Egmont, in 1852. They were long-bodied, fox-eared, sharp-nosed, long-haired, bushy-tailed, yellowish-brown, and dark—almost to black—in colour. They stood about 18 inches high. He branded them as curs. They were evidently lazy, stupid brutes, which never became wild.

Mr Elsdon Best writes to Mr Drummond (*Lyttelton Times*, 11th January, 1913):

Some old Maoris of the East Coast district assert that before Captain Cook's visit there were two distinct breeds of dogs in New Zealand. One was a large dog, with long hair, and lop ears; the other a small dog with erect ears. The first was brought, they say, from Raiatea. This variety was bred solely for food and clothing; it was useless for hunting. The long hair covered the body as low as its knees, and there was a natural parting along the top of the back. The small dog was also introduced from Polynesia, and was useful for hunting the kiwi, weka and parera (grey duck).

The so-called "Maori" dogs seen in the fifties and sixties he believes were crosses. As to the word "pero-pero," he says that whether it was Spanish in origin, or not, it is not improbable that Spanish vessels reached these shores before Cook's visit.

For centuries the Spaniards concealed the results of their voyages very carefully. In the short accounts of their early voyages, the positions of the islands discovered by them were vague and unsatisfactory. The voyage made by Juan Fernandez westward from Chili, and then southward to a land inhabited by white people who made "good woven cloth," may point to a visit to New Zealand's shores. The natives of the more northern islands, unlike the Maoris, did not wear woven material, but used bark cloth, and "white people" might refer to the Maoris, as the Spanish voyagers called all true Polynesians white.

H. J. Fletcher in a late issue of the *Journal of the Polynesian Society* states that the Maori dog was known at the Matapihi Station, Taupo, as late as 1896. The shepherds employed on the station shot a number of dogs, long-haired, bushy-tailed, and of a dirty white colour.

Elsdon Best (April, 1913) gives some statements about the Maori dog as follows:

(1) Captain Mair says that in his youth he saw these dogs trained to hunt by themselves through the kumara plantations for the large caterpillars of the *Sphinx convolvuli*. They were trained to put their noses under the trailing shoots of the vines and to turn the shoots over, in order to expose any caterpillars that might be present. If they succeeded in finding any they devoured them. If the dogs were not watched they ate pieces out of the pumpkins.

(2) Savage, in *Some Account of New Zealand*, published in 1807, says:

as far as I can learn, the natives have no larger animal than the dog, which is a native here, usually black and white, with sharp, pricked-up ears, the hair rather long, and in figure resembling the animal we call a foxhound.

(3) Shortland (in 1856) says:

The natives wore cloaks made from the skins of dogs before Captain Cook's time, and their manner of fabricating such cloaks is particularly ingenious. Moreover, the native breed of dogs still exists in New Zealand, though, perhaps, seldom in its original purity, and is preserved in some places for the sake of its skin. In appearance it is very unlike the European breeds. Its body is long, legs short, head sharp, tail long, straight and bushy. The hair is thick, straight, and tolerably long, varying in colour from white to brown, but it is not spotted.

In a paper written by Dr Hector in 1876 on the "Remains of a Dog found near White Cliffs, Taranaki," he says: "The remains of a dog were found in a hollow tree which was imbedded in a cliff (at a depth of 19 ft.) near the Urenui River." Captain Rowan, the discoverer, and Dr Hector both seem to think that the dog must have crept into the tree, at a comparatively recent date, for though the lignite which occurred in one of the layers above the remains is of great antiquity, the state of preservation of the bones, as compared with the thorough alteration that the vegetable matter of the lignite has undergone, inclines me to believe that the dog remains are of modern origin. But even in that case, the circumstances under which they have been found, and the decayed state of the dentine layer of the teeth tend to refer them to a period further back than any previously obtained.

Dr Hector states (1876) that a bitch and full grown pup were known for several years in the densely wooded country between Waikawa and the Mataura plains, and did great damage among the flocks of sheep, but exhibited such cunning and daring that it was not till after hunting them for two years that they were shot by Mr Anderson, who presented them to the Colonial Museum. Of the smaller specimen both skin and skeleton were taken to the British Museum by Sir Geo. Grey, and the skin of the mother was preserved here, and has been recognised by many old Maoris as a genuine *Kuri* or ancient Maori dog.

In general appearance it resembles a poodle, but it presents characters unlike any other of the many breeds of dogs which we are familiar with. It is a large bodied dog with slender limbs, large ears, and a straight half-bushed tail, wide head, and small pointed nose. Its colour is white, with a black spot on the loins, and a brown spot on the crown of the head, and a few faint spots on the ears. Its nose is black, and its claws are white. The back is covered with hair. The total length is 3 ft., and the height of the shoulder 17 inches.

Taylor White writing in 1889, says:

I consider these dogs entirely distinct from the European dog. For the wild dogs met with on the Waimakariri River in the Alpine ranges of Canterbury during the year 1856, were in colour and markings identical with those found in the Alpine region of the Lake Wakatipu, Otago, in 1860, a distance of several hundred miles apart. There seems little room to doubt that they were an original Maori dog. The fact of their wanting the two tan spots over the eyes mostly seen in European dogs of approximate colour, is a very strong evidence also in favour of this opinion.

The Maori dog has totally disappeared. Mr S. Percy Smith tells me that the last one he heard of was about 1896.

When settlement began European dogs must have crossed freely with the native animal, and many both of the introduced and crossed dogs became truly wild, especially as there were sheep and goats to worry and pigs to chase and kill.

Dr Lyall, who was surgeon on H.M.S. 'Acheron' during the survey of the coast of New Zealand in 1844, in a paper read in 1852 before the Zoological Society of London says of the Kakapo, that:

at a very recent period it was common all over the west coast of the Middle Island; but *there is now a race of wild dogs* said to have overrun all the northern part of this shore, and to have almost extirpated the Kakapo wherever they have reached.

The same thing was practically said by Brunner (1846–1848), who was nearly starved in S.W. Nelson owing to the destruction of the ground birds.

The early settlers could not distinguish between Maori dogs and these half-wild curs. Thus R. Gillies, who arrived in Otago at the beginning of the settlement in 1848, writing in later years says:

For some years after the settlers arrived here, the wild dog was the terror of the flock-master and the object of his inveterate hostility.... They ran from any tame dogs, and tame dogs, as a rule, would follow and attack them with all their master's antipathy.... The bulk of the wild dogs were not domestic animals gone wild, but the true old Maori wild dog.

W. D. Murison, formerly editor of the *Otago Daily Times*, writing at the same period (1877), tells how in 1858, he and his brother took up country in the Maniototo Plains, which they reached by the Shag Valley. The wild dogs were very troublesome. The first was caught by a kangaroo dog (apparently imported from Australia for the purpose of hunting them).

This particular wild dog was yellow in colour, and so was the second killed, but the bulk of those ultimately destroyed by us were black and white, showing a marked mixture of the collie. The yellow dogs looked like a distinct breed. They were low set, with short pricked ears, broad forehead, sharp snout, and bushy tail. Indeed those acquainted with the dingo professed to see little difference between that animal and the New Zealand yellow wild dog. It may be remarked, however, that most of the other dogs we killed, although variously coloured, possessed nearly all the other characteristics of the yellow dog.... The wild dogs were generally to be met with in twos and threes; they fed chiefly on quail, ground larks, young ducks, and occasionally on pigs. On one occasion, when riding through the Ida-burn valley, we came across four wild dogs baiting a sow and her litter of young ones in a dry tussock lagoon. To our annoyance, our own dogs joined in the attack upon the sow, and the wild dogs got away without our getting one of them.... In all, we destroyed 52 dogs between September, 1859, and December, 1860.

Tancred writing of Canterbury in 1856 says: "A few dogs have escaped and become wild in unfrequented parts, where they have become dangerous to the flocks."

The following paragraph appeared in the *Auckland Herald* on 18th November, 1866:

> It is not generally known that about Otamatea and the Wairoa the bush is infested with packs of wild dogs, as ferocious, but more daring, than wolves. These dogs hunt in packs of from three to six or eight. They are strong, gaunt large animals, and dangerous when met by a man alone. Not long since a Maori, when travelling from one settlement to another through the forest, was attacked by three of these animals at dusk, and only saved himself by climbing into a tree, where he was kept prisoner until late the next day. The extensive district over which these packs roam was once well stocked with wild pigs, but most of these have fallen victims to the dogs, and since this supply of food has failed the dogs have ventured after dark to the neighbourhood of native settlements and the homesteads of European settlers, in quest of prey.

G. M. Hassing, writing from Feldwick, Otago (March, 1913), states that wild dogs infested the country about Lake Wanaka in 1860. They were exceptionally plentiful on the western ranges and in the country near the Matukituki River. They became so troublesome that the settlers found it necessary to keep packs of kangaroo dogs to hunt and destroy them. Mr Hassing describes them as of no particular breed, but just coarse inbred mongrels, shy, cautious and cowardly. He adds:

> In 1865, when exploring the Clark and Landsborough Rivers, tributaries of the Haast, I observed numerous tracks of wild dogs, but saw only one of the animals which came out of the bush across the river one evening. It was a large, rough-looking animal like a wolf, and when it caught sight of us it set up a most dismal howl, and plunged into the forest again.

Mr Andrew Wilson, a veteran surveyor, writing (February, 1913) from Hangatiki, about 120 miles south of Auckland, says that the wild dogs which lived in the North Island forests a few years ago had a strain of the original Maori dog in them. He describes two he saw as of a reddish-fawn colour, about the size of an ordinary cattle-dog. As far as his experience went, the wild dogs never barked, but only howled.

Mr J. Hall (May, 1913) says that on the Kaingaroa Plains he found wild dogs, red in colour, with pointed ears, and, when full grown, as large as a small collie. They were usually five or six in a pack. When a person approached they retired to a safe distance, and gave a kind of howl. He never heard them bark, nor did he hear that they ever attacked anyone. On one occasion, however, when the late Mr R.

Mayer, of Otiamuri, was driving along the edge of the plains with his wife, a mob of five or six wild dogs rushed them and jumped at the horses' heads. On another occasion, a young Maori came from a neighbouring pa to see him. In a short time he returned with a terrified look on his face, and stated that a pack of wild dogs were attacking a large calf. Mr Hall, taking his gun, went to the scene, and was just in time to drive the dogs off and save the calf. When travelling over the plains with domestic dogs, Mr Hall noticed that the latter can scent the wild dogs miles off. As soon as they receive the scent, they stand and watch, and their hair becomes bristly almost at once.

My son, G. Stuart Thomson, informs me that wild dogs were at one time so common in Marlborough, and did so much damage on the sheep runs, that packs of hunting dogs were kept and bred for the special purpose of running them down. £5 per head used to be paid for wild dogs.

Mr Elsdon Best wrote that in 1877, the Rev. W. Colenso said in regard to Maori dog-skin garments: "Many a dog-skin mat has he made within the past fifty years of the skins of dogs of the small mongrel breed, before European clothing became common among the Natives."

As settlement proceeded and the country became opened up, wild dogs were gradually exterminated. The only ones which are now met with are curs which have managed to escape from their owners and have taken to rabbit- or to sheep-killing.

Bellingshausen reported wild dogs on the Macquaries in 1820, but it is improbable that they long survived the sealers, who probably originally brought them. As soon as the killing of seals and sea-birds stopped, the dogs probably died out. Captain Musgrave, who was wrecked on the Auckland Island in 1864, discovered wild dogs —like sheep dogs—on the island.

Family Mustelidæ

*Ferret; Polecat (*Putorius fœtidus*). *Stoat; Ermine (*Putorius erminea*). *Weasel (*Putorius vulgaris*)

Nothing in connection with the naturalisation of wild animals into New Zealand has caused so much heart-burning and controversy as the introduction of these bloodthirsty creatures.

The Canterbury Society introduced five ferrets in 1867, and an additional one in 1868. They were apparently not liberated, though the progeny was probably sold to private individuals. In 1873 the

Society had six in Christchurch Gardens. Probably private individuals (dealers) introduced them at all the chief centres, but there is no record.

As rabbits began to increase to an alarming extent, various suggestions were made as to importing what was called "the natural enemy." One authority actually proposed to introduce Arctic foxes, because their fur would be so valuable. When it was pointed out to him that they would probably prefer lamb to rabbit, he replied that as they did not know anything about lambs in their native haunts it was improbable that they would take to eating them in New Zealand. Fortunately his proposal was not given effect to. Meanwhile sheep farmers brought pressure to bear on the Government, and as a result steps were taken to obtain ferrets. Numbers of these were introduced in 1882, and in the following year, Mr Bailey, Chief Rabbit Inspector, recommended the introduction of stoats and weasels. To show the scale on which these recommendations were carried out, I summarise from Mr Bailey's reports for four years as follows:

(*a*) In July, 1883, it is stated that since March, 1882 (15 months), the Agent-General had made 32 shipments of ferrets from London, numbering altogether 1217 animals. Of these only 178 were landed, at a cost of £953. Of 241 purchased in Melbourne, 198 were landed at a cost of £224. Thus the total number landed was 376, and the cost £1177, or £3. 2*s*. 7*d*. per head. The natural increase was 122, but 157 died of distemper. At this period it would seem as if the Government kept a perfect menagerie of these animals. In the same year a substantial bonus was offered to any one who would introduce a certain number of stoats and weasels in a healthy condition.

(*b*) In 1884 he reports: "nearly 4000 ferrets were turned out; 3041 in Marlborough alone, and about 400 on crown lands in Otago." The rest appear to have been sold to private individuals. He also states that "an agent has been sent home to procure stoats and weasels." Mr Rich of Palmerston imported some of these latter in a sailing vessel, but how many I cannot learn.

(*c*) In 1885 two lots of stoats and weasels were received from London, viz. 183 weasels (out of 202 shipped), and 55 stoats (out of 60). Of these, 67 weasels were released on a peninsula on Lake Wanaka of 8000 acres, on which they reduced the rabbits, but by no means exterminated them; 28 weasels and 6 stoats were liberated at Lake Wakatipu; 15 weasels near the Waiau River, Southland; and 8 stoats at Ashburton. The rest were sold at Wellington, Christchurch, and Dunedin.

(*d*) In 1886 the Government introduced two lots, viz. 82 stoats and 126 weasels, which were distributed in about equal lots to the Wilkin River, the Makarora, at the head of Lake Ohau, and on the

Waitaki; and 32 stoats and 116 weasels distributed between Marlborough and West Wairarapa. A private shipment of 55 stoats and 167 weasels was also received for Riddiford's station in West Wairarapa. The localities selected for these animals were those in which rabbits were most abundant. Mr Bailey also reported that "ferrets were turned out by thousands," but the success was only partial.

In this same year a meeting was held at Masterton to consider the administration of the Rabbit Act, and the best means of dealing with the pest. One of the resolutions carried was:

that the introduction of ferrets, stoats and weasels in large numbers is, in the opinion of this meeting, the only means by which the rabbit pest can be successfully put an end to, and that every owner of land infested with rabbits should either turn out ferrets in proportion to his acreage, or contribute to a fund for the breeding and purchase of ferrets, stoats and weasels to be turned out in the district. That the land-owners present form themselves into an association for the purpose of providing the natural enemies.

An Association for the purpose was accordingly formed with this object in view, large sums of money were subscribed and hundreds of stoats and weasels were introduced into the district. Several of the acclimatisation societies took strong exception to the action of the Government and of the sheep owners directly concerned, but as the societies were themselves directly responsible for the rabbits, their protests were ineffective.

However much the introduction of the Mustelidæ is to be deplored, the mischief has been done. Stoats and weasels are common in nearly every part of New Zealand and in some parts are enormously abundant. Ferrets (or the wild form, the polecats) are also met with. The latter do not thrive to any extent in the South Island; it may be that the winters are too severe for them. Probably most of the ferrets originally turned out were white or yellowish; but some shot in the neighbourhood of Dunedin seem to have reverted nearly to the original colour of the polecat.

These animals have not exterminated the rabbit, they do not even seem able in most parts to keep them in check. There is, however, great difference of opinion on the subject. Mr Chas. J. Peters, of Mount Somers, writes about these animals (August, 1916):

Since the stoats and weasels become fairly numerous the rabbits have increased a hundred per cent. and more. I have found weasels' nests both in heaps of fencing material, and also in rabbit burrows. *These nests have always been made out of skylarks' feathers*. I have also found parts of young hares at weasels' camps, but never a sign of a rabbit.

Against this we can place the evidence of an old settler like Mr H. B. Flett of Table Hill, Otago, who states most definitely that he used formerly to keep as many as 16 dogs on his place, and also employed ferrets, phosphorised oats and pollard, and in spite of most strenuous efforts he was only able to hold the rabbits in check. Since the introduction of stoats and weasels the rabbits have become fewer and fewer in number, and now, on his property of 7000 acres, the pest has practically disappeared, except in one corner where trapping is carried on. Where trappers are allowed to work, rabbits increase, for numbers of stoats and weasels are thus destroyed. Mr Flett's experience and opinions are those of many large land-holders throughout New Zealand.

My son, G. Stuart Thomson, has given me this note on the destructive action of the stoat. He says:

> At Lee Stream, in the Taieri district I saw a rabbit paralysed with fright and uttering squeals of terror, and on looking round for the cause observed a stoat fully ten feet away walking deliberately towards the victim. The rabbit was killed by one bite on the neck. A few weeks ago a lady informed me that she had seen a somewhat similar occurrence at Brighton, but in this case the rabbit struggled to the lady for protection, and fell trembling at her feet, while the stoat disappeared.

In regard to any natural enemy it is, of course, absolutely certain that it cannot exterminate, but can only keep in check, the animal it is intended to cope with. If it does more, then its own means of livelihood are imperilled, or it has to find other victims[1]. Thus one direct benefit which stoats and weasels confer is the wholesale destruction of rats and mice which they cause. Indirectly this may explain why certain birds, such as wekas among native species, and Californian quail among introduced forms, have increased of late years in districts where both stoats and weasels abound. It may be that rats are more destructive to eggs and to young birds than even stoats and weasels. The latter certainly will not touch birds if they can get rats. Mr Flett, whom I have referred to above, tells me that 20 years ago rats were a perfect curse about the homesteads, destroying harness, sheep-skins, grain and food, but that since the weasels appeared the rats have absolutely gone. He states he has not seen one about his place for 16 years.

The evidence regarding the destruction of the native avifauna by stoats and weasels is very inconclusive. Imported to destroy rabbits, they have penetrated into regions where rabbits are unknown, and where their food must have consisted exclusively of birds and bush rats (*Mus rattus*). Yet even in such districts there is evidence that native birds still survive in abundance, and there are also cases where birds like wekas, etc. have re-established themselves.

[1] In Taranaki, in March 1917, a litter of nine sucking-pigs was found destroyed one night, apparently either by stoats or weasels.

Mr Richard Norman, Albertown, writes in the *Otago Witness* of 2nd October, 1890:

I think that Mr E. H. Wilmot's experience in the Hollyford Valley, as recorded in the *Witness* a year or two ago, conclusively proves that the imported vermin kill the native wingless birds. He encountered there a ferret warren, and the weka, kiwi, and kakapo were almost exterminated. In the Makarora Valley these used to be plentiful, but since the advent of the stoats and weasels they are very rare, and rabbiting tallies have not depreciated.

Mr Geo. Mueller, Chief Surveyor of Westland, in his report on the "Reconnaissance Survey of the Head-waters of the Okuru, Actor and Burke Rivers" (*Rept. N.Z. Survey Dept.*, for 1889–90, p. 50), says:

Several weasels and ferrets were caught and killed at the Okuru and Waiatoto settlements, within about a mile from the sea-coast. . . . No rabbits were met with until near the Actor, 19 miles from the coast; and they were only seen in numbers at the very head-waters of the Okuru. . . . Meanwhile the Kakapos, Kiwis, and Blue Ducks have nearly disappeared from the district.

Mr Richard Henry, writing from Lake Te Anau in September, 1890, says:

I have known the ferrets to take seven young paradise ducks out of a clutch of ten in 1888, and last year the same pair of ducks only reared two young ones, but away from the lake I have seen larger families. I found two black teal ducks killed by a ferret, though it is seldom any of their work is seen, for they always drag their prey under cover. The black teal are getting scarce.

Mr Henry adds:

I think very few ferrets at liberty survive the winter for want of food.

Sir Thos. Mackenzie has recorded a case in which a weasel killed a black swan; and another which he saw in the Catlins district where a weasel brought down two tuis (*Prosthemadera*) from a tree.

The reverse of this tale is interesting. Mr H. Drummond has accumulated some evidence as to the killing of weasels by wekas (*Ocydromus* sp.). In 1909 Mr Murrell, junr., and Mr Harry Birley described how the wekas had been seen attacking and killing weasels. Mr Murrell witnessed a most interesting fight between them on a path. The weka circled round the weasel, watching a chance to spring in and strike it, which it did, always on the head, finally stretching its opponent out. They both note that native birds were beginning to increase again. In 1916 Mr A. T. G. Symons of Christchurch recorded the fact that wekas were killing weasels.

In regard to the occurrence and distribution of these species at the present time in New Zealand, I have no record of the introduction of the true polecat (*Putorius fœtidus*) into the country; but some eight or nine years ago Mr Thomas Anderton, curator of the Portobello Marine Fish Hatchery, shot two animals, which were too large for stoats, being about eighteen inches long. They were not ferrets, for they were brown coloured. Unfortunately he did not realise the importance at the time of preserving the skin, their smell—for one thing—being so offensive, and so their specific character was not determined. It may be, of course, that they were stoats of unusually large size.

Ferrets are fairly common throughout the country. I was formerly of opinion that this species, which does not in Northern Europe survive the winter unless carefully housed, could not stand the winter in Otago, or indeed in any of the inland parts of New Zealand where the winter is severe. I am informed, however, by trappers of experience, that they survive the Otago winter quite easily. Apparently wet cold is their enemy; and where burrows are warm, they can stand the dry cold quite easily.

Stoats are common from end to end of both islands. Mr Yarborough of Kohu Kohu states that stoats and weasels do not seem to be so numerous now (1916) as they were some few years ago. At that time a great number of these intrepid little animals appeared on the eastern side of Hokianga estuary, and were occasionally observed swimming across the river, which is about a mile wide. For the last year or more they have neither been seen nor heard of.

In the parts of New Zealand where the winter cold is severe, stoats retain their habit of changing their coat in the late autumn. According to Seebohm (*Siberia in Asia*, p. 41), the ermine in Scotland regularly assumes its winter dress in cold winters, and in England as far south as the Derbyshire moors. In New Zealand I have records of white stoats in winter from Burke's Pass, the Mackenzie Country, from the Taieri district, and from Lake Wakatipu, and these not as single instances, but as a fairly common occurrence. Thus Drummond (June, 1913) records the occurrence of a stoat from West Oxford. It was 17 inches long, and pure white in colour, except for the tip of the tail, which is jet black. The stuffed specimen is in the Canterbury Museum[1].

Family Sciuridæ.

Chipmunk; Californian Grey Squirrel (*Tamias striatus*)
Brown Californian Squirrel (sp.?)

About 1906, Mr P. R. Sargood, of Dunedin, liberated two of the former (all that remained out of 12 shipped from San Francisco)

[1] The late Dr Günther of the British Museum was not usually credited with a *great* sense of humour, but when discussing with Dr Chilton of Christchurch the introduction of stoats and weasels into New Zealand, he remarked: "Ach! why did they not send out males only?"

and two of the latter species. They were seen about the Dunedin Town Belt and neighbourhood for two or three years but were not known to increase[1].

Family MURIDÆ.

Maori Rat; Kiore (*Mus exulans*)

A species of rat was one of the four land mammals found in these islands when Captain Cook first visited New Zealand, the others being a dog, and two species of bats. Sir Joseph Banks says in his *Journal* (p. 224):

> On every occasion when we landed in this country, we have seen, I had almost said, no quadrupeds originally natives of it. Dogs and rats, indeed, there are, the former—as in other countries—companions of the men, and the latter probably brought hither by the men; especially as they are so scarce, that I myself have not had an opportunity of seeing even one.

This was not Forster's experience, for in his account of the second voyage of Cook, he says (vol. 1, p. 201):

> "Our fellow voyagers" (Furneaux in the 'Adventure') "found immense numbers of rats upon the Hippah rock (Queen Charlotte Sound), so that they were obliged to put some large jars in the ground, level with the surface, into which these vermin fell during the night, by running backwards and forwards, and great numbers of them were caught in this manner."

Always in reading this account, and considering the facts, I think it highly probable that these rats, spoken of by Forster, were not Maori rats at all, but were black rats (*Mus rattus*). Both Cook and Banks considered the native rat to be rare. The 'Endeavour' was in Queen Charlotte Sound for some days in January and February, 1770, and some of the rats on board were almost certain to find their way ashore. Furneaux arrived in Queen Charlotte Sound in April, 1773, Cook, in the 'Resolution' reaching it in May. Over three years had elapsed between the two visits, and on the second occasion the rats were found to be extraordinarily abundant. The rate of increase of

[1] In *Nature* of 8th March, 1917, the following paragraph appeared: "Sir Frederick Treves, in the *Observer* of 25th Feb., directs attention to the grave results likely to follow from the introduction of the American Grey Squirrel into Richmond Park. Not only has it driven out our own native red squirrel, but it has also spread beyond the confines of the Park into adjoining gardens, working serious damage there. 'They eat everything that can be eaten, and destroy twenty times more than they eat.' 'The buds and shoots of young trees, apples, pears and stone fruits, peas and strawberries are all laid under a heavy contribution. Already it seems the Office of Works has given orders for the destruction of these pests. The order, however, has come somewhat late, for they have already made their way into the open country of Surrey with a steady persistence and in good force. When it has reached the fruit gardens and young plantations of Surrey and Kent, we shall hear more.'

We are evidently in grave danger of having another very practical lesson in the folly of 'acclimatisation,' of which the rabbit in Australia forms a familiar and awful example."

rats is known to be very great, and it seems to me that the animals met with on the second voyage were the progeny of some which got ashore in 1770[1].

The Rev. R. Taylor says that this animal, the Maori rat, was in general size about one-third that of the Norway rat. The Maoris used to make elaborate preparations to catch them, and hundreds would be caught at one hunting. Taylor says the animal is reported to run only in a straight line, and that the Maoris made special lines of roads in order to lead them into their traps, which were baited with miro and other berries; if these roads were crooked, they said the rats ran into the forest at the bends. They fed entirely on vegetable matter and were greatly prized as food by the natives, who also extracted much oil from them. The native rat quickly disappeared before other rats, and imported cats. It was extremely rare 30 or 40 years ago, and it is probably quite extinct now. As, however, the species is common in Polynesia, occasional immigrants may arrive in New Zealand from time to time.

Tancred writing of Canterbury in 1856, says: "the native rat forms numerous burrows, rendering the soil unsafe for a horse." He also says "the rat is being exterminated by the formidable invader the Norway rat."

W. T. L. Travers, writing in 1869, says:

It has been the fashion to assume that before the arrival of Europeans in this Colony, this creature was common, and to attribute its destruction to the European rat, and, indeed, the natives have been credited with a proverb in relation to this point. It is not in effect impossible that the

[1] In *Rats and Mice as Enemies of Mankind* by M. A. C. Hinton (British Museum, Economic Series, No. 8), published in 1918, the following statement occurs: "There have been many attempts to calculate the reproductive potential of rats. For instance, F. von Fischer, in 1872, concluded that the progeny of a single pair might in ten years amount to no less than 48,319,698,843,030,334,720 individuals; Rücker, more recently, has computed the increase of a pair in five years at 940,369,969,152 rats.

Lantz was not so ambitious; for the purposes of his calculations he assumed the rats to breed only three times a year, and to have average litters of ten. Breeding at this rate uninterruptedly for three years, producing sexes in equal numbers, and with no deaths, the progeny of a single pair at the ninth generation would be 20,155,392 rats.

Zuschlag assumed a pair to have six litters of eight in a year; that the young would breed when three and a half months old, then with equal sexes and no deaths the progeny at the end of the first year would be 880 rats.

Although such calculations are purely theoretical, and although their results, in ordinary circumstances, will never be approached in Nature, they are not extravagant, *qua* the power to reproduce, but are based upon moderate and conservative estimates. In proof we may cite Kolazy's record that two females kept by him had twenty-six litters in a space of thirteen months, and produced 180 young—almost double the number assumed by Zuschlag. *We can, therefore, readily understand how the progeny of a few rats introduced to a new country by a ship may, in favourable circumstances, succeed in overrunning the whole country in the space of a few years.*"

ultimate destruction of those which still existed when trade was first opened between Europeans and the natives, long after the Colonization of New *South Wales*, may have been hastened by the introduction of the European rat; but I am satisfied that before that time they had become very scarce, and indeed I have been told by gentlemen who have lived in the Northern part of this island for upwards of forty years, that they never saw a specimen.

My son, G. Stuart Thomson, says: "that the Maori rat was once very abundant seems to be proved by the fact that the Maoris always erected their store houses for food of various kinds on piles, as a protection against the depredations of rats. (I think this was the custom before Europeans landed—cf. Maning's *Old New Zealand*.)" I think, however, that it may have been protection against the black rat which was sought, and that they may have got the idea from early European settlers.

It has been suggested that the disappearance of the Kiore Maori or native rat has led to the diminution, and almost to the extinction of the Laughing Owl (*Sceloglaux albifacies*). Sir Walter Buller says: "The fact that the extinction of the native rat has been followed by the almost total disappearance of this singular bird appears to warrant the conclusion that the one constituted the principal support of the other."

Mr W. W. Smith, writing in the *N. Z. Journal of Science*, says:

The suggestion of Dr Buller...is an important one; and my researches among the rocks at Albury, and experiments with the living birds in captivity, are greatly in support of this. In several of the crevices where I captured them, I found an ancient conglomerate of exuviæ ranging from three to twelve inches thick. From the under surface and through the mass to nearly the upper surface, this conglomerate is thickly studded with Owl's castings, composed entirely of light brown hair (which is unquestionably that of the Kiore Maori) and small bones. The castings more recently deposited among the rocks are composed of elytra and legs of beetles.

*Black Rat (*Mus rattus; Epimys rattus*)

It is impossible to say when the black rat first came to New Zealand, but it probably arrived with some of the first ships which came to the country. I have already suggested that Captain Cook's ships introduced them. Yates in 1835 says: "The natives tell us that rats were introduced in the first ship, by Tasman." Oldfield Thomas (*Proc. Zool. Soc.*, 1897, p. 857) states that "the rats normally inhabiting ships are not, as is commonly supposed, *Mus decumanus*, but *Mus rattus*, and in most cases are the grey variety of that animal, with white belly, though the black form may often be caught in the same ship as the grey." For a long time great confusion existed in the minds of most of those who observed and wrote of rats in this country, between the native or Maori rat (*Mus exulans*) and the black rat. For instance Buller in 1870 in a paper "On the New Zealand Rat," gives figures and descriptions of *Mus rattus*.

The black rat became enormously abundant in the early days of settlement, and used to move about the country in vast armies. The settlers, bush fellers and saw-mill hands of fifty to seventy years ago, have recorded how invasions of them in countless swarms used to move through their district, climbing everywhere, and eating everything that was of a vegetable nature. Oldfield Thomas, in the article already quoted, says: "All the world over *Mus rattus* takes to roofs and trees on meeting its formidable rival *Mus decumanus*, to which it leaves the gutters and cellars[1]."

In 1840 Messrs Dodds and Davis of Sydney established a farming settlement at Riccarton, close to where Christchurch now stands, and sent down James Heriot (or Hariot), as manager, two farm hands, and two teams of bullocks. They ploughed and cultivated about thirty acres of land and secured their crops. But in less than a year they decided to abandon all further efforts. Numberless rats attacked the garnered stores, and the bar at the mouth of the river or estuary proved a sad obstacle to shipping whatever grain had been spared by the scourge of rats.

[1] In his presidential address to the Royal Society of New South Wales on 1st May, 1918, Dr J. Burton Cleland gives some very interesting information on "the Rats that Travel by Sea." I think his remarks on the subject are worthy of quotation in full. "As the old English black rat (*Epimys rattus*), including the Alexandrine variety (*E. rattus alexandrinus*), the Norway rat (*Epimys norvegicus* (*decumanus*)), and the common house mouse (*Mus musculus*), are all subject to plague, it is of considerable interest to see which of these species is most prone to travel by sea. The most frequent traveller of the three would naturally be looked on, other factors being equal, as the most likely introducer of the plague bacillus into unaffected parts. For the purpose of putting this matter beyond dispute, I have had a list prepared of all the rats and mice submitted for examination to the Microbiological Laboratory under my charge, from vessels berthing in the cosmopolitan port of Sydney between April 16th, 1913, and April 14th, 1917. During the period rats or mice were found in fumigation by the Commonwealth Department of Quarantine on 189 vessels, after the accomplishment of 325 voyages.... The ships belonged to all nationalities, though naturally British vessels much predominated, whilst the voyages they made included coastal, interstate and overseas in all directions. On the 325 voyages made by the 189 vessels,

Epimys rattus was present in 293.
2968 individuals were found and submitted, an average per voyage of 9.

Epimys norvegicus was present in 3.
7 individuals were found, an average of ·02 (of these vessels one came from Vancouver, and one from Noumea).

Mus musculus was present in 53.
487 individuals were found, an average of 1·5.

The largest numbers of mice were found on vessels trading with the North Coast of New South Wales, and an undue proportion of such vessels yielded mice, probably as a result of the frequent carriage of fodder, mice were only occasionally found on vessels from overseas."

So writes Dr Hocken in his interesting *Early History of New Zealand*.

Taylor White states that on the west coast (of the South Island) they came in vast crowds, climbing trees, tent poles and ropes, and ate everything. On the shores of Lake Wakatipu they lived under the dead leaves of the Wild Spaniard.

Rutland records how:

in 1856 the district of Collingwood on the western side of Blind Bay was visited by a swarm, and, in 1863, I am informed of a swarm on the Shotover, Otago. Repeated swarms occurred in Picton, in 1872, 1878, 1880, 1884, and 1888....These rat-swarms invariably take place in spring....A few of the animals appear in August; they increase in numbers till November, when all disappear again gradually, as they came. While in a locality dead rats are seen lying about in all directions,—on roads, in gardens, and elsewhere, very few have any marks of violence on their bodies; nor have they died of hunger, since on examination they are generally found fat. In 1884 in Picton, 47 dead rats were found lying together under the floor of the sitting-room (in one house). In another 37 were found dead under the kitchen. The whole town was pervaded with the odour of dead rats. The average weight of full grown specimens is about two ounces. The fur on the upper portion of the body is dark-brown, inclining to black; on the lower portion white or greyish-white. They run awkwardly and slowly on the ground, but run very quickly on the trees. When suddenly startled or pursued they cry out with fear.

"The extremely few females that occur amongst the countless hordes is a fact that shows that if breeding does take place at all during these periods (of travel), it must be on a very limited scale." They do little damage, their food being green vegetable....Though they enter dwelling-houses and barns, it is evidently not in quest of food, as shown by corn and other eatables being left untouched by them." (Rutland adds)—"Among English country people, who have the best opportunity of observing them, it is commonly asserted that in litters of young rats" (? *Mus decumanus*), the males produced outnumbered the females by about *seven to one*.

Meeson describes a plague of these rats in 1884:

Nelson and Marlborough, in other words the whole of the extreme northern portion of the South Island of New Zealand, is enduring a perfect invasion. Living rats are sneaking in every corner, scuttling across every path; their dead bodies in various stages of decay, and in many cases more or less mutilated, strew the roads, fields, and gardens, pollute the wells and streams in all directions. Whatever kills the animals does not succeed in materially diminishing their numbers. Young and succulent crops, as of wheat and peas, are so ravaged as to be unfit for and not worth the trouble of cutting and harvesting. A young farmer the other day killed with a stout stick two hundred in a couple of hours in his wheat field.

Among reasons suggested for the visitation he suggests the pressure of famine: "last summer was very wet, and last winter very cold, the amount of snow lying on the high lands in the interior was very great." Another is the excessive increase in numbers producing an intense struggle for existence. (His conclusions are somewhat at variance with Rutland's who did not think that hunger was an impelling cause.)

"I have examined many of these animals, and have not found a single female. One of my neighbours has examined two hundred of them; and a Maori, at the pa beyond Wakapuaka, one hundred, with the same negative result. Some females have, however, been taken; and in one case they were found breeding." "He is more like a big field-mouse than a Norway rat, and besides being considerably smaller, he is slightly darker in colour, and less malodorous. He climbs trees and flax plants, and is phytophagous rather than carnivorous."

Hutton in 1887 says:

The rat appears to have invaded Picton at the end of March, and to have suddenly disappeared by the 20th April. Old Maoris recognized it as the rat they used to eat in former times, and said that swarming on the low lands periodically was always characteristic of it.... These rats were often noticed climbing trees. In the Pelorus, where they stopped longer, they built nests, like birds, in trees.

Hutton at that time thought the Picton rat was a new species, and he named it *Mus maorium*. He says: "This rat is certainly different from *Mus huegeli*, Thomas, from Fiji; and I should think from *Mus exulans*, Peale, also, but I have seen no full description of that species." Kingsley in 1894 records it as nesting on the branches of small trees, four to five feet from the ground, near Totaranui, and gives examples from Motueka, Riwaka, Collingwood, Nelson and Taranaki. I have myself seen tall thorn hedges at Whangarei full of their nests, large shapeless structures, which at first I thought must be house-sparrows taken to hedge-building.

Marriner reports that he met with grey rats at North West Bay, Campbell Island, which Waite thinks were probably *Mus rattus*. The black rat is at the present time (1916) extraordinarily common about Christchurch; Mr Speight the curator informs me that Canterbury Museum is infested with these animals. A good deal of damage said to be done to orchards by opossums is almost certainly the work of the black rat. In some districts they destroy the native vegetation; and have been found to eat the roots of the larger Umbelliferæ, as *Ligusticum*, *Angelica* and *Aciphylla*; the tubers of *Gastrodia Cunninghamii*; the inflorescence of the kie-kie (*Freycinetia Banksii*) and of the Nikau palm; and the fruit of the native passion-flower (*Passiflora tetrandra*).

* Brown Rat; Norway Rat (*Mus decumanus; Epimys norvegicus*)

This ubiquitous animal very early made its appearance in New Zealand, but there is no record of its arrival. Perhaps every vessel which came to the colony brought some immigrants. In the early days of last century Russell or Kororareka in the Bay of Islands was the chief port of the colony, and rats must have become very abundant there. But, as already pointed out, they were probably mostly black rats. Darwin, who visited the Bay of Islands in the 'Beagle' in 1835, says (p. 428): "It is said that the common Norway rat, in the short space of two years, annihilated in this north end of the island the New Zealand species." Dieffenbach states (vol. II, p. 185) that he never could obtain a native rat, owing to the extermination carried on against it by the European rat.

A. R. Wallace in *Darwinism* says:

> This invading rat (*M. decumanus*) has now been carried by commerce all over the world, and in New Zealand has completely extirpated a native rat, which the Maoris allege they brought with them from their home in the Pacific; and in the same country a native fly is being supplanted by the European house-fly.

The latter statement is quite erroneous. Native flies have been reduced by introduced birds; certainly not by any other insect. During visits to Stewart Island and the West Coast Sounds between 1874 and 1880, I was struck by the abundance of these animals in regions uninhabited and almost unvisited by man. One day the late Mr R. Paulin and I emerged from the bush on the south side of Thule in Paterson Inlet when the tide was low, exposing a wide stretch of beach nearly a mile long. We were very much surprised to find the whole beach alive with rats which were feeding on the shell-fish and stranded animals which the tide had left and exposed. As soon as they saw us they immediately ran for the shelter of the bush. They were literally in hundreds. Rats are also very numerous round the homestead on Campbell Island (Bollons). In 1868 H. H. Travers reported them as very abundant in the Chatham Islands.

A few years ago when a scare arose about the bubonic plague, a feeble and intermittent crusade against rats was inaugurated, but it was, as might have been expected, absolutely futile. While rats are still very abundant, especially about the towns, there is no doubt that the spread of weasels throughout the country has vastly diminished their numbers, especially in the open.

Rats have had a great share in the destruction of the native avi-fauna, and are also responsible for much of the difficulty experienced by acclimatisation societies and private individuals in their attempts to establish game and other birds. But it is impossible to say which species is responsible, or whether both are equally so. It is clear in the preceding references that it has not always been possible to be sure which species of rat was referred to. I have therefore thought it advisable to add in a note some general facts on rats culled from recent sources.

NOTE. In 1869–70 there occurred a great visitation of rats (apparently *Epimys norvegicus*) in the north and north-western plain country of Queensland. The numbers were said to be incredible, and one writer stated that "one rat to every ten square yards in each mile would not represent anything like the numbers." H. E. Longman says that "in Australia, judging from available statistics, the black rat is quite as common as the brown, and is, of course, the species most frequently found in buildings, whereas *E. norvegicus* is characteristically a ground rat." In Mr Hinton's pamphlet already referred to he says, in regard to general habits of rats: "*Rattus rattus* is essentially an arboreal or climbing animal, and it rarely burrows; hence, where infesting buildings or huts, it is found usually in the walls, ceilings, or roof, not in cellars or drains. Although cautious, it does not shun mankind, and it enters into far closer relations with its unwilling host than does the Brown Rat. For this reason it is often the species principally concerned in the transmission of plague. It drinks little, and seldom, if at all, enters water voluntarily. As already mentioned, this is the common rat on ships. In most cases it reaches or leaves the ships by climbing their cables while they are in dock; sometimes it is introduced with grain and other merchandise. Its diet is of a most varied description, but, probably in consequence of its more salubrious station, it is a cleaner feeder than *R. norvegicus*.

R. norvegicus is essentially a water loving and burrowing animal; although far less agile than *R. rattus*, it is a good climber. As compared with the last named species, it is far more voracious and cunning; its greater size and strength, and its much greater fecundity, render it, as far as material prosperity is concerned, a much more formidable enemy of mankind. On the other hand, although it spreads many serious or fatal diseases, it usually exhibits a certain shyness of man, so that, in normal conditions, it is probably slightly less important than *R. rattus* as a carrier of plague."

Prof. P. Chavigny in articles in the *Revue Générale des Sciences* of July 15–30, 1918, on "The Invasion of Trenches by Rats" (condensed in *Nature*, of 19th September, 1918), gives certain interesting facts regarding these animals, viz. both brown and black rats:

(*a*) Starvation kills a rat in about 48 hours.

(*b*) The period of gestation is 21 days, and the minimum time between two litters is 62 days. The female rat may have five litters in a year, and a litter consists of about ten young. A female is capable of producing a litter at the age of 2½ to 3 months. A simple calculation shows that a pair of rats is capable of producing 20,000,000 descendants in three years. In temperate climates reproduction is at a standstill during winter[1].

*The Mouse (*Mus musculus*)

The first notice of the appearance of the mouse in the North Island is in Dieffenbach (vol. II, p. 185), but no doubt it was introduced early last century. When Wilkes visited the Auckland Islands in 1840 the only living creature seen besides the birds was a small mouse. According to R. Gillies, who wrote in 1872:

"it is quite certain that there were no mice in Otago in 1852" (he arrived in 1848), "but a year or perhaps two years after they were noticed in Dunedin first. As soon as the mice appeared, the Rats disappeared....The Molyneux stopped their southern migration for a time, and it was considerably later before Molyneaux Island (Inchclutha) was touched by them."

Taylor White speaks of mice appearing in the Canterbury Plains in the early days of settlement (1855) onwards "suddenly in Thousands." Pastor Wohlers, long a missionary working among the natives on Ruapuke in Foveaux Straits, states that mice were first brought to the island in the 'Elizabeth Henrietta,' which was wrecked there in 1824, and that even as late as 1873 they continued to be known as "henriettas." Mr Philpott writing on 2nd January, 1918, said:

There is a plague of mice in the district west of the Waiau. From Bluecliff to the Knife and Steel near the Big River, and beyond, each hut (the Government huts on the now abandoned telephone track to Puysegur Point) was overrun with them. And not only at the huts, but on the beach and in the dense bush, wherever we went, they were plentiful. At the Hump, near Lake Hauroto, they were as numerous as elsewhere. This prevalence of mice is certainly not usual; I have been on the Hump four or five times since 1911, and last year tramped along the Knife and Steel, and apart from an odd one or two, no mice were in evidence on former trips. One noticeable thing about the mice was their boldness; they were evidently very hungry. The wekas caught many of them, swallowing them whole, head first.

[1] In a letter written from Sydney, N.S. Wales, dated 28th April, 1919, which appeared in *Nature* of 3rd July (p. 345), Mr Thomas Steel states that: "under a creeper in my garden near Sydney, the common snail (*Helix aspera*) was very abundant, and *Mus decumanus* used to devour large quantities; the apex of the shell was always bitten off so that the mollusc could be readily extracted. On the Upper Waikato River, New Zealand, the same rat dives into the water and gathers the fresh-water Unio. On the river-banks the shells are gnawed open and the animal eaten. The shells are always bitten through at the same spot of one valve, but I forget now whether that was the right or left one."

"In Australia at certain seasons a 'cutworm' moth, known as the 'bogong' or 'bugong' (*Agrotis infusa*), swarms in myriads in many places, and is, after the wings have been singed in a charcoal fire, used as an article of food by the aboriginals. These moths sometimes invade the cities and crowd into houses and stores for the sake of darkness. At Melbourne, in a large sugar store, I have noticed *Mus decumanus* collect the moths and eat the bodies, rejecting the wings."

The mouse has never been found very far from the haunts of men in New Zealand. In 1866, during a discussion which arose at a meeting of the Canterbury Acclimatisation Society as to the reported destruction of small birds by hawks, W. T. L. Travers reported "that he had opened a large number of hawks, and in all cases found their food to consist entirely of *Mice* and grasshoppers." At present the mouse is abundant in all settled parts of New Zealand, and is also common on the Auckland, Antipodes and Campbell Islands.

Family CAVIDÆ

Guinea-pig (*Cavia porcellus*)

The only record I have of the introduction of the guinea-pig is by the Auckland Society in 1869, but they have repeatedly been brought in by private individuals and dealers for the last 50 or 60 years. Though they have been frequently liberated, they have never succeeded in establishing themselves anywhere, as the young are mercilessly preyed upon by cats. I had them running nearly wild in my garden in Dunedin for some time, and noticed that violets (*Viola adorata*) growing among grass increased remarkably all the time they were about. The guinea-pigs nibbled the grass very closely, but would not touch the violets.

Family LEPORIDÆ

*Rabbit (*Lepus cuniculus*)

The introduction of the rabbit into New Zealand has produced such far-reaching effects and wrought such changes throughout the country, that it requires more than the sober language of a naturalist to describe them. One thing is quite certain, namely, that it was deliberately introduced into the country. The first definite notice I have found as to the introduction of these animals is in du Petit-Thouars' voyage of the 'Venus' (1838), in which he says (p. 115): "There are still to be found some rabbits imported from New South Wales." The next is in Mr T. Tuckett's diary of his expedition to the South Island, which is printed as an appendix to Dr Hocken's *Contributions to the Early History of New Zealand*. Speaking of the country between the mouths of the Clutha and Mataura Rivers, he writes under date 10th May, 1844: "Palmer has grown wheat and barley as well as potatoes, and has plenty of fine fowls and ducks and some goats.... Returning from Tapuke (Taukupu) we landed on the island, and with the assistance of a capital beagle caught *six rabbits* alive and *uninjured*." He does not say whether any were liberated on the mainland, nor whether it was possible for them to get ashore. Mr H. Travers (February, 1919) says:

> From what I can recollect about the introduction into New Zealand of the black rabbit, or silver greys as they were then called, these were imported by a Captain Ruck Keene, R.N., who had a run, I think, at Kaikoura, as I knew him when he lived in Nelson in the late fifties and saw the rabbits. Other rabbits were imported into Nelson, and were kept as pets—the "French rabbits" as we boys knew them; they were white and foxy coloured. Some of these were turned out at Taradale and increased enormously, but being in a district in which only cattle were run, did no damage. Some years afterwards, these were practically exterminated by a tremendous rainstorm and flood which pretty well destroyed the lot, as it was followed by a snowstorm and the rabbits were smothered in their burrows.

According to Mr Huddlestone, silver-grey rabbits were first introduced into Nelson in or about 1865, but there is no record as to what came of this importation.

Mr James Begg, of Mosgiel, has given me some very valuable information as to the earliest attempts to introduce these animals, and I quote him freely in the following pages. He says: "When Willsher and party settled at Port Molyneux in the early forties they sent to Sydney for rabbits, but whether they obtained them or not, I am unable to say." Perhaps these were the rabbits which Mr Tuckett saw. From early days there was at least one colony of rabbits on the Upper Waitaki. These remained quite local in their habits, and did not increase to any great extent. They were finally overwhelmed by the invasion of the grey rabbit from the south.

Mr Thomas Walsh of Shag Point tells me that these rabbits were turned out by Messrs Julius Bros., on their run at the Rugged Ridges on the Waitaki River, but they never seemed to increase. He also states that the Rev. Mr Fenton, who came down to Dunedin shortly after the commencement of the Otago Settlement—about 1849—brought with him both black and grey rabbits. Some of these were handed over to Mr Geo. Crawford in whose care they increased, and they were distributed from there to various other centres. One lot were liberated on the sand-hills between Invercargill and Riverton; and another lot at Queenstown, the price paid being £1 per pair. These rabbits do not seem to have increased to any extent. At a somewhat later date (1870) when Mr Walsh was at Palmerston he kept long-haired lop-eared rabbits, and turned out a good many of them. He states that the lop-ears quickly disappeared in succeeding generations, though occasionally long-haired ones were seen. They never increased to any extent, however, and only an odd one was afterwards seen. They also were swamped by the southern invasion.

In 1857, Mr John Sutherland (now of Te Kiuti, Auckland provincial district), then a shepherd on the Greenfield Estate, saw some black and white long-eared rabbits between the Tokomairiro and Waitahuna Rivers. These also disappeared in later years. This country was quite uninhabited at the time, though sheep were running on it. Mr H. B. Flett states that a good many rabbits, which had escaped from captivity at Waitahuna, found their way to the open country, and increased to a slight extent, but eventually died out completely. He attributes their extermination to the weka or Maori hen (*Ocydromus*), "which were very plentiful at that time and for some years subsequent. They used to go into the holes and eat the young rabbits. I have seen a weka killing a half-grown rabbit."

The late Mr Telford of Clifton introduced some rabbits, and bred them in hutches till they numbered about fifty. They were then liberated on Clifton, near the banks of the Molyneux, but died out in a short time. This was about the year 1864. Mr Clapcott also liberated some at the old homestead at Popotunoa Station, but they also failed to thrive, and disappeared. It is probable that there were other attempts to acclimatise rabbits, all more or less unsuccessful. Dr Menzies of Mataura also introduced rabbits, and he is usually credited with having been the successful introducer of them to the south, an achievement the credit of which has not been very eagerly sought after. They were liberated on the sand-hills somewhere near the Bluff.

Sir Geo. Grey also appears to have introduced them at about the same date, for in the annual report of the Canterbury Society in 1866 it is said that "an enclosure has been set apart for the Silver-grey Rabbits presented by Sir G. Grey, which have thriven well and increased to a great extent, *and have been distributed to members far and near*." Later in the same year the Society passed this minute: "The suggestion of giving as a reward, for the destruction of hawks and wild cats, some silver-grey rabbits, was approved of." In 1866 the Otago Society liberated 60 rabbits, 23 in 1867, and 18 in 1868, but I do not know whether these came from Britain or from Australia. These are the only records I have been able to secure so far as to the introduction of rabbits into the colony, and they would account for the presence of these animals in Southland, Otago, Canterbury and Auckland. There can be no doubt, I think, that what happened in the south, happened elsewhere at every port where settlement took place, and that private individuals at Nelson, Wellington, New Plymouth and Napier also imported rabbits. But when the animals became a pest, and their increase was recognised to be a calamity to the country, every one was desirous of repudiating the responsibility of their introduction. Thus the framer of the annual report of the Canterbury Society for 1889, not having read the statement in the report for 1866, says concerning "the rabbit, that great scourge to our large runholders,—that the introduction of these cannot be laid to the charge of this society." Similarly Mr Bathgate of Dunedin, in 1897, writes: "It is to them" (the Provincial Government of Southland) "that we are indebted for the presence of the rabbit." Dr Menzies was in these early days Superintendent of the Province of Southland. From 1866 onwards the spread of the rabbits was phenomenal. I quote Mr Begg's account at length:

About the year 1874 they began to make their presence felt in an unpleasant manner. By 1878 they had reached Lake Wakatipu, leaving a devastated country behind them. At the same time they had reached as far east as the Clutha River, and in a few years later had overrun the greater part of Otago as well as the whole of Southland. Those were evil days for farmers, especially for the squatters who occupied large areas of grazing country. The fine natural grasses on which sheep and cattle grazed were almost totally destroyed. Sheep perished from starvation by hundreds of thousands, and it is no exaggeration to say that the majority of the squatters were ruined. In the old Burwood Station the number of sheep fell in one year from 110,000 to about 30,000. This was partly due to heavy snow, but the rabbits prevented any recovery. It is doubtful if the same country to-day carries more than 40,000 sheep. From the year 1878 onwards immense areas of grazing land were abandoned, as the owners gave up the unequal struggle with the rabbits. At first no efforts seemed to have the slightest effect in stemming the invasion, or in reducing the numbers of the rabbits. The wet country in the South suffered equally with the dry lands of the interior, but the former is now showing a power of recovery from the damage done, while in much of the latter the damage appears to be almost irreparable.

In the early days, hunting with dogs, shooting, digging out the warrens, poisoning with various baits, and trapping, were the methods by which farmers tried to rid themselves of the pest. Later, wire-netting fencing, the introduction of stoats, weasels and ferrets, fumigating the burrows with poisonous gases (such as Carbon disulphide and Hydrocyanic Acid) and the stimulus given to trapping by the export trade in frozen rabbits, have been relied upon to reduce their numbers. In the writer's experience, practically no progress was made in reducing the numbers of rabbits till about the year 1895. From that year there has been a steady diminution. For twenty years the rabbit had the upper hand, and though many millions were killed annually, no reduction in their abundance was noticeable. In the last twenty years there has been a steady decrease. Large areas of hill country in the wetter districts are now completely clear of rabbits, though they still persist in favourable situations. In the dry country in Central Otago they are still very troublesome and very vigorous, and their evil effects are there seen on hundreds of square miles of country, once the finest grazing land in New Zealand, now little better than a desert.

Hardly two men will agree as to the cause of the decline in the numbers of rabbits, and I will just state my theory for what it is worth. The grey rabbit, when first introduced, found himself in very congenial surroundings. There was abundance of food and shelter, and the ground was absolutely clean, never having been grazed by rabbits previously. These favourable conditions gave a tremendous filip to the vitality of the rabbits and stimulated their powers of reproduction. They increased at a rate that I believe is not even approached in the worst infested parts of Otago to-day. No efforts at checking them had the slightest effect, and they passed over the country like a prairie fire. After a time the original conditions no longer existed. Food became scarce, the land was foul with rabbits, disease appeared among them, and their fertility decreased. No doubt improved methods of dealing with them hastened their reduction, but I firmly believe that the principal factor in their decrease was lessened fertility, due to the first great spurt to their vitality having spent itself. The decrease first became apparent in the colder and wetter parts of the country. The rabbits abandoned large areas and became concentrated in warm sunny spots. Even in these spots their numbers declined, and from many of them disappeared altogether. In the dry country which is more congenial to rabbits, fertility is still maintained, and may possibly be permanent. The rocky hills round Alexandra may be taken as ideal country for rabbits, and probably this area has suffered more from them than any other part of New Zealand. All known methods of rabbit destruction have had an exhaustive trial there, and have not succeeded. It would seem that in this favourable spot the vitality of the rabbit is not greatly impaired. It would be interesting to try if rabbits could be re-introduced into country where they once swarmed, but which they have subsequently abandoned. I believe that such an attempt would fail.

Opinions different to mine are very widely held. Most men claim the credit of having themselves cleared their ground of rabbits, and the official Rabbit Department staff possibly take the credit to themselves. I should like to believe that to me belonged the credit of having cleared my own place, but my experience leads me to believe that my efforts had little to do with it.

It must not be assumed that every one regards the rabbit as a nuisance. Many a successful farmer of to-day got a start as a rabbiter. The killing of rabbits actually became one of the principal industries of the province. Their presence directly led to the subdivision of large estates, and may have been quite as effective in this direction as all the legislation on the subject.

The introduction of rabbits had a lasting effect on acclimatisation generally. Before their advent partridges and pheasants had become numerous, but they have entirely disappeared in Otago. In the effort to cope with the rabbits, the country was annually sown with poisoned grain. This had a disastrous effect both on native and imported game. Had rabbits not become a nuisance, it is unlikely that weasels and other vermin would have been introduced. These animals are largely responsible for the decrease in the numbers of native birds, and also make the successful introduction of new varieties more difficult.

The initial difficulties in getting rabbits introduced, the terrific success that at last crowned the efforts made, the unexpected ruin and destruction which they caused, and the gradual return to normal conditions, makes the history of their introduction one of the most interesting in the annals of acclimatisation.

Mr H. B. Martin in 1884 states that in various parts of the Auckland district the rabbits have become almost or quite extinct from natural causes; tuberculosis was also believed to be present in the Wairau Valley, where the rabbits were beginning to decrease before the present Act was in force.

In a discussion which took place in the Legislative Council on 4th July, 1883, the Hon. Mr Chamberlain said that rabbits were formerly numerous on Motuihi and Motutapu, and on Flagstaff Hill, but they had now become extinct.

Mr Edgar T. Stead, writing me as late as 25th July, 1919, informs me that:

> in the Wills Valley and the Upper Haast, to the north of Lake Wanaka, the rabbits were at one time,—say ten or twelve years ago,—absolutely swarming. When I was there six years ago I was told that the rabbits were completely gone from the Wills Valley, and I personally observed that they were leaving the Haast. On the flat below the Burke hut there was still a fair number, but above that on the open stretches of river flat there was not one, though there were deserted warrens in all suitable localities. There are many places in Canterbury where rabbits have become scarce in the last ten or fifteen years,—more places still where the case is *vice-versa*,—but, as you remarked in your paper, there were only some races of rabbits that spread badly, and we do not know that the above-mentioned places were inhabited by the virulent races. In the Haast River case we do, for the rabbits had spread over the range from the famous Central Otago stock. It is quite possible that the country going "rabbit-sick" is only part of a cycle, but the subject is well worth investigating.

In the House of Representatives on 1st August, 1883, Captain Mackenzie said that a competent authority assessed the actual loss to the Colony through the Rabbit Plague at £1,700,000 a year. Mr W. C. Buchanan said that the loss for the past ten years was assessed at ten millions sterling. The question of importing a disease from the Falkland Islands was discussed at the same time.

I was at one time under the impression that in this new country, where the causes which kept them in check in their original home were wanting and there seemed to be nothing to arrest their development in any direction, there might arise new varieties of rabbits with modified habits. Particularly did it seem likely that colour variations would thrive unchecked, and the traveller passing through certain districts in Central Otago is certainly surprised at the number of conspicuously coloured animals to be seen. Mr W. H. Gates of Skippers writes me (April, 1916): "As for colour they are of all colours;—grey and white; tan and white; grey, with a black ridge down the backbone; and buff." Other observers speak of the prevalence of black, black and white, and yellow rabbits. But Mr R. S. Black of Dunedin, the largest exporter of rabbit-skins in the Dominion, informs me that while they are of all colours, 95 per cent. of the skins exported are grey. The other colours appeal more to the eye, but they are not so abundant after all. That the rabbits of aberrant colours should survive is not to be wondered at, seeing that in this country there are no foxes, and neither hawks nor owls large enough to tackle a full-grown rabbit. The common harrier-hawk takes a considerable toll of young rabbits, but it is quite unable to keep them in check. In many districts wild cats live mainly on rabbits.

Mr Yarborough of Kohu Kohu tells me (August, 1916) that rabbits became quite common in a district near Kawa Kawa (at the head of the Bay of Islands) many years ago. Recently they have reached the eastern side of the Hokianga River, and it is not unusual to see them occasionally. Then he adds this interesting statement:

> *I have never heard of any rabbit burrows, as they appear to breed among the rocks and roots of trees.* They do not seem to have crossed yet to the west side of the Hokianga River. No complaints have been heard of devastation done by them, and it seems to be doubtful if they would thrive in either our clay lands, or in volcanic areas.

The comparatively heavy rainfall of Hokianga, amounting to some 60 to 70 inches per annum, has no doubt a good deal to do with the comparative scarcity of the rabbit in this part of New Zealand. They are, however, not uncommon near Kaikohe, and do make small burrows[1].

Effect of Rabbits on the Country and Native Vegetation.

The economic waste caused by the vast increase of rabbits in New Zealand is incalculable, and certainly represents a loss in the stock-carrying capacity of the country which probably runs every year into millions of pounds. It is not only that they eat up food which would support some millions more sheep than are at present reared, but they destroy large areas of country, and yield very little return for the

[1] One curious effect of the recent great war has been a phenomenal increase in the price of rabbit-skins. I have not been able to ascertain yet what effect this is having on the rabbit question in Otago, but by the end of 1919 it has become quite impossible to get rabbits for the table. At a sale held in Dunedin in December, 1919, the prices received for skins of winter growth ranged from 155*d*. to 274*d*. for six skins; that is to say, that the highest quality of super-does, as they are termed, brought 3*s*. 10*d*. per skin!

Since this was written prices have altered greatly. In June 1921 the best skins were fetching about 74*d*. per lb., or from 8*d*. to 1/- per skin.

damage they do. The annual export of approximately 3,000,000 rabbits valued at about £70,000 and of some 8,000,000 skins valued at about £115,000 is all the return they give, but it only represents a small proportion of the dimensions of the pest. In all parts where rabbits abound, their destruction entails a heavy expense on the occupiers of the land. There are no data available anywhere to enable one to estimate how many rabbits are destroyed every year, but far more are killed by phosphorus than by trapping. The latter method alone furnishes any statistical data, the former is an unknown quantity, but it represents a very large figure.

Probably the most ghastly exhibition of the work of rabbits is to be found in the grass-denuded districts of Central Otago, parts of which have been reduced to the condition of a desert. It is improbable that this state of affairs could have been brought about by rabbits alone. Before their advent, the runholders who had possession of the arid regions—in which the rainfall probably averages 10 to 12 inches annually, and certainly never exceeds 15 inches—were doing their best to denude the surface of the ground by overstocking with sheep and frequent burning. The latter was resorted to because many of the large tussock-forming grasses—especially such as the silver-tussock, *Poa cæspitosa*—yielded coarse and rather unpalatable fodder, but after burning the tufts, a crop of tender green leaves sprung up, which were very readily eaten. Unfortunately the burning not only got rid of most of the coarse growth of the tussocks, but it also swept off the numerous bottom grasses which occupied the intervening spaces, such as *Agropyrum scabrum*, *Danthonia Buchanani*, *Danthonia semiannularis*, *Triodia Thomsoni* and *Festuca ovina*, which were the mainstay of the depasturing flocks. Even before the rabbits arrived the work of denudation of the grass-covering had been proceeding apace through the causes mentioned. Thus Buchanan, writing in 1865, said: "it is no wonder that many of the runs require eight acres to feed one sheep, according to an official estimate." Mr Petrie thought this an unduly severe estimate, "as in the mid-seventies the sheep-runs of Central Otago were reputed to carry at least one sheep to four acres, and the majority of them carried one sheep to three acres or somewhat less."

Mr Petrie, who reported to the Department of Agriculture on the grass-denuded lands of Central Otago, knows more about this subject than anyone else, and I quote him at some length:

Before the rabbit-invasion began the hill-slopes carried a fairly rich and varied covering of tussock and other grasses, and, except on the steeper rock sun-baked faces, had not been very seriously depleted even in the early nineties. The earlier stages of this depletion may now be seen in several of the Central Otago ranges, as on the spurs of the Rough Ridge and the Morven Hills districts. The northern slopes of the spurs are almost, in many instances entirely, bare of grass, while the southern shaded slopes still carry a fair amount of pasture. The grass covering generally stops abruptly at the bottoms of the valleys, even when these are not worn into water-channels. The vastly greater depletion of the pasture on the northern slopes is easy enough to understand. They are more exposed to the sun and to the frequent violent parching north-west winds; they lose their covering of snow earlier in spring than the southern slopes, and are thus more closely grazed at a critical season for the pasture; and sheep at all times show a preference for feeding on the warmer sunny slopes. When the pasture on the exposed slopes fails, that on the shaded slopes has to feed all the stock that is about, and unless the stocking is reduced to meet the new conditions the remaining grasses are sooner or later eaten out. The desert, with all its problems, is then established.

In this account of how the desert conditions have arisen, Mr Petrie refers only to sheep, because it is the loss in sheep-carrying capacity which is so serious, but later on, after describing a typical specimen of the country, and showing that in inaccessible situations a considerable variety of fairly vigorous grasses live on, he adds:

"This is one of the facts that go to indicate that the extermination of the grasses in this desert country is mainly due to eating out by overstocking, rabbits as well as sheep being included among the stock carried." "The desert and the greatly denuded lands are not wholly destitute of vegetation. In most of their lower areas greyish, flattened, firm, nearly circular patches of scab-weed (*Raoulia australis* and *R. lutescens*) are thickly dotted about the bare ground. Though otherwise useless, these moss-like composite plants help to keep the soil from being blown or washed away, and when old supply, in the decayed centres of the patches, spots with some amount of humus where grass-seeds can more readily settle and grow."

These plants are never eaten either by sheep or rabbits.

In regard to their effect on other native species of plants Mr Petrie writes to me in a letter of 1st May, 1916:

I know that rabbits have done much to reduce the abundance of the Otago Spear-grasses (*Aciphylla squarrosa* and *A. Colensoi* chiefly), probably during times when the ground was covered by snow. When I first visited inland Otago (1874) *Aciphylla Colensoi* was most abundant. In riding about it was almost impossible to deviate from well beaten tracks or roads, because the spines pricked the horses' legs and feet. I know of no evidence that sheep would eat fairly full-grown *Aciphylla* leaves, but young plants must be more or less eaten both by sheep and cattle (as well as by rabbits). Several species of *Celmisia*, notably *C. densiflora*, have been greatly checked, and *C. densiflora* almost exterminated.

Dr Cockayne is now (1921) engaged on an exhaustive investigation.

Captain F. W. Hutton writing me on 23rd March, 1892, said:

As to the extermination of the Wild Spaniards (*Aciphylla*), I believe it to be due to rabbits. When I was in the Nelson district in 1872–3 there were no rabbits on the eastern side of the Upper Wairau near Tarndale, but they were abundant on the western side. Spaniards were abundant on the eastern side, but almost destroyed on the western. The rabbits seemed to burrow under the plant and then eat the root.

Mr B. C. Aston, in ascending the highest point of the Kaimanawas (5700 feet) on 31st December, 1914, found the physiognomy of the sub-alpine scrub, which begins above the beech forest at an elevation of 4200 feet, and consists of *Senecio Bidwillii*, *Veronica buxifolia*, *Olearia nummularifolia*, *Coprosma cuneata* and *Phyllocladus alpinus*, mottled with the brown leaves of a dead shrub. On close investigation this was found to be *Panax Colensoi*, and on still closer inspection with a view to determining the cause of death, we found every bush ring-barked. I have no doubt this was done by rabbits, which after a heavy fall of snow would be driven down from the tussock land to the scrub and forest zone. The scrub only occurs in the gullies, and the beech forest usually has a bare floor with a few *Panax* trees scattered through it. On examination of these we found the same fate had overtaken them. Trees 10–20 feet high of *Panax Colensoi* and *P. Edgerleyi* were found to be dead and ring-barked.

A certain part of the destructive and exterminating work of rabbits on the vegetation in mountain districts is particularly wrought at the beginning of winter. In spring and early summer as the snows melt, the rabbits follow up the mountain side, and are found during the summer at all elevations. I saw them in abundance on the top of Mt Tyndall—nearly 8000 feet—among the snow-beds. When the first heavy snows come on about April, they are driven down in hordes to the lower country, and, as has been told me by more than one resident in the Wanaka district, "they are as thick as locusts, and they eat the ground just as bare as those insects do."

In a good many rabbit-infested districts, particularly in the North Island, these animals have aided very materially in producing a certain amount of erosion and washing down of alluvium, by burrowing extensively in the banks of rivers and small streams. When floods came down these undermined portions were commonly swept away where the firmer banks resisted the impact of the water. Professor Cotton of Wellington considers that some importance can be attached to this agency as affecting the physiography of certain districts and river systems. Cattle, sheep and goats no doubt assist in breaking down such alluvial banks, but rabbits are probably the most active agents in the work. Rev. A. Don writing me in 1901 said:

the rabbits, by so stripping the ground of vegetation and burrowing into the faces of the slopes are converting what were once nice green hill-sides into shingle slopes, because when once the face is so bared and its surface broken, it begins to slip.

Mr Petrie also refers to this process in his report as follows:

> The soil on the grass-denuded slopes, which is by no means infertile, being no longer held together by the roots of plants, is being rapidly removed by wind and rain, and pebbles and angular stones are now closely dotted over great stretches of hillside that not many years ago were covered with soil. On the steeper slopes, indeed, the soil is being rapidly sluiced down into the gullies and thence into the river, and deep, narrow, chasm-like watercourses are being dug out.

* Hare (*Lepus europæus*)

The Otago Society liberated three in 1867, which they obtained from Geelong, Victoria; one in 1869, and three in 1875.

The Canterbury Society got one from Dr Macdonald of the 'Blue Jacket,' and one from Captain Rose of the 'Mermaid' in 1868; and four in 1873 from Messrs Wood Bros.

The Auckland Society introduced two in 1868, and five in 1871. The Nelson Society introduced some hares in 1872, and these increased so rapidly as to become a nuisance in the district. The Southland Society obtained some from the Victorian Society in 1869, liberated two in 1871, and two in 1874, and 40 in 1887. These are all the records I can find of importations from abroad, and considering the casual manner and the small numbers in which they were introduced, the subsequent increase is most remarkable.

The first three introduced by the Otago Society were liberated at Waihola, where two years later they were reported to be plentiful. In Southland coursing was commenced in 1878. Hares soon spread over all the flatter parts of the South Island, mostly about cultivations, and in districts where rabbits were not abundant. They are common from Cook Strait to Foveaux Strait. In the North Island they spread south from Auckland, and the smaller acclimatisation societies assisted to distribute them far and wide. Wellington liberated two in 1874, 14 in 1875, and four in 1876, and in 1885 reported them as "numerous in the vicinity of Wellington and the lower end of the Wairarapa Valley." In more recent years they are reported as in large numbers about Marton; increasing about Pahiatua, and as seen in almost every part of the Eketahuna district. The Taranaki Society introduced them in 1876, and they were reported as thriving in 1884. On Mount Egmont at the present time they are common above the bush line and up to 6000 feet in the summer months.

In 1905 the Waimarino Society purchased and liberated a number, and protected them for two years. Later on they became so numerous that they were declared to be no longer game, and all restrictions about shooting them were removed. Mr E. P. Turner tells me (January, 1916) that they are found all through the volcanic plateau of the North Island from Rotorua to Waiouru.

In no part of New Zealand have they increased to such an extent as in South Canterbury, where they became so abundant that a considerable export trade sprang up, mostly from the port of Timaru. Thus the total number of frozen hares exported from New Zealand in 1910 was declared at 10,744; in 1911 it was 11,418; and in 1912, 7240. I have been told, and it is highly probable, that many more were exported as rabbits.

I am informed that in New Zealand hares usually produce three or four young at once; whereas in England they seldom have more than two. It is also stated that the animals are considerably larger than in Britain. In both cases the statements require verification, but if correct, the superabundance of the food supply is the principal factor. It is probably only true of those districts where rabbits are almost unknown.

In some parts of New Zealand hares tend to become white in the winter season, just as in parts of the old country, following the same seasonal variation as occurs in ferrets, stoats and some other sub-arctic animals. Mr Stead informs me that this is a familiar phenomenon in Canterbury; and Mr E. H. Burn states that they are not uncommon in the Mackenzie Country. My son, G. Stuart Thomson, considers that hares are much more abundant than rabbits in the North Auckland Peninsula.

Order INSECTIVORA

Family ERINACEIDÆ

*Hedgehog (*Erinaceus europæus*)

In 1870 the Canterbury Society received a pair of hedgehogs from Mr D. Robb, purser of the 'Hydaspes,' and in 1871 received one from Mr Nottidge.

In 1885 a shipment of one hundred was made to the Otago Society, but only three survived. These were liberated in a suburban garden, but were very sluggish though the weather was warm. I attributed this to the fact that they had lost their usual season of hibernation. The female died a month or two after arrival, and the two males were allowed to go free. Other hedgehogs must have been imported, for they were found at Sawyer's Bay about 1890.

In 1894 the late Mr Peter Cunningham of Merivale, Christchurch, sent a consignment of wekas home and got 12 hedgehogs out in exchange. They were placed in a pigeon-house, but got out under the wire-netting and escaped. For years nothing was heard of them, but they gradually increased, and are now extraordinarily abundant. Mr Edgar F. Stead of Riccarton says (March, 1916):

> If I hunted through my garden with my dog I could get a dozen now, and I frequently kill them.... They are extraordinarily destructive to chickens, their depredations being readily identified by the fact that they eat their victim's stomach first, whereas a cat eats the breast first, and rats and weasels go for the head and neck. Once a hedgehog starts eating chickens, he will go on until caught or the supply runs out. I know of many cases when a trap set and baited with the remains of a chicken has caught the marauding hedgehog.

An informant at New Brighton tells me (February, 1916) that they are very abundant, and are a pest in the gardens, as they eat the vegetables and dig up the potatoes. They are now (1916) very abundant about Dunedin, and apparently everywhere between Dunedin and Christchurch they are to be found. Mr W. W. Smith introduced two pairs into the Public Park, New Plymouth, in 1913, and they are increasing rapidly.

Among my correspondents, one who hails (40 years ago) from Surrey, England, is a firm believer in the milk-sucking habit of hedgehogs, and warns me that the milking qualities of cows are frequently destroyed by them.

Mr C. Hutchins of Omokoroa, Tauranga, states (April, 1913) that many years ago he found a hedgehog thickly infested with a large blue tick, about the size of a small pea. The hedgehog seemed to be more than half dead, but the ticks were apparently thriving.

Chapter IV

BIRDS

The classification adopted for this class is that used by Dr A. H. Evans, in the *Cambridge Natural History*, 1899.

Somewhere about 130 species of birds have been introduced into New Zealand since the date of Captain Cook's landing, but it is difficult in many cases to distinguish between mere aviary and cage species, and those which it was seriously attempted to naturalise. Besides, of those introduced it is impossible to identify quite a number, and their names may be synonyms of others already recognised. Thus when a society reports that it has introduced so many Indian doves or Indian pigeons, it is manifestly impossible to identify them.

Excluding the common wax-eye, blight-bird or twinkie (*Zosterops lateralis*), which apparently has come in from Australia within the last 60 years and has become extremely abundant, and birds like the Australian swallow (*Hirundo neoxena*), which is an occasional visitor and a migrant, all the species which have been introduced have been purposely brought in.

The following are the only immigrants which have become truly wild: mallard, Canadian goose, black swan, common pheasant, Chinese pheasant, Australian swamp quail, Californian quail, common pigeon, little brown owl, skylark, thrush, blackbird, hedge sparrow, Australian magpie, rook, starling, Indian minah, house sparrow, chaffinch, redpole, goldfinch, greenfinch, cirl bunting and yellow-hammer, a total of 24 species. The record of failures is much greater than that of successes.

Order RATITÆ

Family Struthionidæ

Solomon Island Cassowary; Mooruk (*Casuarius Bennetti*)

Sir Geo. Grey imported some of these birds in 1868, and handed them over to the Auckland Society; but there is no record as to what came of them[1].

[1] Major Bunbury in his report on the proclamation of Stewart Island as Her Majesty's possession in 1840, says: "The cassiowary has also been seen in different parts of the island." The reference is no doubt to the large kiwi, *Apteryx australis*.

Emu (*Dromaius novæ-hollandiæ*)

The Canterbury Society received one from Mr E. Flood of Sydney in 1864; the Otago Society had several in their garden in 1867; the Auckland Society received one from Sir Charles C. Bowen in 1868, and two from Mr F. E. Drissenden in 1871. There is no further record of any of these birds. There are always Emus in the Wellington Zoological Gardens.

The only serious attempt at naturalisation was that made by Sir Geo. Grey who introduced a number into Kawau in 1868, but they all died.

Order ANSERIFORMES

Family ANATIDÆ

No fewer than 25 species of this family have been introduced into New Zealand, but only one—the Australian black swan—has completely established itself, while the mallard and the Canadian goose have been partially naturalised. Domestic ducks appear to have been first introduced by the missionaries, either at the time of Marsden's first visit in 1814 to the Bay of Islands, or very shortly afterwards.

Muscovy Duck (*Cairina moschata*)

In 1865 Captain Norman liberated six of these birds on Adam's Island, one of the Auckland Islands lying to the south of New Zealand. They failed to establish themselves.

English Pochard Duck (♂), Dunbird (♀) (*Nyroca ferina*)

The Wellington Society imported six in 1894, and three more in 1895. Two years later, in conjunction with the Canterbury, Nelson, Taranaki, and other societies, a number more were imported. Private individuals and dealers apparently also brought in several.

The only report of these is a negative one, the Taranaki Society stating in 1902 that "we have not seen anything of the pochard ducks which were liberated in 1898."

Canvas-back Duck (*Nyroca vallisneria*)

In 1905 the Government imported some of these birds, but only two appear to have arrived, and these were handed over to the Wellington Society. There is no further information obtainable about them.

Pintail Duck (*Dafila acuta*)

In 1885 the Canterbury Society received some from the Royal Zoological Society of London, and in 1896 the Otago Society imported some, but in neither case is any information obtainable as to how many were introduced, or what was done with them, and there is no further record.

In 1905 the Government imported a number of birds. Four of these were sent to the Otago Society and were kept for breeding purposes at the Clinton hatchery, but did not increase. Six were handed over to the Wellington Society, and went to their game farm in the Wairarapa. Again there is no further record.

The species is almost exclusively a migrant in Britain, which it visits only in winter. It breeds in the Arctic regions of both hemispheres, and winters in various parts of Europe and North America, India, China, Japan, and Central America. It occasionally breeds in Britain. It is no wonder therefore that it did not settle in New Zealand.

English Teal Duck (*Nettion crecca*)

In 1897 an effort was made by the Wellington, Canterbury, Nelson and other societies to introduce this bird, and several were imported and distributed. But there is no record in any of the societies' reports as to what came of them afterwards.

The teal is a palæarctic species, breeding chiefly in Northern Europe and Siberia, but occasionally in several other countries of Europe. It is a resident of the British Isles, though its numbers are largely increased during the winter season. It has been found as far south in winter as Abyssinia, and abundantly in India, China, Formosa and Japan[1].

Widgeon, Wigeon (*Mareca penelope*)

In 1868 the Canterbury Society received eight young birds from Messrs Nairn and Crawford, who had apparently imported a number. In 1885 some more were received from the Royal Zoological Society of London.

In 1896 the Otago Society received eight birds from London, and these were sent down to the Clinton establishment for breeding purposes. The following year another lot from London were handed over to Mr W. Telford of Clifton, who liberated them in his ponds.

In 1904 the Government imported a number of these ducks. Of these four were handed over to the Wellington Society, and a dozen

[1] It is one of the remarkable facts of the attempts made to naturalise foreign animals in New Zealand that the country formerly possessed a native quail (*Coturnix novæ zealandiæ*) which has been allowed to become extinct; and also that it still possesses a grey duck (*Anas superciliosa*), a small brown duck (*Anas chlorotis*), and a shelldrake or paradise duck (*Casarca variegata*), all fine game birds, and that not one of the societies ever put forth any effort to preserve or protect these birds, except in the way of limiting the seasons for shooting them. In recent years bird sanctuaries have been set aside in many parts of the country, but very little else has been done to increase the supply of native game. Yet the little brown duck is quite as good a bird as the English teal, and well worthy of conservation.

(six pairs) to the Westland Society, which liberated them at Lake Kanieri.

It is very doubtful whether the species will establish itself in this country. It breeds in the northern portion of temperate Europe, extending northwards beyond the limit of forest growth, and occasionally in the British Isles. It winters in Southern Europe, India, South China, Formosa, and Japan; it is only a partial resident in Britain.

Gadwell's Duck, Gadwall (*Chaulelasmus streperus*)

The Wellington Society introduced some in 1894 and 1895; but have kept no record of them.

The species is only a partial resident in Britain, where it mostly winters; breeding in Iceland, North Russia and Central Europe.

Korean Duck (*Eunetta falcata*)

The Wellington Society received some (? date) from Sir F. Sargood, and since 1905 have reared a considerable number. They have apparently no later record of having turned them out.

*Mallard, English Wild Duck (*Anas boscas*)

In spite of the fact that the native wild duck of New Zealand (*Anas superciliosa*) is as fine a bird both for sport and table purposes as any species of the family that can be introduced, the various acclimatisation societies have for many years made continuous efforts to naturalise other species, and notably the mallard.

The Otago Society got a pair in 1867 from the Melbourne Society, and later from London introduced five in 1869, four in 1870, three in 1876, and nine in 1881. Apparently none of these early introductions throve, for there is no record of their increase or distribution, except that some were sent north to Kakanui, and others down to Riverton. In 1896 21 birds were received from London. Of these ten were forwarded to the Southland Society, and the rest were kept at Clinton for breeding purposes. In 1897 another lot was imported and these were handed over to Mr Telford of Clifton to be bred from. In more recent years numbers have been reared in different localities and have been liberated in such quantity that shooting was allowed in 1915. Between 1910 and 1918 the Southland Society have also liberated nearly 1350 birds in their district. They may, now, therefore, be considered to be established in the southern portion of the South Island.

The Auckland Society imported two in 1870, and four in 1886, and kept them in the Domain for breeding purposes. But there is no record of their further progress.

In 1873 the Canterbury Society had 12 in their gardens kept for breeding purposes, but there is no record of any results achieved. In 1897 this society joined with some others in importing a number from London.

In 1893 the Wellington Society imported 19, which they kept in their Masterton enclosures for breeding from; and they also distributed eggs. But stoats and weasels destroyed nearly the whole stock. Some placed in a reserve on Mr Martin's run, Wairarapa, increased rapidly however. In 1904 the Society received four pairs imported by the Government; and in more recent years they have reared and distributed several hundred birds. The species may therefore be considered as established in the Wairarapa, but without careful protection it is not likely that it will increase to any great extent.

In 1898 the Taranaki Society received a number from one of the other societies, but in four years all the birds had disappeared.

Mr Dansey of Rotorua tells me that subsequently to 1906, Mr McBean liberated a number of mallard on Lake Okareka, and that they increased to a flock of about 200. Some of these were presented to the Tourist Department, but there is no information as to the disposal of the birds on Mr McBean's death a few years ago.

In 1911 the Southland Society inaugurated an identification test. They liberated 100 mallards, which were numbered 1 to 100 on leg bands. I have not heard that any of these birds have since been identified.

In 1917 ducks were shot in South Canterbury which were believed to be hybrids between the native grey duck and the mallard.

At Temuka in the Acclimatisation Reserve a pair of hybrids between the mallard and the paradise duck have been reared.

Mr W. W. Smith states that the native grey duck hybridises readily with the domestic duck, and that the hybrids are fertile.

The mallard is a partial resident in Britain, but in many cases is only a winter visitant, nesting in Southern Greenland, Iceland and in Northern Europe. Some of the introduced birds and their progeny may have inherited the migratory instinct.

American Black Duck (*Anas obscura*)

The Government introduced a number of these birds in 1905, giving six to the Southland Society, and four to the Wellington Society. The latter body reared a number of young. They also reared a number of hybrids between this species and the mallard. Whether such hybrids are fertile or not is not stated. As an experiment in acclimatisation it may be interesting, but it is doubtful if it has any scientific value, especially if the results are not collected and published.

I have not heard that the species has been liberated, or if so that the birds have been seen.

American Wood Duck (*Aix sponsa*)

The Auckland Society introduced two in 1867; the Canterbury Society one in 1871; the Wellington Society two in 1894 and four in 1899; and the Otago Society one in 1906. Probably others have been brought in since.

They appear to have been kept as aviary birds in each locality, except in Wellington, where the last four received were liberated. There is no record of increase, except in Christchurch, where they were reported to be thriving in 1908.

Mr Dansey of Rotorua informs me that in 1906, Mr McBean introduced and liberated some Canadian wood ducks (presumably this species) on Lake Okareka, which lies in a basin between Rotorua and Tarawera Lakes. They were seen there for some years, but no young were ever observed.

Mandarin Duck (*Aix galericulata*)

The Auckland Society received one in 1868; the Canterbury Society two in 1871, and a number in 1885 from the Royal Zoological Society of London; and the Otago Society four in 1907. Private dealers also frequently introduced this bird. There is no record of their increase, or of their liberation.

Black Indian Duck

I do not know what species is referred to, but it is probably the tufted duck (*Fuligula cristata*), known to Indian sportsmen as the "Golden-Eye."

In 1870 the Auckland Society received five from the Acclimatisation Society of Victoria.

Mr Dansey informs me that along with the Canadian wood ducks, Mr McBean in 1906 introduced some Indian ducks, which were also liberated on Lake Okareka, but they never increased.

Probably most of the species of geese referred to in this list are more or less migratory species, breeding near or within the Arctic Circle, and wintering in temperate or warm temperate regions.

Egyptian Goose; Cape Goose (*Chenalopex ægyptiaca*)

Sir Geo. Grey brought eight or ten of these birds from the Cape with him in 1860. They bred freely at the Kawau, and many of them crossed over to the mainland. They were not long in spreading through the country, and were found from Te Aute in Hawke's Bay to the Kaipara. Apparently, however, all were destroyed in later years.

The Auckland Society introduced some of these birds in 1869, and kept them in their aviaries for several years, but there is no record as to whether they increased or were liberated.

Sandwich Island Goose (*Nesochen sandvicensis*)

The Auckland Society introduced a pair in 1871, but the record is the same as that of the preceding species.

Brent Goose; Black Brant Goose (*Branta nigricans*)

In 1871 the Canterbury Society received a pair from the Zoological Society of London, but there is no further record of them.

In 1905 the Government imported some from America, and the Wellington Society got a pair, but there is no further record.

*Canadian Goose; Maine Goose (*Branta canadensis*)

The Wellington Society imported three in 1876, and 15 in 1879. These were liberated and were seen for some months afterwards, but ultimately disappeared. It was reported that some of them were killed by paradise ducks.

In 1905 the Government imported a considerable number of these birds and distributed them widely, 11 going to the Southland Society, ten to Otago, a number (unspecified) to Canterbury, and six to Wellington.

The Southland Society liberated their birds at Lake Manapouri, and sent three more there in 1909. Others were sent up to Lake Te Anau, where they were reported to be thriving in 1918.

The Otago Society lost two in the first few years, but after a time sent some to the poultry farm at Milton where young were reared. Ultimately in 1912 some were liberated on Mr Telford's lagoon at Waiwera; and in 1915, 12 were sent to the head of Lake Hawea. These are doing well.

The Canterbury Society liberated a number at Glenmark in 1907–8, and these increased; in 1912 some were sent to Lake Sumner and others to Mount White. Mr E. F. Stead (April, 1916) reports that "they are thriving." At Glenmark they breed freely every year. Those at Lake Sumner are doing well.

Mr Ayson reports in 1915 that "they are doing well in several parts of the dominion."

I am told this species is a migrant in its original habitat, in which case its establishment in New Zealand is rather interesting.

Chinese Goose (*Cycnopsis cycnoides*)

The Canterbury Society imported some of these birds about 1874, when there were eleven in their enclosures; but in 1877 there were only four, and there is no further record.

Common Goose; Grey-Lay Goose (*Anser cinereus*)

The common goose is particularly interesting as the first species of bird which it was attempted to naturalise in New Zealand. When Captain Cook was at Dusky Sound in March–May, 1773, during his second voyage to New Zealand, he liberated some geese. In his *Journal* he says:

> Having 5 geese out of those we brought from the Cape of Good Hope, I went with them next morning to Goose Cove (named so on this account), where I left them. I chose this place for two reasons; first, because, here are no inhabitants to disturb them; and secondly, here being the most food. I make no doubt that they will breed, and may in time spread over the whole country, and fully answer my intention in leaving them.

Dr McNab, commenting on this in "Murihiku," says:

> The non-success of the importation of geese was doubtless due to the depredations of the weka. To show how deadly the weka could be to the harmless goose, the author instances a case which came under his own notice in Dusky Sound. The party disturbed a swan (black swan?) sitting on her nest, and although less than one minute elapsed before they reached the spot, the solitary egg, which proved to be quite fresh, had in that short time been tapped by a weka, and the contents partially extracted. No imported geese could successfully contend with such an ever-present foe.

The weka is no doubt responsible for the failure of many attempts to establish introduced birds, but it is only one of the agents.

The next recorded attempt at naturalisation was made by the Southland Society in 1867, when seven geese were liberated on the banks of the Mataura River. They were duly advertised as protected, but evidently that did not protect them from pot-hunters, for they disappeared.

A more successful attempt was made by the Otago Society, which in 1892 placed a number of goose-eggs in black swans' nests on Lakes Kaitangata and Waihola, Lake Onslow near Roxburgh, and in the Upper Taieri. At first these half-wild geese were shot, but later on they were allowed to increase, till in 1905, permission to shoot them was granted, when they were nearly or quite exterminated. The average (so-called) sportsman is a man out to kill something, and he does not concern himself as to the amount of trouble and expense which has been gone to in order to provide him with the thing to be killed.

Except where strictly preserved the goose has no chance of surviving, for it never becomes truly wild in New Zealand.

In 1887 a flock of about 20 geese frequented the east lagoon on Ruapuke in Foveaux Straits; they were very wild and after being shot at with rifles (though none were obtained) they appear to have shifted their quarters. Mr W. Traill, of Ulva, to whom I am indebted for this information, says that some years later a flock used to be seen about Bench Island, but when persons went to shoot them, they made off to the extensive swamps at the head of Lord's River. These were grey geese, probably belonging to the common species, but specimens have never been secured for identification.

White-fronted Goose (*Anser albifrons*)

In 1905 the Wellington Society received two of these birds, which were imported by the Government. There is no further record. The species is only a winter visitor in Britain; it breeds in the North, mostly within the Arctic Circle.

Oregon Wild Goose; Snow Goose (*Chen hyperboreus*)

In 1877 the Auckland Society received ten from Mr T. Russell, and ultimately liberated them at Matamata. They failed however to establish themselves.

Cape Barren Goose; Australian Wild Goose (*Cereopsis novæ-hollandiæ*)

The Auckland Society liberated two of these at Riverhead, some time before 1869, but they disappeared.

In 1871 the Canterbury Society received two from Mr G. Gould, but no further record of them was kept.

In 1912 the Otago Society received two, and sent them to the Government Poultry Farm at Milton, where several young were reared. From these, four were placed at the head of Lake Hawera in 1915, and others sent down to the Society's hatchery at Clinton. They appear to be doing well by latest reports.

Adelaide Goose

The Auckland Society introduced two birds in 1867, under this name. I do not know what species this is unless it is the maned goose, *Chlamydocen jubata*.

*Australian Black Swan (*Chenopsis atratus*)

This is one of the pronounced successes of naturalisation in New Zealand.

Some time previous to 1864 the Nelson Society introduced seven birds into that district.

In 1864 the Canterbury Society received four birds from Sir Geo. Grey, and liberated them on the Avon; and later two more from Mr Wilkin. In 1866 the same society received one from Mr Mueller, and four more from Sir Geo. Grey.

But the big effort came from the Otago Society, which liberated one in 1866, 42 in 1867, four in 1868, six in 1869, and eight in 1870.

The Southland Society also liberated six in 1869.

The birds quickly established themselves, spreading into all parts of the island from Stewart Island and the West Coast Sounds to Cook's Strait. Their whistling flight is a common sound at night.

Sir W. L. Buller says "the first were introduced into the North Island in 1864." The Auckland Society liberated four in 1867. They were plentiful in the Kaipara district ten years ago. They were also reported to be abundant in the Chatham Islands a few years ago.

Sir W. L. Buller states, and Mr Drummond repeats the statement, that wherever the black swan is found the wild duck (*Anas superciliosa*) disappears.

Sir Thos. Mackenzie records a case of a black swan being killed by a weasel.

White Swan (*Cygnus olor*)

This bird has never gone wild in New Zealand, but is still, as in Britain, always associated with preserved ponds and streams. The Canterbury Society received two in 1866; the Otago Society three in 1868, and one in 1869; the Auckland Society two in 1869 (from Sir Geo. Grey), and 12 in 1871 (from Captain Hutton). Several were also introduced by private individuals and by dealers. It is nowhere common.

Family Megapodiidæ

Scrub Turkey; Brush Turkey (*Catheturus lathami*)

The Auckland Society received two from New South Wales, from Sir Geo. Bowen before 1869, and liberated them at Kaipara, but they failed to establish themselves.

Family Cracidæ

Curassow (probably the Crested Curassow) (*Crax alector*)

The Canterbury Society received two in 1873, and one in 1874; but there is no further record.

Family Phasianidæ

Guinea-Fowl; Pintado (*Numida meleagris*)

The Rev. R. Taylor states that guinea-fowls were first introduced by the early missionaries, who brought them to the Bay of Islands.

The Canterbury Society received a number of these birds from India in the early sixties, from Messrs Guise Brittain and Cracroft Wilson. They presented six to Mr H. Redwood of Nelson in 1864.

The Otago Society introduced 23 in 1867, and distributed them to various private individuals. But they failed to reproduce themselves, and are still extremely rare in Otago, even as poultry. Apparently the winter climate is too severe for them in most parts.

In the North Island private individuals liberated them at several points, but they do not seem to have established themselves commonly as wild birds. I am informed, however, that in the Aberfeldy district, about 40 miles inland from Wanganui, they are not uncommon.

Mr Holman, curator of the Whangarei Acclimatisation Society, tells me that guinea-fowls attack and drive away harrier hawks.

Turkey (*Meleagris gallipavo*)

Turkeys have apparently gone wild in various districts from time to time. Mr B. C. Aston informs me that on crossing the Kaimanawa Range, he came upon wild turkeys on the Erewhon Estate. There are also abundance of wild turkeys on the ranges behind Kaikoura in Marlborough, but they are never far from the homesteads from which they have strayed. Mr T. Hallett states that they were common in Hawke's Bay 30 years ago, but disappeared rapidly after that. Mr Lowe considers that the disappearance of this species from districts where it was formerly common in the North Island was due to the same causes as that of the pheasants. The starlings and other introduced birds ate out their food, especially the insect-life on which the young were chiefly reared. The poults were lost in their efforts to struggle after their parents for food. It was noticed that the young broods became smaller and smaller, until at last they failed to be reared.

An interesting record is that given by Dr Cockayne concerning a flock of turkeys kept by Mr Jas. O'Malley of the Glacier Hotel, Bealey, 20 or 30 years ago. The birds lived in the bed of the Waimakariri River, among the tussocks which covered the land, and harboured immense numbers of grasshoppers. They became absolutely wild and increased to a great extent. The area over which they spread was at an elevation of 2000 feet and over, and was just on the edge of the wet belt, having an annual rainfall of 120 inches or over, and experiencing very severe frosts in winter. Apparently when Mr O'Malley left (or died) the whole flock was destroyed.

Mr H. T. Travers writes me in February, 1919:

At Pakawau, not far from Collingwood in the Nelson district, a settler used to raise a large number of turkeys. I often have, when visiting Collingwood, taken back with me two or three fine young "gobblers," paying 2*s*. 6*d*. each only for them. After a time I was unable to obtain any more, and was informed that in consequence of the great increase in introduced birds, destroying the grasshoppers on which the young turkeys lived, no more could be raised.

They are still common in some localities inland from Wanganui, but it is stated that in these localities, while minahs and Australian magpies are common, starlings are scarcely ever seen.

Pea-Fowl (*Pavo cristatus*)

According to E. Jerningham Wakefield the first peacocks introduced into New Zealand were brought to Wellington in 1843, by Mr Petre, who imported a large assortment of stock and materials in a ship of his own chartering.

The Otago Society introduced two in 1867, and handed them over to some private individual, but no further record was kept of them. In other districts private individuals and dealers introduced them, and, especially in the North Island, they occasionally got into the bush and became wild. Mr Thos. Hallett tells me that they were formerly wild in several places in Hawke's Bay, but disappeared probably from the same causes as the last species. They are still wild in bush districts inland from Wanganui.

Mr W. W. Fulton states similarly that they were formerly numerous in the valleys of the Turakina and Wangahu Rivers.

Common Fowl; Jungle Cock (*Gallus bankiva*)

Captain Cook in 1773 liberated fowls in West Bay, Queen Charlotte Sound, but on visiting the spot on October, 1774, he could not observe any trace of them. But in the account of his third voyage (February, 1777) he says:

All the natives whom I conversed with agreed that poultry are now to be met with wild in the woods behind the Ship Cove; and I was afterwards informed by the two youths who went away with us, that Tiratou, a popular chief amongst them, had a great many cocks and hens in his separate possession.

On 2nd November, 1778, Captain Cook anchored the 'Resolution' at the mouth of Port Nicholson, and gave the natives who came off in their canoes "fowls to take home and domesticate."

I cannot find that these birds established themselves anywhere as a wild species till much later, though it is quite possible that the natives got them from whaling vessels in later years. In 1814, when Marsden came to the Bay of Islands, fowls were brought over from Sydney, and from that date onwards the natives acquired birds and carried them throughout the country. Numbers became wild in the bush. Mr Thos. Hallett informs me that 30 or 40 years ago common fowls, the progeny of birds escaped from the Maoris, were found in many inland parts of the North Island, from Hawke's Bay to Lake Taupo. No doubt they were equally common in other parts, but in recent years these wild fowls have been exterminated.

Mr Jas. Hay says: "there were in the early days many wild fowls in the bush in Pigeon Bay, but whether or not descended from Captain Cook's stock I cannot say." It is much more likely that they were introduced either by whalers or by the early settlers at Akaroa and Lyttelton, as they were not recorded from other parts of Canterbury.

In 1840, during the Ross Expedition, some poultry were landed on the Auckland Islands. Again in 1865, Captain Norman liberated some on Campbell Island. In both cases they failed to establish themselves, though domesticated fowls thrive on the latter island at the present time.

Golden Pheasant (*Chrysolophus pictus*)

The Auckland Society received one from Mr R. Claude in 1867, and imported four more in 1868.

In 1871 several were imported by the Canterbury Society, but there is no record as to what they did with them.

The Wellington Society obtained three in 1891 from England.

The Otago Society received a pair in 1906, which they kept in their Opoho grounds, Dunedin. In 1907 they had 12 birds; in 1908 only five; in 1909 15 and in 1910 30. In the following year they had 35, and then their stock diminished till in 1913 they had only some half-a-dozen, but whether from death or by distribution is not stated. In 1912 a number of hybrids between golden and diamond pheasants were reared, but whether these crosses were fertile or not I cannot find out.

I do not think these birds were liberated at any time; they have, apparently, only been kept as aviary specimens.

Diamond Pheasant (probably Lady Amherst's Pheasant) (*Chrysolophus amherstiæ*)

The Otago Society had two of these birds at Opoho in 1907, but there is no record as to where they came from. They had increased to four in 1910, after which date there is no further mention of them.

*Common Pheasant; English Pheasant (*Phasianus colchicus*)
*Chinese Pheasant; Ringed Pheasant (*P. torquatus*)

These two species must be treated together, because, though separately introduced at several centres, they have apparently interbred to some extent. The first-named species has never established itself so readily as did *Phasianus torquatus*.

The first pheasants introduced into New Zealand were brought from England to Wellington in 1842 by Mrs Wills, a passenger in the 'London,' who landed a cock and three hen birds. In the following year Mr Petre landed some more. I cannot, however, trace the subsequent history of these birds.

In 1845 some English pheasants were liberated at Mongonui by Mr Walter Brodie. These birds, though multiplying round the spot where they were liberated, did not spread much. In later years, up to 1869, they were distributed and turned out at Tauranga, Tolago Bay, Raglan, Kawau, Bay of Islands and Napier.

Mr James Hay states that English pheasants were first imported into Banks' Peninsula by Messrs Smith and C. H. Robinson in the 'Monarch' in 1850. "Mr Robinson gave Mrs Sinclair one pair. She kept them in a wire-house, but one having escaped, the other was let out. Instead of remaining in Pigeon Bay, the birds went straight over the hill to Port Levy in the early spring of 1851." There they increased rapidly, but it was six years before they re-appeared in Pigeon Bay.

In 1853 Sir Edwin Dashwood brought out some English pheasants to Nelson. These birds increased greatly and soon found their way to the Waimea Plains, and in other directions. Two unsuccessful attempts were made later by Mr Henry Redwood, but on each occasion only one bird arrived.

The Otago Society liberated four English pheasants in 1865, 36 in 1866, six in 1867, 12 in 1868, 100 in 1869, 13 in 1870, ten in 1874, and 12 in 1877. There was no mistake as to their intention to establish them thoroughly. In 1871 they are reported as increasing rapidly in Otago; and in 1877 as "abundant from Oamaru to Invercargill; many shooting licences have been granted."

Sir Frederick Weld brought out a number of birds from England to Christchurch in 1865. The Canterbury Society imported four in 1867, and 30 in 1868. In 1871 it was stated that they are "thoroughly established, and needing no further importations."

The Auckland Society introduced seven English pheasants in 1867, and two in 1868.

In 1869 Mr Wentworth liberated a number of English pheasants on the Hokonuis, but I do not know from whence he obtained them.

These are all the introductions of English pheasants (*Phasianus colchicus*) which I have been able to trace.

Turning now to the Chinese species (*P. torquatus*), Captain Hutton states that in 1851 Mr Thomas Henderson imported some direct from China in the barque 'Glencoe.' Two dozen were shipped, but only seven reached Auckland alive, five cocks and two hens. These were turned out at Waitakere. In 1856 Mr Henderson imported some more Chinese pheasants in the schooner 'Gazelle,' of which only six arrived alive. They were also turned out at Waitakere. These thirteen birds, mostly cocks, appear to have been the whole of the Chinese pheasants imported into the province of Auckland. They were lost sight of for several years, then they re-appeared and gradually became more and more abundant in the neighbourhood of Auckland, and were shot in considerable numbers in 1865. They first appeared in the Waikato in 1864–65. In 1869 they were extremely abundant from Auckland through the Waikato and Thames to near Taupo. In the same year they reached Whangarei, but were rather rare further north. In the Auckland Society's Report for 1874–75 it is stated that the "Chinese pheasant is the common bird of this Province."

The Wellington Society imported 20 Chinese pheasants in 1874, and four in 1875.

The above seem to be all the direct importations of this species into New Zealand. But in the early years of acclimatisation a very wide distribution of birds by the various societies was undertaken, and they were spread far and wide. The kind was not always specified; some were Chinese, some were English, and very probably many were hybrids, if it is the case, as commonly believed, that the two species have crossed freely. Thus in 1864 the Otago Society obtained three Chinese pheasants from Auckland, and in 1877, 15 more; seven of these were sent to Oamaru, and five to Tapanui.

The Canterbury Society received three Chinese pheasants in 1867, and in the report for 1871 it is stated that they consider them "to be thoroughly established and needing no further importations." This statement appears to apply, however, both to the Chinese and English species, for in 1869 it was stated that the pheasants were in thousands on the Cheviot Hills Station.

The Southland Society got a large number of both species from other societies in 1869–70, and liberated them in various parts of the provincial district; and in 1874 they liberated four Chinese pheasants at Wallacetown.

Mr T. F. Cheeseman writing to me in August, 1915, says:

> Our pheasant is certainly the Chinese pheasant (*P. torquatus*) with a slight admixture of the English pheasant (*P. colchicus*) in the extreme north of the provincial district. Such enormous numbers of the Chinese pheasants were distributed from Auckland to other parts of the Colony that I am inclined to doubt the existence of the pure English pheasant in the wild state in New Zealand.

There is no doubt that in 1871 pheasants of both species were abundant all over New Zealand, but soon after their numbers began to decrease. Many causes contributed to bring about this decrease, which has continued ever since. The vast increase of the rabbit led to the introduction of phosphorus poisoning, and the grain used was freely eaten by the ground birds. The small birds introduced in 1867 and 1868 and subsequent years, soon increased to an extraordinary extent and ate out the food supply, and lastly, about 1882, the importation of stoats and weasels commenced.

The Otago Society report in 1881 that pheasants are plentiful, but much scattered, and "believed to have suffered greatly from hawks and poisoned grain." In 1882 they "have become very scarce." Poaching was considered to be the principal cause of decrease. In 1885 the "numbers are sadly reduced." In 1890 "pheasants are few and far between"; and in 1892 "pheasants are few and far between, it is very rare to see one." In the same report it is stated that there is no prospect of being able to establish a supply of game, while the present system of liberating stoats and weasels, and of rabbit poisoning is being carried on.

In the Wellington Society's report for 1885 we read: "The number of these birds in this district has greatly decreased of late years." In the Wairarapa district they are nearly extinct, "due to poisoned grain, and the introduction of stoats, weasels and ferrets." In 1888 "they do not seem to be increasing as they should." This is attributed to vermin, poaching, wet weather during the nesting season, and rabbit poisoning.

Captain F. W. Hutton, writing me in 1890, said: "the pheasants are dying off about Christchurch. This is probably due to failure of food; they have killed off nearly all the grasshoppers." The failure of food was the probable cause, but I think it was the starlings and sparrows which mainly decimated the grasshoppers.

Mr J. L. Watson, writing from Invercargill the same year, remarks: "the pheasants became pretty numerous" (in Southland), "but through poaching and poisoned grain, they are now exterminated."

Thus in the South Island, and in those parts of the North Island where rabbits abounded, the history of the naturalisation of the pheasant seemed to be that at first the birds increased rapidly and became very common, then their increase stopped, their numbers decreased and finally in 25 years they were all but exterminated.

In the North Island, in those parts where there were no rabbits, pheasants still continued to be fairly common, but were not nearly so abundant as in the earlier years. For example in Taranaki the report of the local society for 1874 is "pheasants plentiful in the Province." In succeeding years they were largely destroyed, and then breeding and liberating of young birds was resorted to. In 1908 the Chairman (Mr W. L. Newman) at the annual meeting said "he did not think they had a pheasant left."

After a time, most of the societies, undeterred by past failures, and without any proper examination into the causes of the failures, renewed their efforts to stock the country with these birds.

In 1895 the Otago Society obtained seven hens and two cocks from Auckland, and liberated them between Lake Waihola and the coast. Shortly after one was picked up dead, and its crop was found to be full of phosphorised oats. In 1897, 22 birds (out of 32 shipped) were received from London, and appear to have been kept for breeding purposes either at the Milton poultry farm or at the Clinton hatchery. The Society was led to recommence stocking with these birds because "it seems that the trapping of rabbits for export is now carried on to such an extent that much less poison will be laid in future, and if the hawks were reduced in numbers, game birds may increase in our district." It was also recommended that a bonus be offered for the rearing of young pheasants, and in this way "45 strong young birds were secured and liberated." In 1899 another lot of 21 birds was received from England, and was kept for breeding purposes, while over 80 were bred locally and turned out. Yet in 1904 the report is: "An odd old bird is to be seen at times, but no young." Still more recently the attempt is again being carried on, and in the report for 1914–15 we read "the birds liberated during the past two seasons have been seen occasionally and are apparently thriving." In the report for 1919 it is stated that "Pheasants liberated four years ago have not been reported about for some time."

The Southland Society has made very strenuous efforts to stock Stewart Island, where the rabbits have died out, there is no poisoning done, and neither ferrets, stoats nor weasels occur. Five pheasants were liberated in 1869, 16 in 1895, 48 in 1901, 37 in 1902, 36 in 1904, 16 in 1907, 105 in 1909 and 47 in 1910. No doubt some of these survived and may have bred, but Mr Walter Traill—a skilled observer and resident—writing in March, 1916, says "the pheasants did not thrive."

The Canterbury Society began rearing pheasants in 1868, and continued the work for many years, also importing fresh birds from Britain from time to time. Mr Drummond writes as follows:

In 1868 the Acclimatisation Society bred forty birds and sold them to members for £2 a pair. In the tussock-covered land of Canterbury they throve specially well, and the large Cheviot Estate, then held by the Hon. W. Robinson, was soon stocked with them. Mr Robinson spared no expense in preparing for their reception when he arranged for a consignment supplied by the Society. He erected commodious aviaries, ordered that all the cats on the estate should be killed, nearly extirpated the wekas, and had the hawks destroyed at the rate of six a day. The society continued to import pheasants for a considerable time. It bred about a hundred birds in a year, and obtained a fairly good income by selling them to the owners of large estates. It seemed as if pheasants would in a few years spread throughout both Islands and become thoroughly naturalised. After this had gone on for some time the birds received a decided check. Their numbers neither increased nor decreased. They then began to decrease rapidly, and apparently almost simultaneously in many districts. Their complete failure, taking the colony as a whole, is now beyond doubt. In Canterbury and other provinces where they were once exceedingly plentiful they are never seen at all.

The Nelson Society re-introduced pheasants in 1912, and succeeding years, and in 1915 reported them as thriving.

The Wellington Society continued for years to raise broods through private individuals, but the birds did not increase, ferrets occasionally destroying all the birds in an enclosure. In 1897, in conjunction with other societies they introduced 40 from England. In 1905 they received four pairs from the Government. In 1907 they reared 430 at the Game Farm; 347 (English and Chinese cross) in 1908; some 230 in 1910; over 300 in 1912; and no doubt large numbers (not specified) in 1909, 1911 and 1913; and most of these were liberated. But the curator says they are unable to hold their own against stoats and ferrets.

The Taranaki Society were still (1913) "buying pheasants' eggs (which hatched poorly), and rearing them at a game farm." In 1915 the efforts of the Society to rear birds are referred to, but it is stated that these efforts had been unsuccessful. Mr W. W. Smith, one of the best observers and naturalists in New Zealand says, in February, 1916: "they are still fairly plentiful in Taranaki, but formerly were much more abundant."

Mr Peacock in 1913 states that pheasants were very plentiful formerly in the Bay of Plenty district, but are now rare; while blackbirds, thrushes, starlings, goldfinches, green-linnets and Californian quail are now common.

I have given the facts regarding the introduction, spread and subsequent disappearance of the pheasants at very considerable length, because they show what a complex problem the naturalisation of a species in a new country may be. In spite of all the efforts which have been put forth during the last half-century, pheasants are not very common anywhere in the North Island, and are extremely rare in the South. Poisoned grain, ferrets, stoats, weasels, wild cats, hawks, wekas and poachers have all been blamed for this failure, and no doubt all have borne a share in bringing about the present position. But I think Mr Cheeseman's explanation is probably the correct one, namely that the diminution of their food supplies caused by the vast increase of all kinds of small birds has been the chief agent. "These," he says, "have literally starved out the pheasants, which scarcely manage to keep themselves alive during the winter months, and, when the breeding season arrives, they are not in a fit state to reproduce their species." Mr Bell, of Hawera, who strongly endorses this theory, states also that "young pheasants cannot travel far for food, and that they are attacked and destroyed by Californian Quails." Mr J. Grant of Wanganui informs me (1918) that Australian magpies have been seen to attack and kill a hen-pheasant.

What is true about pheasants is also true for all the larger game birds. It is not so much the hawks, ferrets, wekas and owls, as the smaller birds which are the real cause of the diminution and disappearance of the larger species. It is the eating out of the food supply, and chiefly of the insect life which is the main food supply of the young birds.

Pheasants do much damage to crops; they destroy young grass; pull up sprouting maize, attack potatoes, carrots, beans, peas, barley, wheat, and many kinds of fruit.

On the other hand they eat great quantities of insects, as many as 150 crickets having been taken from the crop of one bird. In the Auckland district, when the berries of the ink-weed (*Phytolacca*) are ripe, the pheasants feed largely on them, and their flesh becomes very dark-coloured at that season of the year[1].

[1] The way in which incorrect statements get abroad and are quoted later as facts is illustrated by some remarks in *Animals of To-day; their Life and Conservation*, by C. J. Cornish. This author says: "In Australia, and still more noticeably in New Zealand, the new comers, the most vigorous representatives of the later types of animal, had a clear advantage over the ancient marsupial forms and the wingless birds. The pheasant, which can both run and fly, displaces the New Zealand *Apteryx*, and the rabbit gets the better of the wallaby and smaller kangaroos." The statement about the pheasants is absolutely wrong.

Elliot's Pheasant (*Phasianus ellioti*)

The Wellington Society received three birds in 1895, and sent them to Mr Knowlton of Greytown. In 1897 four more were imported, but there is no further record of them.

Reeves's Pheasant (*Phasianus reevesii*)

In 1897 the Wellington Society, in conjunction with other societies, introduced nine birds, but it is not stated what was done with them.

About 1899 the Wanganui Society imported several specimens of this species (this may indeed be the same as the previous record), and all but two, which had broken wings, were liberated up the Wanganui River. Apparently they disappeared without being recorded again. Of the pair kept in confinement by Mr S. H. Drew of the Wanganui Museum, the male moped after a time and died. Dr Connolly, who examined the body, stated the death was due to tuberculosis. The bird was then sent to Dr J. A. Gilruth, Chief Government Veterinarian, who reported: "the disease affecting this animal is tuberculosis in a most advanced stage, almost every organ being implicated. The nodules in the liver and lungs, when examined microscopically, are found to be filled with masses of the characteristic bacillus."

Silver Pheasant (*Gennæus nycthemerus*)

According to Taylor, 1868, silver pheasants were liberated by the Hon. Henry Walton in the neighbourhood of Whangarei.

The Auckland Society received two from Mr D. B. Cruickshank in 1870, and they were reported as breeding well in 1874.

The Otago Society had four shipped from Hongkong in 1871, but only one cock arrived in Dunedin. In 1905 there were two in the Society's aviary at Opoho.

The Canterbury Society introduced some in 1871 and had four in their gardens in 1880.

Like the golden pheasant, this species has only been treated as a bird for the aviary, and there is no record of their liberation.

Jungle Pheasant (species?)

I do not know what bird is referred to under this name. The Otago Society had four at Opoho in 1907, with no record as to where they came from. In 1910 there was one at Opoho, and one at Clinton. They are not mentioned in any later reports.

Temminck's Tragopan (*Ceriornis temmincki*)

The Auckland Society received a pair from the Zoological Society of London in 1871. One poult was produced in 1874, and there the record ends.

Chinese Quail (*Excalphatoria sinensis*)

The Otago Society received ten of these birds from Mr Winton of North East Valley, Dunedin, whose daughter brought them from China in 1897. They were sent to the aviary at the Camp, Otago Peninsula, for breeding purposes. There is no further record of these birds.

*Australian Quail; Swamp Quail; Brown Quail (*Synœcus australis*) (called Tasmanian Quail (*Coturnix australis*) by Hutton in 1871)

The Canterbury Society received a pair from Lieut.-Colonel White in 1866; they imported five in 1868, and a number in 1871.

The Auckland Society introduced four in 1867, and no fewer than 510 birds in 1871.

The Otago Society imported three in 1868, and nine in 1870. These were liberated at Green Island, south of Dunedin, and were never heard of again.

The Southland Society imported four in 1872; these were liberated at Wallacetown. Of their progeny 25 were liberated on the Awarua Plains in 1911, and in the following year the remaining stock were taken to Mason Bay, Stewart Island. Regarding this last lot, Mr Traill tells me (in 1916) that they did not succeed.

The Wellington Society introduced five in 1875, and 39 in 1876.

The bird is almost unknown in the South Island, but is fairly common in many parts of the North Island. I have frequently been told in certain districts that "Native Quail" occur, and have always found that it is the Australian swamp quail that is referred to.

The Wellington Society report in 1885 that they "are rapidly increasing on the West Coast between Waikanae and Manawatu, and on the East Coast of the Wairarapa." In 1889 it is stated that "they are spreading slowly, but owing to their keeping close on the ground are kept down very much by cats, hawks, and other vermin."

In the 1890 report it is said: "they fall an easy prey to cats, rats, etc. They almost disappeared in some of the clearings in the Forty Mile Bush, where formerly there were large bevies."

On the other hand the Waimarino Society in 1915 report these birds as coming into the district from round about and being protected till 1912. They are now increasing.

Colonel Boscawen informs me that they are now (1916–17) very common in the Auckland district. Mr R. Kemp got four of them in 1907 in the Hokianga district, and on comparing them with skins in the British Museum found that they were slightly different from the typical Australian form. I flushed two near Whangarei in April, 1919.

Mr M. Makai of Paremarema, Auckland, says they are most active agents in the spread of blackberries and gorse in the district.

Mr Peacock reported them as common in the Bay of Plenty in 1913.

Tasmanian Swamp Quail (*Synœcus diemenensis*)

The Auckland Society reported a few in 1869. There is no indication in subsequent reports as to where the birds were liberated, or whether they were ever seen again. Singularly enough, in June, 1916, nearly half a century after the date of their introduction, a specimen, caught at Pirongia in the south of the Auckland provincial district, was brought to the Dominion Museum, Wellington. Evidently then, some examples of the species still exist in the wild state, and are probably mistaken for *S. australis*, though it is fully a third larger than the Australian species.

Californian Mountain Quail; Mountain Partridge (*Oreortyx pictus*)

The Auckland Society introduced three (out of 29 shipped) in 1876. They received nine from Mr T. Russell in 1877, and these were turned out at Matamata. In 1881 a large number was imported, of which 40 were liberated near Lake Omapere, 40 in the Upper Thames district, and about 120 were kept in the gardens. In 1882 a further large number was introduced, and they were distributed all over the provincial district. They never appear to have established themselves, and Mr Cheeseman, writing in August, 1915, says: "it is now many years since they have been observed."

The Otago Society introduced 122 in 1881, and liberated half of them at Gladbrook, Strath-Taieri, and the other half at Mataura Bridge and on Venlaw Station. In 1882 another lot of 64 was imported, and these were liberated at the foot of the Rock Pillar Range. Nothing was ever seen or heard of them afterwards.

Stubble Quail of Tasmania; Australian Quail; Stubble Partridge (*Coturnix pectoralis*)

According to Hutton (1871) this species was introduced into both Auckland and Canterbury.

Robert Kemp says "that they were introduced into the Hokianga district in the seventies, but failed to establish themselves."

Recently (1918) several acclimatisation societies have been desirous of re-introducing this bird, being unaware of the previous unsuccessful attempts, and have obtained permission from the Department of

Agriculture to do so. Reports of its feeding habits have been received from Australia, and are favourable[1].

Indian Quail (*Coturnix coromandelica*)

The Otago Society had four of these birds at their Opoho hatchery in 1907, but there is no record of where they came from, or what was done with them.

Little Australian Quail; New Holland Partridge (*Turnix varia*)

According to Hutton (1871) this species was introduced into Auckland and Canterbury, but there is no other record of them.

Egyptian Quail (species?)

The Canterbury Society Report of 1883 states that: "the Egyptian Quail, which were turned out on the Kinloch Estate, have likewise been seen," but there is no previous reference to these birds. Colonel Boscawen informs me (December, 1916) that his boys bought some tiny Egyptian quail at 3*d*. a pair from a steamer at the wharf (Auckland). They were marked like guinea-fowl. "They got out, and may have lived." This was before the war (1914). I do not know what species is referred to.

Black-breasted Quail (species?)

The Otago Society had two at Opoho in 1909, but again there is no record of whence they came, and what happened to them. Again I do not know what species this is.

Common Partridge (*Perdix cinerea*)

The history of the numerous attempts to naturalise this bird in New Zealand is almost pathetic.

The Nelson Society introduced *eight* some time prior to September, 1864, but there is no record of what happened to them.

[1] In New South Wales this species eats seeds of grass and weeds, and occasionally wheat; also occasionally seeds of *Solanum nigrum*, *Phytolacca*, buttercups and chickweed. Its animal diet consists of army-worms, beetles and plant bugs.

The acting Chief Inspector of Fisheries and Game in Victoria says: "Their food consists of weed-seeds and insects. One of their favourite foods in Southern Victoria is the black seed of the spear grass, but dock-seeds, crickets and a species of weevil have been found in them." The Superintendent of Experimental Work, Department of Agriculture, Adelaide, says: "The Stubble Quail is common here and is becoming increasingly so, but I have neither seen nor heard of damage being done by them. They certainly eat grain, but do not appear to do so unless it falls to the ground, and on the other hand they eat an enormous quantity of seeds of the weeds that accompany the cereals. For instance a rather bad weed with us some years is the rough poppy (*Papaver hybridum*), and when this weed is bad in any district the quail are plentiful, and I have seen young quail raised successfully on the seed of this poppy alone."

The Auckland Society liberated *seventeen* in 1867 (and a covey was seen at Mangere the following season), *twenty* at Howick in 1868, and *nine* in 1871. Three years later it is reported that "the birds do not seem to have multiplied." In 1875, 80 birds were shipped in London, but only 40 arrived, and these were mostly cocks. They were liberated near Lake Takapuna. The report for that year adds: "The failure of all previous attempts, both in this colony and in Australia, is by no means encouraging."

The Canterbury Society received ten birds in 1867, and one in 1868, the latter being the sole survivor of 45 which were shipped. In 1871 they imported 32 brace, and in 1880 a shipment on a large scale was attempted. No fewer than 240 birds were shipped from London, but only 19 arrived in Christchurch, and were liberated. I do not know if any others were imported by private individuals or not, but it was stated in February, 1869, that "partridges may be counted by the hundred at the Cheviot Hills Station." In 1875 Mr R. Bills brought out a number of partridges to the Canterbury Society, and these appear to have been sold and distributed throughout the provincial district. In 1879 another shipment of 25 brace was made, but only a few survived, and these were liberated on the Hororata.

The Otago Society liberated 31 birds in 1869, and 130 in 1871, They commenced at once to multiply and in the country south and west of Dunedin coveys were frequently flushed. In 1877 they were not uncommon from Oamaru to Invercargill. But the introduction of phosphorus poisoning seemed to arrest the progress.

The Society's report for 1881 says they are "much scattered, but suffering from hawks and poisoned grain." In 1882 they "have become very scarce," and poaching is believed to be the principal cause. In 1885 the "numbers are sadly reduced"; in 1890 "the partridges have almost entirely disappeared"; and in the report for 1892 occurs this passage: "Your Council regret that there is every reason to believe that these birds have become extinct."

Another attempt was made in 1896, when the Society received 20 birds from London, and liberated them on the property of Mr R. Charters on the Taieri Plain. They were reported as being seen very often, and doing as well as could be desired. In 1897 23 were received from London (out of 24 shipped); nine of these were sent to Mr Charter's property, and the rest to Mr T. Telford of Otanomomo "(to be liberated after the poisoning season is over)." In 1898 the report states: "The partridges turned out on the Taieri do not seem to have increased in numbers." In 1899, "a few are to be seen at times on the Taieri, but no young broods have been seen."

In 1900 a shipment of 44 (out of 48 shipped) arrived from England, and the birds were liberated in two lots in suitable localities (which were not specified publicly). In the following year we read: "An odd old bird is to be seen at times, but no young ones."

In 1909 still another attempt was made, and several birds (the number not specified) were liberated near Milton, for the report for 1911 says "the partridges liberated in the Milton district *two years ago* are seen at times by the settlers." In 1912, "no recent reports have been received in connection with those liberated near Milton last year." In 1913 we read: "Those liberated near Milton in 1911 have not been seen for a long time, and have probably all been killed by the obnoxious stoat or weasel."

The Southland Society imported 48 partridges in 1899–1900, and liberated them on Stewart Island. Mr W. Traill, writing in March, 1916, says the experiment failed.

In connection with the Southland Society's proposal to introduce these birds, Mr F. Sutton wrote (17th April, 1899):

> My brother and myself brought from England 19 partridges: 12 of them were let go on my farm, and seven on Mr G. Sutton's farm near Winton. This is about twenty years ago, before poison was used to destroy rabbits. They had a few young ones, but soon died out. In (say) five years not one was to be seen. The hawks and cats were supposed to be the cause of their extermination.

In 1889 a private attempt was made in Wellington to introduce partridges, but it was not successful. In 1890 the Society received two cocks and a hen, but the latter died. In 1891 the Society received 15 birds from Lord Onslow, and imported two from England. Unfortunately ten of these birds died suddenly in the following year, and the remaining seven were liberated at the Upper Hutt, but were not seen again. In 1897, with Mr Stuckey, the Society introduced 32, and liberated them near Masterton; and in conjunction with the Canterbury, Nelson and other societies, an additional 16. But there is no record of what came of these. The report for 1900 says: "The partridges liberated at Rangitumau seemed to do well for a time, but have now disappeared." Evidently the Society got some more birds, for the report in 1904 says: "Out of some 35" at the Game Farm "only seven remain[1]."

[1] It is characteristic of the careless way in which acclimatisation society records were often kept, that there is no record of the importation of partridges into Otago in 1909 or 1911, nor of their being liberated near Milton, except the statements quoted above, which were made in later years.

The Taranaki Society in 1894 procured four pairs from some other society, and liberated them in four different localities. In 1902 it was stated that nothing had been seen of them. But in 1904: "it is reported from the Koru district that during the latter part of November or early in December, a covey of partridges had been seen, and indeed one had been killed by a mowing machine passing over a nest."

Mr J. Glessing of Thames stated in 1913 that partridges would have been a success in the Waikato district, but for the abundance of harrier hawks (*Circus gouldi*).

I am not aware of any district in New Zealand where partridges have survived. The causes of their total disappearance are probably the same as in the case of the pheasant, viz. poison, wekas, stoats and weasels, and the great abundance of insect-eating birds[1].

French Partridge; Red-legged Partridge (*Caccabis rufa*)

These birds have been introduced from time to time, but the records of the acclimatisation societies are so hazy with regard to various species, that it is almost impossible to speak with certainty on the subject.

In the Wellington Society's annual report for 1897, we read: "It is reported from the Rangitikei district that red-legged partridges are increasing, and a few are working north into the bush-country." Yet there is no previous record of the introduction of these birds.

The Canterbury Society imported two in 1867, but lost one of them. They also imported 25 brace of "partridges" in 1897, and these are referred to as "Common Partridges"; it is not clear from the reports to which species they belonged.

Mr Ayson writing in 1915 says: "They took a hold well in several parts of the Dominion until rabbit-poisoning commenced, and vermin was introduced to destroy rabbits."

In 1899 18 birds were received from London (out of 20 shipped) and were liberated on Stewart Island. According to Mr Traill, these did not succeed in establishing themselves.

In 1913 the Auckland Society obtained a considerable number in London, but they all died before they could be shipped.

[1] At a meeting of the Tauranga Society (held in 1915) Mr Macmillan suggested that partridge eggs could be brought out to New Zealand (frozen?) and might be successfully hatched out here. "A gentleman residing in Derby had informed him that he had frequently bought New Zealand eggs, which had gone to London and been sold there as fresh-laid Derby eggs. Out of curiosity he had frequently set New Zealand eggs, and from the preserved variety had secured a fair percentage of chicks." It certainly sounds improbable.

Barbary Partridge; Teneriffe Partridge (*Caccabis petrosa*)

The Auckland Society introduced two in 1868, but there is no further record of them.

The Wellington Society in 1892 received 19 of these birds from Mr Hamilton of Teneriffe. Six of them died soon after arrival, and the rest were at once liberated on Kapiti Island. In 1894 a covey of two old and seven young birds was seen.

Hungarian Partridge (*Caccabis saxatilis?*)

The Wellington Society, in conjunction with the Canterbury, Nelson and other societies, introduced 20 of these birds in 1897; but no record is obtainable as to what was done with them except that the Hawera Society obtained three brace in February, 1898, and liberated them.

The Auckland Society in 1912 imported 39, and kept them for breeding, but 20 of them died, and the remainder were liberated, 15 at Kaipara, and four in the Waikato. No report was received of these, and in the annual report for 1914 it is stated: "it begins to be apparent that they are not satisfactory birds to import, there being no perceptible increase."

*Californian Quail (*Lophortyx californicus*, *Callipepla californica*)

This is one of the few species of game birds which have succeeded in establishing themselves in New Zealand.

Mr Huddlestone informs me that they were first brought to Nelson in 1865, but it is not clear where they came from.

The Auckland Society introduced 113 in 1867; of these, ten were sent to the Waikato, and 42 to Nelson in the following year. Fourteen more were received from Mr D. B. Cruickshank in 1870. Sir George Grey also introduced these birds into Kawau, and they increased to such an extent that in later years the Auckland Society was permitted to net hundreds of them for liberation in the provincial district.

The Canterbury Society introduced two in 1867, four in 1868, and a large number (unspecified) in 1871. Later on they purchased 520 birds from Nelson (where they had increased enormously), and turned them out in various localities. In 1883 they introduced 122 more; in 1885 they are reported as being very numerous at Little River. But in 1906 "the Quail seem to be steadily decreasing in most places."

The Otago Society introduced 18 in 1868, and liberated them at Inch Clutha, where they increased, and three years later were reported to be plentiful. In 1871 120 were imported, one half of them being liberated at Waikouaiti, and the other half at Popotunoa (Clinton). The following is the record of the Society in regard to this bird in the years succeeding its introduction: (1881) "they are numerous about Queenstown, also seen in the Clutha district and in the Upper Clutha Valley; numerous about Palmerston"; (1882) "doing well at Queenstown, Goodwood," etc.; (1885) "their numbers sadly reduced"; (1892) "these birds are in considerable numbers in some localities, principally along the coast, where little poisoned grain is used"; (1896) "reported as being on the increase in places where they were not exterminated by the poisoned grain laid for the rabbits, and it is known that the poisoned pollard, which is largely used now for the destruction of rabbits, is less injurious to game birds than poisoned grain"; (1897) "numbers to be seen in the lower parts of the Otago Peninsula, and on the ridges in the neighbourhood of Clyde." They gradually disappeared from the neighbourhood of Dunedin and from both shores of Otago Harbour, and this certainly was not due to poisoned grain. It was more probably caused by the great increase in stoats and weasels. The birds always held their own more or less in Central Otago, and they are reported from 1912 to 1920 as common and increasing in numbers.

The Southland Society liberated two at Wallacetown in 1873, and 29 in 1874, and up to 1890, they increased and were to be seen in considerable numbers; but they have become rarer in later years, and Mr Philpott tells me have completely disappeared from Southland for many years.

The Wellington Society liberated 266 in 1874, and 118 in 1875, but their subsequent reports are very contradictory. In 1885 the statement is made that "the number of these birds in this district has greatly decreased of late years. In the Wairarapa they are nearly extinct, owing to poisoned grain, and the introduction of stoats, weasels and ferrets." In 1886 they are said to be fairly numerous; and in 1889: "they are increasing fast and have taken such a hold, that there is little danger of their extermination by any fair means." In 1904 the Marton sub-committee report that "The Californian Quails, which used to be so plentiful, are disappearing fast, as every succeeding season they seem to be less; even in parts where they were thick, now very few are to be found." In the same season the Masterton sub-committee, writing from a rabbit-infested country, report that "Quail are showing a marked increase." In 1910 they are said "to hold their own fairly against stoats and ferrets. The birds are so numerous in some districts as to be a nuisance. They came into the Waimarina district from outside, and were protected till 1912, they are still increasing (1915)."

In Taranaki they were spreading in some districts in 1874 having apparently worked their way down from Auckland. Then in 1880 the local society imported 60 birds, presumably from Nelson, and they steadily increased, and were reported in 1904 plentiful in all parts. In 1913 they are reported as having become "very numerous in various parts of the district, and farmers complained of the damage done by these birds."

It is, however, in the Nelson provincial district that this species has increased and thriven to the greatest extent. The only record I can find of their introduction into Nelson is the isolated one of 42 birds received from the Auckland Society in 1868. (I may remark that the early records of the Nelson Society are unobtainable.) But there is no doubt that in about ten or a dozen years they became a considerable article of export at one time. At a meeting of the Southland Society in 1897, Mr Whitcombe stated that in Nelson, quails were formerly so numerous that they were tinned by thousands; and Mr Ellis mentioned that when he was at home, a shipment of 20,000 arrived in London in the frozen state from Wellington.

I wrote to Mr John Pollock, President of the Nelson Society, on the subject, and in his reply, dated 13th December, 1915, he says:

> With regard to your enquiry re Quail, I have interviewed Mr Kirkpatrick, and he informs me that he started tinning these birds about twenty-five years ago, and from my own recollection that is about the time they were most numerous in this district. Mr Kirkpatrick informs me that the year he started the industry he was paying as low as 3*d.* per brace for the birds which were, of course, trapped, the price rising each successive year until 10*d.* was reached. At this price it did not pay him to continue, and he accordingly closed down this line of goods. At the same time that tinning was being carried on, large quantities of quail were shipped to the Wellington and West Coast markets. I have myself seen ten or twelve 4-bushel sacks full of birds on the Nelson wharf awaiting shipment. This state of things, however, did not last long. Stoats and weasels made their appearance in the district, having been liberated in the Marlborough district adjoining, and I think there is little doubt that the gradual diminution of the quail ever since may be chiefly ascribed to these pests, assisted by the countless numbers of starlings which now scour the country in search of food.

The Nelson Society report them as quite common in recent years, 1911–15.

Californian quail are regarded by most farmers as somewhat of a nuisance. Mr Drummond states that "At Te Puke, in the Maketu district, quail live largely on clover, taking both seed and the young plants in the bush-clearings." Mr D. Petrie states that in the Waikato they are a nuisance, as they pick up the newly-sown and germinating turnip seed; "one bird was found to have about 130 seeds of gorse in its crop. On being sown, every one of these seeds germinated." This is interesting, as showing the food on which they live, but these birds do not pass seeds; they are ground up before reaching the stomach. Reports from north of Auckland, from Rotorua and from the Bay of Plenty all tell the same tale; the birds are most destructive to young grass and clover seedlings, and in many parts farmers sow poisoned grain with the seed in the hope of checking the nuisance. They are also accused of spreading the blackberry.

At the present time (1916–19) Californian quail are very abundant in the Taranaki district, and are met with on Mt Egmont in summer time up to 6000 feet.

Virginian Quail; Colin; "Bob-White" (*Ortyx virginiana*)

The Wellington Society in 1898 introduced about 400 birds, which were distributed as follows: 80 went to Otago, 40 to Canterbury, 20 to Stratford, 20 to New Plymouth, and the rest were liberated in the Wellington district. In 1899 another large lot was introduced and distributed: 22 to Southland, 46 to Otago, 90 to Canterbury, 70 to Blenheim, 100 to Wellington, 60 to Wanganui, 44 to Stratford, 32 to New Plymouth, 30 to Napier, 56 to Waikaremoana, 6 to Gisborne, and 200 to Auckland.

In 1900 the Otago Society reported that "they were still to be seen in the neighbourhood of where they were liberated, but no young birds have been seen." There is no further record of them.

In 1902 the Wellington Society report: "These birds have so far been a disappointment; reliable information as to their having been seen during the past year is difficult to obtain." The Pahiatua sub-committee report says: "they seem to have disappeared." The Marton sub-committee report them as "doing well, and that they have been seen with young broods."

In the same year the Taranaki Society report: "Virginian Quail are steadily increasing and will, in a year or two, afford good sport." In the 1904 report we read: "Virginian Quail seem to have disappeared."

In 1909 the Auckland Society report "the Virginian Quail have almost disappeared."

Prairie Hen (*Tympanuchus americanus*)

The Canterbury Society imported 17 (out of 28 captured) in 1879 from Topeka, Kansas, and turned them out near Mount Thomas. In 1880 it was reported that "the prairie chickens which were obtained from Nebraska (?) last year, have not been seen or heard of for some time." But in 1885 we read: "About a dozen of the prairie hens, imported about four years ago, have been seen on a farm at North Loburn, in fine plumage and apparently good condition." This is the last record of them.

The Auckland Society introduced 20 of the birds in 1881. In the following year, out of a very large number shipped at San Francisco, about 40 more were received. Half of these were sent to Otago, and (it is said) the remainder were stolen. I cannot find any report of their liberation in Otago. But neither in the north nor the south did the species succeed in establishing itself.

Common Grouse (*Lagopus scoticus*)

The history of the attempts to introduce grouse into New Zealand is nearly exclusively a record of failure to carry the species over the sea. Only two societies appear to have tried to introduce it.

In 1870 five were shipped in London by Mr Larkworthy for the Auckland Society; one survived the voyage, then pined and died. In 1872 he shipped 33 birds; two pairs were landed, but one pair died soon after landing. The other pair, a cock and a hen, were liberated at Matamata. In 1873 Mr Larkworthy shipped another lot (the number is not recorded in the Society's reports), of which one pair arrived. These also were liberated at Matamata, and were heard of for some time afterwards.

In 1871 the Otago Society got Mr R. Bills to attempt to bring some out, and eight brace were shipped in London. When eight days out they all died in the course of a single night. Mr Bills reported that "their legs became suddenly paralysed, immediately after which they drooped away."

Black Grouse; Black Game (*Lyrurus tetrix*)

The Auckland Society in 1873 made an attempt to obtain these birds through Mr Larkworthy, but they all died before they reached the docks in London.

The Otago Society in 1879 through Mr Bills, obtained 80 birds in Scotland, but only 20 reached the ship. "It was found very difficult to take the black cock at all from proprietors' lands in Scotland." Ten survived the voyage and were liberated. In 1882 one was seen near Tuapeka Mouth; and a cock and a hen up the Waitahuna River. There is no further record.

In 1900, 26 were shipped in London, but only a cock and two hens arrived and reached Dunedin. These were liberated in the reserve at the junction of the Leithan and Pomahaka runs; they were seen the following year, but with no young ones, and there the record ceases[1].

Ptarmigan (*Lagopus mutus*)

The only attempt to introduce this species was made in 1897, when the Wellington Society, in conjunction with the Canterbury, Nelson and other societies, had a shipment made in London; but all died on the way out.

Pointed-tailed Grouse (*Pediœcetus columbianus*)

The Auckland Society introduced 22 from Utah in 1876, and they were liberated at Piako. There is no further record concerning them.

Family Rallidæ

Australian Coot; Murray Coot (*Fulica australis*)

The Auckland Society introduced two in 1869, but it is not stated what was done with them.

Sir Walter Buller in the Supplement to *The Birds of New Zealand* (p. 75), describes this as an addition to the fauna, as follows:

> I have to add to the list of New Zealand birds the Australian Coot, a specimen of which was killed in July, 1889, at Lake Waihoia in Otago. *There is no record of this species having been brought alive from Australia*, and, even if it had been, it is difficult to see how it could have reached that remote district.

The species is evidently an occasional visitant from Australia. A second specimen was taken at Kaitangata in May, 1919, and a third at Mataura Island in July, 1919.

[1] In a book of newspaper cuttings belonging to the late Mr A. M. Johnson of Opawa, I found a paragraph taken probably from some Christchurch paper. It is without date, but from the date of some of the adjoining paragraphs, it was probably about 1871. I reproduce the paragraph, as it is interesting in this connection: "In a recent issue we republished from the *Argus* a paragraph stating that some grouse had been successfully brought out from Norway by a Mr Graff, who purposes bringing them on to Otago. A correspondent of the *Hobart Town Mercury* writes on the subject as follows:—'Mr Jalmar Graff, a young engineer, who has arrived by the emigrant ship 'Eugine,' has brought with him from his native place, Frederickshald, in Norway, two pairs of black grouse, a game bird better known to Englishmen as the black cock or moorfowl of the Highlands of Scotland. Because of this bird's peculiar and solitary habits of life, and the fact that it feeds almost entirely on heather, berries and such like food, no attempt had hitherto been made to remove them alive to any distance from their mountain home, much less across the ocean. The grouse were taken as eggs in the wood, and hatched by a common hen. Afterwards they were brought into a cage without a bottom, and every day this was moved on the grass. At first they were fed on ants' eggs, bilberries and red whortleberries, latterly or barley, herbage and birchen-nobs. On the voyage (5 months) they had barley, peas, maize and cabbage. No one else has, to my knowledge, succeeded in hatching and rearing grouse in Norway, and I believe this is the first successful experiment.'" I may add that these birds never came to Otago.

Australian Land Rail (*Hypotænidia philippinensis*)

The Otago Society introduced a pair of them in 1867; but there is no further record of them.

Order CHARADRIIFORMES

Family CHARADRIIDÆ

Golden Plover (*Charadrius pluvialis*)

In 1875 R. Bills brought out some of these birds for the Canterbury Society, and they were apparently among the birds sold and distributed. There is no further mention of them.

The Wellington Society introduced four in 1877, but there is no further record of them.

The Otago Society made an attempt to introduce the species in 1897. Out of 22 birds originally caught, only five reached the ship's side at London. Two were landed at Dunedin and were liberated at Clifton, but at once disappeared.

Seebohm says of this species that

> it breeds in the north temperate and Arctic regions, including the British Islands. Some remain in South Europe to winter, but the majority appear to pass on to North Africa, a few migrating during the winter season as far as South Africa. The Asiatic birds winter in Beloochistan, some of them probably crossing Arabia into Africa.

Lapwing; Green Plover; Peewit (*Vanellus cristatus*)

The Auckland Society in 1872 liberated 36, but nothing more was ever heard of them. Another attempt was made in 1875, when 36 more were shipped for Auckland; only four were landed and three of them died soon after.

The Canterbury Society obtained nine (imported by Mr C. Bills) in 1873, and liberated them, but they were not seen again.

The Otago Society purchased 36 birds in 1897, managed to ship 22, and only landed five in Dunedin. These were liberated at Clifton, and were seen flying about the fields for some time; then they disappeared.

In 1900, 50 birds were purchased, only 14 were shipped in London, and eight arrived in Dunedin. In the following year it is reported that "no young birds have been seen. Those liberated on Goodwood Estate did not survive many days, and some of their skeletons were picked up soon after they were liberated, the birds having been killed either by weasels or by hawks."

The Government introduced a lot in 1904. Of these 35 were handed over to the Wellington Society, and were liberated, but nothing more was heard of them. The Westland Society received

30, which "were liberated on the Upper Kohatahi. They all took wing and landed on the opposite side of Doctor's Creek." Apparently that was the last seen of them.

Apparently also a number were liberated in the Auckland district, for Mr Drummond, writing in the *Lyttelton Times* of 4th January, 1913, states that

a few lapwings were liberated in the Auckland district in 1870, and information supplied to me in 1907 shows that the effort was successful in several northern districts. The experiment has given great satisfaction to the settlers. The birds are credited with killing large numbers of wireworms and grubs in the spring.

I think Mr Drummond was misinformed by his correspondents, who were probably writing about some other bird, for on 2nd October, 1915, he asks: "has anybody seen a lapwing in New Zealand during the past ten years?"

The Auckland Society reported in 1909, that "the Green Plover has disappeared."

This species is a partial migrant in Britain; some nesting there, others only wintering there, and spending their nesting season in Northern Continental and Central Europe.

Grey Plover; Australian Plover (*Squatarola helvetica*)

The Otago Society liberated two in 1867, which were not seen again. In 1881 they obtained eight more, which were liberated on the Lauder Station, Manuherikia. They were observed for some time afterwards, and one was shot by mistake. Then they disappeared altogether.

The grey plover is only a spring and autumn migrant in the British Islands; it breeds within the Arctic Circle, and passes through Central and Southern Europe on migration, wintering in South Africa, India, South China, the islands of the Malay Archipelago, and Australia. In the western hemisphere it is known to winter in Cuba and some parts of South America.

Australian Curlew (*Numenius cyanopus*)

The Canterbury Society received two of these birds from Australia in 1868, but there is no record of what was done with them.

Family PTEROCLIDÆ

Sand Grouse (*Pterocles bicinctus*)

The Wellington Society received two from Mr Hamilton (of Teneriffe) in 1892, but there is no other record.

Pintail Sand Grouse (probably *Pterochlurus alchatus*)

The Otago Society in 1882 liberated eight at the foot of the Rock and Pillar Range; they were not seen again.

Family COLUMBIDÆ

Crested Bronze-wing Pigeon (*Ocyphaps lophotes*)

The Wellington Society introduced four of these birds in 1876, and eight more in 1877. They were seen for some months afterwards, and then disappeared.

The Canterbury Society in 1883 received six from the Melbourne Zoological Gardens; but there is no further record of them.

The Auckland Society introduced ten in 1887, of which five were liberated, one died, and four were retained for breeding purposes.

Bronze-wing Pigeon (*Phaps chalcoptera*)

The Canterbury Society received two pairs from Mr Wilkin in 1864, and introduced four more in 1867. They appear to have bred and increased in the Gardens (there were six in the aviary in 1880), for in 1882 it is reported that "the Murrumbidgee pigeons turned out by the Society have been seen, and are apparently doing well." Again in 1883: "the bronze-wing pigeons from the Murrumbidgee Mountains have been seen lately in one of the numerous patches of bush on the Peninsula, at a considerable distance from where they were liberated." In 1884 the Society received 20 birds from the Melbourne Zoological Society. I can find no further record after this date; they seem to have been exterminated.

The Otago Society introduced six in 1867. Mr F. Deans says: "they were liberated about the Gardens (Dunedin) and probably got killed by cats."

The Nelson Society introduced some in 1867, but there is no further record of them.

The Auckland Society received six from Mr Jas. Williamson in 1867; these were liberated at Kaipara. Some more were introduced in 1869, but none of them was ever heard of again.

The Wellington Society received two Tasmanian birds from Mr Meredith, but the date is not given. There were two reared at the Game Farm in the Wairarapa in 1907; but there is no further record.

Harlequin Bronze-wing Pigeon (*Phaps histrionica*)

The Auckland Society introduced two of these birds from Adelaide in 1869. There is no further record.

Wonga-Wonga Pigeon (*Leucosarcia picata*)

The Canterbury Society received two pairs from Mr Wilkin in 1864, and four more in 1871, from Mr R. B. Johnstone. These were apparently kept for breeding in the aviaries, for in 1872 they had still eight birds. In 1883 seven more were received. There is no record as to whether they were liberated or not, and no further mention of them.

Mr E. F. Stead writes me (April, 1916):

> In the early 90's Mr Peter Cunningham introduced some from Australia to Rockwood, his station near Whitecliffs. I can remember three or four pairs running about the garden there; they rarely flew. But they have all gone. The presence of the weasels would preclude the possibility of establishing them here.

Some were introduced into Nelson in 1867, but I cannot learn what came of them.

The Auckland Society received two pairs in 1868 from Mr Dacre, and in 1870 two more pairs from Sir Julius Vogel. There is no record of what was done with them.

The Otago Society introduced 12 in 1869, but they were kept too long in the aviaries, where many died. The remainder were liberated in the upper part of the Gardens, but were apparently too tame, and were killed by cats.

The Wellington Society imported 12 in 1875, and 22 in 1876. It is not stated where they were liberated, but they were seen for some considerable time afterwards, and some were reported from Wainuiomata. They were not reported again, and apparently completely disappeared.

Ring Dove (*Turtur risorius*)

These birds have been commonly imported by dealers at all the chief ports for the last 40 or 50 years, and passing into the hands of private individuals, were in many cases liberated round homesteads and dwellings. The Canterbury Society kept them for years in their aviaries. The first pair were imported about 1866. In some reports they are called "Ring-doves," in others, "Turtle-doves."

The Nelson Society introduced some of these birds in 1867.

The Auckland Society received five "doves" from Mr Jas. Williamson in 1867; probably this bird.

Mr E. F. Stead writes: "there is a fairly large number of ring-doves in the (Christchurch) domain. People have turned them out in their gardens in various parts of the city, and they stop about and do fairly well, but do not seem to increase." The same is true of suburban gardens about Dunedin.

Turtle Dove (*Turtur turtur*)

According to Hutton (1871) these birds were imported into Auckland and Nelson. They have been confused with the preceding species.

*Common Pigeon; Rock Dove (*Columba livia*)

Pigeons have gone wild in many parts of New Zealand, and I have accumulated a good deal of information in regard to them, especially in connection with reversion to the wild rock-dove type. Mr W. H. Gates of Skippers, near Lake Wakatipu, informs me (April, 1916):

> Fifteen miles down stream from here there are a lot of wild pigeons inhabiting the gorges in all sorts of inaccessible sheltered places. They have a strain of nearly every pigeon tribe. Some make an attempt to "tumble." Some are nearly white with small heads; some biscuit-colour; slate; dark-brown, white breast, with bronze neck and shoulders; and others with many colours mixed.

In Strath-Taieri, wild pigeons occur in thousands along the rocky faces of the Rock and Pillar Range from Middlemarch to Waipiata. Mr W. Sainsbury (April, 1916) says they are of all colours, and apparently of all breeds, and several of them have the "tumbling" habit. One which he brought me was slaty-blue in colour, with the black bars on the wings, but not well defined; black bar at the extremity of the tail, and the outer tail feathers edged with white. The majority of the wing pinions were tinged with red on the inner web. The abdomen and leg feathers were pale slaty-grey. The crop of this bird was quite full of wheat; but I am informed that the sharp winter of Central Otago—when the hills are covered with snow, and the ground is frozen hard for weeks at a time—is most severe upon the pigeons, which fall into very poor condition before the advent of each spring.

In the Galloway Station, Central Otago, where they have gone wild in numbers, they live both in the rocks and in holes in clay banks in the wilder districts.

Forty years ago wild pigeons in thousands occupied the rocks in the Duntroon district, inland from Oamaru, finding shelter in the numerous holes which occur in this limestone region. But shooting in season and out of season, and the eating of poisoned grain, have reduced their numbers to such an extent that they are now only found in hundreds. Mr Alfred Labes of Duntroon, who has looked into this question for me, states that they are now (1916) very timid and most difficult to shoot, and if their nests are disturbed they do not build again in the same locality. It is evident that from time to time tame pigeons join the wild ones, for a tumbler was observed among the latter, and a carrier, with a ring on its leg, was shot among them. In regard to colour, Mr Labes states that most of the birds are of a dark slaty colour, as seen on the wing, but some of lighter hue are occasionally seen. One specimen which was secured showed all the characteristic markings of the wild rock pigeon. Another was nearly black, tinged with slaty-grey on the neck, black tinged with brown on the wings, one distinct black bar on the wings, and a black bar on the tail; underneath the wings the feathers were pure white.

Mr A. Warburton, Teacher, of Cromwell wrote me as follows on 5th March, 1892:

> I have found out rather a strange thing since I came here, viz. that among the rocky mounds of the mountains, which are extremely bare and arid, are growing Kowhai trees (*Sophora tetraptera*), stunted of course, which were unknown in this district until a few years ago. I have noticed that the trees are only found among the loose and disjointed rocky masses. The seeds are evidently bird-carried, the nearest place where Kowhai trees abound being Lake Wanaka, some forty miles distant. The transportation is not the work of a native bird as the trees would have been planted before; consequently some introduced bird must carry the seed. I am inclined to think it is the common domestic pigeon, many of which are living in a wild state among the cliffs and rocks of the Dunstan Range. Of course these birds have not been inhabitants of the district for many years, and I dare say further enquiries will connect their advent pretty closely with the appearance of the Kowhai.

Mr R. D. Dansey informs me that wild pigeons were numerous on the cliffs between Napier and the breakwater; but Mr H. Hill states that they have mostly been driven away, he thinks by harrier hawks.

I have no doubt that these birds occur wild in many other parts of New Zealand, but I have not received reports from any other localities than those referred to.

New Caledonia Green Doves

The Auckland Society received two from Mr Martin of Noumea in 1867, and later four from Captain Stuckey. There is no further record.

I do not know what species of bird is here referred to.

Solomon Island Pigeon

The Auckland Society received a pair from Captain Jacobs in 1870; and another from Mr E. Perkins in 1872. I have no idea of the specific name.

Indian Dove (? *Turtur ferrago*)

The Otago Society had two in their aviary in Dunedin in 1907.

Queensland Doves

The Auckland Society introduced some of these about 1868 Hutton suggests (1871) that they were the little turtle dove or grey-necked pigeon of Australia (*Geopelia cuneata*). They evidently did not establish themselves.

Indian Pigeon

The Otago Society had five in their Dunedin aviary in 1907; but only two in 1908. No one seems to know what species they belonged to, or where they came from.

Java Dove

The Otago Society introduced five in 1867, and after being kept for a considerable time, they were given to the late Mr Fred Jones, and liberated at Green Island. They soon were lost sight of.

The Nelson Society introduced some the same year; but the record is lost.

The Wellington Society introduced eight in 1875, and bred them in the Gardens; but there is no further record.

Again I have no idea what species is referred to.

Moreton Bay Dove

The Canterbury Society introduced four in 1867, and liberated them. But, as usual, no further notice of them is obtainable, nor do I know what bird is intended, though quite probably it is one of the commoner Australian species.

Cape Dove; Harlequin Dove (*Oena capensis*)

Sir George Grey introduced these into Kawau in the early sixties, and according to the Hon. S. T. George, they became very numerous. But it is many years since any were seen.

Squatter Pigeon; Partridge Bronze-wing Pigeon (*Geophaps scripta*)

The Canterbury Society received two pairs in 1866 from Mr R. Wilkin. I do not know what came of them.

Order CUCULIFORMES

Family Psittacidæ

Shell Parroquet; Warbling Grass Parakeet; Long-tailed Grass Parakeet; Budgerigar (*Melopsittacus undulatus*)

According to Hutton this species was introduced at an early date into Canterbury, and was liberated, but failed to established itself.

The Auckland Society introduced and liberated either two or four in 1871, and every bird-dealer has brought numbers into the country for sale.

Mr Holman, Curator of the Whangarei Acclimatisation Society, tells me that Australian Parakeets are abundant at Waitakerei; but I have not been able to find out whether it is this species, or the next[1].

Rose-hill Parroquet; Roselle Parroquet; Rosella (*Platycercus eximius*)

Colonel Boscawen informs me that this species is to be seen occasionally in the neighbourhood of Auckland. It is frequently introduced by bird-dealers, and has evidently escaped or been liberated by private individuals.

The rosella is considered to be a pest in the apple orchards in parts of New South Wales. On the other hand it has been found to feed on the larvæ of blow-flies, which are a much worse pest.

Cockatoo Parroquet (*Calopsitta novæ-hollandiæ*)

The Auckland Society introduced two in 1871 from Adelaide. It is frequently to be found in hands of dealers and private fanciers.

Sulphur-crested Cockatoo (*Cacatua galerita*)

Colonel Boscawen informs me that this species is frequently to be seen on the Waitakerei Ranges, where it appears to have established itself. This species was also reported from Nelson, but Mr F. G. Gibbs informs me that the report arose from one tame bird which frequently flew over the town screeching.

Order CORACIIFORMES

Family ALCEDINIDÆ

Laughing Jackass (*Dacelo gigas*)

The Canterbury Society got two pairs of these birds from Mr Wilkin in 1864, but there is no record of their being liberated. However, in Lady Barker's *Station Life in New Zealand* (p. 16) it is stated that on her voyage from Melbourne to New Zealand in the 'Albion' in 1865 one of her fellow-passengers had a number of birds on board—*chiefly laughing jackasses*.

The Otago Society introduced four in 1866, and two in 1869, and liberated them near the Silverstream. They were seen for some time there, but ultimately disappeared.

Sir George Grey introduced a number into Kawau in the early sixties, but they all died.

[1] In the Report of the Canterbury Acclimatisation Society for 1868 it is stated that a large number of the pretty little Australian love-parrots has been received, which the curator is desirous of disposing of, or exchanging. I do not know what species is referred to.

The Nelson Society introduced some in 1867, but there is no record as to what came of them.

The Auckland Society received one from Dr Stratford in 1868.

The Wellington Society liberated 14 in 1876, and one in 1879. One was seen as late as 1885, but that is the last record.

Colonel Boscawen tells me (December, 1916) that a few laughing jackasses are to be found wild on the east coast of the Auckland district.

This bird in Australia has been found to feed mostly on insects (beetles, grasshoppers, etc.), centipedes and spiders.

Family Strigidæ

Many of the species of small birds introduced by the acclimatisation societies to cope with insects, increased themselves to such an extent as to become a serious pest to the farmers, and steps were taken in some of the districts to introduce a natural enemy in the shape of owls. The following species have been introduced already.

Barn Owl (*Strix flammea*)

The Annual Report of the Otago Society for 1900 contains the following statement:

> Seven barn owls received from London in December (1899) were liberated at West Taieri, Mr Fulton of Ravenscliff supplying them with food at the homestead until they finally left their cages. Some of them returned to their cages for over two weeks. They are now to be seen of an evening flying about the straw stacks at West Taieri, and *as they live almost entirely on rats and mice*, we feel sure the public will protect them.

The italics are mine. The birds were introduced to keep down the small bird nuisance. They do not appear to have established themselves, for they have not been heard of since 1900.

*Small Brown Owl (*Athene noctua*)

The Otago Society in 1906 imported 28 birds from Germany, liberating 14 at Ashley Downs, Waiwera, and 14 at Alexandra; in 1907 39 were introduced and liberated at Alexandra; and in 1908 a third shipment of 80 was received and these were distributed in various parts. In this year also it was reported that several of those introduced in 1906 had reared young broods. In 1909 several fruit growers in Central Otago reported them as having proved already a great boon to their orchards. In 1910 72 birds were received, of which 14 were liberated, and the remainder sold to farmers and orchardists.

In 1916 they are reported as multiplying about Wendon in an old quarry, from which they have displaced a colony of starlings.

At Pyramid Hill they live in rabbit burrows in inaccessible places, and come out freely in the daytime to catch lizards and beetles. It is stated that they do not shun the daylight, but may often be seen sitting at the entrance to their burrows enjoying the sun. Mr W. H. Gates also reports them (April, 1916) as heard occasionally at Skippers. Mr Henry Warden (May, 1916) states that since their appearance in the Wyndham district, the moreporks (native owls) are no longer heard.

Mr A. J. Iles liberated a pair at Rotorua in 1908 (?) and for some time dead sparrows with their heads eaten off were found. After a while the birds disappeared.

Mr A. Gunn of Galloway Station, Central Otago, tells me they are common there in rocks, clay banks, and deserted rabbit holes. They are also reported from the Taieri Plain, and from Kaitangata.

Mr A. Philpott sends me (August, 1916) the following note on this species:

> The little owl (*Carine noctua*) made its appearance near Invercargill in the autumn of 1915. I frequently heard its call, but did not see the bird till September 21st, when I discovered a pair perched in a hole in a dead Kamahi (*Weinmannia racemosa*). The hole was about 25 feet from the ground and faced north, so that the sun (it was 2 p.m.) shone almost directly in. When I approached within about a dozen yards one of the birds flew out, seeming to find no difficulty in guiding its flight in the daylight. After some time the other bird also flew out and perched on a branch overhead. Then it took a long flight to another dead tree and selected a place, not at all in the shade, where it seemed to settle itself to sleep. The trees referred to were on the edge of a bush, and subsequent observations have shown that the little owl does not enter the bush, but keeps on the outskirts or about isolated trees. For this reason I do not anticipate that our native bush birds stand in much danger from this owl. The little owl is much more difficult to approach than the morepork; it evidently sees much better in daylight; but apart from this, it seems to be a more wary bird. Unlike the morepork, the little owl hoots vigorously in the middle of the day,—at least, in the spring. It is now pretty common in this district; a few evenings ago I heard four or five calling in different directions.

Writing in 1918 he adds:

> There can be no doubt that such introduced birds as the sparrow and others which roost about hedges, plantations, and buildings will pay a heavy toll; indeed, I have reason to think that the thrush, the sparrow, and the starling are already diminishing in numbers near Invercargill. Where a pair of owls have established themselves, the evensong of the thrushes and blackbirds gives place to an incessant chorus of terrified alarm-notes.

At Hawera they are also found, according to my son, Dr W. Malcolm Thomson, but there they have not displaced the morepork, which is still to be heard in the neighbourhood (1916).

According to Mr Drummond, some of these owls were liberated in North Canterbury in 1910.

It is evident that these birds have become firmly established in the south portion of the South Island; they are now quite common round Dunedin.

Wood Owl (*Smyrnium aluco*)

Sir Walter Buller says:

In 1873 I sent out from England a pair of Wood-owls (*Smyrnium aluco*). They arrived safely at Napier, and after recruiting their strength were turned out loose in a distant part of the province. They were protected under the Act, but notwithstanding all these precautions, the unfortunate immigrants fell victims to popular prejudice.

Australian Owl, probably the Boobook Owl (*Ninox boobook*)

The Otago Society introduced two in 1866, and liberated them in the bush at Waikouaiti. They were not seen again.

Order PASSERIFORMES

Family ALAUDIDÆ

*Skylark (*Alauda arvensis*)

The introduction of this bird was general throughout New Zealand.

The Nelson Society, sometime before September, 1864, imported 20 larks into the district.

The Otago Society liberated four in 1867, 35 in 1868, and 61 in 1869. Private dealers brought in others.

The Canterbury Society introduced 13 in 1867, and 18 in 1871.

The Auckland Society introduced ten in 1867, and 52 in 1868. By 1873 they were considered to be thoroughly established in the provincial district.

The Wellington Society introduced 52 in 1874, and 56 in 1875.

In 1879 70 were liberated on Stewart Island, and were seen for a time at the head of Paterson Inlet; but Mr Traill informs me (March, 1916) that "none have been reported for years."

In every part of New Zealand they increased rapidly, and spread throughout the whole country, but they confine themselves to cultivated districts, and are not found in the bush or on open mountain country, though Dr Hilgendorf's statement on the following page modifies the last paragraph.

Next to the sparrow, the skylark is considered by farmers to be the most destructive of the small birds which have been introduced into New Zealand. They are particularly destructive in spring, when they pull wheat and other grains out of the ground just as they are springing. They also uproot seedling cabbage, turnip, and other farm plants. In the Foxton district pea-growing is quite impossible, owing to their depredations.

Several observers note that skylarks sing from a perch. Potts, writing in Canterbury in 1884, says:

> In the old country I never observed a Skylark in full song when perched. This habit is not very infrequent here. Taking up a position on a post or rail, gently turning from side to side, now and then with a slight movement of the wings, it indulges in song as joyous and powerful as when ascending in spiral circles skyward. At Sumner it has been observed singing whilst on the ground.

He also records great variation in the coloration of the eggs:

> whitish or grey-yellow, profusely speckled with brown of various shades; dull greyish with a green tinge, freckled or mottled with an ashen-brown; rich brown, abundantly marked with darker shades highly varnished; pale dull pink, profusely speckled with reddish brown.

Mr H. Watts, of Maungatua, Otago, states that the skylark has mated freely with the native pipit (*Anthus novæ-zealandiæ*), and he considers that the hybrid is the mischievous bird. He states that the hybrid bird rarely rises to any height when singing, and that when on the wing he only utters a few expressionless notes; then he makes a horizontal flight to some distance, or alights upon a post and continues his song. Mr Watts is a good observer and keen student of nature, but I cannot obtain any corroboration of his views, and am doubtful whether the two species are able to hybridise.

Mr Drummond (July, 1916) climbed Mt Leadhill, south of Collingwood, and notes in his diary:

> It is surprising to see so many larks in this desolate misty region, amongst rocks and boulders, where the food supplies must be poor. On all sides below us there are valleys throbbing with life, and further on are plains and meadows, yet these birds are spending their time here. With the exception of a few small but beautiful sub-alpine flowers, such as Celmisias and Sundews, they afford the only relief to the dreariness of this frowning mountain side.

The presence of the lark is, of course, proof that the food they are dependent on, viz. seeds and insects, was not poor. Hilgendorf found the nest of the skylark with eggs among rocks at an elevation of 5000 feet in the Canterbury Mountains. He says of larks: "Skylarks in this district are almost purely insectivorous; in agricultural districts, poisoned grain scattered over a field of sprouting wheat kills more larks than sparrows[1]."

Wood Lark (*Lullula arborea*)

The Auckland Society introduced five in 1872. There is no record as to what came of them.

Family Turdidæ

*Song Thrush (*Turdus musicus*)

Somewhere about 1872, the Nelson Society introduced five of these birds. They disappeared for many years, and then reappeared later. The probability is that the earlier lot failed to establish themselves, and that later the district became stocked by an immigration from some other part.

The Otago Society introduced two in 1865, four in 1867, 49 in 1868, 48 in 1869, and 42 in 1871. There was no mistake as to the determination of the Otago settlers to have their favourite song-bird—the "*Mavis*"—established in New Zealand. It shows, too, the hardiness of this bird in confinement, that Mr J. A. Ewen shipped the above 48 in London, in 1869, in charge of Mr R. Bills, and

[1] The skylark is a resident in the temperate regions, but the Arctic birds migrate in autumn to South Europe, North Africa, North-west India, and North China. Seebohm (*Siberia in Europe*, p. 257) gives a most interesting account of this migration, as observed by him in Heligoland: "In the afternoon it was a calm, with a rising barometer; in the evening a breeze was already springing up from the south-east. I called upon Gätke, who advised me to go to bed, and be up before sunrise in the morning, as in all probability I should find the island swarming with birds. Accordingly I turned in soon after ten. At half-past twelve I was awoke with the news that the migration had already begun. Hastily dressing myself, I at once made for the lighthouse. The night was almost pitch dark, but the town was all astir. In every street men with large lanterns and a sort of angler's landing-net were making for the lighthouse. As I crossed the potato-fields birds were continually getting up at my feet. Arrived at the lighthouse, an intensely interesting sight presented itself. The whole of the zone of light within range of the mirrors was alive with birds coming and going. Nothing else was visible in the darkness of the night but the lantern of the lighthouse vignetted in a drifting sea of birds. From the darkness in the east, clouds of birds were continually emerging in an uninterrupted stream; a few swerved from their course, fluttered for a moment as if dazzled by the light, and then gradually vanished with the rest in the western gloom. Occasionally a bird wheeled round the lighthouse and then passed on, and occasionally one fluttered against the glass like a moth against a lamp, tried to perch on the wire netting and was caught by the lighthouse men. I should be afraid to hazard a guess as to the hundreds of thousands that must have passed in a couple of hours; but the stray birds which the lighthouse men succeeded in securing amounted to nearly three hundred. The scene from the balcony of the lighthouse was equally interesting; in every direction birds were flying like a swarm of bees, and every few seconds one flew against the glass. All the birds seemed to be flying up wind, and it was only on the lee side of the light that any were caught. *They were nearly all skylarks*. About three o'clock a.m. the migration came to an end or continued above the range of our vision." The date was the 12th of October. I am not aware of any migratory tendency in the skylarks which are now naturalised in New Zealand.

that every one was landed alive in Dunedin. They established themselves at once.

The Canterbury Society landed 36 in 1867, 24 in 1868, and a third lot in 1871 The Society's Report for the latter year states that "they have not increased so well as expected, and it is much to be feared have been killed by cats." The large amount of native bush in the neighbourhood of Dunedin was, no doubt, more favourable for their protection and increase than the comparatively open country of North Canterbury.

In 1875 a further lot was brought in by Mr Bills, some of which were sold, and others liberated in the Christchurch Gardens. It was, however, more than 20 years before thrushes were thoroughly established there.

The Auckland Society introduced 30 in 1867, and 95 in 1868. They established themselves at once.

The Wellington Society introduced eight in 1878.

In the Otago Society's Report for 1881, it is stated that "thrushes, we are glad to find, are becoming more plentiful in the neighbourhood; *they are blamed for destroying fruit.*" Apparently this was thought by the writer to be a habit specially acquired in its new habitant. At the present day thrushes are found from one end of New Zealand to the other in enormous abundance. They are responsible, along with blackbirds, for continual and serious depredations in orchards. Before their introduction fruit of all kinds could be grown in the open, but as they began to increase it became impossible to grow small fruit, especially, without protection. Netting has had to be resorted to by all small growers, while in large orchards, guns, supplemented by owls, cats, crippled hawks and gulls, have to be employed to keep the depredators at a distance.

Against this must be placed the fact that they eat a great quantity of insect life, and of land mollusca (snails especially). The latter they destroy in the orthodox manner by dropping them on to rocks, stones and hard roads; and on the sea-coast they also eat periwinkles, leaving heaps of broken shells at the spots where they drop their victims. In New Zealand, as in Europe, earth-worms are their favourite food, but these all belong to introduced species. Mr Drummond quotes a Hawke's Bay correspondent as follows:

For about 130 days in the year, until well into January, a thrush has come to my farm morning after morning. Over an area of about 300 square yards he collects worms and takes them to his mate, sometimes carrying two or three at a time. I have watched him frequently, and from 7.30 a.m. to 8 a.m. he takes about fifty worms. I think I underestimate it in putting it at two hundred worms a day.

Philpott writes (1918):

The song-thrush does not appear to penetrate far into the big forests, nor to spread into unsettled areas. In the coastal forest of Fiord County they are seldom to be heard, though plentiful enough about the settlements of Tuatapere and Papatotara. Nor does the bird favour the mountains; I do not think I have ever heard one above the bush-line (about 3,000 feet). They are certainly absent along the upper limit of the Titiroa Forest (Hunter Mountains), and I have no record of meeting with them on the Longwood tops or the Hump.

The effects produced on the native and introduced vegetation of New Zealand by the introduction of thrushes and blackbirds have been very marked in at least one respect. The indigenous flora of New Zealand contains an exceptionally high proportion of plants with succulent fruit, amounting to approximately 16·55 per cent.

In Britain about 5 per cent., and in Australia 9 per cent. of the whole flora have succulent fruits. The introduction of fruit-eating birds such as thrushes and blackbirds, which in the case of small fruits swallow them whole and so distribute the seeds, and in the case of large ones like plums and apricots, carry them off to some distance where they can pick off the flesh and leave the stone, has led to a considerable increase in succulent-fruited plants. A considerable proportion of the indigenous birds of New Zealand are frugivorous, and it is their prevalence which, no doubt, accounts for the abundance of indigenous succulent-fruited plants. But the advent of the thrush and blackbird has increased this feature, though the former does not penetrate far into undisturbed forest. For example, in the Town Belt of Dunedin, a wooded area in which the vegetation is protected from all grazing animals, there has been a marked increase in the numbers of individual plants of *Fuchsia*, *Coprosma*, *Melicytus*, *Muhlenbeckia* and other berry- and drupe-bearing genera. Along with this, certain introduced plants, such as gooseberries, currants, brambles, raspberries, cape fuchsia (*Leycesteria*), but above all the elderberry (*Sambucus*) have spread through the native vegetation. The last-named plant in particular threatens to crowd out everything else, and a considerable sum of money is spent each year in eradicating it.

In great parts of New Zealand, the Blackberry (*Rubus fruticosus*) and the sweetbriar rose are most obnoxious pests, and thrushes and blackbirds are to some extent responsible for their spread. This question of the distribution of succulent-fruited plants by thrushes and similar birds is of especial interest to naturalists in New Zealand, and I have summarised a good deal of the evidence which Kerner has given on the subject, especially that relating to plants which are now found in these islands. Thus Kerner in *Flowers and their Unbidden*

Guests (p. 29) states that thrushes "are made ill by the *Phytolacca* berries, which many other birds feed on without injury." Apparently the statement was based on the case of one individual bird which was unwell after eating some of the fruit, for it is repeated again in his larger work on the *Natural History of Plants*, where he says: "a song-thrush sickened after eating berries of *Phytolacca*." Now this plant, the common ink-weed or poke-weed, is very common in the warmer parts of New Zealand, and Mr Cheeseman informs me that thrushes eat the fruit freely.

Kerner also states that when the fleshy fruits of *Berberis* (Barberry), *Ligustrum* (privet), *Opuntia* (prickly pear) and *Viburnum* (Laurustinus, etc.), all of which have seeds exceeding 5 mm. in diameter, were introduced into the crop of thrushes, along with other food, the pulp passed into the gizzard, but all the seeds were thrown up. "The seeds of fleshy fruits which were greedily devoured were thrown out of the crop if the stones which they inclosed measured as much as 3 mm." Now barberry is certainly spreading in the bush reserves near Dunedin, and is distributed either by thrushes or blackbirds.

He also found that of the fruits and seeds which passed the intestines of the thrush, no less than 85 per cent. germinated. In most cases the germination was retarded in comparison with seeds not so treated. But in the case of a few berries, e.g. *Berberis* and *Ribes* (currants and gooseberries), it was hastened. The seeds of such plants as grow on richly-manured soil (e.g. *Amaranthus*, *Polygonum* and *Urtica*) after passing uninjured through a bird's intestine, produced stronger seedlings than did those which were cultivated without such advantages. The time taken by seeds to pass through the alimentary canal of a thrush was very short, half an hour in the case of the elderberry (*Sambucus*), and three-quarters of an hour with seeds of *Ribes*. The majority of seeds took from one and a half to three hours to perform the journey. Small smooth fruits of *Myosotis sylvatica* (forget-me-not), and *Panicum diffusum* (a grass) were retained for the longest period.

The habits of thrushes have not altered appreciably in their new country. Their nests are of similar construction to those found in Britain, and they are lined with mud or cowdung. They breed in September and October, and I have seen the fledgelings in the end of the latter month, and the beginning of November. They usually breed again later in the season.

They commence to sing, in the South Island at least, in the month of May, that is at the commencement of winter. The earliest date I have noted is a record from Dr Brittin of Papanui, who heard one in Christchurch on 24th April.

At one time I thought, with Sir Walter Buller, that albinism was on the increase among thrushes in New Zealand, but as the result of long observation I am compelled to think this is not the case. Any thrush showing a tendency to develop white feathers seems to be a marked bird, not only by man, but by other birds, and even by other thrushes, and they do not appear to have a happy time.

Thrushes have found their way to the Chatham Islands, a distance of 450 miles east-south-east of Cape Palliser.

*Blackbird (*Turdus merula*)

The Nelson Society introduced 26 blackbirds about 1862, but there is no record as to their success at the time.

The Otago Society liberated two in 1865, six in 1867, 39 in 1868, 21 in 1869, and 70 in 1871. Ten years later we read they "are now exceedingly numerous and we regret to say are found to be rather partial to cherries and other garden fruits."

In *Station Life in New Zealand*, p. 16, Lady Barker, writing of her voyage from Melbourne to New Zealand in 1865, says:

> Ill as I was, I remember being roused to something like a flicker of animation, when I was shown an exceedingly seedy and shabby-looking blackbird with a broken leg in splints, which its master assured me he had bought in Melbourne as a great bargain for only £2. 10*s*. 0*d*.

The Canterbury Society received two in 1865 from Captain Rose of the 'Mermaid' who also sold "a number of songbirds" to the Society for £18. I regret to say there is no record of these "songbirds," to enable us to identify them. In 1867 the Society introduced 46, and in 1868, 152 blackbirds. In 1871 the Report states of them, as of the thrushes, that "they have not increased as well as expected, and it is much to be feared have been killed by cats." In 1871 Mr R. Bills brought a further consignment of 62 to the Society, and many more were introduced in 1875.

The Auckland Society introduced eight birds in 1865; about 30 in 1867, and 132 in the following year, when they were "considered to be thoroughly acclimatized." In 1869 a further large consignment was liberated. It is rather singular that in the far north, Whangarei and further north, blackbirds are rare or altogether wanting, while thrushes are common.

They were liberated on Stewart Island in 1879, and are seen every breeding season near settlements.

This is now one of the commonest of our introduced birds in very many parts of New Zealand.

Mr Philpott (1918) says that:

> unlike the thrush the blackbird is to be found in the heart of the big

bushes. I have met with the bird wherever I have gone, and found it as common on the Hunter Mountains at 3000 feet elevation, as in the bush near Invercargill. I have no records of the thrush occurring in Alpine forests. The spread of succulent-fruited plants is probably accomplished to a greater extent by blackbirds than by any other species.

They can evidently hold their own very well among the native avifauna, for Mr L. J. Phillips of Kaitoke states that on several occasions he has seen two or three blackbirds set on and kill a tui (*Prosthemadera*).

Kerner states that the blackbird is much less fastidious in regard to its food than the thrush. When fed in confinement, it swallowed even poisonous fruits like those of the yew, and never rejected a single fruit that was mixed with its food. Of the fruits and seeds which passed through the intestines, 75 per cent. germinated.

The blackbird has found its way to the Chatham Islands, which are distant 450 miles from the nearest point of New Zealand, and are increasing there, and scattering seeds of such noxious weeds as the blackberry.

Mr Drummond also is responsible for the statement (in 1907) that they "have taken up their residence on the lonely Auckland Islands." They are about 290 miles south of the Bluff, but only 230 miles from the south end of Stewart Island. The prevalent winds, however, would sadly impede the passage of a bird bound southwards.

In Europe there are migratory races both of thrushes and blackbirds, and it is quite possible that some of the birds introduced into New Zealand may have belonged to such races.

Robin Redbreast (*Erithacus rubecula*)

The Nelson Society attempted to introduce robins in or about 1862, but only one bird arrived.

The Auckland Society introduced three in 1868, three in 1871, and three in 1872.

The Canterbury Society introduced a number (not specified) in 1879, and the report for that year states that "the old familiar shrill note may be heard in the Society's grounds morning and evening."

The Wellington Society liberated ten in 1883, and three years later one was reported to have been seen in Happy Valley.

The Otago Society liberated 40 in 1885 at Fulton's Bush, West Taieri; and R. Bills, who brought them out, sold another 40 to private individuals. In 1886 some 20 more were imported and liberated at the same spot. They were scarcely ever seen again, but in 1891, Mr A. C. Begg reported one in a Dunedin suburban garden.

The cause of failure was never understood by the Society, but Mr A. Binnie, who was Mr Bills' assistant at the time, assures me that all the birds which were imported were cocks—which is a possible explanation, seeing that Mr Bills brought them out to sell. In 1879 the Otago Society received two (out of ten shipped from London) and in 1900, one (out of eight shipped); these were liberated on Otago Peninsula, but were not seen again.

In explanation of the failure of these birds to establish themselves, I am more inclined to favour the idea that birds of migratory races were brought out, for bird-catchers frequently make their best catches of birds which are gathering preparatory to starting on their journeys.

Nightingale (*Daulias luscinia*)

An attempt was made by the Otago Society in 1871 to introduce these birds, and a number of them were shipped from London, but they all died when a few days out.

The Auckland Society had exactly the same experience in 1875, none of those shipped surviving the passage.

The Canterbury Society repeated the experiment in 1879, when one was landed in Christchurch, and died soon after.

The nightingale is purely a migratory species in Britain, and any attempt to naturalise them in New Zealand was foredoomed to fail.

*Hedge-Sparrow (*Accentor modularis*)

In Dr Arthur Thomson's *Story of New Zealand* published in London in 1859, it is stated that "Mr Brodie, the settler who introduced pheasants, sent out, in 1859, 300 sparrows, for the purpose of keeping the caterpillars in check." I cannot verify the statement; Mr Brodie lived at Mongonui, and there have never been hedge-sparrows in North Auckland.

The Auckland Society introduced one in 1867, two in 1868, seven in 1872, 19 (out of 80 shipped) in 1874, and 18 in 1875. The nests were first observed in 1873, and the bird soon established itself.

The Otago Society liberated 18 in 1868, and 80 in 1871.

The Canterbury Society liberated nine in 1868, and 41 in 1871. Mr Drummond says (1907):

> It was Captain Stevens who brought the first hedge-sparrow to the colony, and it is claimed to the Southern Hemisphere. It came in the 'Matoaka' together with the first house-sparrows. It was the only survivor of a consignment. For a long time it was an object of interest in the Society's grounds in Christchurch, many people journeying to the gardens to see the stranger.

I do not know where Mr Drummond got his information, certainly not from the annual reports of the Canterbury Society. Nor does the pretty story tally with those told of the introduction of the house-sparrow (*q.v.*). The fact is that a number of those who were concerned with the introduction of the small birds in the early days of acclimatisation activity did not know a hedge-sparrow from a common sparrow, and while in later years it was quite creditable to have been concerned with the introduction of the former bird, no one is inclined to claim any credit for the latter.

A number of hedge-sparrows were brought to Christchurch in 1875, some of which were sold, and the remainder liberated in the Gardens.

A number were liberated in Hawke's Bay in 1876 by Mr Walter Shrimpton, but Mr Guthrie Smith says they are not known at Tutira.

The Wellington Society introduced four in 1880; 26 in 1881; and 20 in 1882. This species has now become very widely spread throughout New Zealand. It is the one bird against which no word of complaint has ever been raised. It is not met with in undisturbed bush country, but, according to Philpott (1918), is equally at home in the smaller areas of bush, in the suburban garden, and in the shrubby groves at 3000 feet on the mountains. The majority of nests are built quite low down, often practically on the ground, so that the prevalence of stoats or cats is probably a controlling factor in the increase of the species. On the other hand it is extremely common in suburban areas near Dunedin, where cats also abound.

The value of the bird was demonstrated to orchardists in Central Otago at the beginning of 1919. In February, very heavy rain caused an extraordinary outburst of vegetation in the gardens, and this was followed by an invasion of the green fly (*Aphis*). The outlook for some crops was very serious, till a great number of hedge-sparrows appeared in the orchards, and in a very short time cleared off the whole of the pest in the most perfect manner.

In Otago the note of this bird is occasionally heard in winter, but it begins to sing regularly in August, and nests are found from September onwards. Some County Councils rather foolishly pay for the eggs of hedge-sparrows. There is no excuse for this, for every boy knows the eggs, and would not take them at all unless a price was offered for them.

Whitethroat Warbler (*Sylvia cinerea*)

The Auckland Society introduced two in 1868, but they were not heard of after liberation. In 1874 another attempt was made to introduce them to the colony, and a considerable number were shipped, but all died on the voyage out.

Black-cap Warbler (*Sylvia atricapilla*)

The Auckland Society introduced five in 1872.

Both of the above-named species are summer visitants in Britain, and it was folly to attempt to naturalise them in New Zealand.

Family Hirundinidæ

Australian Swallow (*Hirundo neoxena*)

This species is an occasional visitant to New Zealand. Sir Walter Buller, in his introduction to his *History of the Birds of New Zealand*, says:

> In March, 1851, a flight of the Australian Tree-Swallow appeared at Taupata, near Cape Farewell; ten years later they were observed again at Wakapuaka, near Nelson, and a specimen obtained; and after a further lapse of fully twenty years another flight, from which a specimen is now in my possession, appeared for several days in succession in the outskirts of Blenheim.

In 1888 Mr W. W. Smith observed and recorded them from the neighbourhood of Timaru. In 1901 numbers of them appeared at New Brighton near Christchurch. I have been told of their occurrence since at Whangarei and in the neighbourhood of Auckland, but have no authentic information on the subject.

I would not mention the species among introduced birds, were it not for the action of various acclimatisation societies in regard to them. In 1874 the Auckland Society made the futile experiment of obtaining some eggs and placing them in two nests, of a sparrow and a chaffinch respectively. Needless to say the attempt failed. Even had the foster-parents succeeded in hatching out the young, they would not have supplied them with the right kind of food.

About 1915 several of the acclimatisation societies proposed to subscribe £5 each for the introduction of this migratory species; fortunately for them the project was not carried out.

Family Laniidæ

Australian Shrike

I do not know what bird this is, for there are many species of Australian shrikes. The Wellington Society liberated 14 in 1877, and 15 in 1878; but there is no further record of them.

Australian Magpie-lark; Mud-lark; Pee-wee; Pied Grallina (*Grallina australis*)

The Agricultural Department introduced a number of these birds from Sydney, and liberated them on the west coast on the North Island, where they promptly took to building nests. I do not know the date of this attempt, but apparently it did not succeed, for no on seems to know anything about the birds since.

Dr Cleland says that in Australia "this bird occasionally feeds on maize and wheat obtained near fowl-yards, etc., but it is doubtful whether it touches crops. It is also found to eat plague-locusts, grasshoppers, cockchafer larvæ, etc. It is one of our foremost useful birds." In their stomachs, in addition, there have been found moth larvæ, mole crickets, ants, small flies, and occasionally grass-seeds.

*Australian Magpie; White-backed Crow-Shrike (*Gymnorhina leuconota*)

The Canterbury Society liberated eight birds in 1864; four in 1866, and 32 in 1867, all from Victoria. They also received some 18 from Tasmania. In 1870 Mr E. Dowling imported a large number from Tasmania, and these were liberated on Mr Moore's station at Glenmark. The Society liberated 24 more in 1871. The birds soon established themselves in the provincial district, and are now fairly common. Of late years they have spread south of the Waitaki and as far south as the Horse Ranges.

The Otago Society introduced three in 1865; 20 in 1866; 32 in 1867; 20 in 1868; and six in 1869. At first it seemed as if they were doing well, for they began to build nests at Inch-Clutha, and in the vicinity of Dunedin. But from some unexplained reason—(Mr Deans thought they were shot or taken by boys)—they entirely disappeared, though now coming in again from the north.

The Auckland Society introduced ten in 1867, and one in 1870. But Sir George Grey introduced a number into Kawau probably at an earlier date; they very quickly became numerous, and spread to the mainland.

The Wellington Society introduced 260 in 1874.

These birds are fairly common in many parts of the North Island, from Wellington to north of Whangarei, but their numbers vary a good deal. Mr W. W. Smith tells me that they are not so abundant in Taranaki now (1916) as they were some years ago. Inland from Wanganui, on the edges of the unbroken forest, they are very common. T. H. Potts records (in 1873) how this bird defends itself successfully against the native quail-hawk (*Falco novæ-zealandiæ*) by throwing itself on its back, striking out with beak and claws and shrieking most wildly. Mr J. Grant of Wanganui informs me (1918) that the Magpie has been seen to kill a fantail by a direct blow on the body, then it stuck its bill into the little victim and carried it away. This bird has a wonderfully fine flute-like song; both the male and the female sing, and they begin their concert even before sunrise. They are readily tamed and are very sociable in confinement; but they are apt to drop their beautiful song, and take to imitate all sorts of domestic sounds. I knew of one in Dunedin which could bark like the dog, and mew like the cat; but its favourite amusement was to sit on the fence and call the fowls together.

During the mouse-plague in Victoria in 1905, in some districts, crowds of magpies were seen to follow the plough, and catch and swallow every mouse that was unearthed. In one case 150 to 200 magpies were seen following one plough and no mice got away.

Family PARIDÆ

Titmouse; Blue Tit; Tom-Tit (*Parus cœruleus*)

The Canterbury Society in their report for 1874 state that "these have been imported in considerable numbers." There is no previous record of these birds unless they are included in the unspecified birds introduced in 1871 in the 'Charlotte Gladstone.' They must have died out, for there is no further record.

Family CORVIDÆ

*Rook (*Corvus frugilegus*)

In 1862 three rooks were introduced into Nelson, and stayed about there for a few years, when they disappeared. The popular belief was that they left for Canterbury, when others were introduced there.

It would appear that rooks were introduced about the same time into Canterbury, for in a press cutting dated April, 1870, recording the presentation of a single specimen to the Acclimatisation Society, it is added: "The rooks first imported into the province by Mr Watts Russell, *some years ago*, were all killed by cats."

The Auckland Society introduced two in 1869, and 64 in 1870. At first it seemed as if these birds would not succeed in establishing themselves, for the Society's report for 1872 states that:

> eight nests were built in the Gardens, but unhappily a night review of the volunteers took place just as incubation commenced, when the firing caused the majority of the rooks to forsake their nests, so that only three small broods were hatched. In January a severe epidemic broke out

amongst them which destroyed several. Dr Wright, who examined two of the dead birds, stated that the disease was identical with the epidemic then prevalent amongst domestic poultry. The dead birds were unusually fat, and the survivors refused the food placed for them on the outbreak of the disease.

In 1873 several pairs built and nested; but in the following year's report we read: "they are not doing so well. The young have died," and it is suggested that the climate is too hot for them.

The increase of these birds has been slow in the North Island; they spread to Hick's Bay, Hawke's Bay and Lake Taupo. Mr H. Guthrie reports: "A colony of rooks has for long existed in Puketapu. In favourable years colonies start—one, for instance, in Petane,—but the rook does not do well in this part."

The rook seems to have a bad time in Hawke's Bay, and pastoralists and fruit-growers alike blame it for many evil things it is supposed to do. They conveniently ignore the good. In 1917, at a meeting of the Fruit-growers' Association at Hastings, a member stated that the rooks "were doing considerable damage to walnuts, amounting to some hundreds of pounds." Later on, farmers complained that

> the rooks have acquired the habit of attacking lambs and full-grown sheep, and the losses in some parts of the district are becoming serious. The birds not only attack flocks in the daytime, but also during moonlight nights, and one farmer near Farndon has lost scores nightly. The rooks attack the throats of the sheep, and wethers can be seen in paddocks with open wounds. One was seen with the head completely severed, with the exception of the spinal column. The birds also eat the flesh right down the middle of the back, rendering the skin quite useless.

This is the sort of newspaper paragraph that gains credence in the country, but is absolutely incorrect. One farmer states that he had 30 acres of sprouting oats completely uprooted by rooks. Against this unvouched-for evidence I have information from several well-known men in Hawke's Bay, who certify that the birds are at no time a nuisance. One gentleman, connected with the Tomoana Freezing Works, suggests that several hoggets were dying or dead, and the crows seeing them with ticks on them, picked the latter away. He ridicules the idea that they could hurt sheep. Mr H. Hill, lately Mayor of Napier, and formerly Senior Inspector under the Education Board, sums up the case as follows:

> Rooks may pull up wheat now and then, but only to discover a worm, and the question is whether the balance is not in favour of the rook under the circumstances. Surely a bird cannot be expected to live and benefit man without obtaining a part of his maintenance from what it helps to preserve. The fact is, the farmer expects his crops to be protected from all insect pests without cost or responsibility on his own part.

The Canterbury Society got a large number (36) shipped from London in 1871, but only five survived the voyage, and these were liberated in the Gardens at Christchurch. In March, 1873, 35 more were liberated in the Gardens. Writing me in 1890, Captain Hutton said: "these birds are well naturalised about Christchurch, but do not now increase much; possibly owing to poisoned grain." They are fairly common now (1916) south of Christchurch, but are strictly localised, and have hardly spread from the spot where they were originally liberated. They are very destructive to the grass-grub (*Odontria striata*) in Canterbury.

Mr A. H. Cockayne (April, 1919) says that they used to be most abundant in Dean's Bush at Riccarton, Christchurch, but are now rare. Also that two years ago there was a large rookery near Hastings, Hawke's Bay, which has since been abandoned. Mr Graham, Manager of Weraroa Government Farm, informs me, however, that Rooks are very common in Hawke's Bay.

Jackdaw (*Corvus monedula*)

The Otago Society had some of these birds (the number not specified) in their depot in 1867; but there is no further record of them.

The Canterbury Society received one from Mr McQuade of the 'Mermaid' in 1868. But they evidently obtained some more, for in the Report for 1872 it is stated that "three of the jackdaws have remained about the Society's Gardens since they were liberated." In a newspaper cutting of 1871, received from the late Mr A. M. Johnston, it is stated that two jackdaws are at Prebbleton, and two at Kaiapoi. Mr E. F. Stead, writing in April, 1916, says: "there are certainly no jackdaws wild in Canterbury at present."

Family STURNIDÆ

* Starling (*Sturnus vulgaris*)

The Nelson Society introduced 17 starlings about 1862, but have lost their record.

The Otago Society imported and liberated three in 1867, 81 in 1868, and 85 in 1869.

The Canterbury Society introduced 20 in 1867, and 40 in 1871.

The Auckland Society introduced 12 in 1865; 15 in 1867, and 82 in 1868; while the Wellington Society's record was 60 in 1877; 90 in 1878; 14 in 1881; 100 in 1882; and 34 in 1883.

Besides all these, great numbers were introduced by private enterprise. The increase of this species was phenomenal. Mr C. Hutchins writes (November, 1913):

> when I arrived in Napier from England in 1875, there were only four starlings in the town. They increased rapidly and took possession of the limestone bluff that looks out over the bay, boring into the softer veins of limestone. After eleven years they were there in hundreds of thousands. The bird has few, if any natural enemies.

Mr H. Hill (of Napier) considers that the comparative disappearance of the bird from the cliffs about Napier, where it and the wild pigeon (*Columba livia*, i.e. tame pigeons gone wild) used to be extraordinarily abundant, is due to their being driven away by the harrier hawks. If Mr Hill's view is correct, it is probably due to the fact that the hawks were after the pigeons, and the starlings suffered through association with them. In open country I don't think hawks are a serious enemy to the starlings.

These birds are abundant in most parts of the country, and in favourite spots, where they congregate in numbers, the noise they make when roosting can be heard, literally for miles. Mr W. W. Smith of New Plymouth writes me:

> Every evening tens of thousands of starlings perform their cloudlike gyrations around and above the island of Moturoa which is clearly seen from this hill. Every person who sees them compares them to rapidly moving clouds. It is truly magnificent to see them; they form densely black cloud-like masses.

Mr W. Best of Otaki reports (April, 1912): "that the starlings in that district make long daily flights (of 30 miles?) to and from their roosting place to their food." These only occur in the summer and autumn, after the breeding season, and the birds fly at sunset and before sunrise. Mr Mahoney of Tuparoa describes how they come in thousands from some feeding ground near the coast to roost on the trees behind the house. Mr Johannes Anderen says they roost on Kapiti and fly to the mainland daily for food.

Mr H. J. Fowler of Marton writes (June, 1912): "The daily evening migrations began about seven or eight years ago. At first the flocks were small and infrequent, now they pass in battalions. On calm evenings the air is filled with the rushing sound of their wings." He has seen flocks a mile long, all the birds flying in line, like soldiers marching in ranks. These appear to be made up of flocks rising at intervals across the country and uniting in the air. He estimated the numbers at hundreds of thousands. In Marton the birds fly south-west, and it is stated that they go to the Manuka scrub on the coast. At one place a piece of native bush, about four or five acres in area, was used as a roosting place by the birds, and so great were their numbers that they were killing the bush with their droppings. They had to be driven away by firing guns at roosting time. Another roosting place was at Martinborough, where they occupied an avenue of bluegums:

In going to those places they seemed to converge from all parts of the compass, and their cries made a great roar as they settled themselves for the night. With regard to their returning, they seem to go back soon after daylight, and in twos and threes flying low. It seems that they work up country from field to field during the day, until the time for making homeward arrives, when they rise and fly straight for their roosting place, the flocks gradually increasing, as they draw nearer home by the addition of other flocks. As a rule, when once in full flight, nothing in the shape of a ploughed field will tempt them down, and any stragglers that may stop, seem to be uneasy and soon rise and follow the others. Starlings follow the binders in clouds for the caterpillars, but they are not observed on the ploughed fields to any great extent.

This last statement may be correct for Rangitikei, but is not so for Otago and Southland, where they may be seen following the plough in considerable numbers. Mr A. Philpott, writing me in April, 1892, says:

The Rev. J. G. Wood states that "when a flock of starlings begin to settle for the night they wheel round the place selected with great accuracy. Suddenly, as if by word of command, the whole flock turn their sides to the spectator and with great whirring of wings the whole front and shape of the flock is altered. No body of soldiers could be better wheeled or countermarched than are these flocks of starlings, except an unfortunate few who are usually thrown out at each change." I have watched flocks of starlings arriving at their roosting-places very often, and in one case only have I seen anything resembling the company evolutions referred to by Mr J. G. Wood. They appear to arrive in flocks of large and small numbers, and immediately on arriving drop down wherever they can find a perch. Perhaps this is an instance of an altered habit. It is also possible that only the first flock to arrive wheels and circles in the manner described.

Soames' Island in Wellington Harbour in pre-war days was a night retreat for starlings, which used to resort to it in immense flocks. Since it became a place for interned German prisoners the birds have largely abandoned it, on account of the number of people about.

The effects produced on the insect-life of the country by starlings, and through that on the vegetable and other animal life, is incalculable. They have nearly destroyed the grasshoppers which used formerly to be so abundant, and many other groups of insects must have suffered equally. They also remove great quantities of ticks from sheep, and cattle, and help to keep insect pests from them.

Indirectly they are credited by many observers with having exterminated pheasants, partridges, introduced quail, wild turkeys, wild fowls, etc., from many districts, by having so eaten out the insect food, that these larger birds are now unable to rear their young broods. They have driven the Indian Minah out of all the southern towns where formerly they were established.

In many places they are accused of being fruit-stealers, attacking not only small fruits, but also pears, plums, and peaches, and some of my correspondents have thought this was a new trait developed in their new surroundings. But it is familiar enough in the northern countries from which the starlings came. There is a well-known passage in Rabelais' *Gargantua* in which it is stated that "at this season the shepherds were withdrawn from the hills in order to keep the starlings off the grapes." I have frequently seen them in this country feeding on the rather hard white berries of the cabbage tree (*Cordyline australis*).

In the *Otago Witness* of 2nd October, 1890, J. H. E. of Anderson's Bay, Dunedin, writes: "Last season my jargonelle pears were alive with starlings, the pears eaten by scores, and the leaves and fruit in a disgusting state from their droppings."

Mr Philpott several times found their gizzards full of the berries of the broadleaf (*Griselinia lucida*).

Mr J. Drummond states that Mr D. L. Smart of Napier, who formerly lived at Tuakau, on the banks of the Waikato, found that great flocks of starlings in the late autumn visited a large kahikatea (*Podocarpus dacrydioides*) forest in order to feed on the berries.

It may be noted here that in many parts of New South Wales, cherry-growing has become an impossibility, owing to the persistent attacks of starlings.

Mr P. J. O'Regan considers that starlings (as well as blackbirds and thrushes) are responsible for the diminution in the number of native pigeons, as they eat the berries of pine trees, *Fuchsia* and *Aristotelia*, which form part of their food. (This may be partially true, but pigeons subsist on other materials, and in summer their crops will be found quite full of the leaves of the kowhai—*Sophora tetraptera*.)

On 18th August, 1890, I wrote to the *Otago Witness* asking certain questions on acclimatisation matters, among others as to whether starlings were eating poisoned grain, as was stated to be the case in Southland. Mr Richard Henry, writing from Lake Te Anau, was inclined to think they did, but his opinion was based on the fact that he found three dead starlings in the first week of rabbit-poisoning, and none before or since. Mr Richard Norman, Alberton, replied:

> In this district the starlings roost in thousands in the blue-gum trees in winter time, and are fighting, scratching, and screeching for positions

all night long, and if they are disturbed, the rustling of their wings as they rise sounds like distant thunder. Frequently some are found dead on the ground, and it is generally concluded that they have perished from the cold, and not through the effects of phosphorised grain. The cats eat their bodies without harm.

Mr George Green, Broad Bay, wrote: "The only poisoning done here is intended for the sparrows, and I have never heard of starlings taking the grain; but they have developed a taste for the elder-berries."

Mr H. Watts of Maungatua stated that the starlings do not touch poisoned grain, though they had developed a strong predilection for red currants. They were commonly found among the bushes, and one which was shot contained a large number of berries in the crop.

Mr Thomas M'Latchie, Owaka, wrote: "Poisoned grain has been laid for rabbits for two or three months in my paddocks. Flocks of starlings have been busy amongst the grass, but I have never seen one of them touch an oat."

Mr A. Philpott writing to me in July, 1916, says:

This bird is certainly less plentiful than it was twenty-five or thirty years ago. Possibly the want of suitable nesting-places may have something to do with it. So much bush has been cleared away since the starlings' greatest abundance, that it must be somewhat difficult now to find suitable hollow trees, for the bird does not appear to penetrate deeply into the bush. In 1914 I found it building in a mass of ivy on a cabbage tree (*Cordyline*) in the centre of Invercargill, and last year at Wyndham I found several broods in masses of *Muhlenbeckia*. The bush tree there is chiefly matai (*Podocarpus spicatus*), a tree that does not provide many holes suitable for nests.

Two years later he states that the bird is not nearly so plentiful as it used to be, and attributes the change to the decrease in the number of suitable nesting places, on account of the disappearance of old forest trees. A North Canterbury farmer writing to Mr Jas. Drummond in August, 1910, states that starlings are very destructive to humble bees, and he has repeatedly seen them catching these insects and taking them to their nests.

Mr B. C. Aston tells me that starlings frequently imitate other birds, which are new to them. He has noticed them in the neighbourhood of Wellington imitating both the Californian quail and Australian magpies.

Hilgendorf says that in the Cass district they leave the houses to the sparrows, and build among the rocks and tussocks[1].

[1] In connection with the blow-fly pest in sheep, in Australia, Dr Cleland says of the starling (and sparrow): "Though useful to a slight extent, they do much more harm than good. Neither apparently plays any definite part in controlling the blow-fly pest." "The stomachs of seventy-three of the introduced birds were examined. As regards the vegetable food, wheat grains were found in a few and

*Indian Minah or Myna; House Myna (*Acridotheres tristis*)

This species appears to have been introduced in the first instance in all centres by private individuals, and by a few of the societies in the early seventies. One of the most remarkable things about them is their increase after their first introduction, and then their subsequent diminution, and—in some districts—their ultimate disappearance. The latter appears to have been due, either directly or indirectly, to the starlings, the increase of the latter coinciding with the decrease of the farmer.

In 1870 Mr F. Banks introduced 18 of these birds, which he termed *Indian Minaul birds*, from Melbourne, where they had been acclimatised for some time, and presented them to the Canterbury Society. Writing in 1890, Captain F. W. Hutton said: "A few used to be about Christchurch, but they have disappeared before the starlings." Mr Stead, writing in 1916, says: "In the early nineties there were a few minahs nesting in some houses on the North Belt (Christchurch), but there are now none left, and there have not been any for fifteen years at least."

Some were imported into Dunedin in the early seventies by Mr Thomas Brown. They used to build in the First Church Steeple and on one or two houses in the neighbourhood, but they had all died out or were driven away before (1890).

Mr F. G. Gibbs of Nelson says (July, 1916):

Minahs were imported in the seventies. I remember that they were very plentiful in the streets when I arrived in 1877, but a few years later

fruit in one. This result, however, does not by any means indicate clearly the destructive tendencies in the direction of vegetable food, as the accessibility of such food must be considered at the time the bird was shot. Unquestionably starlings feed greatly on cultivated fruits and on cultivated grains during the season when these are available." "As regards the insect food of these seventy-three birds, we found that locusts or grasshoppers were present in five, wireworms in two, cutworms in thirty-four, flies in four, psyllids in one, and scale (?) in one. The cutworms were found in most of the starlings obtained in the Wagga district, these having been shot while this pest was present." "Flies were found in four. These could not be identified as blow-flies. It is, however, likely, though not proved as yet, that the starling does destroy a few of these insects. As indicated by the list of insect foods, the starling can unquestionably play a useful purpose in the direction of destroying insect pests.

Summed up, it may be stated that the starling does marked harm to fruit gardens and that it does some harm to crops, but that it does some good in destroying certain insect pests, such as cutworms, when these are present in abundance and perhaps other food is scarce. The starling has spread very extensively over Australia, and it is a prolific breeder. Moreover, it interferes with the breeding-places of many of our useful insectivorous birds. It is also so wily and so hard to approach that it will never be possible to eliminate it from Australia, or even to diminish materially its numbers, whatever human means are adopted to attempt this. Its virtues are unquestionably less than its defects, and no encouragement whatever should be given to its appearance in any part of the country. On the other hand, any discouragement offered is likely to have little effect."

they disappeared. As they were very tame, they were shot down by boys in large numbers, and may have been exterminated in this way.

The Wellington Society introduced 30 birds in 1875; and 40 in 1876; they are not now common about Wellington, but are to be found up the coast to Wanganui and throughout Taranaki, where they are fairly common; also at Wairarapa. In and about Napier they are in thousands. Mr W. W. Smith says (1916): "though now common in Taranaki, it is said to be less numerous than it was twenty years ago." The Agricultural Inspector for New Plymouth in 1903 blamed this species as the chief cause of the spread of the blackberry. This is a manifest error, for not only is the bird mainly insectivorous, and not to any great extent a fruit-eater, but it is also almost confined to towns, and builds mostly on houses. On the other hand Mr Drummond (May, 1910) says "they are very destructive to apricots, apples, pears, strawberries and gooseberries."

Mr Mahoney of Tuparoa (May, 1912) says they are quite common in the neighbourhood of buildings, and are very destructive to fruit and to grass-seed, quite as much so as yellow-hammers and sparrows. In April they attacked and cleared off most of the late peaches. He also describes how starlings dispossessed a pair of minahs from the ventilator of the school, where they had built their nest for many years. The minahs took themselves off to a willow-tree. Mr F. P. Corkill of New Plymouth also reports how starlings have displaced minahs in the town. Mr H. J. Fowler of Marton states (June, 1912) that minahs follow the plough, as many as a dozen or more together, all day unweariedly, and pick up abundance of grubs.

Australian Minah (*Myzantha garrula*)

In Hutton's *Catalogue of the Birds of New Zealand*, published in 1871, it is stated that this species was introduced into Canterbury and Nelson from Victoria. The early records of the Canterbury Society do not mention them, and those of Nelson are lost. This is the Australian bird known as the noisy minah or miner.

The Otago Society liberated 80 in the neighbourhood of Palmerston in 1880, and they were occasionally seen for two years, and then disappeared. The late Mr Deans, curator of the Society, said these birds were quite different from *Acridotheres tristis*, the Indian minah.

Mr Huddlestone states that Australian minahs were introduced into Nelson in the seventies, that they flourished for a time, but have now (1916) disappeared. I think he is referring to the same species as Mr F. G. Gibbs describes as plentiful in Nelson, and that they were Indian minahs obtained from Australia, in many parts of which they are very common. The Wellington Society liberated 184 birds in 1874; eight in 1876; 12 in 1877, and 20 in 1878. A colony was seen for a time at Taita. The Canterbury Society purchased 200 pairs from Mr Bills in 1879, and liberated them in various localities. I have a very strong suspicion that these two lots also were Indian minahs, caught in Australia—where they are now very common—and brought over to New Zealand by Mr Bills. Unfortunately those who knew the facts are all gone, and it is now impossible to verify my suspicions. But there are no Australian minahs now in New Zealand, whereas there are great numbers of the Indian species in certain districts.

Family Zosteropidæ

Wax-eye; White-eye; Gold-eye; Blight Bird; Silver-eye; Twinkie (*Zosterops cœrulescens*)

If this bird is truly indigenous in New Zealand, then it is a southern form which has recently increased and migrated northwards, but it is more likely to be a comparatively modern natural introduction from Australia. Buller considers it is an indigenous species, but it seems to me the record he gives is against this hypothesis. Captain Howell states that he first noticed the birds at Milford Sound in 1832. In the fifties they were recorded by I. N. Watt, then Resident Magistrate at the Bluff, as coming apparently from Stewart Island, and all migrating northwards.

They did not appear north of Cook Straits till 1856, when they were suddenly abundant, and were called "blight-birds," because they destroyed quantities of the "American blight" (*Schizoneura lanigera*). They only remained for about three months, from June to August, and then disappeared completely. They appeared again in Wellington in 1858, and after that became permanent residents. They were recorded from Nelson in 1859. In 1861 they were first observed by the natives in Hawke's Bay, when the name given to the bird by the Maoris was *Tau-hou* or the Stranger. They were recorded by Colenso in Napier in 1862, and by the natives on the Upper Wanganui in 1863. In 1865 they were observed at Auckland, and by 1868 they had penetrated to the most northerly part of the North Island.

They are stated by Mr A. Shand to have appeared in the Chatham Islands about 1856 and 1857.

They are extremely abundant now, and come right into the very heart of the towns in winter, but in the early summer months they move away out into the country for the breeding season.

At their first appearance in settled districts, their visits were made in the winter months, and they were hailed as valuable insectivorous birds by orchardists and gardeners. When they became permanent residents they discovered the potentialities of fruit, especially plums and pears, and now they are looked upon as great robbers in the fruit season. The good they do for ten months of the year probably far outweighs the toll they exact during the remaining two.

In 1913 large flocks of them visited Akaroa in the autumn and punished the orchards. In 1914 very few were to be seen.

In July, 1910, Mr H. Boscawen reported finding seven white-eyes caught on the sticky seeds of *Pisonia Brunoniana*.

Mr B. E. Collins of Takapou, Hawke's Bay, reported them (1914) as visiting flowering currants and tritomas when in flower.

These little birds are more mercilessly attacked and destroyed by larger predatory birds than any other species, perhaps because they seem less suspicious of enemies than introduced finches, and even the small indigenous birds. They are frequently found killed by hawks, kingfishers, long-tailed cuckoos and shining cuckoos, as also occasionally by others[1].

Family MELIPHAGIDÆ

Australian Bell-bird (*Manorhina melanophrys*)

The Wellington Society introduced two in 1874, but have no further record of them.

Family TANAGRIDÆ

Scarlet Tanager; Cape Cardinal (*Pyrangra rubra*)

The Auckland Society introduced two in 1868. They immediately commenced to breed, and in 1869 it was stated that they were not rare in the vicinity of the Gardens. They did not, however, succeed in establishing themselves.

[1] Dr Cleland says of this species in Australia: "The stomach contents of fifty-five Silver-eyes have been examined. Forty-five of these contained vegetable food, chiefly fruits of various kinds. Thirty-two contained insect food. Amongst the insects occasionally eaten were cabbage-moths, froghoppers, psyllids, thrips, aphids, black scale, and plant bug. During the fruit season there is not the slightest question that the Silver-eye does a very considerable amount of damage to orchards. By feeding on the fruits of such pests as blackberries and lantana, and passing the seeds in their droppings, Silver-eyes act as potent disseminators of these and other plants. However, during the season when fruit is not ripe they apparently serve a definitely useful purpose in destroying certain insect pests. As energetic measures adopted for the destruction of Silver-eyes have never yet been successful in materially reducing their number in any locality, there is little likelihood, whatever action be taken, of eliminating this bird from any particular part."

Family Ploceidæ

Australian Wax-bill; Sydney Wax-bill; Red-browed Finch (*Ægintha temporalis*)

The Otago Society introduced four in 1867, and the Auckland Society four in 1871; in neither case is there any further record.

Java Sparrow; Rice-bird; Paddy-bird (*Munia oryzivora*)

The Nelson Society introduced a number in 1862, but they are not mentioned again.

The Auckland Society obtained six from Captain Forsyth in 1867, but there is no report as to what came of them.

Nutmeg Sparrow; Cowry Bird (*Munia punctulata*)

The Auckland Society received eight from Queensland from Miss Wright in 1868; but there is no further record of them.

(Captain Hutton in 1871 referred this species to Gould's *Estrelda temporalis* (= *Ægintha temporalis*), which is the red-eyebrowed finch; the temporal finch, or red-bill of the Australian colonists.)

Chestnut Sparrow; Rockhampton Sparrow; Chestnut-breasted Finch (*Munia castaneithorax*)

In Judge Broad's *Jubilee History of Nelson* (published in 1892) in a list of birds imported and liberated up to September, 1864, are included six Australian sparrows, which are most probably this species.

The Auckland Society liberated 25 of these birds in 1867, and state in the report of the following year that the "Australian sparrows are considered to be thoroughly acclimatized." Two more were liberated in 1871. Then they disappeared.

The Canterbury Society obtained 12 from Mr Wilkin from Sydney in 1864. Nothing more was recorded of them.

Diamond Sparrow; White-headed Finch; Spotted-sided Finch (*Steganopleura guttata*)

The Canterbury Society introduced a number, apparently in 1864, for the report for 1866 says: "the greatest success has been attained by the little Australian diamond sparrow, which may now be seen in flocks." They have never been heard of since.

The Wellington Society introduced 12 in 1874, but there is no further report. The Nelson Society also introduced them.

(According to Hutton (1871) this was the diamond bird of Australia (*Pardalotus punctatus*).)

Sir George Grey told Sir Walter Buller that out of nearly a hundred diamond sparrows which he liberated on Kawau, very few survived the ravages of the morepork (the small native owl—*Ninox novæ-zealandiæ*).

Family ICTERIDÆ

Californian Starling; Red-winged Starling (*Agelaius phœniceus*)

The Auckland Society introduced two in 1869. They did not increase.

Meadow Lark (*Sturnella neglecta*)

The Auckland Society introduced two from California in 1869; but they did not increase. (This species in Hutton's 1871 catalogue is called *Sturnula ludoviciana*.)

Family FRINGILLIDÆ

Firetail Finch (*Zonæginthus bellus*)

The Auckland Society received two from Captain Coppin in 1870. The Wellington Society introduced eight some time before 1885, but there is no further report concerning any of them.

Zebra Finch; Chestnut-eared Finch (*Tæniopygia castanotis*)

The Wellington Society introduced 12 some time before 1885, but again there is no report about them.

Numbers of ornamental finches and small cage birds of many species have been introduced from time to time for private aviaries, and have been bred in confinement. Mr F. L. Hunt of Ravensbourne, Dunedin, has been a particularly successful rearer of these beautiful species, but when liberated they never succeed in establishing themselves. Apart from climatic reasons, they readily fall a prey to cats and other enemies.

*House-Sparrow (*Passer domesticus*)

The responsibility for the introduction of the common sparrow is very generally shared by the men who were so active in acclimatisation work 50 years ago, as the following record shows. They introduced it in all good faith, and congratulated themselves on their success. But it is amusing to observe how averse the implicated societies are to-day to accept this responsibility. There has grown up, too, by way of explanation, a certain amount of myth about the business; but the facts are incontrovertible. For example, Mr Drummond in 1907 writes:

> The story is that the (Canterbury) Acclimatisation Society ordered twelve dozen hedge-sparrows from England. The order was placed with Captain Stevens of the 'Matoaka,' who submitted it to a bird-fancier at Knightsbridge. Either the fancier or the Captain blundered, and the latter

took on board thirteen dozen house-sparrows, which are generally known by the common name of "Sparrow." He was very attentive to them on the voyage out, believing that they were the valuable hedge-sparrows which the colonists were anxious to secure. Most of them died, however, and when he reached Lyttelton in February, 1867, only five were left. The officers of the Society, realising that a mistake had been made, refused to accept the strangers. The Captain then took them out of their cage, and, remarking that the poor little beggars had had a bad time, set them at liberty. They flew up into the rigging and remained twittering there for some time. The members of the Society went below to look at other birds. When they reached the deck again the sparrows had flown. The birds stayed about Lyttelton for three weeks. Then they disappeared, and when next heard of they were at Kaiapoi, about twenty miles distant, where, at the end of 1869, they were reported as being "particularly numerous."

This story was evidently current in Canterbury, for at the annual meeting of the Society in Christchurch in 1885, the Chairman, the Hon. J. T. Peacock, said: "The Society used to give bonuses to captains of ships for bringing out small birds. One captain brought five sparrows, which the Society refused to purchase, and which that captain let go himself. From those five, *the whole of the sparrows in the Province had, he believed, sprung*."

In the report for 1889, under the heading "The Sparrow," the same Society are responsible for this statement: "we most deliberately deny ordering or introducing this questionable bird, but we well remember the devastations made by the caterpillars and grubs previous to their advent."

In 1895 Mr A. Bathgate of Dunedin wrote: "I believe our (Otago) Society turned out one or two, but the sparrows came to us from Christchurch."

Now for the actual facts.

According to Sir Walter Buller the Wanganui Society introduced sparrows in 1866, and these were therefore the first brought into the country. But the Nelson Society forestalled Wanganui, for they succeeded in bringing in *one* sparrow in 1862.

The Canterbury Society in 1864 printed a list of prices which they offered to immigrants for each pair (cock and hen) of birds, viz.:

	£	s.		£	s.
Black Cock or Grouse ...	10	10	Robins	1	10
English Partridges ...	5	0	Wrens	1	10
Common Thrushes ...	2	0	Grey Linnets ...		15
Blackbirds	2	0	Green Linnets ...		15
Skylarks	2	0	*House-Sparrows* ...		15
Rooks	2	0	Hedge-Sparrows ...		15

The same Society liberated forty sparrows in 1867, and the annual report for 1871 states that they are "thoroughly established and need no further importations[1]."

The Auckland Society in 1864 also gave a list of the prices offered by the Auckland Provincial Government to immigrants bringing out various birds, per pair, cock and hen.

	£	s.		£	s.
Black Cock or Grouse ...	5	0	Blackbirds	1	10
(Hares)	(5)	0	Robins	1	10
English Partridges ...	3	0	Wrens	1	10
Rooks	3	0	Wheatears	1	10
Nightingales	3	0	*House-Sparrows* ...	1	10
Song Thrushes	1	10	Hedge-Sparrows ...	1	10

This Society liberated 47 sparrows in 1867, and in the annual report for 1868, "consider them thoroughly acclimatised."

I am indebted to Mr E. D'Esterre, Editor of the *Auckland Weekly News*, for unravelling some ancient history regarding the introduction of the sparrow into Auckland. In reply to a representative of the *Auckland Star*, Mr T. F. Cheeseman is reported to have said:

> Sparrows were not introduced by the Acclimatisation Society, although that body is credited with having brought them here. I can speak with certainty on that point, because I was here before the small birds were introduced. It was the late Mr S. Morrin and Mr T. B. Hill who introduced the first lot of sparrows and distributed them.

Mr D'Esterre saw Mr Cheeseman later (June, 1916) and was informed that Mr T. B. Hill, who came out in the 'Morning Star' in 1861, resided at Auckland, and that he brought out the sparrows either for or in conjunction with Mr Sam Morrin. He then wrote to a correspondent of his, a Mr P. T. Hill, also resident at Raglan, to find out his namesake if possible. This gentleman, who disapproves very strongly of the sparrow, sent a cutting from the *Raglan County Chronicle* of 8th July, 1915, with a report of a Farmers' Union Meeting at which the question of poisoning small birds came up. "Mr Taylor said by getting rid of the small birds insect pests would be increased, and would probably be a far worse trouble than the birds. The President thought it was an unwise thing to disturb the balance of Nature. Mr T. B. Hill agreed that it was best to leave the birds alone. *He had introduced sparrows and sold them at 10s. each in Auckland*." Eventually Mr D'Esterre got into communication with

[1] The late Mr Bills used to narrate how he trapped the sparrows for the Canterbury Society with large folding nets in the streets of London in the early mornings, and how the Londoners were surprised that any country should want such birds. He explained that the caterpillars in New Zealand were so numerous and large, that the farmers had to dig trenches round their houses to trap and bury the voracious creatures, lest after eating up all the crops, they should turn to and eat up the farmers themselves. Mr Bills' statement can be taken *cum grano salis*.

Mr T. B. Hill himself and received a letter dated 1st July, 1916, in which he says:

As I see it is going the round of the papers that I and the late Mr Morrin introduced the sparrows I shall be glad if you will contradict this. I don't think for a moment my friend Mr Cheeseman made the statements intentionally, but as I believe he was Secretary of the Auckland Acclimatisation Society at the time, of which Council I was an individual member, I think if he refers back he will see it was the Auckland Society that introduced them. With many other birds they were sold by auction by Mr S. Jones at his Auction Mart, and the House-sparrow was the favourite. I was, I think, the largest purchaser at *One pound per pair*, and I successfully acclimatised them to my building, with Mr Soppet's Flour Mill adjoining, in Freeman's Bay, and soon had all the sparrows others had brought down in my yard and flying in and out of the window of my room, where I kept several confined. I had people come to me for birds to replace what they had lost at the price I paid for mine. The first breeding season they proved a great nuisance in filling the spouting and other places in the Bay with their nests. So many of us then were not yet acclimatised ourselves that when we woke in the morning hearing the little "cheer-up, cheer-up," it made us fancy we were back in the old country again. I certainly sent them to friends in the country who were anxious to get them. There now you have the whole history as far as I am concerned.

In 1865 the ship 'Viola' from Glasgow arrived at Auckland, and landed *two* sparrows out of six dozen which were shipped.

In 1864 the Nelson Society imported a number of Sparrows, but only one was landed alive. In 1871 six were introduced, and were liberated at Stoke, where they soon increased.

The Otago Society liberated three in 1868, and 11 in 1869.

The sparrows very quickly increased in all parts of New Zealand until they became a very serious pest. But while farmers rail at them to-day, it has to be remembered that at the time of their introduction crops of grain and grass were threatened with absolute destruction by the hordes of grubs and pigeons. Mr Drummond in his very able pamphlet on "our feathered immigrants" (1907) has summarised the case for and against the sparrow, as far as New Zealand is concerned, with the balance very much against. Nearly every county council and agricultural association in the country wages war on him, by selling poisoned grain to the farmers, and offering bonuses for eggs. Yet he continues to thrive and flourish.

Mr R. E. Clouston writing (July, 1916) of the Gouland Downs in the Nelson district, which is noted for the abundance of its native bird life, and where thrushes, blackbirds, skylarks, and particularly redpolls are common, says that in all the years he was there he only saw about two sparrows. Mr Philpott observes that while the sparrow is abundant in cultivated country it does not penetrate far into the bush. Sparrows are found and are increasing in the Chatham Islands, which are distant 450 miles from the nearest point of New Zealand, and where they are self-introduced.

Mr T. W. Kirk gives much important information as to its rate of increase in his note on the breeding habits of the sparrow in New Zealand. The following is a summary of his facts:

The breeding season begins in spring, the first brood appearing in September, and the last in April. There are never less than five eggs in a nest, but usually six or seven. Incubation lasts 13 days. The young are fed in the nest for eight or nine days, then return to it for two or three nights and afterwards shift for themselves. In five instances fresh eggs were found in the nest along with young birds, and the author thinks that the young birds do the chief work of incubation of succeeding broods. In at least one instance, marked birds reared in September were themselves breeding at the end of March.

Calculating from nests which were watched, the author thinks that the average annual increase is five broods of six each, and this is a low estimate. Allowing for deaths at the rate of one-third of the whole annual increase, then one pair will produce 11 pairs at the end of the first year; 121 pairs at the end of the second year; 1131 pairs at the end of the third; 14,641 pairs at the end of the fourth; and 161,051 pairs, or an actual increase of 322,100 birds in five years, without taking into account: (1) the early broods which are themselves breeding; (2) the fact that more than five broods are probably hatched in a year; and (3) that often more than six eggs are hatched at a time[1].

Food of the Sparrow. The average farmer's opinion on this subject is valueless; he only sees the harm that is done at sowing time and in harvest, and concludes—on very imperfect evidence—that the bird is only a grain-feeder. Mr T. W. Kirk says: "I have myself dissected fifty-three birds, taken at all seasons of the year, and am forced to admit that the remains of insects found in them constituted but a very small portion of the total food." Unfortunately Mr Kirk does not say where he took the birds which he examined. He himself dwells in or near a large town, and the chances are that a considerable amount of the food of the sparrows would be from households, grain from horsedroppings, etc. Mr Kirk's communication was made to

[1] Some idea of the abundance of small birds in the farming districts of New Zealand may be gathered from such facts as the following. One trapper in the Rakaia district received a cheque for £54. 9*s*. 3*d*. for the month of July, 1918, which represented 17,429 heads. The Ashburton County Council commenced buying birds' heads on 17th June, 1918, and by the middle of August had paid away £495. 17*s*. 8*d*., representing 158,681 small birds, chiefly sparrows and skylarks.

the Wellington Philosophical Institute in 1878. In the discussion which followed, Mr W. T. L. Travers said that "His experience led him to believe that their principal food was insects. The cicadæ especially are caught in hundreds by them" Sir Walter Buller says:

If the sparrow is fond of ripe grain it is still fonder of the ripe seeds of the variegated Scotch Thistle. This formidable weed threatened at one time to overrun the whole colony. Where it had once fairly established itself it seemed well nigh impossible to eradicate it, and it was spreading with alarming rapidity, forming a dense growth which nothing could face. In this state of affairs the sparrows took to eating the ripe seeds. In tens of thousands they lived on the thistle, always giving it the preference to wheat or barley. They have succeeded in conquering the weed. In all directions it is dying out.

I have myself watched sparrows hawking for moths and crickets, and have observed them feeding on the fruits of meadow plantain (*Plantago major*) and of dandelion (*Taraxacum densleonis*).

Dr Hilgendorf points out that "so purely a grain-eating bird as the sparrow feeds its nestlings for about six weeks on nothing but insects."

During the Farmers' Union Conference in May, 1918, one delegate stated that during the preceding season his grain crop of 40 acres was black with caterpillars, as many as three or four being on each head. Then sparrows attacked them, and in a short time not a caterpillar was to be seen.

Sparrows are often very destructive to flowers in gardens, picking them to pieces for no ostensible cause. Primroses, violets, crocuses are the most commonly attacked, and these are all spring-flowering. The habit is recorded from several parts of New Zealand.

The native hawk (*Circus gouldi*), kingfisher (*Halcyon vagans*), long-tailed cuckoo (*Urodynamis taitensis*) and shining cuckoo (*Chalcococcyx lucidus*) are all credited with catching and destroying sparrows[1].

[1] Dr Cleland, writing of the food of the sparrow in New South Wales, says: "One hundred and twenty-seven sparrows were examined, the majority of them coming from Richmond, New South Wales. Sixty-four were found to feed on wheat and maize. Various grass seeds were found in others. Occasionally they have been found to feed on white ants, cabbagemoth larvæ, cutworms, locust, blowflies and aphids. The large amount of grain eaten far outweighs any value that the sparrow may have as an insectivorous bird during the period when such grain is available, but during other seasons of the year it probably plays a mildly useful part." (It is to be noted that the 127 birds referred to by Dr Cleland were shot in May. Had sparrows been examined in September to November, the results would almost certainly have been different.) One of the birds examined contained 400 millet seeds, besides maize and other grasses. It is due very largely to sparrows about Sydney, that several species of Cicadas are almost extinct.

Mountain Sparrow; Tree Sparrow (*Passer montanus*)

The Otago Society liberated two in 1868. The Auckland Society also liberated three in 1868, and nine in 1871. None of these was heard of again.

* Chaffinch (*Fringilla cœlebs*)

The Nelson Society introduced 23 of these birds between 1862 and 1864, but kept no record of them afterwards.

The Canterbury Society liberated 11 in 1867, and five in 1868; and three years later reported that they are considered to be "thoroughly established and to need no further importations." In 1871 a further lot were introduced.

The Auckland Society liberated several in 1864, 45 in 1867, and stated the following year that they were thoroughly acclimatised. But they introduced 68 more in 1868, and a considerable number in 1869.

The Otago Society liberated 27 in 1868, six in 1869, and 66 in 1871.

The Wellington Society liberated 70 in 1874, 36 in 1876, 20 in 1877, and a few more in subsequent years.

Private individuals and dealers introduced them also at all the principal centres.

This bird is common throughout both the islands, and very abundant in some parts, especially from Taupo northwards. Even up to the present time, some county councils in grain-growing districts (e.g. South Canterbury) are giving bonuses for their destruction.

Their occurrence in Otago has been rather curious, and has puzzled observers a good deal. For example, about Dunedin they became fairly common a few years after their introduction, and then nearly altogether disappeared. I attributed this to their eating poisoned grain, for their scarcity dated from the time that this method of destroying rabbits came into use. But since this method was abandoned in favour of pollard-poisoning, and poisoning generally was substituted for trapping, other small birds have increased very considerably, but the chaffinch is still a comparatively rare bird.

Mr A. Philpott states that while the species is not common near Invercargill, where it began to appear in 1910, it is abundant at Queenstown, and common in the Longwood, Waiau and Titiroa (Hunter Mountains) forests. It is especially abundant in these localities on the upper limit of the bush (about 3000 feet).

An attempt was made in 1879 to establish them in Stewart Island, when 70 were liberated. They were seen for a time at the head of Paterson Inlet, but Mr Traill tells me (1916) that "none have been reported for years."

Mr T. H. Potts states that where lichens are scarce, chaffinches frequently used fragments of paper in nest-building.

Bramble Finch; Brambling (*Fringilla montifringilla*)

The Canterbury Society liberated two in 1868, and six (?) in 1871. But the annual report for 1873 speaks of them as having been imported in considerable numbers.

The Wellington Society liberated three in 1874, and one in 1877, and reported them in 1885 as having been seen.

Seebohm (*Siberia in Europe*, p. 120) describes this as a migratory species, which winters in the British Isles but mostly in Central and Southern Europe, occasionally crossing the Mediterranean. It breeds throughout the northern portions of the palæarctic region, at or near the limit of forest growth.

Linnet (*Linota cannabina*)

The Nelson Society introduced seven linnets in 1862, but have lost all record of them.

The Otago Society liberated two in 1867, and 18 in 1868.

The Canterbury Society liberated 20 in 1867, three in 1868, and a number (not specified) in 1869. In 1875 another consignment was introduced by Mr R. Bills, and these were either sold and distributed, or liberated in the Christchurch Gardens.

The Auckland Society liberated eight in 1865, 14 in 1867, 20 in 1868, and a number (not specified) in 1869. Five years later the annual report states that "they are thoroughly established in Auckland Province." I cannot understand this statement, unless the report refers to the greenfinch, a bird which is sometimes called the green linnet.

The Wellington Society liberated 22 birds some time before 1882, for it was reported in that year that two were seen in the Porirua district.

The failure of this species to establish itself in New Zealand is one of the most inexplicable problems in animal naturalisation. Some allied species have become very common, but the linnet disappeared soon after liberation.

The bird is a partial migrant in Britain, and I have suggested that this may be one explanation. This, however, is discounted by the fact that those introduced were brought to the country at so many different times, and were probably obtained from very different localities.

Mr Edgar F. Stead, writing me in April, 1915, says: "It is possible that the birds imported were all, or nearly all of one sex. If they were caught during certain seasons of the year, nesting, and perhaps migratory, they may have been nearly all cockbirds."

Twite; Mountain Linnet (*Linota flavirostris*)

The Nelson Society imported some of these birds in 1862, but only two reached the colony.

The Otago Society liberated 38 in the Dunedin Botanical Gardens in 1871. Mr F. Deans, the curator, informed me in 1890 that he "never saw or heard of them again." This species is a partial migrant in Britain, and I have little doubt that the birds once set free started on a voyage of discovery.

*Red Pole; Redpoll (*Linota rufescens*)

The Nelson Society were the first to attempt the introduction of this species, but only *two* arrived of those shipped in 1862.

The Otago Society introduced ten in 1868, and 71 in 1871. Mr F. Deans reported in 1890 that "some of these were seen, but I do not think they increased." The fact is they migrated from the immediate neighbourhood of Dunedin to the high open ground a few miles away.

The Canterbury Society liberated 14 in 1868, which, according to their report, migrated in a body and settled at Timaru. In 1871 they liberated 120 more. In 1875 Mr R. Bills brought out a number, some of which were sold and distributed, and the rest were liberated in the Christchurch Gardens.

The Auckland Society introduced one in 1871, and 209 in 1872, and these were liberated in various districts south of Auckland.

The Wellington Society introduced two in 1875.

This species is not commonly seen about the towns or in thickly settled districts, but is abundant in both islands, especially in open upland country at moderate elevations. It is common in the back country near Dunedin.

Mr A. Philpott says: "I first saw this bird in Invercargill in 1909. Since then it has become very common, but it appears to leave this locality during the winter. I have noticed it in numbers, feeding on the seed of the Toe-toe (*Arunda conspicua*) and resorting to the coastal sandhills, where flocks may always be found feeding on the seeds of *Juncus* and other plants. It occurs commonly about the upper edge of mountain forests," that is at an elevation of about 3000 feet.

The abundance of redpolls is shown from the following:

In March, 1911, Mr T. H. Jones of Christchurch caught 70

in one pull of the net near New Brighton; Mr C. Bills of Dunedin said he could complete an order for a thousand of them within a fortnight; while in Southland as many as 500 were taken in a day.

During the summer these birds eat quantities of the green fly (*Aphis*); while during grass-seed harvest they subsist mostly on seeds of grasses.

*Goldfinch (*Carduelis elegans*)

The Nelson Society introduced ten goldfinches in 1862.

The Otago Society liberated three in 1867, 30 in 1868, 54 in 1869, and 31 in 1871.

The Auckland Society liberated 11 in 1867, and 44 in 1871.

The Canterbury Society liberated 95 in 1871, and a number in 1875.

The Wellington Society liberated one in 1877, 52 in 1880, 22 in 1881, and 103 in 1883. The birds appear to have at once established themselves at all the centres, and to have quickly spread.

They are now extraordinarily abundant in all parts of New Zealand. They occasionally eat grain and seeds of other cultivated plants, but chiefly confine themselves to seeds of thistles and small seeds. Mr W. W. Smith records them from Ashburton as assisting to spread throughout the district such plants as knapweed (*Centaurea nigra*) and Scotch thistle (?) (*Onopordon acanthium*).

Mr A. Philpott (1918) says these birds are never found far from settlement. Yet it is interesting to note that they have found their way to the Auckland Islands, 230 miles to the south, and the Chatham Islands, 470 miles to the east, and are now established in both these outlying regions.

The goldfinch is a bird with a good reputation, and I have heard few complaints of any harm it does, either to farmers, orchardists, or gardeners. But in the *Otago Witness* of 31st October, 1892, one strawberry grower complained bitterly that this species committed great ravages on the growing fruit by picking out the seeds, and thus completely destroying the berries. I do not know whether other growers suffered in the same way.

Young goldfinches are not unfrequently killed and eaten by the long-tailed cuckoo and the native kingfisher.

Siskin (*Carduelis spinus*)

The Wellington Society introduced two in 1876. The Canterbury Society report that several were liberated in 1879, "and they have taken up their quarters in the plantations around Hagley Park."

Mr Edgar Stead (1916) says: "I have never seen a Siskin in Canterbury."

On the other hand Mr W. W. Smith informed me in April, 1919, that these birds are wild about New Plymouth. Mr Smith is one of the most careful and observant naturalists in New Zealand, and it is most unlikely that he has made any mistake about this. These birds may be the progeny of those already referred to, or they may be descended from other introduced birds, for dealers in all the centres continually import siskins, along with canaries.

The species occasionally nests in Britain, but is mostly a winter visitor, usually nesting in Norway, Mid-Sweden and Russia.

*Greenfinch (*Ligurinus chloris*)

The Nelson Society introduced five greenfinches in 1862, but kept no further record of them.

Mr J. Drummond (1907) says:

the first greenfinches about which I have been able to secure any information, were liberated in Christchurch in 1863, where a pair were purchased at auction for five guineas. They soon nested, but the only occupant at first was one little greenfinch. Before the warm summer days had passed, however, a second family of five was reared, and in the following winter a flock of eight was seen daily. In the next year, late in the autumn, more than twenty were flushed from a little patch of chickweed, and it was not long before the birds had spread so widely that their note became a well-known sound in Canterbury.

In another paper he gives the date of the introduction into Canterbury as 1866, but there is no record in the Society's reports.

The Auckland Society liberated several in 1865; 18 in 1867, and 33 in 1868, and in that year "considered them to be thoroughly acclimatized."

The Otago Society liberated eight in 1868.

The bird is now particularly abundant in all the settled parts of the country, and is most destructive to the ripening grain crops. It is also most destructive to fruit in certain fruit-growing districts, especially in Central Otago. Mr G. Howes in a note on "Fruit Destruction by small Birds in Central Otago" (*Trans. N.Z. Inst.* vol. XXXVIII, p. 604) says that green linnets attacked the apricots when the fruit was forming, the cherries while in flower, and later in the season the peaches and plums. It was to combat the serious pest to orchards which these and other small birds had become, that the small brown owl (*Athene noctua*) was introduced, and the result has been a great diminution in the numbers of fruit-eating birds.

The late Mr T. H. Potts recorded that in winter it feeds largely on the seeds of pine-trees (*Pinus pinaster*), which it picks out of the fir-cones. It certainly also feeds on the seeds of many weeds, and probably spreads some of these.

Mr T. A. Philpott states that the species is much less common in Southland now (1918) than it was 25 years ago. He attributes this as probably due to the wide adoption of dairy-farming and stock-raising in preference to grain growing.

This species has found its way to the Chatham Islands.

Bullfinch (*Pyrrhula europæa*)

According to Captain Hutton six bullfinches were introduced into Nelson. But Mr F. G. Gibbs says: "Bullfinches have never been acclimatised in Nelson."

In 1875 Mr R. Bills presented the Canterbury Society with a pair of these birds.

Mr H. Guthrie-Smith also, as quoted by Mr J. Drummond, says: "I saw a pair of bullfinches on one occasion in manuka country. Two friends on whom I can rely have seen bullfinches." This was in Hawke's Bay. While there is no record of any of the societies having introduced them, there is no doubt they have been frequently brought in by dealers.

Mr H. Hill (1916) speaks of it as a regular visitant to Napier and says it attacks the birds and flowers of the peach, nectarine and apricot, sometimes disappearing altogether. It has also been doubtfully reported from Taranaki.

In many of the recorded cases of bullfinches being seen, the birds when sent to experts proved to be chaffinches. There should be no possibility of mistaking such a conspicuous bird.

Reed Sparrow; Reed Bunting (*Emberiza schœniclus*)

The Otago Society liberated four in 1871, but there is no further record of them.

*Cirl Bunting (*Emberiza cirlus*)

The Otago Society liberated seven in 1871. They seem at once to have increased and spread, and Mr W. W. Smith in 1916 reports them as common in Taranaki. They occur in flocks along the coast, at Hawera, etc. I have noticed that some dealers and bird-catchers do not know the difference between these and yellow-hammers. In October, 1915, I saw a large cage of so-called yellow-hammers which had been taken in the neighbourhood of Dunedin; several were cirl buntings. It is difficult to distinguish between the young of the two species. In 1879 18 were liberated in Stewart Island, but they failed to establish themselves. Their occurrence is very erratic. At one time they increased to a very considerable extent in Otago; then they seemed to become quite rare. Now they are more common again. Mr Drummond reports something of the same kind as occurring near Christchurch.

If all the cirl buntings now in New Zealand are descended from the seven originally liberated in Otago, the case is certainly a very interesting one.

*Yellow-hammer (*Emberiza citrinella*)

The Nelson Society introduced three of these birds in 1862, but kept no further record.

The Auckland Society introduced eight in 1865, four in 1867, five in 1868, a number (unspecified) in 1869, 16 (out of 148 shipped) in 1870, and 312 in 1871.

The Canterbury Society introduced one in 1867, and 34 in 1871.

The Otago Society introduced eight in 1868, and 31 in 1871.

They quickly spread all over New Zealand, and to day are common from Foveaux Straits to the extreme north of the North Island. In 1879, 32 were liberated in Stewart Island, but they have not been seen there for years.

They are destroyed wholesale as noxious pests in all grain-growing districts, a price being put on their heads, and their eggs being purchased by thousands by the county councils. Mr W. W. Smith, writing from Taranaki, where they are very numerous, in February, 1916, says: "When at Rangiotu Camp on August 2nd, last, I observed these birds there in hundreds, quite tame, subsisting on bread, etc., thrown out from the soldiers' mess." Mr Philpott states (August, 1916) that they were not uncommon in Southland 30 years ago, but are now very rarely met with. Within the last few years (1918) it has become rather more plentiful. Lieut. Cox in 1910, recorded yellow-hammers as occurring in the Chatham Islands, to which they had found their own way across 470 miles of ocean.

Ortolan; Ortolan Bunting (*Emberiza hortulana*)

The Wellington Society imported three pairs in 1885, and they were liberated near Otaki. They bred in the following year, and the report says "there is now a small flock." That is, however, the last that has been heard of them.

Canary (*Serinus canarius*)

No serious attempt has ever been made to naturalise this species, but I have heard of several private efforts in this direction. Several bird-fanciers have bred canaries in their aviaries, and have given the birds freedom to come and go as soon as the eggs were hatched. Apparently, however, the domestic cat is an insuperable obstacle to their establishment. The canary is such an artificial, domesticated and closely-inbred species, that it would apparently take some generations of birds to acquire habits tending to its self-preservation.

In concluding this list of birds, it may be mentioned that the Otago Society many years ago introduced and liberated what were termed diamond-eared finches and jager birds. The former may have been the diamond sparrow (*Steganopleura guttata*) referred to previously. But I have not the slightest idea what the last named bird is. Neither species was heard of again.

Chapter V

REPTILES AND AMPHIBIA

Class *REPTILIA*

Order LACERTILIA

Family Lacertidæ

English Scaly Lizard (*Lacerta vivipara*)

Mr T. W. Kirk (in 1886) reports the occurrence of several specimens on the Tinakori Hills, Wellington, probably introduced with some cases of plants for the Botanical Gardens. I am not aware that they increased, or were seen again after the first year.

Sub-Class *CHELONIA*

Family Chelydidæ

Australian Fresh-water Turtle (*Chelodina longicollis*)

Representatives of this species—popularly known as Australian tortoises—have been frequently imported into New Zealand by dealers and private individuals. In 1889 a large number were brought to Dunedin for the Fisheries Court of the South Seas Exhibition, and at the close of the Exhibition were sold and distributed. Two of these were given to Mr A. M. Johnson in October 1890, and he had them for some years.

None of these animals at the Dunedin Exhibition was ever seen to eat, though they were offered all kinds of food.

Mr G. Howes states that one or more fresh-water turtles were seen about the Waihopai River in North Invercargill about 1903.

Japanese Tortoise

Mr Cheeseman informs me that "a number of these small tortoises were procured from Japan for the Auckland Exhibition in 1913, and at its close about ten remained. These were placed in one of the greenhouses in the Auckland Domain Gardens to be kept there through the winter, but when spring arrived none of them could be found. They had evidently escaped through some unknown opening." I have not seen them at all, and do not know what species they belonged to, though possibly it was either *Clemmys japonica* or *Damonica reevesii*.

One or more turtles or tortoises were seen about Hawera in Taranaki in 1915.

Class *AMPHIBIA*

Order ANURA

Family Hylidæ

*Australian Green Frog (*Hyla aurea*)

The Auckland Society introduced two in 1867, and in the following year received several small lots from Sydney. They increased quickly and are now abundant all over the North Island.

The Canterbury Society received some frogs in 1867 from the Hobart Acclimatisation Society, and some tadpoles in 1868 from Mr Alport of Hobart and from Mr W. L. Hawkins.

The Southland Society received some spawn (from Hobart?) in 1868, which was hatched out at the Wallacetown Ponds. Frogs soon were carried to various places on the Southland Plains, but they did not thrive and had all disappeared by 1890.

A similar experience was met with in Otago. About 60 frogs and tadpoles were obtained from Napier in 1888, and were liberated in a marsh. They were seen about for a few days, and then all disappeared. It is possible that in both cases wild ducks of some kind or other accounted for them all. Later on others were brought down from the north and liberated in Otago and Southland.

Green frogs are now common throughout the South Island. In Westland, where they were introduced from Nelson about 1896, they have largely displaced the small brown frog (*H. ewingii*), which was established on the west coast some 20 years previously. Marriner says: "From what I have seen of the *Hyla aurea*, it would find the small brown frog very eatable, and if it does not stop at eating its own kind, there is very little chance of it sparing the small strangers."

Mr Dansey informs me that frogs were introduced into Rotorua by Captain Gilbert Mair about 1878, and that now (1916) they are numerous everywhere in the district.

I well remember the uneasiness and consternation in the native village, upon some native excitedly reporting his having seen a peculiar *ngarara* (reptile) in a pond near the lake, and describing that it had fingers and toes and swam like a human being. Dread was expressed at the idea of swallowing young ones while drinking water, that they might grow inside to gnaw away at their stomachs. Others ascribed the bringing into the district of such reptiles, as the doing of some evil-minded European to wipe out the natives and secure their lands.

Mr A. C. Yarborough of Kohu-Kohu states that frogs were abundant in 1884 on the east side of the Hokianga, and appeared on the Kohu-Kohu side in the following year. They are not very numerous in the district at the present time (1916), but at Kaikohe they are extremely common.

The remarkable power of vitality possessed by the green frog—*Hyla aurea*—is shown by the following. In December, 1884, my wife and I left Dunedin for a tour in the North Island, our house being in charge of a housekeeper during our absence. In the front hall there stood a large ornamental fern-case, filled with local ferns (*Asplenium*, *Hymenophyllum*, *Trichomanes*, etc.), and in this case lived a full-grown green frog, which, however, seldom showed itself. We were absent about six weeks, returning at the end of January, 1885. The ferns had not been watered in our absence, with the result that all the filmy and more delicate ones were dead. My wife was too disappointed to start re-filling the case, and it was carried out to a lumber room in an outside shed, and left there for over a year. In March, 1886, Mrs Thomson thought she would like the case re-stocked with ferns, so I personally set to work to empty out the old, perfectly dry material, and incidentally said that I would search for the skeleton of the frog. I could not find it anywhere, but in the bottom of the case, under one of the largest dead ferns, was a lump of clayey soil about four inches in diameter, quite dry externally. On breaking this up I was intensely surprised to find the frog, looking very much as it was when we last saw it—15 months before—and perfectly cool and moist. I at once put it into a glass vessel, shut a common house-fly in, when the frog immediately came to attention, and caught and ate the fly. It fed quite freely afterwards and lived for some months, when it perished by a singular accident. Its little glass-house was left standing on my microscope table in a window facing the midday sun. A large bull's eye condenser stood on the table near the window, and this unfortunately focussed the sun's rays on to the glass-case, and when discovered half an hour after the unfortunate frog was dead. The ball of clay in which the frog was found after its 15 months' imprisonment was not, as far as either of us could remember, in the case originally. We both thought that the animal had in some way or other gathered it together as a protection, but how it managed to get inside the ball and apparently leave no external aperture, I cannot explain. It seems to me that the incident throws some light on the stories which one occasionally reads in newspapers about frogs being found inside of rocks and stones. Our frog was not in a rock, but it was inside a remarkably hard piece of clay, and yet it managed to breathe and retain its moisture for that long period of time.

Australian Brown Frog; Whistling Frog (*Hyla ewingii*, var. *calliscelis*)

This little frog was introduced into New Zealand in a curious manner. A Mr W. Perkins brought some over from Tasmania in a bottle in 1875, and liberated them in a drain in Alexandra Street, Greymouth. From there they spread up the Grey River to Ahura (24 miles) on the south bank, but do not seem to have got to the north side. They also spread south for a few miles.

In 1878 Mr F. E. Clarke read a paper before the Westland Institute on "Notice of a Tadpole found in a Drain in Hokitika." He says: "No frogs or frogs' spawn having been introduced nearer to the West Coast of New Zealand than Nelson and Christchurch, it is puzzling to conjecture in what manner the little stranger arrived." He was evidently unaware that about three years before he wrote his paper *Hyla ewingii* had been introduced into Greymouth. But the frog was probably introduced into Hokitika by some person who carried the spawn or the tadpoles; as it is scarcely or not at all found between the two centres. About 1900 Mr James King, of Hokitika, brought some of the frogs from Greymouth to Hokitika, and they increased for a time. But Mr King informs me that they are now very rare, if not extinct, at Hokitika, being apparently displaced by the larger *Hyla aurea*.

The little whistling frog is one of the commonest frogs of Eastern Australia and Tasmania. Mr J. J. Fletcher of Sydney says that though a true climbing frog it has, at least in Australia, altogether or nearly lost the arboreal habit. Mr A. P. Harper of Greymouth, however, informed Mr Marriner that he had "personally seen these frogs climbing over blackberry bushes at a height of from six to eight feet above the ground."

Australian Climbing Frog (*Hyla cærulea*?)

In 1897 a consignment of six dozen climbing frogs was obtained by the Agricultural Department from Mr J. Stein of Sydney, and 71 arrived alive. Mr T. W. Kirk said of them in his report in 1898:

> this frog is similar to the ordinary common frog, so common in many parts of New Zealand, except that it has a very considerable advantage over that species in that its toes are provided with suckers, which enable the animal to climb trees and houses in search of insects. In Sydney I have seen these frogs at the top of a wall four stories high.

Neither Mr Kirk nor Mr Stein can identify the species, but Mr McCulloch of the Australian Museum thinks it was probably *Hyla cærulea*, from the above account of it.

Mr Kirk informs me that some of these frogs were liberated in Hawke's Bay:

at Greenmeadow Vineyard and Orchards, some on the Frimley Estate, and a few at the Hastings Racecourse. In the Wellington district a portion of both this and a subsequent consignment (imported in 1899) was released in the Wellington Botanical Gardens and in the orchard of Mr G. A. Grapes, Paraparaumu. At Auckland, the Island of Motuihi was chosen. Some were also released in Queen's Gardens, Nelson, by Mr Kingsley, who liberated them on my behalf. Both attempts to establish these frogs proved failures, for neither the liberators nor myself have ever seen them since, although I personally have carefully searched each locality several times since. In addition to the localities already mentioned, some were liberated on the Government Experimental Station at Moumahaki; the result was exactly the same as in the other places, total disappearance.

It seems possible, however, that some of them have survived, and may yet turn up in unexpected localities. Mr J. Killen of Whangarei, when at Kaikohe about 1913 or 1914, saw and held in his hand a small green frog which was quite different from the common *Hyla aurea*. Unfortunately the specimen was not preserved, and so cannot be identified[1].

Family RANIDÆ

European Brown Frog; Grass Frog (*Rana temporaria*)

In 1864 Mr A. M. Johnson imported 30 frogs to Canterbury from Great Britain, presumably of this species. He kept, however, no record of them, and it may be that they were lost soon after arrival[2].

In the report of the Canterbury Society for 1868 the following paragraph appears:

The old original frog which was imported into the colony by Mr Murray-Aynsley, and which at one time drew a concourse of 300 visitors to the Acclimatisation Gardens in one day, is supposed to have been swallowed by a stray swan.

Family BUFONIDÆ

Common Toad (*Bufo vulgaris*)

The Canterbury Society received 12 toads from the Hobart Acclimatisation Society in 1867. There is no record whatever of their success or failure.

[1] Mr Huddlestone of Nelson states that tree frogs were introduced there in a warden case in 1856. It is, of course, impossible to say what species they belonged to, and there is no record of the survival and increase of the animals.

[2] Mr Huddlestone states that large edible frogs were introduced into Nelson in 1867, with the idea of providing food for wild ducks. Probably the common water-frog (*Rana esculenta*) was the species experimented upon. They do not appear to have been seen again after liberation.

In 1893 toads were introduced into Gisborne. Mr W. Chambers writing to the *Poverty Bay Herald* says:

On the 1st March thirty-nine toads were shipped by the Kaikoura, and when the boxes were opened here on 23rd April thirty-four were found in first-rate order. The five that died were most probably hurt before shipment. The plan adopted was to put the torpid toads in boxes filled with wet moss, which were put in the cool chamber and kept at a temperature of about 35° F. on the voyage.

Mr Chambers has since supplied the following information to me (17th June, 1918):

To allow them to recuperate before turning them out, I placed them on the lawn under a large wire-netted cage, and fed them with worms, etc., for a few days. This ended in disaster, as one night the rats got in and killed half of them. I then turned the survivors out near an old tunnel on the edge of a swamp. We saw one occasionally for a year or two after, when they disappeared completely. I don't think they bred, as I kept a good look-out for tadpoles and young toads. Two of them, a male and a female, I kept in a glass cage for about two years, and when I judged the breeding season had arrived (from their behaviour), I gave them access to a large pan of water, but without result. These also came to an untimely end, both dying on the same day from a surfeit of crickets.

Mr Chambers was under the impression that these toads were natterjacks (*Bufo calamita*), but Mr H. N. Watson of Horowhitu, Palmerston North, who was with Mr Chambers at the time of their arrival, has given me some further information on this interesting importation. Writing to me in February, 1919, he says:

The Toads were sent out by Mr Ralph Arthur, a brother of the late Mrs W. Chambers, who lived, so far as I remember, near Newton Abbot in Devonshire. They were collected by the boys of the neighbourhood, and I think twopence each was given for them. They were the common English Toad and not the Natterjack. They were packed in damp moss in tin boxes with some perforations in the lids, and were kept in the cool chamber of the steamer. There were about one hundred, and only three were dead on arrival; their death was probably caused by rough handling when captured. On arrival they had all paired and were spawning. It was impossible to separate the males from the females; there were a few more of the former than of the latter. The females were all large and reddish; the males were much smaller and greenish. There were masses of spawn in the boxes, and this we collected and put into a basin, but I think it had not been fertilised, for none of it ever showed any signs of germination. We changed the water at intervals and kept it in the sun, but as the month was about April, we should not have been able to rear the tadpoles if any had hatched. We selected five (three females and two males), and the rest of the consignment (about ninety), so far as I remember, were liberated in a swamp at Repongaere Station near Gisborne. They swam away and that was the last seen of them. There were a great many moreporks there, and lots of rats at the station buildings not far away, and I think that these were probably the cause of the disappearance of the Toads, especially as the land surrounding the swamp was eaten bare by horses, so that the toads would find very little cover when they were travelling in search of food at night.

I got a couple of glass-cases made, and in one of these I put a male and a female, which I kept for a couple of years, while Mrs Chambers had the remaining three in the other. They were fed on flies, which were eaten in great numbers. On one occasion we gave a "Maori bug" to one of them, and its disgust after swallowing it was very amusing. Being well fed they frequently changed their skins. Also being kept in the house, they never became torpid, but fed freely all the winter. But they never showed any signs of breeding. Ultimately Mrs Chambers turned her specimens out in the same place as the others; while my pair were allowed to escape during my absence for a short time in 1895.

The discrepancy in numbers between these two accounts is due to the fact that both gentlemen were writing from memory, more than 25 years after the events narrated, and that apparently neither of them noted these events at the time.

Considering the ease with which these animals were carried, and their hardiness in confinement, it is rather remarkable that no other attempts have been made to introduce them into New Zealand.

Chapter VI

FISHES

The classification adopted for the fishes which have been introduced into New Zealand is that of Professor G. A. Boulenger in the *Cambridge Natural History*.

Class *PISCES*

Order TELEOSTEI

Family Clupeidæ

Herring (*Clupea harengus*)

To the popular mind the introduction of the herring into New Zealand waters is the most desirable form of acclimatisation work which could be undertaken, as it is considered that its commercial value to this country would be so great. The success which has attended the introduction of certain species of Salmonidæ has led unthinking persons into the belief that it should be quite easy to introduce other species of desirable fish, and as the majority of people are unthinking, even if they do not come under Carlyle's famous dictum, it is not to be wondered at that the introduction of the herring has been frequently urged. A fairly full report of the efforts which have been made, and of the difficulties which have to be overcome has recently been, or is about to be, published in a Bulletin of the New Zealand Science and Art Board on "The History of the Portobello Marine Fish Hatchery and Biological Station."

It is therefore only necessary here to state the facts and summarise the history of the attempts as briefly as possible. Herring and other soft-scaled fishes cannot be transported alive at any stage of their existence. Handling is generally fatal to them. Therefore the only plan left open is to convey the ova. The eggs are adhesive, and under normal conditions are deposited on stones, gravel and other objects at the bottom of the sea in comparatively shallow water. The eggs hatch in about 16 days, the time being shortened or lengthened according to the temperature of the water; cooling causing retardation. The problem then was in the first place to retard development for a period of at least 50 days. If this difficulty can be satisfactorily overcome, there are several others which have to be met.

The first attempt was made in March, 1886, and was an ill-considered experiment, for no provision whatever existed for dealing with the ova had they arrived at the Colony. The obtaining and shipping of the ova was entrusted to Professor Cossar Ewart of Edinburgh, who obtained between two and three million eggs from the Ballantræ beds off the coast of Ayrshire, Scotland, where the famous Loch Fyne herrings spawn. The ova were placed in large carboys and in wooden boxes with glass slides, which were fixed in stoutly made barrels filled with sea-water, and in this way were conveyed to Plymouth and placed on board the 'Ruapehu.' Professor Ewart had designed a special apparatus so constructed as to preserve through the entire voyage a steady quiet flow of pure sea-water over the eggs at an equable temperture of 33° Fahr. So far the experiment succeeded admirably, but owing to lack of foresight in the cooling apparatus, the pipes which were to supply the chilled water, instead of being surrounded with ice, were led directly through the refrigerating chamber. The result was that the water froze in them, none reached the ova, and by the time Madeira was sighted, all were dead.

The second experiment, also unsuccessful, was made by the Government in 1912. The Portobello (Dunedin) Marine Fish Hatchery was opened in 1904, and early in its history I had looked into the question of the introduction of the herring. The question of the retardation of the hatching of the ova again appeared to be the principal difficulty, and I entered into communication with Dr Fulton, Scientific Superintendent of the Scotch Fishery Board, with the object of getting experiments conducted at the Dunbar Hatchery to test this. Owing to the change from Dunbar to the Bay of Nigg, Aberdeen, and to pressure of other work, the matter was allowed to lapse, and it was not till 1908, this time at the instance of the New Zealand Government, that a series of experiments was commenced by Dr H. C. Williamson. The outcome of these experiments was that retardation of hatching of the ova for a period of 50 days, which we considered was the time required, was only successful to the extent of a small fraction of one per cent. Less than one in ten thousand survived such a long chilling. Both Mr Anderton, Curator of the Portobello Hatchery, and I considered that under the circumstances it would be a waste of public money to proceed further with the attempt to introduce the herring by this means. However the Government decided to make the attempt, and Mr Anderton was sent to the Old Country in 1912, to carry it out, while at the same time he took charge of a large shipment of turbot, lobsters and crabs. An ingenious apparatus for the conveyance of the eggs, and for the cooling and aeration of the water, was placed on board the Shaw, Savill & Albion Co.'s S.S. 'Waimana,' and was thoroughly tested beforehand. The vessel called in at Plymouth, where Dr Williamson had secured about 60,000 ova adhering to glass plates. These were duly placed in the apparatus, and the voyage was commenced on 12th January. A fairly uniform temperature of 35°·5 Fahr. was maintained. On 24th January when the equator was crossed the ova were fairly clean, and the outline of the embryo could be easily distinguished. Some dirty water got into the boxes at Cape Town, and the sediment was removed from the eggs by means of a camel hair brush. By this time the chord and eyes were visible in all the live eggs. On the 6th February the plates were still in very fair condition, and then a mass of rust and sediment was forced through the pipes, and the eggs were thickly coated with sediment. The experiment was abandoned on 14th February when all the ova were dead owing to the state of the water. The ova were fertilised on 10th January. The majority contained live embryos on 6th February, 27 days after fertilisation, and some still contained live embryos on 12th February, 33 days after being fertilised. None of the ova hatched out.

The full report is worth studying, and it conveys a good idea of some of the difficulties encountered in this department of acclimatisation work.

Family SALMONIDÆ

The rivers and lakes of New Zealand contained originally a poor and rather sparse fish-fauna. It consisted of the grayling (*Prototroctes oxyrhynchus*), found mostly in clear rapid rivers, and a fine sporting fish; the smelt (*Reptropinna richardsoni*), common in rivers and lakes; several species of *Galaxias*, a mud-fish (*Neochanna apoda*), only found in the west coast rivers of both islands, two species of eel (*Anguilla*) and a lamprey (*Geotria chilensis*).

The kokopu, a name corrupted in the south of the South Island to cock-a-bully (*Galaxias kokopu*), was sometimes popularly called trout; it is a fat, sluggish fish which lurks under logs and stones, furnishes no sport, and is not particularly good to eat. The fish known as the minnow (*Galaxias attenuatus*) is, as its name implies, a small fish. According to the late Professor Powell, "White-bait is the fry of this species," but the facts want working out[1].

The common eel (*Anguilla aucklandii*) is always with us, and is a very valuable food-fish, if people only knew it. The lamprey makes an annual visitation up the rivers in the spring months, usually about October.

[1] In an interesting article on "Some Trout Fishing in New Zealand," which appeared in *Blackwood's Magazine* for March, 1918, pp. 365–77, Mr A. R. Chaytor states that New Zealand white-bait is the larval stage of the common eel. Anyone who has looked into the question of the development of the Anguillidæ knows that this is a quite mistaken idea.

Though there are altogether about a dozen species of indigenous fresh-water fish, yet to the early settlers of the colony, the rivers and streams seemed singularly empty. There was no sport for the angler, unless he happened to live near a stream where the grayling abounded, and except eels—which were abundant—there was practically nothing of an edible character to be found. It is not to be wondered at, therefore, that the minds of the colonists early turned to the idea of importing such fish as they knew would provide both sport and food. The success of attempts to introduce species of Salmonidæ into Tasmania and Victoria encouraged the hope that it would be possible to do the same in this colony, and in the sixties and seventies a systematic importation was commenced, not only by the Government, but by all the principal acclimatisation societies. Over a dozen species, or, counting varieties, about 17 kinds of Salmonidæ have been introduced into this country. The success of several of these has been phenomenal, but the failure of others to establish themselves has been inexplicable. The majority of the streams and lakes of both islands are now stocked with species of Salmonidæ, and an interesting problem has arisen, namely, are the species remaining distinct, or is there a tendency among the allied forms to hybridise and produce a generalised type? Another interesting problem faces both the angler and the naturalist. In a number of streams which are now heavily stocked with trout, the native aquatic fauna has been nearly exterminated, and the question of future food supply has been raised. Are new species of animals to be introduced, or can any method of renewing the indigenous fauna be devised?

I have dealt with the particulars of the introduction of the various species in some detail, but here I propose first to give some facts regarding the general question, and the early attempts to bring Salmonidæ into Australia and New Zealand. These show the difficulties which had to be faced in the early days, when transport was by sailing ships, and refrigeration was unknown. Mr J. Murdoch contributed an article, I think to *The Field*, but I have failed to find the date, on "The Introduction of Trout into Australia and New Zealand," from which the following facts are gleaned:

The first attempt to introduce ova into Australia was made in 1852 by Mr Borcius, who however failed, and lost £300 in the experiment.

In 1854 Mr Youl began to study the subject, and made many experiments. No hope of success was held out by the most experienced pisciculturists, and Mr Youl was told that he might as well try to fetch England to Australia as to carry spawn to it in moss. Mr Edward Wilson, President of the Victorian Acclimatisation Society, associated with Mr Youl and some influential colonists in obtaining £600 by subscription. Thirty thousand ova were shipped in the 'Sarah Curling' on 25th February, 1860. The ice-house consisted of two rooms, one within the other, lined with lead; the space between was filled with powdered charcoal; a filled water-tank over the ice-house with a pipe leading into it allowed a gentle and continuous stream of water to pass over the ova as they lay in swing trays. The passage was long, the 15 tons of ice gave out, and the last of the ova was found to be dead when the ship was 68 days out.

Mr Youl then visited the fish-breeding establishments of Scotland and Ireland. He also made a series of experiments to test the vitality of ova at a low temperature. The experiments proved three, heretofore, unknown facts. First, that a continuous stream of water is not essential to the preservation of vitality; secondly, that partial deprivation of air is not fatal; and thirdly, that light is not essential. After these experiments he felt assured of success if a sufficient supply of ice could be preserved throughout the voyage.

In January, 1864, he again tried. Boxes were made (of inch pine) measuring twelve inches by eight inches by five inches, with perforated top, bottom and sides. At the bottom was first spread a layer of charcoal, next a layer of ice, then a nest of carefully washed moss, and on this spring cushion were deposited the ova. Over them was laid a covering of moss, then a double handful of broken ice, and the whole was saturated with iced water and screwed down. One hundred and eighty-nine boxes, containing 100,000 salmon and 3000 trout ova, were packed closely on the floor of the ice-house, and upon them were piled blocks of ice to the height of nine feet. The 'Norfolk' sailed on 21st January, 1864, and arrived in Melbourne on 15th April. The State of Victoria retained 4000 salmon ova, of which it is said 400 were hatched. The remainder were sent to Tasmania by a Government steamer. They were taken to the Derwent River, and placed in the hatchery provided. Mr Ramsbottom estimated that there were 30,000 salmon and 500 trout ova living. On 4th May the first trout was hatched, on the next day the first salmon, and by 25th May there were 300 trout and 700 salmon. At the end of 1865 the surviving salmon were allowed to enter the sea.

Of the 300 trout many died; about 30 were liberated in the River Plenty, while only six pairs reached maturity and spawned in the ponds. Their progeny have been liberated in many rivers and streams of Tasmania, Victoria, and New Zealand.

In 1866 Mr Youl brought out 87,000 salmon, 15,000 salmon-trout, and 500 brown-trout ova. The result of this shipment was 6000 salmon and 900 salmon-trout being hatched from the 30,000 living ova which arrived. Other shipments followed, including brook-trout.

These early Australian experiments are interesting as showing the difficulties encountered, and, in the case of brown-trout, the remarkable success which ultimately attended the first introduction into Tasmania.

The question of the probable merging of various species into one generalised form is an interesting one. Those who have discussed it in the colonies have not perhaps very clear ideas as to what constitute specific distinctions. Indeed one might go further and say that very few biologists have yet attained to clear views on this subject. Most species are founded on structural characters, some of which appear to be very mobile, while physiological characters—which may be and often are the dominant factors in differentiation—are extremely difficult to estimate. Thus *Salmo salar* may be considered to be a distinct species, structurally and physiologically, and though healthy hybrids are readily produced with *S. trutta* and *S. fario*, there is no proof that these hybrids are fertile. Indeed the question does not seem to have been tested. On the other hand it seems to me doubtful whether *S. trutta* and *S. fario* are distinct species or only varieties of one somewhat variable species. They cross freely, and it is only in reference to this crossing that there has arisen the question of a generalised type occurring in New Zealand. There is no question of the probable crossing of species of *Salmo* with those of *Salvelinus* or *Onchorhynchus*, these latter genera belonging to totally distinct types; and while their species may be fertile within generic limits, they are not likely to hybridise at any time with species of *Salmo*.

Mr A. J. Rutherford considers that the commingling of so many species and varieties of the genus *Salmo* in our streams will result in the establishment of one generalised type, which, with some disregard of the laws of scientific nomenclature and publication, he calls *Salmo trutta novæ-zealandiæ*. I quote the following passages from a Memorandum (undated) which he furnished to the Otago Society.

Though these islands lie in the latitude of the Levant, Italy and Switzerland, with widely different ocean surroundings, we introduced various forms of Salmonidæ from higher latitudes in the Northern Hemisphere, such as the Loch Leven Trout, Scotch Burn Trout, Sea Trout, etc., and fondly imagined that these varieties would retain their characteristics in their altered environment. The fallacy of this idea has been abundantly proved, and after making some slight study of the forms of trout found in the North Italian Lakes, such as Como and Maggiore, I came to the conclusion that they are almost identical with the form of *S. trutta novæ-zealandiæ* found in Wakatipu, Wanaka and in South Alpine New Zealand Lakes. The trout in our rivers are also much more akin to the Italian and Swiss varieties than to the British types, and in my opinion it matters little whether you liberate a Loch Leven trout, a Scotch burn trout, or an Italian trout. He soon loses his identity and becomes a distinctive type (*novæ-zealandiæ*), a quasi-Italian-antipodean variety. I think there is inherent in most of the forms of Salmonidæ a wandering habit, and that in search of a wider range, health and better food, they soon learn the advantages of sea-going habits, some varieties more readily than other. The border line between sea trout and brown trout is very fine in this country and depends really on environment.

In several hatcheries in this country hybridising various forms of Salmonidæ has been purposely carried out, and the fry seem just as healthy as those of pure-bred varieties. For example, at the Opoho hatchery (Dunedin) some 700 hybrids, produced by fertilising brown-trout ova with milt from a male salmon (*S. salar*), were reared, of which 650 were placed in the Waitati, a small stream a few miles to the north of Dunedin, and 50 were retained in the ponds. The former disappeared, the latter—which were strong and lively fish—were watched for a few months, and then their identity apparently was lost. There was no further record of them. This sort of experiment is futile, unless carried out with the definite object of finding out whether such fish grow to maturity, and are able to perpetuate their hybrid characters, or whether naturally reared fish taken in our streams resemble them. No systematic experiments, such as have been carried out at the Howietown (Scotland) fish hatchery, have ever been conducted in New Zealand. Until careful and exhaustive work in this direction has been undertaken, it is impossible to speak with certainty as to the natural crossing of all the imported varieties in this country, and the emergence of a generalised type.

Fish have been repeatedly taken in our streams which have puzzled all parties in the colony, and they have been submitted in many cases to expert opinion in the Old Country. I give a few examples to show how difficult the question is. Regarding a fish which was taken in Nelson Harbour in 1881, Sir James (then Dr) Hector said:

a careful examination shows that it must be classed as a true sea- or salmon-trout, although, as has been found invariably to be the case in Otago specimens, it presents a certain admixture of the characters of the many species into which the sea-trouts from the various rivers in Europe have been subdivided.

In 1889 the Canterbury Society forwarded a fish to Dr A. Günther, who reported on it as follows:

(1) The fish is most decidedly not a salmon. You can always distinguish a salmon by the large scales on the tail; this is an invariable characteristic. Your specimen has *thirteen* to fourteen scales between the adipose fin and the lateral line; a salmon has *eleven*, very rarely *twelve*. The maxillary

is much too long for a salmon of this age. (2) Your specimen is a migratory trout. If it had been caught in England or Scotland, I should not have hesitated to call it a *Salmo trutta*, a sea-trout. (3) Your specimen is not a common trout (*Salmo fario*), not having the vomerine teeth in a double row, and differing from it in usual minor points. (4) There is one other probability, that it is a cross between a salmon parr and *S. fario*. You, knowing the history of the introduction of the salmonoids into the Selwyn River as you do, will be able to judge whether such an assumption is possible. I am unable to distinguish between these crosses and the migratory trout; I have seen some bred in captivity. Whatever the fish may be, one thing you may rely upon, that it is not a young salmon. Also recollect that sometimes the common *Salmo fario* wanders into salt water, and then assumes a silvery and very deciduous coat.

In 1891 Mr Tanner, Hon. Secretary of the Southland Society, sent home three fish from the Aparima River, which were supposed to be from the salmon fry liberated in that stream in February–March of 1890. They were from one and a half to two pounds in weight. They were submitted to Dr Günther, who reported on them as follows:

These specimens are most assuredly not salmon (*Salmo salar*), neither are they the brook trout (*S. fario*). They are a kind of sea-trout (*S. trutta*), looking extremely like the Irish White trout. But the different kinds of migratory sea-trout are so closely allied to each other, that it is almost a matter of impossibility to give an opinion on artificially reared fish, or their offspring.

The Field of 9th January, 1892, commenting on this statement says: "This leaves the question precisely where it was, and will confirm the opinion of those who insist that the acclimatised trout in Tasmanian and New Zealand waters acquire a distinct character of their own."

On 12th August, 1893, *The Field* published a letter from Mr Tanner in which he says:

It is remarkable how the trout in New Zealand assume the habits and aspects of sea-trout, whenever they are found towards the mouths of rivers near the sea. They first make for the estuaries, then for the open sea, and have been traced for more than twenty miles along the coast. Our trout ova were obtained from Tasmania, which, we believe, was supplied from the tributaries of the Thames in England. Is the explanation possibly that the Thames trout was originally a migratory fish, but, being prevented from going to sea, lost its migratory instinct, which has been recovered in this country by the descendants under new and favourable conditions? If this is so, it is a very interesting fact. Generally the trout in the Oreti resemble most the white trout of Ireland.

Sir James Hector made the same suggestion at a meeting of the Wellington Society held in 1896, when he said:

From enquiries he had made, he had satisfied himself that these fish were descendants of sea-going trout which had once been in the habit of issuing out of the mouth of the Thames, but which were prevented, by the state of that river of late years, from doing so.

The problem is very interesting from the point of view of the naturalist. All that I can do in the way of elucidating it is to give as full details as I have been able to obtain of the various species and varieties which have been introduced into the country, and of the dates and amounts of these introductions, so that any future investigators may be furnished with a record of facts.

Atlantic Salmon (*Salmo salar*)

For nearly half a century attempts have been made to naturalise this fish in New Zealand waters, and untold sums of money have been expended in the undertakings, as the following record shows, indirectly. Fish have been hatched by the million, and liberated in a great number of the rivers both of the South and North Islands. Glacial streams, rivers from the great lakes, rivers from the Canterbury mountains, rapid streams, sluggish streams—all have been tried. In several cases the same river has been stocked with young fish for many years in succession. In many cases salmon have been reared from the egg, have been kept in confinement till they spawned, and their fry have been liberated—always in the same stream—for a succession of years, by the hundred thousand. The fish have grown well to a certain age in our waters and have then gone to sea in a normal manner, just as they do in European streams, but from that point they are lost. With the exception of two identifications recorded below, not a single authentic instance has been recorded of their return from the sea to the rivers. The fish has absolutely failed to establish itself. Our record is the same as that of Tasmania. As W. Saville Kent, the Queensland Commissioner of Fisheries, said in 1872:

> the attempts to stock Tasmanian streams with the true salmon have utterly failed. The young fish have thriven magnificently until their departure for the sea as smolts, at which stage they have simply vanished from human sight, the warm seas of the South being too enervating for them.

It is extremely difficult to suggest any explanation of the facts. Mr Kent's explanation is almost certainly wrong. It does not apply at all to the south and east coast of the South Island of New Zealand, the region where most of the salmon have been liberated. A southerly current sets up this coast to the east of Stewart Island and Otago, and the temperature of the sea at all seasons of the year is lower than that of the seas round Britain and Ireland, or on the west coast of Norway.

It is almost inconceivable that they have perished at sea. Other species of Salmonidæ thrive in the sea and grow to a great size, periodically returning to the rivers to spawn. It has been suggested that the fish has changed its habits and that it spawns at sea, but there is not a trace of evidence in favour of such an improbable theory. It has also been suggested that the fish migrate to other shores, but if so—where?

What has been wanted all along in this work of acclimatisation of new species in New Zealand has been some sort of scientific supervision and co-operation. Every centre and society went on its own way, independent, as a rule, of every other. There never has been in the country an organised fishery department. The result has been waste of money and effort right along the line. Had experiments in fish-marking been carried out systematically from the commencement of operations, it is probable that ere this we would have been in possession of information as to our missing salmon. Until some such regular work is undertaken the subject will remain a mystery. It may not be solved even after fish-marking has been undertaken for years, but the strong probabilities are that light would be thrown on the problem. Up to the present nothing has been done to trace the fish when they go to sea.

The following extract from a letter written by Mr Youl to the Superintendent of Otago before any salmon were introduced, is not only very interesting, but it may be in part the explanation of the problem. Unfortunately I cannot find the exact date:

> May I beg of you on no account to permit the Brown Trout to be introduced into the Molyneux or any of its tributaries, until you have got the salmon fairly established in them. They are the greatest enemies the Salmon can have. I can compare them to nothing, but wolves in a flock of sheep. Again and again I have warned Dr Officer, of Tasmania, of the danger of admitting these voracious fish into any stream suitable for Salmon before the Salmon are established therein. I am sorry to observe that so many of the Provinces of New Zealand have introduced these Brown Trout before they have got the salmon. Depend upon it, for every £10 spent by these Provinces in this way, they will, in those rivers where they have placed them, have to spend £100 to successfully introduce the king of fishes.

Introduction. The first attempt at the acclimatisation of the salmon in New Zealand was made by Mr A. M. Johnson, who put 600 young fish on board the 'British Empire' bound from London to Canterbury in 1864. Snails, water-lilies and weeds of various kinds were placed in the tanks; contrivances for aerating the water were provided; the tanks were provided with a frame-work case, with double cane matting, which was kept constantly wet throughout the tropics in order to keep up evaporation and lower the temperature.

In spite of all the care exercised, however, the experiment was unsuccessful.

The next attempt was made by the Provincial Government of Otago which took action in 1867, with the result that 100,000 ova were placed on board the 'Celestial Queen' in January, 1868. Most of these were from the River Tay, but some came from the Severn. The ship reached Port Chalmers on 4th May, and the 300 boxes of eggs were at once sent down to the Waiwera Ponds at Kaihiku, which had been specially prepared for them. The Waiwera is a tributary of the Clutha. In 77 of the boxes all the eggs were dead, but out of the remaining 223 boxes about 8000 healthy eggs were taken. On 20th May a flood filled the boxes with mud covering the eggs just as hatching seemed about to begin. On 28th May some began to hatch out, and in a few days, Mr Dawbin, the curator, had between 500 and 600 young fish. In his report he says:

"In about ten or twelve weeks all of the fish had made their way into the tank at the end of the shed and thence dropped into the feeding pond. It was some time before they began to show again, and although I supplied them with plenty of food I am inclined to believe that they mostly lived on the natural food in the water. It was not very long before two or three began to appear round the edge of the pond, grown wonderfully; and after a time, by regular feeding, I could collect little mobs of them at every corner of the pond. I have now fish five, six, and, I believe, one or two seven or eight inches long, and thick in proportion, which in a few months will be ready to go to sea" (this was written on 31st May, 1869), "and if they return safely it might be possible to get ova from them and hatch fish enough to stock a river well. I have only seen three dead since they went into the ponds, which, however, are so large and deep that I am not able to keep so accurate an account of them as I should like. The size and depth of the ponds, however, is advantageous in this respect, that the fish are kept supplied with abundance of natural food. The water of the Waiwera, like, I have no doubt, that of all the New Zealand rivers, seems to be admirably adapted for salmon, but the river itself I do not consider good for breeding. The bottom is rocky, and although there are said to be gravel banks high up, I should be afraid of the floods, which here occur in the winter and spring. The supply of food in the river is most abundant."

A small lot of these ova (ten boxes) were hatched out in Mr Duncan's ponds on the Leith, just about the mill, but I can find no record of what was done with the fry. They probably escaped into the Otago Harbour[1].

[1] "In an article in the *Field* of Jan. 26, 1878, we are told that there were 500 young salmon fry in the ponds out of the 'Celestial Queen' shipment. When they were fifteen months old, and ranged from 12 to 15 in. long, the services of Mr Dawbin were dispensed with by the commissioners, they having appointed a gentleman who seems to have had some influence with the Government, and on whose land the ponds were situated, but who was totally ignorant of the treatment the

The second shipment came to Port Chalmers in April, 1869, in the 'Mindora.' (In the Southland Society's pamphlet, *Acclimatisation in Southland*, published in 1915, this vessel is called the 'Minerva.') The passage occupied 133 days, and the ova never hatched out. Canterbury received 700 ova, Southland 7000, while 100,000 were retained in Otago. There was some hope of them at first, for Mr Dawbin in the letter just quoted from stated that he had about 2000 good eggs on 31st May, but nothing came of them.

In 1871 the Southland Society obtained about 3000 ova from Mr Frank Buckland in furtherance of an experiment. The eggs were packed in bottles surrounded by saw-dust and presumably by ice, and were despatched by sailing ship to Melbourne; they were delayed in transit to Southland, and none hatched out.

In 1871 the Auckland Society made an attempt to introduce salmon from England *via* San Francisco. The experiment failed in consequence of the long detention on the Pacific Railway, and from lack of attention to the ova. The transit occupied 100 days, instead of less than 50, as was anticipated. The ova were presented by the Duke of Northumberland.

In 1873 the 'Oberon' brought 120,000 salmon ova to Port Chalmers, of which 95,000 were sent down to the Makarewa ponds, Southland, and the remaining 25,000 went to Canterbury. Between the packing of the ova and the unpacking at the Southland ponds, 114 days elapsed, and 85,000 were dead. The remainder were placed in the hatching boxes, but only 300 fry hatched out. Most of these died, and the remaining 96 were removed when one year old to a pond near the Aparima River. In April, 1875, they were about seven inches long and healthy; in June they were carried by a flood into the river.

fish would need. Mr Dawbin's offer to continue his services gratuitously for a term of six months was refused, and he was instructed to hand over his charge to the new-comer. This was too much for one who had devoted his time day and night for fifteen months to the care of the fish. The new-comer's incapacity would almost inevitably have resulted in their destruction; or, if this had not happened, he would have claimed whatever success might accrue. Impressed with the con-conviction that he was doing the best thing possible in the circumstances for the colony, Mr Dawbin chose a night when a slight fresh was coming down, opened the gratings, and allowed the prisoners to escape into the river. It is not our province to defend Mr Dawbin, but we would ask the commissioners why the circumstances which led up to this are suppressed in their reports, and the colonists whether they approve of the arbitrary substitution of an inexperienced manager for one who had abundantly proved his ability and deserved public confidence? Since the above events the magnificent breeding ponds on the Waiwera have gone to ruin, as we are informed." (Arthur Nicols, *Acclimatisation of the Salmonidæ at the Antipodes*, p. 49.)

At p. 87 of the same work, Mr Nicols states: "About the middle of 1874 a salmon grilse weighing more than three pounds, was taken in the river Molyneux, no doubt the offspring of a pair of the 500 smolts liberated in that river in 1869 by Mr Dawbin."

Of those which went to Canterbury, only 38 fish were obtained. Some were kept in the ponds and appear to have been lost. With others an experiment was tried.

A large cage was made, which was anchored in the River Avon a little below Victoria Bridge; in the cage were placed on 3rd Nov. eleven of the largest salmon. They remained there sixteen days, during which time they throve well. On 19th Nov. the cage was raised and floated down to a spot in the Avon below New Brighton, where at high tide the water is brackish. The Garden Committee have three times visited the spot. On the last occasion (24th Dec.) the cage was raised and the eleven fish examined; they were in good health and had increased in size considerably. It was calculated that one of them was a foot in length.

The reports of the Society do not contain another word about this experiment, and this is characteristic of the isolated and discontinuous manner in which most of the societies work. Lack of continuity of effort has nullified many of their experiments. In 1875 the 'Timaru' brought 300,000 ova to the Bluff, after a passage of 105 days, but many of the ova had been collected 30 days before the vessel sailed. No fish hatched out. According to the report of the Otago Society, Mr Howard of the Wallacetown Ponds liberated (on their behalf?) 1400 young fish in the Aparima River, but I cannot find a definite record of this, nor is it quite clear what lot of eggs they came from.

In 1876 the 'City of Durham' brought to Melbourne 90,000 salmon ova, which were transhipped, mostly to the Bluff, a few boxes going to Canterbury. The Southland ova were placed in the hatching boxes 69 days after sailing, and about 87 days after having been taken from the parent fish. From 25,000 to 30,000 were apparently healthy, but of those it appeared that about two-thirds were not fecundated. In October about 1500 fry were liberated in the Aparima River. It is quite possible that this is the lot referred to in the preceding paragraph.

The Canterbury Society state that only 175 fry hatched out of the boxes which went to Christchurch, but in the following year they placed 181 fish in the Ashley River; and in 1878, 240 were placed in the Heathcote River. These were seen afterwards, and were from 12 to 14 inches in length. There is no record as to where these fish came from.

In 1878 the 'Chimborazo' brought 45,000 ova to Melbourne. These were transhipped and reached Invercargill on 19th March, commencing to hatch out on 4th April. Altogether about 2500 fry hatched out, of which 1700 were placed in the Aparima. Again the Otago Society report that 2500 fry were placed in the Aparima

River in 1878; the Southland figure is 1700; the one reports the total number hatched, the other the number liberated.

Regarding these shipments Mr W. Arthur, then Hon. Sec. of the Otago Society, wrote in 1880:

Mr Howard has informed me that the ova of the salmon turned out in the Aparima in 1874, 1876 and 1878, came originally from the Tweed, Tyne, Ribble, Hodder, Lune, Avon and Dart Rivers. Yet who can say from which of these rivers the ova were taken which eventually hatched at the Wallacetown ponds?

In 1881 Mr C. C. Capel of Footscray, Kent, sent out 100,000 ova to the joint order of the Otago and Marlborough Societies; but all the eggs were dead on arrival.

In 1883–84 Sir F. D. Bell, the Agent-General for New Zealand, and Sir James Maitland, the eminent pisciculturist, were working together to send out salmon ova to the colony, and a shipment, of which I cannot find particulars, was forwarded in 1884. Sir Francis writing to the Colonial Secretary on 30th October, 1885, says:

I had taken the greatest pains all through 1883 and 1884 to interest many people in this country, eminent for skill and experience in pisciculture, about sending ova to the colony on the supposition that the spasmodic experiments which had been going on for so many years were to be superseded at last by a systematic and persistent action on the part of the Government itself, extending over some seasons at any rate. The first experiment of sending out ova in a "moist-air chamber" and at a regulated temperature was made in the steamship 'Ionic' in January, 1884; and it is hardly open to doubt that this method was not only in itself a right one, but, in fact, the best that had till then been devised. Further experience, however, had shown that the first expense of that method would not have to be repeated. A shipment of trout ova privately made by Sir James Maitland had brought out most valuable information, showing how cheaply as well as safely ova could be got out under certain conditions; and when the reports came home of our shipment by the 'Ionic,' Sir James Maitland wrote to me that he had no doubt whatever of "perfect success next season, as we had now the key to the whole problem, namely, the period which ought to lapse between spawning and packing, and could insure the success of every egg we sent."

Unfortunately neither the Government of New Zealand nor those interested in the subject in the colony were informed of the exertions which Sir F. D. Bell and Sir James Maitland were making in Britain.

Meanwhile there was a good deal of dissatisfaction felt at the poor results of all the attempts hitherto made to introduce salmon, and at the request of several societies, Mr S. C. Farr, Hon. Sec. of the North Canterbury Society, went to Britain in December, 1884, and succeeded in getting 198,000 from the Tweed. Mr Farr had previously complained that not ten per cent. of the eggs hitherto received had been fertilised, so he took care to see that this lot was fully fecundated. Forty-four days after sailing, and 90 days after being taken from the parent fish, the ova were landed at Lyttelton, when 117,000 eggs were found to be alive.

Two cases went to the Napier and Wellington Societies, one each to Otago and Waitaki, and two to Canterbury. I have failed to find what results were achieved by the Napier and Waitaki Societies. The Wellington Society liberated 4600 fry in the Hutt River; the Otago Society liberated 3900, presumably in the Waiwera; while in the Canterbury Society's ponds about 21,000 fry were hatched out. Of this number, more than 3000 were born with curved spines, and about 6000 more were sickly and attenuated; these soon died. Owing to the long drought and the high temperature of the water during the season the number was reduced to about 7000. Of these, 1000 were liberated on 23rd December, 1885, in the Temuka River; and on 30th December 1000 in the Opihi. They varied in length from three to five inches. On 6th January Mr Farr started with 1000 parr for the Clarence River, about 105 miles from Christchurch, but owing to the excessive heat in the carriage (96° F.), 178 died *en route* and others sickened. So on arrival at the Perceval River, the sickly ones were picked out and put in there, where they soon revived and went into a deep pool. The remaining 725 were turned into the Clarence River. On 25th January 500 were liberated in the Ashburton River; on 2nd February 1000 in the Rangitata; on 23rd February 1000 in the Hurunui; and on 7th May 200 into the Selwyn. The Society evidently were under the impression that by scattering the fry over so many streams the chance of survival in one or more of these was increased.

It was claimed that between 1868 and 1878 no less than 824,000 (I think the correct number is 772,000) ova were shipped from Britain and that only 3996 fry survived to be turned out; while of Mr Farr's shipment of 198,000 nearly 120,000 arrived, and about one-half of that number were liberated. It is interesting to note at this point that Mr Farr also tried to bring live fish out to the colony. He brought 50 parr in a tank of fresh-water, and these were carried safely till near the Cape, when some one tampered with the water-supply, and all the fish perished. Up to this date, as was stated in a letter addressed by the Waitaki County Society (Dr H. A. de Lautour, President) to Sir Julius Vogel, the Commissioner of Trade and Customs, the efforts at acclimatisation were running in at least four different channels, viz.: (1) efforts by Government, conjointly with the societies; (2) efforts by the Government independent of the societies; (3) efforts by the societies themselves, conjointly and independently; (4) efforts by the Agent-General independent alike of the societies or the Government. The Government now took the matter in hand themselves as far as importation was concerned, and prepared for shipments on a larger scale, utilising the societies for the hatching of the ova and the distribution with the colony. In January, 1886, the Agent-General shipped over 200,000 ova (most of which came from the River Tay) by the 'Ionic,' which arrived in Wellington on 21st March, and they were distributed to the various societies on the following day. I have been able to follow up those sent to the Wellington, Nelson, North Canterbury, Otago, Southland, and Lakes Societies. The Wellington Society received one box, from which about 8000 fry hatched out. These were liberated in the Hutt and Manawatu Rivers, and a small lot in the Ruamahanga. The Nelson Society hatched out approximately 12,000 fry, but I do not know the subsequent history of this lot. The Canterbury Society received four boxes with approximately 22,000 eggs in each. In one box which was outside the cold chamber, all the eggs were dead. From the others about 41,000 fry, many of them rather small, were hatched out. At the end of the first month there were 37,000, but at the end of the second only 11,000 were left alive. Of these 4000 were sent to the Waitaki Society, 1500 to South Canterbury, 1500 to Geraldine, 1000 to the Taranaki Society, 1000 to the Hawera Society, and 300 were placed in the Selwyn River.

During this year reliable reports were received of young salmon having been repeatedly seen in the Opihi, and some that had been accidently taken measured from 9 to 12 inches in length. The Society kept a considerable number in their ponds, but in May a flood washed 400 parr into the Avon, where they appeared to be doing well.

The Otago Society got 12,000 fry from their consignment, of which number they liberated 9000, presumably in the Waiwera.

The Southland Society hatched out over 12,600 fry, of which a little over 9000 were alive in June. In July 5500 were turned into the Aparima, while owing to a sudden melting of snow over 3000 more escaped into the river.

The Lakes Society hatched out about 14,000 fry which were liberated in Lake Wakatipu.

In 1887 three large shipments of ova arrived in Wellington. The 'Kaikoura' with 160,000 from the Tay, Forth and Tweed, and the 'Doric' with 330,000 from the same rivers arrived in January, and the 'Tongariro' in February with 120,000 ova from the Tweed, and 100,000 from the Rhine.

(*a*) Of the 'Kaikoura' shipment, 120,000 went to the Clinton Hatchery (Otago), and 40,000 to Southland. Of the six boxes which went to Clinton, one contained "Forth" ova, and the others "Tay" ova; they averaged about 50 per cent. good eggs, and hatched splendidly. The Southland lot hatched out well and yielded nearly 15,000 fry.

(*b*) Of the 'Doric' shipment, ten boxes were handed over to the Otago Society for the Clinton Hatchery, and about 70 per cent. of the eggs were good. Five boxes went to Southland, containing nearly 66,000 good eggs, and from them nearly 52,000 fry were hatched. One box went to Oamaru and from it between 4000 and 5000 fry were hatched out.

(*c*) The 'Tongariro' shipment did not turn out so well. Five and a half boxes went to Opoho (Dunedin) and yielded about 65 per cent. of good eggs. One box with 9000 ova went to Oamaru, and yielded between 3000 and 4000 fry. The Canterbury Society received nominally 50,000 Rhine salmon, but only 3620 good eggs were found, from which 3250 fry were hatched out. The Wellington Society also received the same quantity of Rhine salmon ova, and about 5000 fry were hatched out. Some of these were sent down later to the Opoho ponds.

Out of the 430,000 eggs received by the Otago Society, Mr Deans, the curator, calculated that 270,000 fry were hatched out, but the Society's report for 1888 states that only 270,000 eggs were good on arrival.

They hatched out about the beginning of April. During September, October and November, 98,000 young fry were liberated in the upper waters of the Aparima. As the Southland Society have turned out about 60,000 in the same neighbourhood, we trust that at last the acclimatisation of the salmon may prove to have been accomplished.

Faith was still strong in the minds of all interested in this work that acclimatisation of salmon would soon be accomplished.

The Canterbury Society report of 1887 states:

250 fish of the 1886 hatching were liberated in the Lower Selwyn in February, and in September 1000 parr. These have been turned into that river with a view of establishing what are termed in America "land-locked" salmon; and having the advantage of this river flowing into Lake Ellesmere, the water of which is brackish, we have great hopes of success. It is two and a half years since the first Salmon was turned into the Lower Selwyn by your secretary, and that they have bred there is verified by Dr Anderson of Sydenham, capturing two when netting for live bait. They were returned to the water, and the doctor reported the fact to your secretary. Now, as none of these fish when liberated were less than three inches in length, and those taken by the doctor were less than two inches, this is, we think, sufficient proof that they have been hatched in that river. This evidence is matter for congratulation; and the desire of your council is that the

Government should next session bring in a bill to prevent the wholesale netting in that lake for a few years in order that this noble fish may be thoroughly established. In three years the fish will not only be fully developed, but established in such numbers as to remove all doubts about the experiment.

This paragraph is worth reproducing, because it shows the readiness with which enthusiasts jumped to conclusions on very meagre evidence, and also advocated a course of action which would deprive the public of a plentiful supply of flounders for some years, and cut off several fishermen from their means of livelihood.

In this same year (1887) the Taranaki Society received 810 salmon fry from Christchurch, and liberated them in three streams in the Mt Egmont district.

In the following year the Waitaki Society liberated 150 smolts, and between 4000 and 5000 (Tay) fry in the Ferry Creek, a tributary of the Waitaki; and 200 (Tweed) smolts in the Kakanui River.

In January, 1889, the 'Arawa' left London with 150,000 ova from the River Forth, and reached Wellington early in March. Three boxes with (nominally) 53,300 ova went to Southland, and yielded 38,000 good ova. Presumably the fry were liberated in the Aparima, but there is no record. Five boxes with 97,000 ova went to the Clinton Hatchery, and contained about 95 per cent. of good eggs, which hatched out very well.

In February, 1889, the 'Aorangi' left with 483,000 ova, and reached Wellington on 24th March and Port Chalmers on the 29th. These were taken from Tweed salmon. Ten boxes with 170,000 ova went to the Opoho Hatchery, ten with 182,000 to Clinton, and seven with 128,000 to Invercargill. Mr Deans estimated that from the two Otago lots about 320,000 young fry were hatched out. Of this number some 250,000 were liberated in the Aparima.

In January of this same year, Mr J. B. Basstian of Dunrobin reported having seen young salmon in the Aparima. On 1st March Ranger Burt saw a number in the same river both in the smolt and parr stage, evidently waiting for the first fresh in the river to go down to the sea. Two good specimens were obtained, and the prohibition of netting in the Riverton estuary, which had been enforced when fry were first put into the river, was renewed.

In 1891 the Otago Society report that "nothing has been seen of these fish after their return from the sea. The estuary of the Aparima has been netted several times, but without success."

In 1892 it is stated that:

periodical trials with the seine-net have been made in the estuary of the Aparima to ascertain whether the salmon liberated in 1887 and 1889 were preparing to ascend the river for spawning. Numbers of small fish ranging from one-half to two pounds were caught. Some were sent to Dr Günther, who pronounced them to be Sea Trout.

In an article on the Aparima experiment in the *Southland News* of 20th January, 1891, it is stated: "The first lot of Salmon was put into the Aparima in 1873, and the last in March, 1890, the total being 494,000." It was pointed out to the Marine Department that the people in the district wanted the river made available for anglers. The reply received was that "the Government were determined to keep the river closed, until it had been definitely ascertained that the salmon experiment was a failure."

From time to time reports of salmon being met with were received. Thus in *The Field* of 19th December, 1896, a letter appeared regarding a fish which was caught near Oamaru Harbour and was sent to the editor by the Waitaki Acclimatisation Society. In this he says: "This was submitted to Mr A. Boulenger of the Natural History Museum, and stated by him to be a true *Salmo salar*. It was between 8 lbs. and 9 lbs. in weight." The angling editor of *The Field* commenting on the fish said:

I have examined dozens and scores of frozen fish sent from New Zealand as Salmon, which, in spite of their size and appearance, were trout of some kind. This is the first of these forwarded specimens which has been to my mind thoroughly satisfactory, and which, from a superficial inspection, would lead me to say with confidence,—"This is a salmon."

Somewhat later in date is the following statement from the report of the Otago Society in 1900–1:

We have assurance of an old experienced salmon fisher, Baron Bultzingslowens, who visited our shores during the last fishing season, that he caught a true grilse in the Waiau last February. He says: "The main object of these lines is to tell you that one of the 4 lb. fish was a true grilse, and not a trout. I am too old a salmon fisherman, and have landed too many hundreds of grilse and salmon, not to know the difference between a grilse and any kind of trout. There is to me not a shadow of doubt about that fish being a true grilse. Had it been possible to send you the fish I should have done so.

To resume the record of introduction. In 1895 the Wellington Society received 200,000 ova from the Government ex 'Kaikoura.' These were in bad condition and only about 20,000 hatched out, of which 3000 were weakly. Of these 500 fry were retained in the ponds, but did not grow well. Apparently a lot more arrived, but it is most difficult to trace them. The only record I can find is that 10,000 ova were sent to the Southland Society, 1500 to Westport, 1500 to Greymouth, 1500 to Hokitika, 1000 to Buller, and 1000 to Marlborough;

but I can find no report as to their success or failure. In 1901–2 two lots of ova arrived from Great Britain, 150,000 by the 'Gothic,' and 50,000 by the 'Paparoa.' The numbers of fry hatched out from these were respectively 51,200 and 25,500, but amongst them was a considerable proportion of deformed fish. There must have been very considerable loss among these, for in the following year, only 42,806 one-and-a-half-year-old fish were liberated in the Hakataramea River, and 4200 were retained in the ponds. Whether the latter were allowed to escape or whether they died in the ponds is not stated, but in 1904 there were only 230 two-year-old fish left. The report of the Marine Department for 1904–5 states that:

> Several fish, believed to be salmon, have been caught at the mouth of the Waitaki River. A gentleman, recently from Scotland, states that he caught one of the fish, which weighed $4\frac{1}{2}$ lbs. and that it was undoubtedly a salmon in appearance and taste. Although the taste was not so pronounced as that of Scotch salmon, still the flavour was fine and quite different from that of trout.

This, of course, is not very convincing evidence.

In the following year (1905) 55 three-year-old fish were liberated, and 131 retained in the ponds. In 1906 seven four-year-old fish were liberated, and 50 retained; and in 1907 11 five-year-old were liberated, and 43 retained in the ponds.

In this same year the Canterbury Society obtained 50,000 ova from the Canadian Government, and these hatched out well, some 47,000 fry being found in the boxes. In the Society's report for 1908 it is stated that there are 20,000 yearlings in the races; and in the following year, 11,500 fish were liberated in the Selwyn, but nothing more was ever heard of them.

It is noteworthy in this, and indeed in most of the reports, how the numbers of fish in the ponds dwindled year by year. None of the societies offers any explanation of the fact. The losses were probably due to eels and the kingfishers, and in a less degree to shags. The latter birds are most destructive in rivers and lakes, but are always somewhat shy of coming too near human dwellings, and the ponds were nearly always placed in proximity to the latter.

At the beginning of 1908, Mr C. L. Ayson went to Canada for ova, and returned with 150,000 eggs. They were scarcely "eyed" when packed, and a heavy loss was anticipated. They reached Wellington on 1st April and were at once sent to Lake Anau, which was reached on the 5th, when it was found that 140,000 eggs were in good condition. This attempt was a new departure on the part of the Marine Department, which determined to select the Waiau River as a suitable spot to hatch out salmon ova for several successive years.

The small stream flowing into the Upokororo River was selected as the best place in which to liberate the young fry. At the end of 1908 Mr L. F. Ayson, Chief Inspector of Fisheries, was sent to England for more ova, and shipped 500,000 by the 'Turakina.' Of these about 400,000 came from the Tay, and the rest from Irish rivers. This lot arrived early in 1909, and was at once despatched to Te Anau, where 447,000 fry hatched out, which were placed in the Upokororo River. Mr Ayson followed in the 'Rakaia' with another 500,000, made up of 55,000 from the River Test in Hampshire, 120,000 from the Dee in Wales, and about 340,000 from the Rhine, near Trier. The last lot were rather too far advanced when shipped. Seven cases of ova (approximately 350,000 eggs) were taken to Lake Te Anau, and the fry liberated in the Upokororo, and three cases (150,000 eggs) to Hakataramea. From this last lot, 103,440 Rhine salmon and 6900 English salmon fry were liberated into the Hakataramea River.

In 1910–11 another shipment of ova came out in the 'Ruahine' in charge of Mr C. L. Ayson, 400,000 eggs being obtained from the River Wye, and 600,000 from the Rhine. These were all taken up to Lake Te Anau, where over 930,000 good eggs were unpacked, the Rhine ova again showing much the heavier loss. As the young fry hatched out, they were liberated in the lake. During this year also 10,274 one-and-a-half-year-old fish were liberated in the Hakataramea. In 1911–12, only 181 three-year-old Rhine salmon, and 49 three-year-old Atlantic (English?) salmon were liberated in the same river.

Pond-bred Salmon. Various attempts have been made to retain salmon in the ponds and to rear fry from them, in the hope that even if the fry from imported ova would not return in due time to the rivers into which they were originally turned, those from locally reared fish would do so. This expectation has not, however, been realised in a single instance. I have collected a good deal of the available information on the subject. In 1887 Mr A. J. Rutherford stated that from each shipment received by the Wellington Society, a few young salmon had been retained in the ponds, so as to test the possibility of rearing in our waters a land-locked variety. Unfortunately the Society does not seem to have kept a separate record of the fish in the ponds, the number of ova taken, or of fry reared from them. In 1888 the Canterbury Society obtained 5240 eggs from some of the fish imported by Mr Farr in 1885, and from these 5000 fry were hatched out. It was claimed for them that they were the first ova taken from imported salmon in the Southern Hemisphere. Unfortunately the whole lot were subsequently lost by disease or accident. About the same time the Otago Society commenced to rear fry from pond-bred salmon, and continued the experiment for some years. In

1889 there were some 116 four-year-old fish at the Clinton ponds, and from these some ova were obtained, from which 300 fry were hatched. These were ultimately liberated in the Waiwera. In 1890 some 14,000 eggs, hatching out about 10,000 fry, were produced, and of these 8000 were placed in the Upper Mataura, and 2000 in Lake Ada, Milford Sound. In 1891 over 20,000 ova were produced, of which 1000 were sent to the Wellington Society, and the remainder were hatched at the ponds. Of the resulting fry 7000 were liberated in the Waiwera, 7000 in the Owaka, while 3000 were sent to Milford Sound, and were liberated in Lake Ada. About 800 were retained in the ponds. The Society's report says: "Those in the Waiwera have done well, and have been seen in large numbers up to ten inches long." They disappeared during a fresh in the river in the beginning of November. In 1892 it is noted that the fish continue healthy in their confinement, although they do not grow to a large size. Some 20,000 ova were got from them, about 17,000 of which were hatched and liberated in the Waiwera. The report for 1895 says:

The last of the stock of Salmon which had been kept in the Society's ponds for the last nine years were liberated early in the Spring, as they were attacked with a fungus disease. These fish for the last five years produced over 20,000 ova annually, from which 15,000 to 20,000 fry have been liberated every year, which did well in our streams, and could be observed going down to the sea in the smolt stage; but it cannot be said that a real salmon has been got after its return from the sea.

In succeeding years some fish were always retained in the ponds. Thus we read in the report for 1900:

There are 176 fish, six to eight years old; and 260 three-year fish (from imported ova) in the ponds. *The old stock fish are not doing well with us. They were reared from pond-reared salmon got from imported ova, and do not seem to have the same vitality as those reared from imported ova.*

Some doubt has been expressed as to whether the fish in the Clinton ponds were salmon at all. This would tend to throw suspicion on some of those who were concerned in sending out the original ova, or on those who were in charge of the Otago Society's ponds—an utterly unworthy suggestion. The following statement is, therefore, of interest. The London correspondent of the Dunedin *Evening Star*, writing in November, 1892, of this suggestion, that by some blunder the ova sent out to the Otago Acclimatisation Society were not salmon eggs at all, but those of trout, says:

Fortunately the fish which resulted from these eggs were not all liberated. Some were kept in ponds, and in consequence, have degenerated until they have certainly become not unlike trout. To settle the question finally, some of these pond fish were sent home from Otago to Mr Tegetmeier, of the *Field*, who promptly forwarded them to the British Museum. Here they were carefully examined and declared to be undoubted salmon, though degenerated.

A summary of the attempts to introduce this species of fish into New Zealand waters is here given to show the continuous nature of the effort. The total number of ova introduced during the last half-century, beginning in 1868, was 4,813,000 or close on 5,000,000. In addition, some 120,000 eggs were obtained from pond-reared fish.

The total number of fry liberated, at various stages of growth, including those obtained from pond-reared fish, is approximately 2,620,000. These have been turned out into the following rivers or their tributary lakes and rivers:

South Island

Waiau	1,767,000	Selwyn	13,250
Aparima... ...	494,000	Heathcote	240
Mataura... ...	8,000	Avon	50 (?)
Owaka	7,000	Ashley	180
Clutha	146,900	Perceval	200
Leith	500	Clarence	725
Kakanui... ...	200	Hurunui	1,000
Waitaki	162,000	Nelson, Marlborough, Grey, Buller and Hokitika Rivers	15,000
Opihi	1,500		
Temuka... ...	1,500	Lake Ada, Milford Sound	5,200
Rangitata ...	1,000		

North Island

Ruamahanga	400	Streams in Taranaki ...	2,800
Hutt ...	8,400	Hawke's Bay	?
Manawatu ...	3,800		

Salmon Trout; Sea Trout (*Salmo trutta*)

In 1868 the 'Celestial Queen' brought 1500 ova of sea trout to Port Chalmers (on 2nd May), but I cannot trace what came of them, as to whether they went to the Society's ponds at Opoho, or to Mr Duncan's ponds, which were on the Leith just above the mill. If they were hatched it is probable that the fry got into Otago Harbour.

In 1870 the Otago Society received some ova of this species from Tasmania, and from it obtained 140 fry. These were liberated in a pond communicating with the Shag River, except 20 which were liberated in the Water of Leith.

The early reports of the Southland Society are not obtainable, but in a history of the Society recently (1916) compiled by Mr A. H. Stock, it is stated that in 1870, "154 ova were received from Tasmania and hatched out well." These fry were retained in the ponds for breeding purposes, and in September, 1874, 1100 ova were obtained from them:

The resultant fry were turned out into the New River. Of the 50 fish retained at the ponds (probably adults), nothing further is recorded, but probably they were included amongst "the old fish to be turned out into the Makarewa" in June, 1875.

Mr W. Arthur the Hon. Secretary of the Otago Society, writing in 1881, says:

I have tried to find from what river in England the original ova sent to Tasmania came, but the Secretary of the Salmon Commissioners there assures me that he cannot possibly find any record of this fact.

In 1871 the following appears in the annual report of the Otago Society:

Sea trout have been many times caught in fishermen's nets on the coast, particularly within Otago Harbour, but no reliable instance has been established of the capture of this fish in any stream or river.

If these fish were the progeny of those brought over from Tasmania in 1870, they must have been very small and few in number. The first record of trout in Otago was in 1868, but no sea trout were included. In those early days there was no doubt expressed, such as arose later in regard to fish caught in sea-water, as to the difference between sea trout and brown trout.

In 1873 the Canterbury Society obtained 300 ova from the Tasmanian Society, from which several fry were hatched out. Writing of this shipment (1895) Mr A. M. Johnson said:

On opening the box at the Christchurch Gardens a large portion were found to have hatched and died in the moss during transit by small steamer. The remaining good eggs, a few hundred, were hatched by myself as then Curator of the Society. The young fish were longer, thinner and more active, but appeared much more delicate than the common trout. A pond with spawning race through which the whole of the water in the Gardens flowed was especially prepared for their reception, and into which about 50, all that were reared were liberated. These, at about four years old, made some nests and deposited their eggs.

In 1874 Captain Hutton exhibited at a meeting of the Otago Institute a sea trout caught in Otago Harbour, and stated that another capture had been recently made. I was myself dredging a good deal in Otago Harbour then and in subsequent years, and the fishermen at that period always distinguished between two kinds of trout, which were not unfrequently taken with the seine-nets and which they distinguished as "Salmon" and "Trout." The former were almost certainly sea trout, and the latter brown trout.

In 1881 the Otago Society report says: "The Council has been made aware of the capture of many Sea Trout in Otago Harbour by fishermen during the year." And again in 1885, it is stated: "Salmon trout have not been found during the past year in any river, and their occurrence in salt-water is not so common as in former years. It is to be feared, therefore, it has been netted out, as a species" (a very improbable suggestion it seems to me).

In 1881 a trout was taken in Nelson Harbour, and was submitted to Dr Hector, who identified it as a true salmon trout (*S. trutta*). This fish was taken near the mouth of the Maitai River, and proved to be a female, 25 inches in length, which had just spawned.

In 1884 the Wairarapa Society received three trays of ova from the Government, and from these they obtained 2300 fry. Of these 500 were put in the Ruamahanga, and 700 into tributary streams, 500 in the Hutt River, and 350 in the Makakahi at Eketahuna. The remainder were retained in the ponds. I do not know where these ova came from, or whether any other ova were imported at the same time.

In the same year the Wellington Society received 900 ova, which were imported from Scotland for the Otago Society, and placed 750 fry in the Makora River. In 1888 there is a statement in the annual report as follows: "Our young Salmon trout have yielded us 2500 fine healthy fry, and these are, we believe, *the first taken from the fish and hatched out artificially*." If this means the first salmon trout reared in New Zealand from imported fish, then it only emphasises the want of knowledge and co-operation which existed between the different societies, for the Southland Society claimed to have done this 14 years earlier.

A very sharp correspondence took place in Canterbury in 1895 as to whether there were any sea trout in New Zealand. I mention it here, because many people still claim that there are none and never have been any, and consider that the question was settled in the negative by this controversy. It was initiated by a Mr W. H. Spackman who wrote to the *Christchurch Press* on 22nd June, as follows:

> I find it stated that 7000 salmon trout were liberated in the Waimakariri, and 1000 sold. Can the Garden Committee tell me whence these were obtained? Are they from ova direct from England, or were they from fish in the Gardens; and if so, had these so-called Salmon trout ever been to sea before spawning? I think this matter should be cleared, as I was not aware there were any real salmon trout in the colony.

This drew forth a letter from Mr S. C. Farr on 1st July:

> Salmon trout were brought here by me from Dunedin, as a gift from the Otago to the Canterbury Acclimatisation Society, about thirteen years since (? 1882), and they bred in the Society's Gardens for several seasons

under my own notice, the stock always being kept in one pond, without the slightest sign of degeneration.

Mr Spackman thereupon wrote to the Otago Society, and Mr Deans, the curator, replied from Clinton, 24th July, 1895:

The only salmon or sea-trout that this Society has ever had were brought from Tasmania by Mr Clifford in 1870. The ova numbered about 140. They were hatched at Opoho, and subsequently in the fry stage were removed to a pond on the bank of the Shag River. Some years ago, probably ten or twelve, the Government were getting out some ova from Sir Jas. Maitland, and it was reported that they were salmon trout, but Sir Jas. contradicted that and said that he was not aware of ever having sent salmon-trout ova.

This produced another strong letter from Mr Farr who stated (Aug. 2nd) that he attented a meeting of the Otago Council when a resolution was agreed to that a certain number of "Salmon Trout" fry should be delivered by Deans to him at the railway station:

I brought them without loss (to Christchurch), and delivered them at the Society's Gardens. They were put into a small race, and were subsequently transferred to a pond, kept as a distinct fish, and from them ova were taken for some years previous to my removal from the office of Secretary.

Mr Farr then quotes from the Otago Society's reports, and also the statements of Captain Hutton and Dr Hector already given, and then from the reports of the Canterbury Society as follows:

"In 1888 there were 580 salmon trout in the ponds." "In 1889 there were nine boxes of Salmon trout in the fish house." "In 1890, 8000 salmon trout were distributed." "In 1891, 9200 salmon trout distributed." "In 1892, 6000 distributed." "In 1894, 8000 sold and distributed." "During the years mentioned Mr Spackman was a continuous member of the Council, and I believe took part in the distribution of the young fry on more than one occasion, as Salmon trout, without once raising the question as to their species."

This shrewd thrust at Mr Spackman did not, of course, affect the question, and that gentleman again wrote to Mr Deans and drew from him another letter in which he expressed his opinion that we have no salmon or sea trout. It also brought Mr A. M. Johnson into the fray, who concluded his letter by saying: "it appears, however, we are all misled by mistakes about the identity of these fish." The outcome of all the discussion was only to show that the fish which the Canterbury Society had been keeping and distributing as salmon trout were probably brown trout. It did not affect the facts that the Otago, Southland and Canterbury Societies all imported salmon trout ova from Tasmania in 1871 and 1873, that mature trout were taken in the coastal waters and identified in 1874 and 1881, and that two further importations of ova were received from the home country in 1884.

In 1895 Mr A. P. O'Calloghan of Timaru caught a fish in the Opihi, which was sent to Dr Günther, who stated in a letter of 2nd April to Mr H. A. Bruce:

The fish mentioned in your letter of Jan. 23rd reached me to-day. It is without question a genuine Sea-trout. It is a great beauty and fatter than I have ever seen a Sea-trout or Salmon, showing it must have had abundance of food, and grown up under the most favourable conditions. It has been stated (erroneously in my opinion) that Salmonoids change their specific characteristics when transplanted from the Northern to the Southern Hemisphere. The specimen sent by you is strong evidence that no such change has taken place in your New Zealand Salmonoids.

In my opinion it is only evidence that the fish was derived from the egg of a pure salmon trout.

Carp Trout (Great Lake Trout?) (*Salmo trutta*, var. *lacustris carpione*)

In 1887 the 'Tongariro' brought to Wellington for the New Zealand Government, a large shipment of salmon ova, and along with it 25,000 Rhine brook-trout ova, 25,000 alpine-char ova, and 25,000 carpione-trout ova. "The latter is said only to be found in the Lago di Garda." Unfortunately the boxes arrived in very bad order, and no marks were placed on the trays to indicate to which particular kind the ova belonged, and there was nothing to indicate which was which. The societies to which this lot of eggs were given were asked to keep each tray apart, in order that the various kinds of fish might not be mixed.

The Canterbury Society received 12,000 ova supposed to be of this variety. Only 23 hatched out and all died.

The Wellington Society received the larger portion of the shipment but the eggs were in a very bad condition, and only about 550 fry were hatched out. Fifty of these were sent to the Auckland Society for use as a parent stock, but there is no record as to what was done with them.

About 30 were received by the Otago Society, and were placed in a pond at Opoho. The report for 1892 says:

These fish are growing, but they produced no ova this year. They do not appear to have settled down to a proper spawning season, as some of them are found to be ripe during the whole year. Some were stripped, but the ova were found to be in a very bad state, and only survived a few minutes. The report for 1893 states that thirteen fish, six years old, and 105, one to two years old, are in the ponds at Opoho. It is not stated where these last came from, and there is no further record.

The Wellington Society kept a number of the fry, and in 1889 they spawned, and about a thousand fry were obtained. I do not know what came of them, but Mr A. J. Rutherford, the Hon. Sec. of the Society, stated that "Unfortunately, owing to a want of knowledge of its habits, we have lost the breed."

*Brown Trout (*Salmo fario*)

The naturalisation of this species of trout in New Zealand waters is the most successful piece of acclimatisation work undertaken in this colony. It has exceeded all expectations. It has not only stocked the streams and rivers with the finest of sporting and edible fishes, a reputation which it shares with the rainbow trout, but it has brought numerous sportsmen to the country, and made it known far and wide as a paradise for anglers. It has also given to the coastal waters of the dominion the finest of food-fishes. It is true that the restrictive laws passed in the interests of the acclimatisation societies, and which are still in force, prevent the trout in the sea from becoming available, as they ought to be, as a food supply for the people. But it is quite possible to safeguard the trout in the streams, and yet enable those in the sea to be taken like other free-swimming fish, and then the public will get the benefit.

The first attempt to introduce brown trout into New Zealand was made by the late Mr A. M. Johnson of Opawa, Christchurch, who did actually ship 600 young trout in London in 1864 by the 'British Empire,' but a careless deck-hand dropped a lump of white-lead putty into the tank (this was afterwards found at the bottom) and killed all the fish.

In 1868 and the following year, Mr Huddlestone, on behalf of the Nelson Society, introduced trout ova from Tasmania, but the record of this work has been lost.

Mr Johnson claims that he was the first introducer of this fish into the country. In a letter written by him to the Minister of Public Works on 6th February, 1878, he says: "I may also add that the English brown trout was first introduced into New Zealand at my expense." This shipment was one of 800 ova from Tasmania, and it appears from the reports that they were brought out for the Canterbury Society in 1867. Mr Johnson brought them across from Hobart, but unfortunately there was so much friction existing for years between him and Mr Farr, the Hon. Sec. of the Society, that it is difficult to get at the facts. Of these 800 ova, only three hatched out. The report states that "it was not long before one of them was lost (escaped into the Avon). The two remaining proved to be male and female, so we concluded that even from these our rivers would in a few years be stocked." The faith of the pioneers was charming, and in this instance was justified. The following is Mr Johnson's own account of this incident.

A tremendous flood, the highest ever known in Canterbury, submerged the Gardens, and although most of the stock was saved, *the three Trout were washed out into a swamp* leading to the river, and appeared hopelessly lost. With a faint hope of their recapture, a spawning race was prepared near their rearing home, and at the season *two of the lost trout were seen and secured*. They proved to be male and female, and from these a supply of ova was obtained annually. By 1876, the Society had received about £100 from the sale of young trout, and many thousands had been liberated in Canterbury Rivers; all the progeny of those two fish.

In September, 1868, Mr G. P. Clifford brought over to Dunedin from Tasmania about 800 brown trout ova. They were packed thinly in well-washed moss in four boxes, which were kept cool with frozen snow. The voyage lasted nine days. Forty-nine dead ova were removed on arrival, and the rest were placed in covered boxes on a bed of small gravel, over which ran a small stream of filtered water about an inch and a half deep. During the time the fish were hatching the temperature varied from 40 to 55° Fahr., averaging 46°. The first fish hatched out on 28th September and the last on 29th October. The total number hatched was 729. The ova were not artificially impregnated, because the spawning was nearly over in Tasmania, but were obtained from the ridds made by the fish in the race at the Plenty ponds. The pond accommodation consisted of an oval pond 12 by 8 feet, and from a few inches to about 2 feet deep. The fry were liberated in various streams in the provincial district.

In distributing the young fish they were carried in an ordinary fish-kettle, 15 by 9 inches, and 9 inches deep, and they were mostly carried by hand.

Our last attempt to take trout from Dunedin to Queenstown—a distance of 208 miles, over rough bush roads, with at times a bad supply of water—proved a failure. The time occupied was four days. Out of 55, the number that left Dunedin, 25 were carried successfully a distance of 170 miles.

The Southland Society received a small lot of ova, and liberated about 200 fry. The Canterbury Society also received a lot from the same source, and 545 fry were hatched out and distributed in various streams. In the report written at the close of 1869 it is stated: "Trout may now be considered as established in the Province."

In 1869 Mr Clifford brought another lot of ova from Tasmania, from which 1000 fry were obtained and mostly distributed; and in 1870 a third lot yielding 1084 fry.

From these three shipments most of the trout of this species now found in New Zealand have come, for they not only throve in every stream into which they were placed, but quickly came to maturity and spawned so freely, that it became easy to distribute them

As to the origin of these fish, Mr W. Arthur, who investigated the subject more carefully than anyone who has written on it, states that the brown trout in Tasmania were descended from three lots from England. "Of these, Mr Francis Francis sent one from the Weycombe, Bucks, and another from the Wey at Alton, Hants, and Mr Buckland sent one lot from Arlesford on the Itchen, Hants."

These appear to be the only shipments made to the South Island

In 1870 the Auckland Society received 1000 ova from Tasmania, but only 60 fry hatched out. In 1872 a large quantity was brought over from which many were hatched and were sold for distribution throughout the district. Again, in 1873 and in 1874, further lots were introduced, all from Tasmania, and all hatched out well, but the reports give no record either of the number of eggs or of fry.

The northern streams are apparently too warm for brown trout, for Mr Cheeseman writing in May, 1880, says: "I am sorry to say that we have no evidence to prove that trout exist in any of our streams at the present time." Mr W. Arthur in 1881 says: "The acclimatisation of Trout does not seem as yet to be a success in the province of Auckland." Evidently he thought the summer temperature too high for this species, for he adds: "A gentleman just arrived from Victoria has assured me that the trout in that colony are fat, sluggish and give no sport when caught with rod and line." No other society appears to have got eggs or fry from outside the colony; the others got their stocks from the south.

Thus in 1874 trout were liberated by the Wellington Society in the Kaiwarrawarra Creek, the Hutt River and the Wainui-o-mata. From the first they disappeared, and in the latter they keep to the higher waters. The reason assigned at the time for this was that they got more congenial food higher up, but I am inclined to think that it was the cooler water in summer which they preferred.

Probably all the Hawke's Bay brown trout—with the exception of 300 which were imported from Christchurch—were originally brought from Otago in 1876.

In 1877 the Wanganui Society received their first consignment of fry, 300, from Mr A. M. Johnson of Opawa.

In 1878 Nelson received 200 young trout from Christchurch, and the Marlborough Society reared 700 fry from ova obtained from Otago.

In 1878, 1879 and 1880 the Grey Society received trout from Otago.

Since 1880 there has been a constant interchange of ova and fry going on throughout both islands, and enormous quantities of both have been sent out, especially from the hatcheries of the Otago, Southland, Canterbury, Westland and Wellington Societies. It is impossible to estimate the numbers which have been dealt with, but it is safe to affirm that the various hatcheries throughout the dominion have handled over 50,000,000 young fish. This was up to the end of 1916 only. The Southland Society are responsible for about 8,000,000, Otago 24,000,000, Canterbury between 3 and 4,000,000, Westland over 4,000,000, and Wellington nearly 7,000,000. But there are altogether 28 societies in the country, and all, or nearly all, have hatcheries in operation, and have been distributing trout for many years past. In addition to these, several million ova have been collected at the Hakataramea Hatchery for distribution, some thousands going as far as the Transvaal. Between 1916 and 1921 another 14,000,000 ova and fry have been distributed.

An experiment in the carriage of frozen ova was made in 1886. A box of ova was placed in the freezing-chamber at the Victoria Docks, London, and kept at a temperature of 18° Fahr. At the end of a month it was found that, although most of the eggs had been killed, a large proportion were alive apparently uninjured. To test this matter further a box of ova was placed on the 'Ionic' which arrived in Wellington in March, 1886. The following is Mr A. J. Rutherford's report on it:

The extremely interesting experiment of sending a box of trout ova in the refrigerator is, I regret, a total failure. The sawdust round the inside box was dry, and the box exceptionally well packed. Within, the moss was frozen into a solid mass, the trays all being stuck together; and on opening a layer it was evident that the ova had been frozen to death. There was no sign of life, and the appearance presented was like layers of light-yellow transparent unfertilized ova, with one side of each egg slightly fallen in. A coating of hoar-frost surrounded each egg. The animal matter was in good condition, and what looked like traces of yellow dead fish could be seen in many of the ova.

I tried several experiments, such as thawing very slowly in iced water, thawing in the air, i.e., but could detect no sign of vitality with a glass. The ova turned opaque at once on being placed in water, but the indentation in the side swelled out and each egg resumed its proper shape. There are about a dozen ova that have not turned opaque, and I have left them in a hatching-box to see if there is any possibility of vitality.

I think that this experiment has demonstrated plainly that the intense cold evolved in the freezing chamber is fatal to life in ova, even when well insulated and protected, as in the case of the box I received.

Rate of Growth and Food Supply. When brown trout were first liberated, the rate of growth was phenomenal, and this, according to Mr W. Arthur, was "due entirely to new and abundant food, and it may to some extent be to new water, also to the constitution or stock of trout." I think it most probable that the food-supply was the most important factor. The streams originally abounded with insects and insect larvæ (including various flies, as ephemerids, gnats, caddis-flies, etc., grasshoppers and beetles), mollusca, crayfish and other crustacea. In many streams, and these are the streams which have lasted out best as fishing streams, there were also countless shoals of minnows, smelts and other fish. At first the growth was enormously fast, then after a time the food supply gave out, and the big fish began to eat the smaller ones, and gradually the lakes and streams became more or less depleted.

The extraordinary increase of imported birds which dates from their first importation about 1868, and which was synchronous with the increase of the trout, has also made a very great difference in the food supply of the imported fishes. Grasshoppers, which were remarkably abundant in 1868, are now comparatively rare, and this is chiefly due, no doubt, to the increase of the starling; but smaller insects, not so conspicuous to the ordinary observer, have suffered equally. In this way it is certain that the insect-life of the streams has been greatly reduced; while the trout ate the larvæ, the birds fed largely on the mature insects. In 1870 crayfish (*Paranephrops*) abounded in nearly every stream; and I could collect quantities of shrimps (*Xiphocaris curvirostris*) and amphiphods (*Paracalliope fluviatilis*). The crayfish are now rare, and the other crustacea are scarcely to be found in any stream into which trout have been placed. One of the problems which now faces those interested in keeping the lakes and streams stocked with well-grown trout is that of finding and maintaining a suitable food supply. The cultivation of suitable aquatic plants (*Potamogeton*, *Myriophyllum*, etc.), and of insects, mollusca and crustacea, will be as much the work of a hatchery as the hatching and rearing of the fish themselves.

Mottram states that in New Zealand, on one occasion, the stomach of a fish (*Salmo fario*) was filled with *Spirogyra*, Link; subsequently it was proved that the fish took the weed in order at the same time to capture a small Trichopterous larva. The yellow bloom of the furze, *Ulex europæus*. Linn., was also taken on account of a small grub, probably one of the Tineina.

On March 30, 1911, on Lake Okeraka, New Zealand, the stomach contents of a trout were four grasshoppers, two cicadas, and three short pieces of stick of about the same length and thickness as the grasshoppers.

This is stated to show that the fish mistakes these things for the insects on which it is feeding at the time.

In the appendix to his work on New Zealand Neuroptera (published in 1904), Mr G. V. Hudson gives several tables showing the contents of the stomachs of trout. Their principal food appears to be caddis-worms.

Another interesting fact he brings out is that the larvæ of the larger species of dragon-flies destroy considerable numbers of very young trout-fry.

Mr W. Arthur has investigated the rate of growth more fully and carefully than any other New Zealand writer on the brown trout. He states that the average growth in all Otago streams between 1878 and 1883 was 1·53 lb. per annum. The lowest recorded was in the Otaria 0·751 lb., and the highest is Lake Hayes, 3·5 lb.

The first fish in the Shag River were liberated in 1868, and the first taken were in 1874; a male weighed 14 lbs., and a female 16½ lbs., representing an average annual growth of 2⅓ and 2¾ lbs. respectively. In the Leith the average increase at first was 1½ lbs. per annum; in the Lee Stream 1 lb.; in Deep Stream 1⅓ lbs.; and in the Upper Taieri 1 lb. These were not marked fish, and the average is based on the assumption that they were among the first fish liberated. In the Shag and the Leith there were great numbers of smelts found in addition to what occurred in the other streams.

The other three are inland streams, mostly running in rocky gorges.

In the Avon and Cust Rivers in Canterbury the average yearly growth at first was 1¾ lb.

Apparently in the large lakes of both islands the growth was much more rapid, but the records were not carefully enough kept to be quite trustworthy. Thus Mr A. J. Iles states that the largest brown trout taken out of Lake Rotorua weighed 27½ lb., "and was netted about eighteen years ago, some four or five years after the brown trout were first liberated." There is good reason to believe that the rate of growth in the early years of stocking was phenomenal, but the record is not definite enough to be accepted.

Mr W. Arthur points out, however, that similar rapid growths have been recorded elsewhere, and gives this example:

Mr J. V. Harvie-Brown, of Dunipace, stocked a loch in the north of Scotland, which had no trout in it at all. In two years they multiplied and attained a weight of 4½ lbs. So soon, however, as the number exceeded the food supply, or in two years, they fell off in condition, colour, etc., and latterly were not worth catching. Like cases have occurred elsewhere at home.

Size of Brown Trout. Anglers are notorious for the extent to which their scales and yard measures stretch when they are recording their catches. But acclimatisation societies and anglers' clubs usually keep pretty accurate records. It is therefore surprising and amusing to read the following paragraph from the 18th Annual Report (1883) of the Otago Society, where certain fish are reported (the italics are mine):

One taken at Lake Hayes, *said to have been* 60 lbs. in weight; two *seen* in the Clutha River, below the mouth of the Lindis, *estimated* at 80 lbs. each by Sergeant McLeod; and one from the Mararoa, which *weighed* 42 lbs. As no *Salmo fario* over 30 lbs. seems ever to have been taken in English waters, the above weights *must be received with caution*.

As one friend suggests, apparently fish grow much faster out of water than in it.

The biggest trout are found in the lakes and in the sea; Lakes Wakatipu, Wanaka and Hawea have yielded many fish up to 25 lb. in weight, and trout of 20 lb. weight and upwards are abundant in the sea, but are very difficult to catch. They are, however, occasionally taken in moki nets. The prohibition against taking, having in possession, or selling trout without a special licence, prevents any accurate record from being kept of these big sea-fish. It does not pay fishermen to take out a licence, because the catch is too erratic, but the fish are constantly taken both in set nets and seines, and they are nearly always so injured about the gills that if thrown back into the sea, as the law demands, they are almost certain to perish. Therefore a great number of them are taken, hidden, and sold surreptitiously, not as trout, however, but as "canaries." I do not know how this popular term has arisen, but it is in common use. A law which tempts men to do illegal actions is a bad law, and should be swept out of existence. Other means must be devised to protect the streams from being depleted of fish, such, for instance, as prohibiting seining altogether in certain areas.

The following records of actual catches of brown trout are taken from the lists annually published by the Otago Society.

Waipahi River

J. P. Maitland	9 fish	34 lb.	Nov. 11th, 1885
W. Carlton	17 "	28 "	Nov. 8th, 1889
Geo. Steel	6 "	27¼ "	
D. A. Purvis	17 "	38 "	1896
J. Nelson	17 "	35 "	
Many fine baskets got up to		40 "	1896

These were nearly all got with bare fly. The reports for 1890–92 state that:

The Waipahi still holds the premier position as a fly stream. It is now yearly attracting the attention of tourists from England, Victoria and the North Island.

Pomahaka River

C. Williams on five trips took	$139\frac{1}{2}$ lb.	1890
Wm. Fraser one day (18 fish)	53 ,,	
T. E. Brown 6 hrs (12 fish)	52 ,,	

The Pomahaka gives, perhaps, a better day's sport with the minnow than any other river in our district, baskets of 50 to 60 lbs. have been taken in one day by Dunedin anglers.

Records for 1890–91

Clutha River (above Cromwell): one rod took 90 lbs. in one evening.

Mimihau and Otaria Rivers: baskets from 30 to 40 lbs. frequently.

M. Lowrie this season in the Mimihau took 241 fish averaging 3 lbs. each.

The Upper Mataura is now splendidly stocked, and baskets of 30 to 40 lbs. can easily be obtained on a good day, fishing with the natural cricket and grasshopper as a lure.

H. Schluter, fishing at the mouth of the Waitaki, in Oct., 1888, took three fish weighing $69\frac{3}{4}$ lbs.

The Waitaki Society reported in 1890 that at Waitaki North (in one day?) 169 fish were taken weighing 1123 lb. or $6\frac{2}{3}$ lb. each; the heaviest were, one of 14 lb., one of 13 lb., two of 12 lb., four of 11 lb., and eight of 10 lb. The smallest fish taken weighed 3 lb. The best baskets were two of eight fish weighing 60 lb. and one of six fish weighing 50 lb.

These records only apply to Otago rivers, but similar records are available for other streams all over the areas stocked with brown trout. Thus the largest brown trout taken from Rotorua Lake, according to Mr A. J. Iles, was $27\frac{1}{2}$ lb.; the largest from Lake Taupo was $25\frac{1}{2}$ lb.; but Mr C. P. M. Butterworth states that fish of 29 lb. have been taken from Lake Taupo. Mr W. P. Cotter tells me (July, 1916) that in Lake Hawea specimens weighing 26 lb. have been taken, and that though the fish have been in the lake for 25 years, there is no sign of deterioration.

Mr W. Arthur gives the following interesting facts about this species in New Zealand. The trout spawn (in Otago) from 20th June to 4th August, a half-pound fish giving about 400 ova; a seven pound fish about 6000. The eggs hatched out in 78 days. At Opoho during the winter the temperature of the water averaged 42° F.; and in the hatchery from 42° F. to 52° F. The best temperature for hatching is 48° F. The young fry average $1\frac{1}{2}$ inches long in six weeks, and 3 inches in a hundred days. They carry best when from 1 to $1\frac{1}{2}$ inches long. The only variations which he considers the brown trout to have developed in the new country is that the spawning season is about two months later (relatively) than in England, and the duration of hatching about 14 days longer.

Mr Dansey tells me that brown trout fry obtained from the Tauranga Society were liberated in Lake Rotorua in 1889, where they multiplied and throve wonderfully in spite of the huge flocks of shags that then infested the Lake, and which were quite capable of swallowing with ease a ¼-lb. trout, and of the immense numbers caught by the natives in nets at the mouths of streams; specimens up to 22 lb. were not at all uncommon. Those caught in the Lake had a muddy flavour; their principal food appeared to be young cray-fish.

The shags soon learned how to catch them. From Kawaha Point, on a calm sunny morning, I have watched a mob of over 200 shags away out on the lake, suddenly take wing, light again on the water near the western shore where it is shallow for a considerable distance out. They would spread themselves out in a long line—at apparently correct intervals apart—and all swim quietly towards the shore, the ends of the line gradually bending inwards. Suddenly, as if by some given signal, the whole line would dive, and every shag reappear with a trout in its beak. These tactics were only undertaken when the sun was at a certain altitude to, I suppose, cast the shadow of the birds at a certain angle on the sandy bottom, and thus drive the fish towards the shore into shallower water. The sight impressed me very much at the time.

Mr Dansey states that brown trout afford little sport to anglers except in unpleasant weather. He further adds that:

Brown trout are now only occasionally seen or caught in the Rotorua Lake. They commenced to disappear after the introduction of the Rainbow. Some ascribe this disappearance to the fact that the two species spawn at a different time; for the Rainbow, being a much stronger and more active fish, disturbed the Brown when thus engaged, and the ova failed to be fertilised. I have never seen a cross between a Brown and a Rainbow Trout; but I have between a Brown and a Fontinalis, with the red spots enlarged to the size of a threepenny piece[1].

[1] The relative merits of different kinds of trout for inland waters are thus recorded by Ernest Phillips in *Trout in Lakes and Reservoirs* (p. 36). "I have a record of a reservoir in which 6000 fish were put down, all two years old. There were 2000 brown, 2000 Levens, and 2000 rainbows. The next season we started fishing, knowing there were 6000 trout to go at. The season's catch was 450 brown trout, 301 Loch Levens, and only 85 rainbows. The brown trout and the Levens were much alike, many of them up to 1 lb. each, and a few over, but all the rainbows were 1 lb., several reached 2 lbs., and a few were actually 2½ lbs. It will be seen, therefore, that the brown trout provided the best and the most consistent sport. Rainbows gave the heaviest fish, but they were erratic and disappointing. They would be on the feed for a day or two and then vanish from view, and it was no uncommon experience for a whole week to elapse and not a single rainbow be returned to the keeper's list, though fifty or more of the other two varieties were caught in the same length of time. As for Loch Levens, I believe it is a fact that they do not grow to as great a weight as brown trout or rainbows, and that a fish of 4 to 5 lbs. is a monster. At any rate, searching through another keeper's book and taking a period of five years to allow for good and bad seasons, I find that only 60 Levens were killed over 2 lbs. as against 225 brown trout, and 74 rainbows."

*Lochleven Trout (*Salmo fario*, var. *levenensis*)

It is perhaps owing to its isolation, and consequent in-breeding in a small Scotch lake, that this variety of *Salmo fario* is a more delicate fish, and more difficult to transport than the common and dominant variety. This characteristic seems in some respects to have been overcome in fish reared in New Zealand, but it was apparently in evidence in the case of ova brought from Scotland.

In 1882 a shipment of ova was made by Sir James Maitland on behalf of the Otago Society, but they were all dead on arrival in Melbourne. A second shipment in the following year from the same source shared the same fate. A third attempt made in December, 1883, was more successful, and the ova were divided between the Otago and Wellington Societies, 1700 fry being hatched out at Opoho Ponds, and about 800 at Masterton. Both societies distributed a portion of their stock, and kept a portion in their ponds for breeding. From these, great numbers of fry have been distributed right throughout the South Island, and in the North Island from Wellington to Mt Egmont.

In 1887 the 'Tongariro' brought 40,000 ova, half of which went to the Wellington Society and the rest to Canterbury. The former lot hatched out about 15,000 fry, and the latter less than 10,000. In 1889 a further lot of 27,000 ova arrived by the 'Aorangi' in Wellington.

Altogether nearly 700,000 fry of this species have been distributed throughout New Zealand, of which over 470,000 came from the Otago Society.

*Scotch Burn Trout (*Salmo fario*, var. *samardii*)

In 1885 the Otago Society received 15,000 ova from Scotland, from which only 1700 fry were hatched out. The mortality among these was so great that there were only 490 survivors at the end of the year. Fifty of these were sent to Mr Pillans of the Lower Clutha, and 40 to A. M. Johnson of Opawa. The rest were retained in the ponds. Fifty more were distributed in 1886, and the balance were kept for breeding purposes; 700 fry being liberated in 1887, and 14,300 in 1888–89. The number annually distributed rose to 154,000 in 1897–98, but has varied from 50,000 to 120,000 in subsequent years. The total number of fish distributed from the Otago Ponds to the end of season 1919 (31st March) has been over 2,000,000. The Southland and Canterbury Societies have reared and liberated a few thousand; as did the Wellington Society till about 1898, since when they seem to have devoted all their attention to rainbow and brown trout.

*Lake Blagdon Trout (*Salmo fario*, var.)

This fish is only a brown trout, and it is questionable even if it is varietally different from the common form of *Salmo fario*. Lake Blagdon is an artificial reservoir in the heart of Somersetshire, England, which supplies Bristol with water. It was stocked originally with brown, Lochleven and rainbow trout, and more of the latter have been taken in the lake than any of the others. The fish grew at a phenomenal rate, and in 1904 the reservoir sprung into celebrity among trout fishers in Britain owing to the size of the fish which were taken, the average weight for that year being 5 lb. 6 ozs. each. Individual fish were taken up to 8 and 9 lb. each. (Though not stated, it is probable that the great average weight was raised by the rainbows, which grew more rapidly than the others.) The reputation of these fish drew anglers to the lake from all parts of the kingdom, so that the waters were very heavily fished, and the average weight fell. But the record still remains unbeaten by any other water in Britain.

The Otago Society introduced ova of the brown trout from Lake Blagdon some years ago, and began to liberate fry and yearlings in 1908. Since then up to 1920 they have liberated altogether some 950,000 fish in Otago waters.

Alpine Char (*Salmo* (*Salvelinus*) *alpinus*)

Among the many kinds of fish which Mr A. M. Johnson of Opawa attempted to bring out to New Zealand in 1864, were a number of char. He does not indicate the species, except that the fish were European, and not the American *Salmo fontinalis*. They therefore almost certainly were alpine char, which were obtainable at various ponds in England at the time of shipment. Mr Johnson lost nearly all his fish from lead-poisoning, due to the carelessness of one of the sailors.

In 1887 a shipment of 25,000 ova was brought to Wellington by the 'Tongariro,' but (as is stated at p. 211) it was mixed up with two other lots of ova, and none of the trays was labelled. The shipment was also in very bad condition, smelling offensively, and with a large number of the ova dead. All the eggs supposed to be of this species were handed over to the Wellington Society. Rutherford wrote on 31st May, 1887: "the small white ova, supposed to be Alpine Char, were in a very bad condition, and only about twenty sickly fish hatched out, three of which are still alive." Mr Ayson, Inspector of Fisheries, who was then in charge of the Wellington Society's ponds, says: "The few fish which were hatched at the Masterton Hatchery were put into one of the deepest, coldest ponds, but they did not thrive well and died off within twelve months."

German Lake Trout (*Salmo* (*Salvelinus*) *umbla*)

In 1868 the Otago Society received 6000 eggs of this species by the 'Celestial Queen.' These were taken to the ponds at Opoho and hatched out there, but I cannot trace their subsequent history. This is the "Ombre chevalier" of the Swiss lakes.

*American Brook Trout or Char (*Salmo* (*Salvelinus*) *fontinalis*)

Mr A. M. Johnson of Opawa, who came out to Christchurch in 1864, claimed that the first *Salmo fontinalis* were brought out at his expense. I have no means of verifying the accuracy of this statement, but he certainly received a considerable stock of eggs from New York (*via* San Francisco) in March, 1877. From these a large stock of fish was obtained, and they were sold to many parts of the colony. Thus, in 1883, the Canterbury Society purchased 100 fry from him. In 1884 the Auckland Society obtained a small lot from him, and liberated them at Western Springs; and in 1885 the Otago Society got 50 fry, which were kept in their ponds and were reported to be doing well.

In 1877 the Auckland Society received 5000 ova from Mr T. Russell of San Francisco. From these 400 fry were hatched out, of which half were liberated in a tributary of the Waikato near Cambridge; and half in the Kaukapakapa stream, Kaipara.

In 1880 the Wellington Society placed 250 fry in a tributary of the Hutt River, but there is no record as to where they came from. In 1881 they liberated a further lot of 220 in the same stream and these were obtained from Christchurch—presumably from Mr Johnson. In the same year they placed 900 fry in the Makara and Ohariu streams.

In 1883 the Canterbury Society received 25,000 ova in January from Mr R. J. Creighton of San Francisco, and in February a second lot of 10,000. Unfortunately all the eggs in both shipments were dead.

In 1884 the Auckland Society received 30,000 ova from San Francisco, but again all were dead.

In 1887 the 'Kaikoura' brought 30,000 ova from the Solway Fisheries to Wellington, but only about 4000 arrived in good condition, for they travelled in the cool chamber and not in one of the insulated cases. About 2500 healthy fry resulted, of which between 500 and 600 were retained at the hatchery at Masterton, and the rest were distributed in various streams.

These appear to be all the direct shipments received from home or America; the other societies obtained their supplies from Christchurch chiefly.

Thus in 1885 the Otago Society received 400 young fry from the Canterbury Society. These were placed in the boxes at Opoho, but serious mortality set in, and at the end of two days only 20 survived. The stock in the ponds increased however and spawned, for in 1886 1400 fry were turned out, in 1887 over 18,000, and the numbers went on increasing. The Society still liberate several thousand each year, and from the date of their first introduction till 31st March, 1915, have sent out to many streams in Otago about 800,000 fish. In the *Otago Daily Times* of 10th June, 1891, the following statement appeared:

The fish at Opoho and Clinton are attacked by a disease, which Dr Scott (Professor of Anatomy in Otago University) considered closely corresponded with cancer in mammals. Mr Deans stated that while confined to *Salmo fontinalis* in the Otago Society's ponds, it was similar to one which attacked the Rhine trout in the Wairarapa ponds. Dr Scott said it was a fatal and malignant spreading tumour in the throat.

In 1887 the Southland Society received ova or fry, presumably from Christchurch, and for a few years turned out a few thousand in the Oreti and various tributary streams. Altogether only some 33,000 fry have been liberated in Southland.

The Taranaki Society also got ova from Canterbury in 1887 and liberated 600 young fish in one or two streams.

The Canterbury Society up to 1915 had liberated about 175,000 young fish. The Wellington Society up to 1899 had liberated about 700,000.

Salmo fontinalis has not thriven as was expected by those who introduced it, for it is a smaller species than most of the others and is not able to compete against them.

Mr Stevens of the Clinton Hatchery tells me that there are a number of small streams in Central Otago in which this species thrives, but there are no brown trout among them. He says:

I have liberated thousands of these fish—both fry and yearlings—but seldom hear of any being caught by anglers. I have no hesitation in saying the tendency is for these fish to disappear from streams already stocked with Brown Trout.

In the upper waters of the Hedgehope stream which rises in Bushy Park Station in Southland this species is found in abundance, but it is in sole possession. It is also found in a small stream called the Back Creek on the east side of the Blue Mountains in Otago.

The Hawera Society liberated 5000 fish in their district in 1890, but resolved in the following year to get no more, but to devote their attention to brown and rainbow trout.

A small number were liberated in the Rotorua district in 1890. In the Horohoro stream, a small tributary of the Waikato, they did so well that in about five years they afforded capital sport, many fish of 4 lb. and even 5 lb. in weight being secured with the minnow. Unfortunately some years later a flood of exceptional severity swept everything before it down the valley of the Horohoro, including nearly all the trout, and so few were subsequently caught or seen, that in 1899 the stream was restocked with rainbow trout by Mr Dansey.

Mr Wilfred Howell of Cave, near Timaru, sends me the following interesting facts about *S. fontinalis* in South Canterbury (Aug. 1916).

This fish was turned out in the Pareora a good many years ago. In this river there is a dam for the Timaru water-works, which it is impossible for fish to climb, as there is no fish-ladder. *Fontinalis* were, as far as I know, turned in both above and below this dam. Below the dam, with the exception of three holes just above the falls, there is no sign of these fish. The habit of the fish is apparently to go up stream as far as they can go. At the head waters, up the hills, all the bush creeks are full of them. In fact all streams running into the Pareora have them after they reach the bush. I have caught them up to 3 lbs. weight, but they mostly run about 8 or 9 inches. They also occur in the Hinds river about thirty or forty miles above Ashburton, but also only in the head-water creeks, and only when there is plenty of cover. In my opinion these fish will not spread on account of their habit of always climbing up stream.

I am of opinion that they occasionally cross with the Brown Trout; I have caught fish that gave me the impression that they were certainly hybrids.

As far as sport is concerned I think that *Fontinalis* are a failure; they are too easily caught. They will take almost anything from a red leaf pulled through the water to a minnow. They will take meat, part of another fish's gill, and any fly with much red hackle on it. When hooked they have only one good run in them, and then you can pull them wherever you want them.

[1] In the *Bulletin* of the U.S. Fish Commission for 1887 (vol. VII), in a paper on "American Fish cultivated by the National Fish Culture Association of England," by W. Oldham Chambers, the following statement is made regarding this species: "It is with reluctance that we omit from this list the American brook trout, *Salmo fontinalis*, which has had an excellent chance of asserting its qualifications for introduction into our group of *Salmonidæ*, but has failed to do so, except in confined waters. Its first appearance in this country was heralded with jubilant anticipations; its capacities for rapid growth were hailed as a good omen, and its gorgeous dress and graceful form won golden opinions from all piscatorial classes, who willingly paid large sums of money for what was then considered the coming trout. Gradually, however, its true character appeared, and now it is universally regarded as a fish not to be depended upon. No authority rebuts the evidence forthcoming as to its suitability to British waters, if inclosed, nor as to its value as an addition to our fresh-water fish. The sole cause, and a very grave cause it is, for its denunciation is that it escapes from those places where it is turned in. Before finally discarding this unique char it behoves us to question more closely than we have yet done its habits, instincts, and the nature of its native home, in order to render it full justice. Probably the waters in which it has been placed have not been suitable, and this assumption certainly seems justified by the fact of the fish wandering as it does. The question naturally arises as to where it goes. Does it find suitable places in its wanderings? Does it descend to the sea, or does it pine and perish for lack of natural conditions? If death explains the mystery, which is hardly likely, we have at once a solution; but if not, it is difficult to say what has become of the thousands

Mr C.J. Peters also states that this species was liberated in streams in the Mount Somers district about 1880, where they throve remarkably, and all the creeks are at the present time well stocked with them.

*Rainbow Trout (*Salmo* (*Salvelinus*) *irideus*)

In 1883 the Auckland Society introduced a quantity of ova from which about 4000 fry were hatched out. These were distributed in four streams in the neighbourhood of Auckland.

In 1884 another and larger importation was made. According to Mr Cheeseman this shipment was referred to as American brook trout, and the name was kept up in the two subsequent reports. It was not till 1886–87 that it received its proper name. Mr Cheeseman states (Aug. 1915) "I believe that the whole of the wild stock of Rainbow Trout in New Zealand has been derived from the Auckland Society's introductions."

The Auckland Society have liberated many millions of fish but their annual reports do not enable one to ascertain with any approximation to accuracy the total number. It probably exceeds 10,000,000.

The Canterbury Society received ova from Auckland in 1885, and of the fry reared, distributed some and retained others as breeding stock. Up to 1915 they had liberated in various streams about 1,200,000 fish.

The Wellington Society got a number of young fish from Auckland in 1891; many of these were deformed. Some 900 of them were placed in a rearing-box into which a large eel managed to find its way, and only left 12 alive when it was discovered. In the following year a number more were obtained, and distribution of the fish throughout the provincial district was commenced, and carried on vigorously for nearly 12 years, rearing in the ponds being discontinued in 1905. This was owing to a disease[1] of the gills which attacked the breeding stock in 1903, and increased to such an extent as to discourage the authorities from breeding any more. Meanwhile the Society had liberated nearly 2,800,000 fish.

In 1895 the Otago Society received from the Wellington Society 5000 ova which hatched out only moderately well on account of being obtained from immature fish. Ultimately 1500 were liberated in the Waipahi, and 400 retained for breeding purposes. During the last 20 years they have liberated over 500,000 fish. Among other localities Lake Hawea is particularly well stocked with them.

turned out into our English streams. In America the brook trout is regarded as a home-loving species—therefore it seems somewhat likely that we have not yet provided the domestic comforts to which it is habituated. The suggestion, at least, is worth studying, and the Association still has these fish under culture, not being convinced of their unsuitability for inclosed waters."

[1] Mr Deans of the Opoho Hatchery, Dunedin, considered that this was similar to the cancer which attacked *Salmo fontinalis* in the Otago Society's ponds.

The Southland Society began to liberate fry in 1900, but appear to have discontinued in 1904, after some 40,000 fish had been liberated, mostly in the Makarewa.

The Hawke's Bay Society began distributing fry in 1900, and liberated up to 1915 about 750,000 fish.

The Westland Society, which liberated altogether over 500,000 fish, in 1907 declined an offer of eggs from the Tourist Department "on account of rainbow trout failing to do well in this district."

Some of my informants have stated that this species cannot hold its own against the brown trout, for A. J. Rutherford who knows the North Island trout streams very well, says (in 1901): "the only stream I know which is well-stocked is the Tahuna-atara stream between Rotorua and Taupo, which is full of them, and contains no brown trout."

The President of the Southland Society states:

> there is only one authenticated case of a rainbow trout having been found (in Southland), and that was in a poor starved condition. Fishermen attribute the destruction of these fish to the brown trout.

Mr A. C. Henderson, Hon. Sec. of the Waimarino Society, reports (1915) that about 1900, two settlers, Messrs Nathan and Robertson, turned out a number of brown trout in the Makotuku stream but reported them as all dead. In 1903, however, some fairly large fish were found to be in the stream, and they have increased steadily since:

> At first this increase of brown trout was viewed with satisfaction and the Society went on liberating both rainbow and brown trout until 1908, when it became evident that in the streams thus stocked, the rainbow trout gradually disappeared and the brown trout increased. Profiting by this experience the Society does not now liberate rainbows in the same stream where there are brown trout. The southern streams of the district, with one exception, have been given over to brown trout.

The opposite opinion is held by Mr Bell of Hawera, who informs me that rainbow trout dominate and are too strong for brown trout. The former go up the streams and occupy the head waters, while at the river mouths, brown trout chiefly are found.

Mr C. P. M. Butterworth states that in the Tongariro River which runs into Lake Taupo, he has caught only brown trout in a certain pool, when the river was in flood and muddy, and the very next day in the same pool, when the water was clear, has taken only rainbows. He is of opinion that the rainbow prefer the lake and only move up stream on the approach of the spawning season. They give very much better sport than the brown trout.

In many Otago streams both brown and rainbow trout will be taken in the same stream in the same day. In Lake Hawea the same thing happens. This is the evidence of Mr McIntosh, President of the Otago Society (in 1916); but against this Mr Steven, Curator of the Clinton Hatchery, states (12th June, 1916): "I know of no stream in the South Island in which brown and rainbow Trout thrive together."

On the other hand Mr W. P. Cotter informs me that in Lake Hawea both brown and rainbow trout have been taken in the same net and are known to favour the same creeks and spawning grounds.

In netting for trout in Lake Hawea in July (1916) only brown trout were taken in the lake itself, 150 fish in one haul. But in a trap set in Timaru Creek which runs into the lake, ten rainbows were captured.

Mr J. King of Hokitika says that rainbows are difficult to keep in confinement after two years of age, as they are exceptionally liable to gill-disease.

Mr Dansey, who was largely responsible for stocking Lake Rotorua with this species, distributed most of the fry as soon as they were free of the yolk sac, as they carried better and stood more knocking about and in greater numbers at that stage than at any other.

Size of fish and rate of growth. Mr Iles states that the largest fish taken out of Lake Taupo weighed 21½ lb. and that they are frequently netted up to 20 lb.

Mr W. P. Cotter informs me that in Lake Hawea, fry were first liberated about November, 1911, and that in less than two years later rainbow trout of 10 lb. weight were taken in nets set along the shallow beaches. He adds:

The fact of trout of this weight having been secured caused a discussion as to whether they had entered the lake from the sea, as it was not considered possible that such growth could be attained in the time mentioned. Against this theory it is merely necessary to state that, roughly speaking the lake is nearly 200 miles away from the sea, the winding Molyneux the only connecting link, that no rainbow trout had been netted in previous years, and that in practically virgin waters for one particular species trout might thrive for a brief space beyond the expectations of the most sanguine.

The following appeared in the Dunedin *Evening Star* in Aug. 1913:

A 13 lb. male rainbow trout, suicidally trapped at Lake Hawea during ova-stripping, has been sent to the Otago Acclimatisation Society and will be stuffed and placed among the office trophies. The rainbow trout, as fry an inch long, were put into Hawea five years ago, and the finding of such a well-grown and healthy specimen is proof that the fish are getting good food and thriving in this lake. Up North the rainbow is reckoned a splendid sporting fish, and the flesh is excellent eating.

Mr A. C. Henderson states that "in one virgin stream (in the Waimarino district) we liberated 200 fry, and two years afterwards the average fish taken was 3 lbs. in weight." Mr C. P. M. Butterworth informs me that fry hatched in 1914 were liberated in

Lake Onslow, a large artificial lake or dam near Roxburgh, in the following year. In 1916 one was caught which weighed 3 lb. 14 oz. There is reason to believe that the growth in the larger lakes at the beginning of stocking was much more rapid. Mr Dansey states that some caught in the Waikato River weighed from 7 to 8 lb. within four years from the liberation of the fry[1].

Speaking of rainbow trout in South Canterbury, Mr Wilfred Howell says:

In the rivers they are not doing much good, as they seem to go down to the sea soon after being turned in, and no big fish ever come into the rivers from the sea. In the lakes, however, especially Lake Alexandrina in the MacKenzie country, they are doing very well indeed. Some were put in there five years ago as yearlings, and last month (July, 1916) three were caught, the largest 17 lbs., the smallest 14 lbs.

The increase of these fish in the inland lakes of the North Island was so great that in 1913 the Government decided to take over the administration of the inland fisheries, and, in co-operation with the acclimatisation societies throughout the dominion, to endeavour to improve the condition of affairs. In Rotorua and Taupo the fish had deteriorated greatly, and measures were at once taken to reduce the number in the lakes. This was done in Rotorua by means of barriers in two of the largest rivers, and by netting traps and drag-nets in other streams. In Taupo all were taken by means of traps and nets.

For the three years ending 31st May, 1916, the total number of ill-conditioned fish taken and destroyed was as follows:

	1913–14		1914–15		1915–16		Totals	
	Number	Tons	Number	Tons	Number	Tons	Number	Tons
Rotorua	18,271	19·3	14,941	20·7	25,243	22·5	58,455	62·5
Taupo	2,830	4·6	12,779	27·0	15,674	27·9	31,283	59·6

The total number of good fish taken out and sold for the same period amounts to the following:

	1913–14		1914–15		1915–16		Totals	
	Number	Tons	Number	Tons	Number	Tons	Number	Tons
Rotorua	25,851	15·9	28,460	18·2	35,464	20·5	89,775	54·6
Taupo	6,243	11·0	11,574	22·3	16,137	22·5	33,954	55·8

[1] Ernest Phillips, author of *Trout in Lakes and Reservoirs*, says (p. 37): "Rainbows are no use at all for rivers. They disappear down to the sea very soon after they have been liberated. It might be thought that the Rainbow would find it hard to exercise this faculty for getting away from lakes and reservoirs and ponds, but it is apparently just as easy for a stock of Rainbows to disappear from a fenced and walled-in reservoir as it is from an open river. Until this was discovered there had been a great run on Rainbow trout, and thousands upon thousands had been turned down in municipal reservoirs. But when reservoirs had been stocked with countless numbers, and, after two or three years they were drained and found to be empty, a reaction set in."

The marketable value of these fish for the three years named was £1865, £1513, £1217.

So that out of those two lakes there have been taken in these three years 213,467 fish weighing 232½ tons.

To still further keep up the fisheries and to renew the stock already in the lakes the Rotorua Hatchery has sold or distributed within recent years, 3,330,000 ova and 1,896,000 fry.

With the object of improving the condition of the fish some 250,000 native shrimps (*Xiphocaris curvirostris*) were caught in the Waikato River in 1908, and were liberated at suitable places in the lakes. In 1909 another 185,000 were liberated, and another large quantity in 1912. Since the Department of Internal Affairs took charge of the fisheries, 280,000 shrimps have been brought from the Waikato and placed in Lake Rotorua, and a further 110,000 in Lake Taupo; all in sheltered places. I question very much whether this expensive mode of feeding the trout has any permanent value. If the shrimps are placed in waters to which trout have no access, but from which they can escape into the trout-frequented areas, then they might form a permanent food supply. Otherwise they will simply be eaten up as quickly as they are liberated. Some more scientific method of feeding and of conserving the food supply should be adopted.

Mr C. Chitty of Cambridge (according to Mr Jas. Drummond, Aug. 1914) says that in the Waikato, the native grayling used to ascend the river in thousands every spring. They were not seen after 1875, though the mullet continued to ascend. He blames the rainbow trout, but as a matter of fact these fish were not introduced into the Auckland district before 1883.

*Mackinaw Trout; Great American Lake Trout (*Salmo* (*Cristivomer*) *Namaycush*)

In 1906 Mr L. F. Ayson—at the request of the Tourist Department—brought a case of eggs of this species from America. They were hatched out at the Christchurch Society's Hatchery, and 4000 fry were liberated in Lakes Pearson and Grassmere. Another lot of 4000 were taken over to the west coast, with the intention of placing them in Lake Kanieri, but Mr Jas. King of Hokitika reports (July, 1916), that owing to the carelessness of the curator then in charge of the hatchery they were all lost. Those in the Canterbury lakes have been caught by anglers during the last two or three years. Mr E. F. Stead, writing in April, 1916, says:

> The Mackinaw Trout are apparently thriving in Lake Pearson, as several have been caught weighing about 10 lbs. As the lake is but little fished, this would indicate that there must be a fair number of these fish there.

Lake Tahoe Trout (*Salmo clarkii Tahoensis*)

The Auckland Society received 30,000 ova from Mr T. Russell in 1878. About 3000 fry hatched out, but only 1000 survived. Of these, part were placed in Lake Omapere, part in Lake Waikare, and a few in the Onehunga Springs. No one has any knowledge about them to-day; they do not appear to have established themselves.

There is some confusion about the identification of this fish and the succeeding species. The name I have given above is that furnished by Mr Ayson, Chief Inspector of Fisheries, who says that it is a species which runs up to 20 or 30 lb. weight in Lake Tahoe and other lakes in the Sierra Nevada. He adds: "No other result could be expected from these fish when turned out in water like Lakes Omapere and Waikare, and the Onehunga Springs."

Black-spotted Trout (*Salmo henshawii*)

The report of the Auckland Society for 1885–86 states that ova of this species were received from San Francisco, "which were reared to maturity in the Society's ponds, producing fry, some quantity of which was liberated. They either disappeared, or merged with the Rainbow Trout."

No one seems to have taken the trouble to look after the fish, once it had been successfully introduced.

*Californian Salmon; Quinnat Salmon; King Salmon; Chinook (*Salmo quinnat*; *Onchorhynchus tschawytscha*)

The quinnat, like the brown trout, has been a great success in acclimatisation work. The species is now thoroughly established on the east coast of the South Island, and its range is being very steadily increased.

The Hawke's Bay Society was the first to take steps to introduce this fish into New Zealand, and through Dr Spencer F. Baird, Chairman of the United States Fishery Commission, a shipment was despatched to Napier in 1875. Unfortunately it never reached its destination. The steamship having it on board went direct to Sydney, and failing to obtain a fresh supply of ice there, it was found on the trip to Auckland that the ova had begun to hatch out. To save them from total loss, Mr J. C. Firth took a portion of the eggs and placed them in the Auckland Society's ponds; the result was that out of 20,000 ova which arrived, about 10,000 were placed in the Waikato and the upper tributaries of the River Thames. The remainder were placed in the hatching-boxes; about 1450 fry were forwarded to the Thames, Wairoa and Tauranga districts, though the loss, due to the heat of the season, was very great. About 1000 fry were retained in Auckland.

In 1876 a very large shipment arrived, which included 84,000 eggs for the Government, a large parcel (the number not specified) for Auckland, 60,000 for Napier, and a quantity for the Canterbury Society.

The Government supply was sent to Southland, where Mr Howard reported 10th March, 1877:

> The salmon were a most unqualified success; very nearly 18,000 have now been turned out, and about 200 kept for observation. All those turned out have been taken as far as possible up the Oreti, and placed chiefly in the five rivers at Lowther.

I do not know what was done with the 60,000 ova for Napier, the records appear to have been lost.

The Auckland Society hatched out about 20,000 fry, of which 10,000 were placed in the Waikato, 3000 in the Tuakau, 2000 in the Mahurangi River, and 600 in the Southern Wairoa. A thousand sent to the Whakatane River were lost in transit. Apparently the Wellington Society received 400 fry from this lot (or from Napier), and these were liberated in the Hutt River, seven miles from the mouth.

The Canterbury Society received 100,000 ova, which it was stated hatched out about 90 per cent.; but they only distributed some 20,000, which were liberated in the Waimakariri, Hurunui, Rangitata, Little River, Rakaia, Avon and Ashley.

In 1877 a big order for ova was sent to America, but owing to some bungling between the societies (which were always trying to act independently) and the Government, only 11 boxes arrived at Auckland, instead of 20 which were expected. Some were retained, and the others distributed to Nelson, Greymouth, Canterbury, Otago and Southland.

The Auckland Society received 100,000 ova, and distributed them as follows: 40,000 in the Punui River in the King Country, 8000 in the Thames, 7000 in a small stream near Wairoa North, and 43,000 in the Mangakahia River. About 95 per cent. were said to have hatched out. But this placing of ova in the rivers was rather a stupid procedure in face of the constant menace from eels and shags. Presumably the Auckland Society did not have proper ponds for dealing with large quantities of eggs.

The Wellington Society again received some fry, about 1700, in this year, though it is not stated where they came from, and liberated them in the Hutt, in the Manawatu River in the gorge, 35 miles from the sea, in the Wairau, 15 miles up, and in the Wanganui, ten miles from the mouth.

The Nelson Society received 25,000 and placed them in the Motueka and Wairoa Rivers. The Marlborough Society appear to have received 500 fry, but where from, I cannot find. Probably they are the same as are referred to in the preceding paragraph as having been liberated in the Wairau.

The Grey Society received a box of ova, presumably 25,000 eggs, and these were placed in the Grey River.

The Canterbury Society received 50,000 ova, and hatched out between 30,000 and 40,000 fry. Of these 10,000 were placed in the Waimakariri, 10,000 in the Rangitata, and smaller lots in the Shag, Hurunui and Heathcote. In 1880—three years later—three fish were caught in the Waimakariri, weighing 8 lb., $5\frac{1}{4}$ lb., and $4\frac{1}{2}$ lb. respectively. These were considered by many to be true quinnat salmon, but to make sure it was resolved to get a true quinnat from America either in spirits or in ice. However, with the lack of continuity which characterised so much of the work of the acclimatisation societies, this was never done, and the identification was not made.

The Otago Society shipment appears to have numbered 50,000, and it hatched out pretty well, for 13,000 fry were liberated in the Kakanui River, and 18,000 in the Waipahi.

The Southland Society received 100,000 ova, and placed 35,000 fry in the Oreti, 18,000 in the Makarewa (a tributary), and 10,000 in the Waipahi. Howard stated that the young fish were exceedingly healthy and strong.

On 1st February, 1878, the Colonial Secretary writing to the Governor states:

the half million salmon ova which arrived in November last have been successfully hatched and distributed to the different rivers of the colony; and that, owing to the extreme care with which the ova were packed, about 95 per cent. hatched out. In addition to the half-million sent at the request of the Government, an equal quantity has been sent to the various Acclimatisation Societies without charge.

In 1878 the Auckland Society imported 100,000 ova, and these were deposited in the tributaries of the Upper Thames, where numbers of young fry were seen. Those placed in the Thames and in the Waikato in 1876 and 1877 had also been seen, and the report concludes that "the full stocking of both these streams is now little more than a question of time." As a matter of fact the fish were never heard of or seen again.

In 1880 the Wellington Society liberated some 4600 fry in the Hutt River, probably from Auckland ova.

Nothing further in the way of introducing ova was done for many years. All these early experiments failed, and though an occasional doubtful fish was taken, as recorded below, the species did not succeed anywhere in establishing itself.

The Canterbury Society's report of 1885 states that a fish found dead in the Avon in February, 1884, was "identified by Dr Bean, Ichthyologist of the Washington Museum, as a Californian Salmon."

In its issue of 21st August, 1895, the *North Otago Times* congratulates Mr George Dennison of Hilderthorpe, on being the first angler to capture a properly identified true salmon in New Zealand waters. The salmon was taken in the Waitaki River. This paragraph was evidently based on the following correspondence. *The Field* of 20th July contained an account of four fish sent from the Waitaki Acclimatisation Society to the Editor, who passed them on to Dr Günther and Mr Boulenger—Nos. 1, 2, and 3 were identified as belonging to *Salmo fario*. Of No. 4 from the Waitaki River they say: "The specimen (length 29 in.; girth 15 in.; weight 9¾ lbs.) was a female with well-developed ova; it was not the English *Salmo salar*, but undoubtedly an American species, but which one has not yet been decided." In *The Field* of 27th July, Dr Günther further writes:

In the editorial note (to previous letter) you assume that one of the specimens sent to you and examined by myself is *Salmo quinnat*, commonly called "California Salmon." This specimen differs so much from the others of the same consignment, in the form of the head and its component parts, in the shape of the body and tail, as well as in coloration, that I must consider it to have a different origin than the other specimens, which, in fact, I regard as beautifully grown specimens of *Salmo fario*. It is certainly not a *Salmo salar* (as I think I demonstrated to Mr Tegetmeier), neither is it *Salmo quinnat*, which has a many-rayed anal fin, and is readily recognised. Being told that Californian Salmonoids had been introduced into the Waitaki waters, I consider it probable that that specimen might represent one of the numerous species of *Salmo* of the west coast of America, with which I am very imperfectly acquainted.

After a lapse of several years, during which no importations were made, the Government took up the matter seriously and a continuous policy was entered on.

Early in January, 1901, a shipment of 500,000 quinnat-salmon ova was received from California, a gift from the United States Fish Commission, and of this a portion was sent up to Hakataramea, and the balance to Lake Ohau. In the following year 23,600 yearlings were liberated in tributaries of the Waitaki, and 20,000 retained in the ponds. In 1903 12,000 twenty-months old, and 20,000 twenty-six-months old, were liberated in the Hakataramea River.

In January, 1904, 300,000 ova were received from the United States as a gift, and 98 per cent. hatched out. But evidently there were far more received, for during the year, 5000 three-and-a-half-year-old quinnat were liberated in the Hakataramea, while in 1905 the numbers set free were 448 four-year old, 12,000 one-year old, 224,252 eight-months old, and 162,613 three-months old. In December, 1905, a fish believed to be a salmon was caught in the Waitaki and submitted to Sir James Hector, who said it was a true salmon grilse, probably belonging to *Onchorhynchus quinnat*, but he was unable to determine the species with certainty at that early stage.

In 1906 another shipment of 500,000 ova was procured from the United States, Mr L. F. Ayson going to San Francisco for them. Half of these were taken to Lake Ohau, and 245,000 fry hatched out, which were liberated as soon as they absorbed the yolk sac. The other half were equally successful at Hakataramea, 224,833 fry hatching out. In addition to the foregoing there were liberated in the Hakataramea River 73 five-year-old, 12,587 two-year-old, and 53,378 one-year-old fish.

On 6th June, 1906, Sir James Hector received a fish from Hakataramea of which he wrote: "The fish sent is a true Pacific Salmon (*Onchorhynchus quinnat*), being a female of about 16 lbs. weight." On 29th June he reported on three more fish from the same river. One was a male, 25 in. long, weighing 6 lb., the second a female 22 in. long, and 5½ lb. in weight; and the third (probably only a three-year-old fish) was 17 in. long, and only 1½ lb. in weight. He thought it probable that all these fish, which were all in bad condition when received, belonged to *O. quinnat*. Both the Secretary of the Waitaki Society and the Collector of Customs at Oamaru stated that fish, supposed to be salmon, had been taken in the Waitaki and in Oamaru harbour by fishermen; those identified by Sir James Hector being of the number. These captures seemed to establish the fact that the fish were now returning to the river to spawn, and that the naturalisation of the species was secured.

In 1907 Mr Ayson again went over to San Francisco and brought back 500,000 ova, which reached Hakataramea on 8th April; and from these 482,000 fry hatched out. During the year 62 three-year-old, 21,282 two-year-old, and 224,647 one-year-old fish were liberated in the Hakataramea River; and later in the year 290,000 fry.

The report of the Marine Department for 1906–7 states that:

> this year, fish which are undoubtedly Quinnat salmon have been caught in the Hakataramea River, up which they are going to spawn; and the Manager of the Salmon Station reports that he has seen large numbers of them in the river.

During May and June the manager obtained 30,000 ova, *the first taken in New Zealand from these fish*, from which about 25,000 fry were obtained; of these 17,000 were liberated in the river. In addition to the few fish which were stripped, numbers of salmon were seen spawning in the side streams of the Waitaki from Station Peak to some distance above Kurow; also in the Ahuriri River higher up, in the Ohau, Haldane and Gray's Hill Creeks, and in the Mary Burn.

In 1908 103 four-year-old, 173 three-year-old, 18,937 two-year-old, and 166,851 one-year-old fish were liberated in the Hakataramea River; while 2000 were placed in the Selwyn River by the Canterbury Society. In this year more and larger fish ran into the Hakataramea, and 78,400 eggs were obtained.

In 1909 43 four-year, 199 three-year, 611 two-year, and 14,624 one-year-old fish from imported ova were liberated, together with 8000 one-year-old and 51,000 three-months-old fish from ova procured from river fish. The number of ova collected during the spawning season from running fish was 238,000.

During the year a 5 lb. quinnat salmon was caught near the mouth of the Rakaia River. This may either have come from the Waitaki, or have been one of the 200 fish which the Canterbury Society liberated in the Selwyn River in 1907.

In 1910 only 210,000 fry were liberated from the Waitaki, owing to the very dry summer which preceded the spawning season, and the low state of the rivers. Of these, 32,000 were reared at the station for liberation in the Hakataramea; 25,000 ova were sent to Tasmania, and 150,000 to the hatchery at Kokotahi, Westland. From this last lot of ova about 145,000 fry hatched out and were liberated in streams flowing into the Hokitika River, the stream which the Department decided should be stocked with this fish: 70,000 being put into the Harris Creek, 50,000 into Murray's Creek, and 25,000 into Duck Creek.

There were liberated into the Hakataramea River, 126 three-year-old, 821 two-year-old, 23,854 one-year-old, and 22,700 fry from the season's ova, and into the Seaforth-MacKenzie River 3000 fry.

The record for 1911–12 is taken from the Marine Department's report:

The largest run of Quinnat salmon which has yet taken place came up the Waitaki River last spawning season. They were found spawning in the main river itself, from a few miles up from the sea to where it branches off at the junction of the Ahuriri, Ohau, Pukaki, and Tekapo Rivers. Large numbers were seen spawning in these four large tributaries, and in the case of the Ohau and Pukaki they had run right through the lakes at the heads of these rivers, and were found spawning in the rivers beyond. These fish spawn in much deeper and heavier water than trout, and are therefore very difficult to capture for spawning purposes, as only a very small percentage of the fish which come in from the sea run up the smaller streams, such as the Hakataramea and Gray's Hills Creek, to spawn. The number of eggs collected last season was 240,000. These were disposed of as follows: 25,000 were sent to Tasmania, 157,500 to the Hokitika River, 3000 salmon fry to the Seaforth-MacKenzie River, and the balance were retained at the Hakataramea Hatchery. This season the Manager reports a good run of salmon spawning in the Tekapo, and the collection of eggs for this season is now proceeded with. During the year there were liberated from the Hakataramea ponds: 137 three-years old, 1011 two-years old, 8317 one-year old, and 12,426 four-months old fish.

Mr Ayson's report for 1912 says:

In point of numbers the run of salmon which spawned in the Waitaki River and its tributaries last season was quite equal to any of the previous years. The average size of the fish was, however, larger, and a peculiarity of the run was the very large percentage of male salmon which were captured. In other seasons the fish taken were about equal sexes, but last season nearly twice as many male fish were taken as females. Had the percentage of females been equal to other seasons nearly double the quantity of eggs would have been collected. The total quantity of eggs taken was 237,000, and these were disposed of as follows: 27,500 were supplied to the Tasmanian Government; 190,000 sent to the West Coast; 7500 retained at the Hakataramea Hatchery: 12,000, the loss during incubation. The salmon-eggs sent to the West Coast hatched out very well, and the young fish were planted in tributaries of the Hokitika River. It is interesting to note that a number of the young fish were taken in whitebait-nets in the tideway of the river during the early summer, showing that they maintain in this country the same characteristics of going to sea at an early stage of their existence as they do in their native country.

The following fish were liberated from the station in October, 1912, 503 three-years-old, and 567 two-years-old.

In 1913 251,000 ova were collected in the tributaries of the Waitaki River, of which 150,000 were hatched at the Department's hatchery at Kokatahi and liberated in the tributaries of the Hokitika River, 25,000 were sent to Tasmania, 45,000 were liberated at Hakataramea, and the fry of 20,000 were kept in the ponds at Hakataramea. Several thousand were hatched at Taupapa for the fresh-water aquarium at the Auckland Exhibition.

The following is from Mr Ayson's report for 1914–16:

A succession of floods in the Hakataramea River during the month of May, 1914, interfered seriously with the collection of salmon-eggs. The nets were washed out several times, and most of the salmon escaped up-stream and spawned in reaches of the river in the gorges. The manager at Hakataramea reports a heavier run of fish than the previous season, but owing to the unfavourable river-conditions the number of eggs collected was less. On account of the large number of salmon which escaped up the Hakataramea during the floods, the river was heavily stocked with the salmon-fry hatched from the natural spawning. In the late summer and autumn thousands of fine strong healthy fish were to be seen in every pool. The total number of eggs collected for the season was 243,000, which were disposed of as follows: 25,000 were supplied to the Tasmanian Government, 145,000 were sent to the West Coast for stocking the Hokitika River; 53,000 were hatched out at Hakataramea.

During the year 41,000 three-months-old fry, 19,254 yearling salmon, 580 two-year-old and 36 three-year-old fish were liberated.

During the angling season it was reported that salmon were caught with rod and line at the mouth of the Waitaki and Rangitata Rivers, and also that they were frequently taken with hook and line off Timaru and Oamaru by persons fishing for sea-fish.

The run of spawning salmon during the present season (1915) in the head-waters of some of the main tributaries of the Waitaki is undoubtedly the heaviest since the fish first commenced to run up from the sea. When recently in the Upper Waitaki district I was told by men who have lived for a number of years near the lakes and rivers in that region, and who are in the habit of observing the spawning every season, that there are more salmon and larger fish than any previous season. Mr Macdonald, manager of Ben Ohau Station, said that for some years he had watched the salmon spawning in the Ohau River, and this year from its outflow from the lake to its junction with the Waitaki River (a distance of about eighteen miles) he had never seen so many fish.

When I arrived at Benmore Station after leaving the Ben Ohau Camp, Mr Sutherland (manager) told me his head shepherd and musterers had returned a few days before from the head of the lake and reported hundreds of large salmon spawning in the Dobson River, so I went on to the head of the lake the same afternoon to see for myself, and ascertain whether it would be possible to get any eggs. Mr Fraser, the shepherd in charge there, provided me with a riding-horse and accompanied me to the Dobson the following morning. We examined the river from its junction with the Hopkins to where the camp joins it, a distance of about eight miles. The statement of the Benmore shepherds with regard to the fish I found to be practically correct. We saw a number of large fish in every pool—we counted as many as fifteen in one—and large spawning beds every chain or so as far as we went. From the appearance of the fish, the number of spawning beds, and the number of dead fish on the shingle beaches, it was evident that the spawning was about finished for this season. We saw some very large fish: two spent dead fish measured 42 in. and 42½ in., and I estimate the average size of the fish we saw at from 20 lb. to 25 lb. I may say that I inspected the Dobson at the end of the spawning season of 1911; then I saw from thirty to fifty fish and a number of spawning beds. This season I estimate there are well on to ten times as many, and much larger fish.

The Marine Department's (Mr Ayson's) report for 1915–16 states:

Last spawning season 251,000 eggs were collected; the most of these were taken at the Hakataramea Salmon Station. Two up-country collecting stations were worked, viz., Gray's Creek, on the Tekapo branch of the Waitaki River, and the Twizel River, a tributary of the Ohau branch. A large number of male fish were netted at Gray's Creek station, but all the females were spent fish and we were not successful in getting a fair quantity. Owing to the very low state of the Hakataramea River throughout the spawning season comparatively few fish ran up, and the quantity of eggs collected did not come up to expectations. The rack across the river, built on the American principle, was effective in stopping salmon from getting past, and all which came up were caught.

This season we decided to work the Dobson River, one of the rivers which flow into the head of Lake Ohau, and the men commenced operations there about the 20th April, and notwithstanding the difficulties experienced in working such a heavy river they have to date been very successful in getting eggs. The run of salmon on the Hakataramea is by far the heaviest that has been experienced. The river this season is carrying a good body of water, and as it is discharging directly into the main branch of the Waitaki, all the salmon which come up from the sea have to pass its mouth, and as the condition of the Hakataramea is so favourable, a good many fish enter its mouth and find their way up through the nets. The first run of fish this season was fully two weeks earlier than in any previous season, and another unusual feature is the large number of unripe fish which have been taken. If the salmon continue running for another ten days at the rate they have been doing, the collection of eggs taken from fish netted in the Hakataramea alone will exceed half a million. From my own observations and from reports from the Tekapo and Pukaki Rivers it would seem that there is an exceptionally heavy run of fish in the Waitaki and all the tributaries this season.

Salmon have been caught freely by anglers at the mouths of the Waitaki, Rangitata, and Rakaia Rivers this season, and information is to hand to the effect that large numbers have been caught by hook and line off the Timaru Breakwater. Last spawning season a large number spawned in the Rangitata River and its tributaries. All this goes to show that the salmon are fast making their way into the large snow-fed rivers north of the Waitaki. One of the reasons why the Waitaki River was chosen in the first instance for the salmon was because of the northerly set of the ocean current along the east coast, so that by stocking the Waitaki all the rivers north of that would in time be stocked by the fish being carried northward.

The success of the efforts to establish quinnat salmon in New Zealand is mainly due to the zeal and continuous energy of Mr L. F. Ayson, Chief Inspector of Fisheries.

The fish has spread south as well as north along the coasts of the South Island, and in 1917 was reported as being commonly taken by line fishermen off Otago Heads, and as moving down to Foveaux Strait.

The early attempts to introduce this fish apparently all failed, and it is interesting to summarise the dates of these attempts, and the rivers or districts which were stocked. They were: (1875) Thames, Waikato, Wairoa district, and Tauranga district; (1876) Tuakau,

Mahurangi, Mangakahia, Punui and Hutt, Napier district, Southern Wairoa, Manawatu, Wanganui; Grey, Wairau, Hurunui, Waimakariri, Rangitata, Heathcote, Shag and Oreti; (1877) Northern Wairoa, Mangakahia, Punui and Hutt; Wairau, Motueka, Hurunui, Waimakariri, Heathcote, Rangitata, Shag, Kakanui, Waipahi and Makarewa; (1878) Upper Thames; (1880) Hutt. Since 1901 they have been placed almost exclusively in the Waitaki and its tributaries, though a few were put in the Selwyn and the Seaforth-Mackenzie; and (in 1913) some hundreds were liberated in the Leith and Waikouaiti streams. More recently the Hokitika River has been chosen as the west coast stream to be stocked.

In a letter written some years ago asking for information as to the failure to establish the quinnat salmon in the rivers of New Zealand, Marshall McDonald, Commissioner U.S. Department of Fish and Fisheries, said:

We have experienced the same difficulty in attempting the acclimatisation of this species upon our eastern coasts; all experiments having failed completely after expending a large amount of money, and being tried on a scale of magnitude and under a variety of conditions sufficient to test fully the capabilities of our eastern streams in this direction. We have attributed the failure to the different temperature conditions prevailing in the rivers of the east and west coasts at the spawning season, which is from July to September. The streams of the west coast at this period, fed as they are by the melting snows in the mountains at the head of the large rivers, present a relatively low temperature which invites the ascent of the salmon in obedience to the natural instinct which pervades the entire family to move from warmer to colder waters in seeking their spawning grounds. On the east coast at this season of the year our rivers are warmer than the adjacent seas, and we have concluded therefore that the failure to enter our streams is due to the higher temperature conditions prevailing in them. This is probably true in regard to your own waters. The summer temperature of the Pacific Coast streams in which the salmon enter at the season of spawning rarely reaches 60° F. During the season on our eastern coast the temperature rises to at least 70° F., and sometimes reaches a maximum of 80° to 85° F.

Mr Ayson writes me (August 17th, 1915) in regard to this communication:

Quinnat salmon begin spawning about 1st April, and are finished by the end of May. In America there are two distinct runs, which are called the summer and winter runs. Marshall McDonald in the report you quote evidently referred to the summer run. The winter run commences well on in October, and finishes in December. The Quinnat eggs with which we stocked the Waitaki were all from the winter-run fish, and it is interesting to note that we have only a winter run of spawning salmon, so far, in the Waitaki; which would go far to show that eggs taken from winter-run fish in America only develop winter-run fish in this country.

The time of running in Southern Alaska, according to Dr Bean, is from May till August; at North Sound, the northern limit of its known migration, it is early in June. There can be no winter run in the far north of America, when the rivers are blocked with ice. From Mr Ayson's statement it would seem probable that all the ova received in New Zealand were from rivers which were open in the winter months. I have no record of size of the quinnat salmon captured in New Zealand. In Alaska rivers (Yukon, etc.) they average about 20 lb., but have been recorded up to 100 lb. and more.

I am told by some anglers who have caught the quinnat salmon in New Zealand waters, that the fish is a very inferior one for the table, being coarse and dry. It would be unfortunate, but a quite possible occurrence, that an inferior race has been introduced. Though so many separate shipments of ova have been received and hatched, it is possible that all those now running in the rivers of the east coast of the South Island are derived from one lot.

(See Appendix B, p. 557.)

*Sock-eye Salmon; Blue-back Salmon (*Salmo* (*Onchorhynchus*) *nerka*)

In 1901–2 a shipment of 500,000 ova of this species was sent from Canada to New Zealand *via* San Francisco. It arrived in the colony in bad condition, only 160,000 being good when unpacked, and there was a large percentage of deformed fish among those hatched out. Of these, 5000 fry were liberated in tributaries of the Waitaki, 91,200 in the streams flowing into Lake Ohau, while 20,000 were retained (on 30th June) in the hatchery at Hakataramea. In the following year 10,000 fry eleven months old, and 1500 sixteen months old were liberated from the ponds into the Hakataramea River. In 1903–4, 5981 fish two-and-a-half years old were liberated in the river, and at the end of the year (31st March) there were estimated to be about 2000 three-year-old fish left in the ponds. Of these, 1273 were liberated the following year, and by the 31st March, 1905, there were still left about 216 four-year-old fish. There must have been some loss in the ponds, for 34 were liberated next season, and only 18 remained in confinement.

On 22nd May, 1906, Mr Chas. L. Ayson, Manager of the Hakataramea Hatchery, wrote as follows:

While cleaning the pond-net which I have set at the mouth of the Hakataramea River to-day I caught on the top side of the net a fish about 16 in. in length, and which would, if in proper condition, weigh about five pounds. This fish is undoubtedly a sock-eyed salmon (*O. nerka*), which has been up the river for the purpose of spawning and was returning down stream. The fish was in a dying condition, being greatly covered with fungus. I now have it in formalin at the station. This, I think, should now set at rest all doubts as to them returning from the sea to spawn.

In the following year some fish believed to be sock-eye salmon were caught in Lake Ohau and sent to Sir James Hector, who reported as follows (4th May, 1907):

These fish are without doubt young sea-run specimens of the blue-back salmon, sock-eye (properly "Saw-qui") or red fish of Fraser River, and the krasnaia ryba of Japan—two males and two females. These particular fish are so much out of condition that they are not fit either for food or sport; yet, had they been allowed to mature, in the course of a few weeks they might have produced about 2000 fertile eggs, which would have been quite sufficient to stock Ohau Lake.

The sizes of these fish were respectively: males, 19 in. and 42 oz.; 18 in. and 36 oz.; females, 28 in. and 28 oz.; and 23 in. and 23 oz.

Mr L. F. Ayson, Inspector of Fisheries, writing to me in September, 1916, says:

An occasional sea-run Sock-eye has been taken in the spawning seasons. The last was a pair which I caught in the Twizel River (a tributary of the Ohau) in May, 1915, when collecting Quinnat salmon eggs. A number of them, however, have remained in Lake Ohau, and have run into the creeks at the head of the lake every season in the months of March and April to spawn. These lake salmon are not plentiful and are dwarfed in size, the average weight being under 2 lbs. This lake habit of some of the Sock-eye is not peculiar to New Zealand, as in their native home in British Columbia, a number remain in the lakes in the same way.

Dr Bean says of this fish in the Alaska Rivers that they average from 7 to 8 lb. in weight, though individuals are occasionally seen up to 15 lb. They run up the American rivers from April to June or July.

Canadian land-locked Salmon (*Salmo Sebago*)

In 1905 Mr L. F. Ayson, Chief Inspector of Fisheries, who went to America for eggs of quinnat salmon, brought over to New Zealand a case of eggs of the land-locked salmon for the Southland Acclimatisation Society. About 10,000 ova were received at the Wallacetown Hatchery, and these hatched out well. Owing to some accident, all of the fry with the exception of about 100, escaped into a race, leading into the Makarewa. The remainder were probably placed in Lake Te Anau, which was the destination originally intended for all, but the Society's statement closes with the words "No further record."

The species is incorrectly named *Salmo ovanniche* in the Southland Society's report.

White-fish (*Coregonus albus*)

In December, 1876, a case with 125,000 ova was despatched from San Francisco (from U.S. Fish Commission) for the Government of New Zealand, and arrived in Auckland on 29th January, 1877. Owing to there being no ponds at Wellington, the eggs were sent on to care of the Christchurch Society, where they were delivered on 3rd February. They began to hatch at once, for Dr Hector writing to the Hon. Spencer F. Baird on 9th February says: "The Secretary reports that over 200 young fish have come out, and says they are three-quarters of an inch long (five days old), very transparent, with bright yellow eyes, are very lively, and appear to be doing well." On 22nd March, Mr Farr writes that owing to water overflowing the boxes, all but six or seven of the fry were lost. It was hoped that as they were washed out into the race, some of them would turn up again, a hope that was not realised.

The Auckland Society received a box of ova in 1876, presumably from the same shipment, but only nine fry hatched out. It is very difficult to tell whether the Society got a supply on their own account previous to the Government's shipment. There was such a stupid want of co-operation on the part of many of the Societies, and the Auckland one especially seems to have been a sinner in this respect. It is in reference to this lot that Mr Dansey writes me on 28th June, 1916, as follows:

While I was in charge of the telegraph station at Te Ngae in the East shore of Lake Rotorua in 1876, some white fish, brought over from America by the late Joshua Firth of Matamata, were turned out into Te Awahua stream on the North West shore of Lake Rotorua, an exceptionally cold stream. I never heard of anything having been seen of them afterwards. Had they survived they would not have escaped notice, as there was at that time, and for some years afterwards, nothing in the lake but inanga (minnow), bull-heads and fresh-water crayfish.

A box of the Government shipment was sent down to the Southland Society's ponds at Wallacetown, but the eggs were dead and quite undistinguishable on arrival.

On 14th February, 1878, another shipment of 500,000 ova arrived in Auckland, and was distributed. One box with 50,000 ova was retained and hatched in Auckland; they turned out very badly, and were practically all lost. A second box with 50,000 ova was kept in Auckland (presumably in ice) till 19th April, when it was sent on to Mr A. M. Johnson of Opawa, Christchurch. Mr Johnson received them on the 23rd, and reported them "as all hopelessly bad, with the exception of three."

The Canterbury Society received 100,000 ova, but only acknowledged 20,000. From these only 12 fish hatched out and eight survived. These were placed in a stream running into Lake Coleridge.

The Waitaki (Oamaru) Society received a box of 50,000 ova, but there is no report as to what was done with them.

The Otago Society also received one box, and were more successful than any of the others, about 1000 young fish hatching out "which throve very well at the breeding ponds." Mr Arthur, writing on 10th July to Dr Hector, says:

The last I know of them is, that Deans started with the whole lot for the Wanaka, before they had reached that age and size which we all agreed to be most prudent before turning them out. He got as far as the Teviot, but they had nearly all died, except one or two which were liberated in a lagoon communicating with the Clutha.

The remaining four boxes with 200,000 ova were taken by Dr Hector himself to the Bluff by steamer, and conveyed as rapidly as possible to Lake Te Anau. By special train to Lumsden, and travelling all night in an American wagon, Te Anau was reached by 3 p.m. on 23rd March, and the boxes were unpacked.

Out of the four boxes of ova three were almost completely destroyed by the growth of white fungus, and the young fish, which had evidently been hatched out for some time were reduced to a pulpy jelly. In the fourth box in which there was only a slight growth of fungus, a considerable number of the ova were found in sound condition, and hatched out rapidly as they were transferred to the trough.

Mr S. Herbert Cox reported from Te Anau to Dr Hector on 20th February: "The whitefish are doing very well. They are all hatched out, and are feeding well, they will, I presume, be let loose in the lake about Saturday, if it is calm enough." There is no further record.

The Nelson Society received an earlier consignment of ova which was brought over at the expense of Mr John Kerr of the Lake Run, but I cannot learn the date of this importation. "The ova were placed in a creek running into Lake Rotoiti, and hatched out well." None have ever been caught, but Judge Broad in his jubilee history of Nelson says: "it is believed that they exist in the lake in considerable numbers." Mr F. G. Gibbs, Chairman of the Rotoiti Domain Board, writing on 3rd February, 1917, says:

I have frequently made inquiries about the whitefish, but I cannot find any one who has seen any trace of them. I well remember the ova being taken up to the lake, and I also remember that the local newspapers shortly afterwards reported that the young fish had been seen. If they really were seen, which many doubt, they have since completely disappeared, probably exterminated by eels, which are very abundant in the lake.

In 1879 the Auckland Society imported 500,000 ova. Most of

these were placed in Lake Taupo, but small lots were distributed to Lakes Okataina, Titikapu and Tarawera, and to the Awahou Basin discharging into Lake Rotorua. About 50,000 were placed in the hatching-boxes in the Domain, but failed to hatch out.

In 1880 another shipment of 1,000,000 ova was made by the Government, and distributed to various societies.

The Auckland Society received 50,000 ova. "The fish hatched out very well indeed, but the temperature being 65° F., they died day by day, and in a few days all but two had died."

The Napier Society also received 50,000 ova on 19th January. The hatching commenced the same day, and about 200 were hatched out, but by 30th January all were dead but 12.

The Nelson Society received 250,000 ova, but they were kept too long in Auckland before being forwarded, and reached their destination in a bad and stinking condition, many of them apparently already hatched out and dead. Only about 40 or 50 fish hatched out alive, and "with the exception of some eight or ten, these young fish quickly died off; those that were alive were put into one of the ponds, where they appeared to thrive." The Secretary adds: "I regret to say they suddenly disappeared. I do not think they died, as they were constantly looked after; and they were large enough to be seen in the pond, as the water was quite clear."

The Christchurch lot amounted to 300,000. Hatching commenced on 20th January, and the whole of the young fish—estimated at 50,000—were hatched out by the 29th.

> Fungoid disease, however, made its appearance among them, although every precaution was taken to insure success, and daily the numbers were rapidly diminishing. On 24th Feb. the whole of the fish—numbering about 25,000—were liberated in Lake Coleridge. After watching them for a few seconds we noticed that they took a spiral course to the depth of about eight inches, then dived suddenly downwards and were lost to sight in the deep azure water.

They were liberated from a boat at a distance of about half a mile from the shore. "The temperature of the water in the lake was taken, and to our astonishment was found to be 59° at a depth of fifty feet, and 60° at the surface."

Mr A. M. Johnson of Opawa received two boxes (100,000 ova?), from one of which only 28 young fish were obtained, and from the other many thousands hatched out. This was on 29th January. On 4th May Mr Johnson writes to the Colonial Secretary stating that

> after seven weeks of the time of hatching, the numbers continued to visibly diminish daily, in spite of every care and precaution, till the total number

left cannot now be as many hundreds as there were thousands; those fish liberated in ponds full of crustacea and insect life appearing to share the same fate as the ones in deep and protected races.

I can find no further report of this experiment.

The final lot of 250,000 was sent down to the Bluff, which was reached on 19th January, met by Mr Deans, curator of the Otago Society, and conveyed to Queenstown the same day. "They commenced to hatch at once, but died shortly after. As there seemed no chance of saving them, it was decided to turn them out at Beach Bay, about eight miles from Queenstown, the latter place being infested with trout and perch. All, with the exception of from 1200 to 1300 (ova), were hatching out in the cans while going up the lake, and seemed quite lively when turned out. I regret to say I believe quite one-half of the ova have gone bad." By 28th January the remaining ova were reduced to about 800 or 900, and about 40 live fish, and these were liberated at Halfway Bay, between Queenstown and Kingstown. The temperature of the water was 56°.

The only subsequent reports of this 1880 shipment come from the Canterbury Society. The annual report for 1885 states that "we are credibly informed that shoals of the White Fish, placed by us in Lake Coleridge, have been recently seen there." In 1886 it is said "we have received several reports of the White Fish having been seen in Lake Coleridge, some of large size, and that they are multiplying has been proved by some young ones being washed ashore in a gale." In a letter received from Mr Edgar F. Stead in April, 1916, he says: "No whitefish have ever been caught in Lake Coleridge, nor have any skeletons of them been found."

In 1884 the Nelson Society received about 1,000,000, of which 300,000 were placed in the hatching-boxes, and the remainder were put in Lake Rotoiti. There is no report of any of these, for the Nelson Society have lost their records; but in 1901, Mr A. Rutherford was of opinion that the fish were then in Lake Rotoiti.

Though all previous attempts had so far failed, it was determined to try again to introduce this desirable species, and two shipments were made to the Government from San Francisco in 1886–87.

On 13th February, 1886, 1,000,000 eggs from the U.S. Hatchery at Northville, Michigan, were shipped by the 'Alameda' which arrived at Wellington on 12th March. On opening the boxes the contents were found to be putrid, and this appeared to be due to carelessness on board the steamer, as the ova had been "packed with the greatest care and in the most approved method."

On 15th January, 1887, the 'Alameda' again took a shipment of

1,500,000 white-fish eggs, and arrived in Auckland on 5th February. They arrived in excellent condition, but according to the letter of the Minister of Marine on 26th February to the U.S. Commissioner of Fish and Fisheries, though "the percentage of bad eggs on being unpacked was less than one per cent.," yet "on being placed in the water the ova did not separate freely, and on the second day nearly fifty per cent. was dead."

Ten trays were handed over to the Nelson Society on the 6th, but again I can find no record of what was done with them. This lot represented 500,000 ova. One tray (50,000 eggs) was sent to Clinton. About two-thirds of the eggs were bad, but a considerable number of fry were hatched out. The mortality, however, was so great that the rest, about 1000 in number, were turned into a large pond. One or two were seen later, but the rest disappeared. On February 25th, 1889, the pond was emptied and one specimen was found, which died the following day. It was 12 inches long, 3 in. deep and weighed close on 11 oz.

A small lot from this tray was taken to Opoho, Dunedin, where over 200 fry were hatched out, but some died and the rest disappeared.

The remaining 19 trays, containing 950,000 ova, reached Queenstown on 8th February. The following report by Mr Davidson, curator of the Lakes District Society is given in full:

The ova were placed in the boxes on the 9th, the temperature of the water being 47°, lowered from 50° by ice. I was able to keep the temperature at 47° for two days with ice; after this, when the ice was finished, the temperature remained at 50°, and never rose higher. Some of the fry were moving in the boxes on the 10th, but the greater portion died in the egg, not more than 50,000 hatching out. When unpacking the ova it was found that too much pressure had been used, making the ova stick together in one mass; the ova, however, looked perfectly healthy, and were all alive, but it was impossible to separate them. If the ova had not been so far advanced there would have been a much greater chance of success. When the fry were fifteen days old, I observed the sac absorbed on most of them. I liberated about six thousand in Lake Wakatipu on 28th February. On 5th March about twenty thousand were liberated in Lakes Wanaka and Hawea. I then began to feed those remaining on bullock's blood. They appeared to thrive well on it for a time—say—for about a month; after that they appeared to be not thriving so well. I therefore liberated the whole of them in Lake Wakatipu about 31st March. The fry have been seen on several occasions, and are doing very well apparently, being 1½ in. long. I consider they are established without a doubt this time. I have liberated quite fifty thousand in healthy condition.

This was written on 1st June, 1887; there is no further record.

Nothing further was done in the way of attempting to introduce white-fish till 1904, when Mr L. F. Ayson, Chief Inspector of Fisheries, went to San Francisco and brought back with him 2,000,000 ova; half of these were taken to Lake Kanieri and half to Lake Tekapo.

Mr L. F. Ayson, writing in September, 1916, says:

In 1904, on my recommendation, the Marine Department decided to make a systematic effort to introduce this fish, and hatcheries were established on Lake Tekapo, in the Mackenzie country, and Lake Kanieri, on the West Coast, and these hatcheries were equipped with the proper hatching jars. The eggs of whitefish cannot be successfully hatched in the ordinary trout boxes. In the American and Canadian hatcheries a special jar is used, and in the attempts to hatch out the earlier shipments of eggs imported, in the ordinary trout boxes, I am afraid very few of the young fish came to maturity. Two million eggs were imported each year from 1904 to 1907. Each shipment of eggs arrived in first-rate condition, was successfully hatched out, and the fry liberated in the lakes mentioned.

Writing on 22nd May, 1906, Mr Chas. Ayson, manager of the Hakataramea Hatchery, says:

While at Lake Tekapo in January last I was informed by two different persons that they saw on different occasions at the bridge where the Tekapo River flows out of the lake, a strange fish, and from the description given me I am inclined to think that the fish seen were whitefish.

And Mr L. F. Ayson on 25th May states:

I may say that reports are current at Lake Kanieri similar to those mentioned by the manager about Tekapo, viz. that strange fish have been seen, and from the description given resembling whitefish. At Kanieri Lake these fish are reported as having been seen in the shallow water near the foot of the lake.

Early in 1907 Mr Ayson brought the last shipment of 2,000,000 ova from San Francisco. Half of these went to Lake Tekapo, arriving at the head of the lake where the hatchery was situated on 9th March. The other half went to Lake Kanieri on the 11th; the eggs were in the latter case just 43 days from hatchery to hatchery.

In his letter already referred to Mr Ayson says:

These fish are not easily seen, and the only way to prove whether they have taken a hold or not is by having the lakes tested with deep set nets. So far this has not been done.

Reports have reached me several times of late of schools of fish being seen in both these lakes—fish, which, from the description given, do not resemble trout. When I was on Lake Kanieri in the beginning of August last (1916) I saw fish shewing on the surface of the lake several times which were quite different in their movements from trout.

Lake Herring; Cisco Herring; or Lesser White-fish
(*Coregonus artedi*)

In 1907 a small shipment of the eggs of this species was brought over from America by Mr L. F. Ayson, Chief Inspector of Fisheries. They were brought partly because of their value as a food-fish, and partly to increase the food supply for trout in Lake Rotorua and other lakes in the Thermal district. About 40 per cent. of the eggs died *en route*. The remainder were taken to Rotorua to be hatched and liberated. Apparently nothing further has been heard of them since.

Family CYPRINIDÆ

*Carp (*Cyprinus carpio*)

In 1864 Mr A. M. Johnson shipped 200 carp in London in the 'British Queen' bound to Lyttelton. The experiment was, however, unsuccessful, all the fish dying during the voyage. In 1870 Mr E. Dowling imported a number of Chinese and Prussian carp into Canterbury.

In 1867 the Auckland Society introduced 114 Prussian carp. Of these 12 were placed in Takapuna Lake.

In 1868 a number of fish were shipped by the 'Celestial Queen' for Otago, but none survived the voyage.

In 1881 the Otago Society obtained six fish, but I do not know where from. These were placed in a dam at Waihemo, but this burst, and the fish were washed away.

In February, 1911, Mr E. T. Frost reported that carp were very common in the Waikato district. They are red, golden, white and black, red and black, and white. The Maoris eat them in great numbers but find them too bony.

Carp were liberated at Lake Mahinapua on the west coast

Mr W. W. Smith tells me that they are common and of large size in Taranaki.

Mr R. D. Dansey of Rotorua, who has given me a great deal of most interesting information regarding introduced animals in that part of New Zealand, tells me that carp are very plentiful in Rotorua, and are called "Morihana" by the Maoris. He gives the origin of this name as follows:

I was present when in 1873 a small number of carp were first liberated in Lake Taupo by Sub-Inspector H. Morrison of the Armed Constabulary, then stationed at Tapuaeharuru. They had been brought up from Napier in a billy. Members of the Constabulary had been purposely stationed at intervals of several miles along the track from Napier to Taupo, a distance of 90 miles, and the billy and its precious contents was passed on from man to man till it reached Tapuaeharuru, where the fish were liberated near the outlet of the lake. All hands and the cook from the redoubt proceeded to the spot to see the liberation, and many natives came across the Waikato River to see the new pakeha fish. There was great cheering as the little carp swam out from the bank. The natives called them then and there "Morihana" after Captain Morrison, and they are still only known by the natives in the Taupo and Rotorua districts by this name.

In 1880 five of us subscribed £1 each and commissioned "Jack Loffley" to bring a billy of young carp down from Taupo, where by that time they had become exceedingly numerous. They were duly liberated at the mouth of the Utuhina Creek and in a small lagoon emptying into the Lake, where they multiplied at an enormous rate. The Maoris did not like them, considering them too full of bones and dangerous for their children. Ere long a lucrative trade in gold-fish sprang up between the Ohinemutu Maori children and visitors. Carp frequenting the thermal waters along the southern shores of Lake Rotorua soon turned a bright red or white, some partly red and partly silver. The children became adepts at catching them with their hands among the reeds and rushes, up to a quarter of a pound weight or more.

In June, 1916, at a meeting of the Arawa tribe in Rotorua, it was decided to send a telegram to the Hon. W. H. Herries, Minister for Native Affairs, protesting against a recent Government notification forbidding the catching of *carp* in Lake Rotorua, and pointing out that the Maoris were thereby deprived of a food supply which they had enjoyed for the last 30 years.

The Canterbury Society received a number of silver carp from Sydney in 1868; I do not know what was done with them.

Golden Carp; Gold Fish (*Cyprinus carassius*)

The first attempt to introduce goldfish into New Zealand was made by Mr A. M. Johnson, who succeeded in bringing a few alive (the only survivors out of a large and varied assortment of fish) in the 'British Empire' in 1864. These were landed at Lyttelton.

In 1868 the Canterbury Society received a number from the Acclimatisation Society of Melbourne.

I do not think these fish are specifically distinct from the ordinary carp[1].

[1] The following paper on the "Rapid Growth of Carp due to Abundance of Food," by J. H. Brakeley, is taken from vol. VII, *Bulletin* of U.S. Fish Commission, 1889:

"The European carp in becoming naturalised in this country has changed its habits in several important particulars. Instead of hibernating for several months with its nose in the mud, as in Europe, here it does this for a very short time, if at all, even as far north as the Middle States. The eggs hatch here in from four to seven days, according to the temperature of the atmosphere, while in Europe it requires from twelve to twenty. Here it readily takes the bait when skilfully

Japanese Minnow (*Pseudorasbora parva?*)

A number of minnows from the rice-fields of Japan were imported to the ponds at Opawa by the late Mr A. M. Johnson. These probably belonged to the species named above. Several were distributed, but I have no word of their subsequent history.

Gudgeon (*Gobio fluviatilis*)

In 1864 Mr A. M. Johnson shipped a number of gudgeon on board the 'British Empire' for Canterbury, but none of them survived the voyage.

In 1868 another attempt was made by Mr Frank Buckland, a number of fish from the Thames, presented by Mr S. Ponder, being shipped to the Otago Society by the 'Celestial Queen.'

This shipment was also unsuccessful.

Barbel (*Barbus vulgaris*)

Several specimens of barbel were included in Mr Johnson's unfortunately unsuccessful experiment, made in 1864.

Bleak (*Alburnus lucidus*)

Mr Johnson also shipped some bleak by the 'British Empire' in 1864, but all died on the way out. The tank in which these and several other species of fish were carried was lined with slate, and so divided by perforated partitions that fresh water flowed freely through it; there were also contrivances for aerating the water. The whole was surrounded with a framework case, with double cane-matting, which

presented, while it is said not to bite at the hook in its native land. So, in becoming Americanised, it has become quite a different fish in habit, if not in form.

The rapidity of growth, too, which characterised many of those distributed by the U.S. Fish Commission during the first four or five years, seemed to foreshadow another important change of habit. It was supposed that the waters of this country were more favourable for its development than those of its native land. But in this, I fear, we are doomed to disappointment. Further experience has shown that this remarkable growth of which we hear so much, and of which there are many examples on record, was due to the abundance of food with which the carp were supplied, rather than to other causes. The small number furnished by the Government to each applicant—usually not over twenty—were frequently placed in large ponds, and often at the close of the first summer the fish had reached a weight of from one to two pounds apiece, and by the end of the second summer from four to five pounds, and in some instances their growth far exceeded this. But now, since they have multiplied so that we can fully stock our ponds, their growth is much less rapid. In the autumn of 1884 the writer placed a little over 2500 carp, then one summer old and much larger than their parents when received from the Fish Commission, in a five-acre pond. In the following autumn they were found to average about eleven ounces each; and last autumn, being the close of their third summer, they fell a little short of a pound apiece, and this, too, with the number in the pond reduced about one-fourth. In another pond of about half the size the growth was no more rapid."

was kept constantly wet in the tropics. Troops of snails, water-lilies and weeds of various kinds were also introduced, partly for food and partly to assist in aeration. A lump of white lead in one of the tanks seems to have poisoned nearly all the fish before it was discovered.

*Tench (*Tinca vulgaris*)

Among the fish shipped from London for Lyttelton by Mr A. M. Johnson in the 'British Empire' were several tench. This was in 1864. Unfortunately all the fish died on the voyage out.

In 1867 the Canterbury Society received some live fish from the Hobart Acclimatisation Society, and in a report issued in 1871 it is said "they have successfully multiplied."

In the following year the Southland Society received some from Mr Morton Allport of Tasmania.

In 1868 Mr Frank Buckland shipped a number for Otago by the 'Celestial Queen,' but none reached their destination. They got on well for some weeks till one day one of the ship's boys

who was changing the water for the fish, got them into a bucket of fresh-water and emptied it over the ship's side, instead of so doing with a bucket containing the stale water that had been drawn off.

In 1869 the Otago Society liberated 18 in the Ross Creek Reservoir, Dunedin, but the report does not state where they came from. In 1880 the Society sent 30 to Otekaike and 30 to Elderslie, both in the Oamaru district. In 1887 the Elderslie ponds were overhauled and cleaned, when great numbers of tench were distributed throughout the district.

They are to be found in a few localities throughout South Canterbury in ponds and dams, as at Cave and near Timaru. But they do not seem to occur in any waters south of the Oamaru district. A good many are to be met with near Hokitika and other localities on the west coast.

They were formerly introduced into the Rotorua district as food for trout, but Mr Dansey tells me there are certainly none there now.

Rudd; Red-eye (*Leuciscus erythrocephalus*)

This was another of the species which Mr Johnson endeavoured to introduce in 1864, but unsuccessfully.

Dace (*Leuciscus leuciscus*)

Mr Johnson had a number of dace in the shipment of 1864, but none survived the voyage.

Roach (*Leuciscus rutilus*)

Shipped by Mr Johnson by the 'British Empire' in 1864, but all died on the voyage.

Minnow (*Leuciscus phoxinus*)

A number were shipped by Mr Johnson in 1864, but all died on the voyage out.

Family SILURIDÆ

*American Cat-fish (*Pimelodus cattus*)

In 1877 the Auckland Society received 140 fish from Mr T. Russell, and placed them in St John's Lake. They were lost sight of for a time, but reappeared in considerable numbers in 1884, and it was stated that they were evidently increasing fast. In 1885 they were caught in hundreds, and were sent to many parts of the provincial district. At the present time (1916) they are plentiful in Lakes St John and Takapuna.

In 1885 some 30 fish were sent down to Wellington and placed in a Mr Perry's pond at Petone, but no one to-day seems to know anything about them. Probably about the same time Mr A. M. Johnson of Opawa obtained some from Auckland, and had a few there till recently (1916).

Mr Jas. King of Hokitika informs me that some were liberated in Lake Mahinapua on the west coast, but he does not know when. In the annual report of the Westland Society for 1904, the secretary states that "Mr T. Green, of the South Spit, showed me two nice American Catfish, weighing 3½ lbs., caught accidently while fishing for eels in Mahinapua Creek. Mr Green considers the lake and creek to be full of them." Mr Ayson tells me that at the present time (1916) they are plentiful in Lake Mahinapua.

Mr W. W. Smith informs me (April, 1919) that they are abundant in ponds about Ashburton, probably obtained from Mr Johnson in the eighties.

Family GASTEROSTEIDÆ

Stickleback (*Gasterosteus aculeatus*)

In 1885 Mr S. C. Farr obtained 36 sticklebacks from the Brighton (England) aquarium, and shipped them by the 'Kaikoura' to the Canterbury Society, but they all died in the tropics.

In 1892 Mr Clifford brought a number of these little fish out to Dunedin for the Otago Society. Some of them were kept for years at the Opoho Hatchery, and some were transferred to Clinton; but both lots seem to have disappeared, which perhaps was not to be regretted.

Some were sent to Mr Johnson of Opawa, who kept them close in his aquaria, and wrote stating what an undesirable importation they would prove if liberated in our rivers. I do not know what came of any of these, but I am not aware of any sticklebacks being in any of the New Zealand waters at the present time.

Family PERCIDÆ

*Perch (*Perca fluviatilis*)

The late Mr A. M. Johnson claimed that he first introduced perch into New Zealand; he arrived in Christchurch in 1864 from the Old Country. His first shipment of 200 fish per 'British Empire' in that year was, however, unsuccessful.

In 1868 three perch were received by the Otago Society from the Hobart Society, arriving in July in the 'Swordfish,' and these were turned into the Ross Creek Reservoir, which supplies Dunedin with water. In September of the same year Mr Clifford landed 19 more from Hobart, and these were placed in the same reservoir. In 1870, 18 more were landed. These fish increased and were spread far and wide through Otago, viz. to Lawrence, Gore, Clydevale, Kaitangata, Otekaike, Elderslie, Tapanui, Waikouaiti, Waihemo, etc. They were also sent to Ashburton, to the Canterbury Society, and to Nelson. The Otago Society's report for 1891 says: "These fish are becoming very numerous; Kaitangata Lake and Lovell's Creek are simply swarming with them." In 1892 the report is: "Perch are still on the increase. Some have been caught weighing as much as 5 lbs."

Also in 1868 the Southland and Canterbury Societies received perch—the number not specified—from Mr Morton Allport of Hobart. The annual report of the latter society for 1871 says: "they have successfully multiplied and no further importations are needed."

In 1883 Mr Shury of Ashburton reported to the Canterbury Society that "perch in large numbers could be seen in some streams on the Wakanui road," and the report of the following year shows that they were extremely abundant in the district.

In 1877 the Wanganui Society imported about 50 dozen perch from Ballarat, Victoria: "They were put into canvas bags filled with water and slung on frames on board ship. They arrived in capital order." A second consignment was not so successful, about half dying.

In 1878 the Wellington Society got about two dozen from the preceding Wanganui shipment and placed them in the Wellington Reservoir. In 1886 they were very numerous, and several lots were placed in lagoons in the Wairarapa, and in lakes near Otaki.

In 1885 the Hamilton Domain Board obtained 100,000 ova from the Canterbury Board, and liberated the fry in the Waikato district.

In 1887 the Taranaki Society obtained a number from Mr Johnson of Opawa, and Mr W. W. Smith, writing from New Plymouth in February, 1916, says: "They are common in the district; though introduced many years ago, I have not seen any large specimens."

Mr Jas. King informs me that they were also liberated in Lake Mahinapua, on the west coast.

Family SERRANIDÆ

Gippsland Perch (*Percalates colonorum*)

In 1868 Mr A. M. Johnson imported a number of these fish, and kept them in his ponds in Opawa, Christchurch.

Family OSPHROMENIDÆ

Paradise Fish (*Polyacanthus opercularis*)

In 1908 Mr A. M. Johnson imported some of these aquarium fish from Japan for his tanks at Opawa.

Gourami (*Osphromenus olfax*)

In 1869 Captain Tobin of the 'Sea-Shell' attempted to bring a number of these fish from Mauritius (where they have been acclimatised) for the Canterbury Society. They did not survive the voyage, however.

Family PLEURONECTIDÆ

Turbot (*Psetta* (*Rhombus*) *maxima*)

A successful attempt to introduce this species into New Zealand was made in 1913. The shipment was made on behalf of the Government and was under the care of the late Mr T. Anderton, curator of the Portobello Marine Fish Hatchery, Dunedin[1].

[1] The only previous attempt to carry live sea-fish across the equator to southern waters appears to have been that of the New South Wales Fishery Department in 1902, under the superintendency of the late Mr H. C. Dannevig. On that occasion 722 plaice, 28 black soles, *four large turbot* and four large brill were shipped at Plymouth on the 'Oroya' on 21st June. *The turbot* and the brill died before the voyage was half accomplished, the last being taken out on 11th July. On arrival at Fremantle on 24th July, 581 plaice and 23 soles were alive. The 'Oroya' reached Sydney on 2nd August, and the surviving fish (560 plaice and 23 soles) were liberated in an enclosure at the Maianbar fish-farm in Port Hocking, situated about one and a half miles inside the Heads. Owing to inadequate preparation no suitable permanent enclosure for the fish was secured; the result was that during the intense heat of summer the whole stock of plaice died. Had the attendant in charge of the station opened the sluices and allowed the fish to escape into deeper and cooler water, they might have kept together and spawned at a later date, instead of being lost altogether. The soles appear to have survived the first summer.

Seven hundred young turbot, varying from one to a little over two inches in length, were caught in the surf at Whitsand Bay, some 20 miles from Plymouth, in September, October and November, 1912. Of these, 400 died while being retained for shipment in the tanks at the Plymouth Marine Biological Station. The remainder, 298 in number, were placed in a tank on board the 'Waimana' on 12th January, 1913, and Otago Harbour was reached in March.

The survivors, 195, were placed at once in the tanks of the Portobello Marine Hatchery, and by scrupulous attention to cleanliness and feeding, their growth and healthy condition have been phenomenal. On 19th May, 1916, more than three years later, the fish numbered 182, only five having died and eight having been liberated in the harbour. Before leaving Plymouth, Dr Allen strongly recommended that at least 75 per cent. of the fish should be liberated immediately on arrival in New Zealand. However, local knowledge of the ground-feeding habits of so many of our indigenous fishes, and of the possibility of the majority if not all, of these small fishes being devoured by such species as the red cod, ling and groper, led to the decision to retain the whole lot, if possible, in the tanks until they had attained to such a size as to guard them against much risk of capture. Owing to the low temperature sometimes experienced in the winter months, care was taken to slightly heat all the water passing through the tanks, and it was accordingly not allowed to fall below 42° F.

As the fish had increased so much in size—many of them measuring as much as eighteen inches in length—and were crowded in the tanks, and on account of the time, labour and expense in feeding them, it was resolved to liberate a large proportion of them.

Accordingly on 19th May, 128 fish were placed on board the S.S. 'Invercargill' by Mr Anderton, and were liberated during the night in a previously selected bay (Tautuku Bay), where it was considered they would be safe from trawlers, and from most of their natural enemies. The fish were liberated as rapidly as possible at one spot, in about ten fathoms of water on a clean sandy bottom, and from their schooling instinct—which was very marked during their confinement—it was anticipated that they would tend to keep together.

The remaining fish still showed no signs of spawning, and as some of them were 22 inches long, another large batch was turned out in the same locality as the previous lot on 1st September, 1917. The temperature of the water at the time of capture of the young fish in the English Channel averaged 56° F., and this was the temperature of the water in Tautuku Bay at the time of liberation of the fish in May, 1916.

There are now 14 fish left in the Portobello Hatchery Tanks, some of them just 24 inches in length, but they show no signs of spawning. I am inclined to think that the difficulty of getting most kinds of sea-fish to spawn in confinement is due to the shallowness and consequent lack of pressure of the water in which they are kept. At the spawning season all the large flat-fishes of indigenous species move out of shallow bays and estuaries, and it may be that a pressure of 30 to 40 fathoms of water assists the fish in the extrusion of the ova. We have noticed at the Portobello Hatchery that native flounders (*Rhombosolea plebeia*) taken at spawning time and placed in our ponds can hold up their ova for weeks.

It is impossible to say whether the turbot has or has not become established in New Zealand waters. Even if any succeed in spawning their progeny are not likely to be in evidence for some years.

Chapter VII

MOLLUSCA

Class *GASTROPODA*

Order PECTINIBRANCHIA

(Family Muricidæ, see Appendix B, p. 558.)

Family Volutidæ

Ericusa sowerbyi, Kiener

An example of this Australian mollusc was picked up in Evans' Bay, Wellington, by Miss M. K. Mestayer some years ago. She suggests that "it may have come to New Zealand adhering to the bottom of some ship, and may possibly have been knocked off by the vessel being put on the Patent Slip." I have not found or heard of any other examples being found in New Zealand waters, but there s no doubt that, from time to time, marine organisms are so introduced.

Family Conidæ

Conus marmoreus, Linn.

Early in 1917 the lighthouse-keeper at Farewell Spit picked up a living specimen of this species, which he gave to Captain Bollons of the S.S. 'Hinemoa.'

Miss Mestayer says of this genus:

> In his Catalogue of the Marine Mollusca, 1873, p. 23, the late Captain F. W. Hutton recorded two species of *Conus* as belonging to New Zealand, *Conus zealandicus* sp. nov. and *Conus distans* Hwass, N.Z. (Cumming). The type of *Conus zealandicus* is in the Dominion Museum.

This species was founded upon a single specimen from the Bay of Islands. It has since been identified as *Conus anemone*, Lamarck, from Australia, by Suter, while in Tryon and Pilsbury's *Manual of Conchology* it is placed as a synonym of *Conus aplustre*, Reeve. In 1882 Mr Justice Gillies stated that he had a single specimen of *C. aplustre* from the Bay of Islands, but suggests that this and other shells picked up in the same locality were perhaps dropped from some South Sea whaler.

Order PULMONATA

With one exception all the mollusca introduced into New Zealand belong to the Pulmonates, known popularly as slugs and snails. With the exception of the first named on the list, they have been unwittingly introduced, with plants, garden stuff, etc. In spite of introduced enemies (birds, hedgehogs, etc.) they are extraordinarily abundant now in many parts, and are a great pest in gardens.

The following list is compiled by the eminent New Zealand malacologist, the late Mr Henry Suter, and is practically the same as is given in his *Manual of New Zealand Mollusca*, published in Wellington in 1913 (p. 1071).

The classification of Families is that adopted by the Rev. A. H. Cooke in the *Cambridge Natural History*.

Family LIMNÆIDÆ

Lymnæa stagnalis, Linn.

Onehunga Springs, Christchurch, introduced as food for trout in the River Avon (Suter). These were originally introduced from England in 1864 by Mr A. M. Johnson, late of Opawa, who brought them out as food for the fish he was endeavouring to introduce.

Suter thinks that the Tasmanian water-snails introduced by Johnson in 1868 were most likely the same species. It is very abundant about Hobart and was described by the Rev. J. E. Tenison Woods as *Limnæa tasmanica* in *Proc. Roy. Soc. of Tasmania*.

Hutton stated in 1881 that this species was abundant in the River Avon, below the Christchurch Botanical Gardens. It was also recorded before 1890 from springs at Onehunga, near Auckland, and is now known from streams in Taranaki.

Lymnæa auricularia, Linn.

An empty shell was found near Wanganui (Suter). According to Kew, eggs of this species have passed unharmed through the digestive system of swans.

Family TESTACELLIDÆ

Testacella maugei, Ferussac

Originally described as *T. vagans* by Hutton who thought it was an indigenous species.

Found in gardens in the vicinity of Auckland (Suter).

The first specimens obtained were got by Mr W. W. Smith at Ashburton.

This "snail-slug" is a native of South-west Europe, and it was first noticed between 1812 and 1816, in Britain.

Family Limacidæ

Agriolimax lævis, Müller

A cosmopolitan slug (Suter). In 1867 Mr Fereday stated that he had seen ten common English slugs on one cabbage in his garden near Christchurch, and he used this as an argument for the introduction of birds, such as thrushes.

This was probably the species he referred to. It appears to be common in New Zealand.

Agriolimax agrestis, Linn.

Common in meadows, fields and gardens (Suter); Auckland, Wellington, Taranaki, Nelson, Greymouth, Christchurch, Dunedin, etc.

Suter writing in 1917 says:

> In 1887 I was living on a ten-acre clearing in the Forty-Mile Bush, surrounded by native bush. This clearing had been laid down in grass about ten years earlier, and was used for feeding horses. Everywhere *Agriolimax agrestis* was common, but these slugs never penetrated the native bush. They evidently must have been brought to that place with the grass-seed, and no doubt in the egg-stage. In a similar way introduced slugs were brought to Campbell Island. They were, if I am not mistaken, a variety of the common *A. agrestis* or *A. lævis*.

Musson and Hedley were both of opinion in 1890 that *A. lævis* was indigenous in Australia; but the latter in 1892 considered that all the species of *Limax* described as native to Australasia are referable either to *L. maximus*, *flavus*, *gagates*, or *agrestis*, all believed to be introduced by man from Europe.

Limax maximus, Linn.

Reported from Dunedin by Captain Hutton, and from Taranaki by Mr W. W. Smith.

(In Tasmania this slug is found to be infested with a mite, possibly the same as is found in England under similar circumstances. Mr Hedley says of it: "should it prove to be identical with the parasite attendant on the European mollusc, this fact would argue that the animals migrated in the adult stage, and not in the eggs." He further states that the species of *Limax* (all introduced) in Australia, have far outstripped their shell-bearing relatives.)

Limax flavus, Linn.

This species is rather common now, and has been reported from Dunedin, Greymouth and Taranaki. Mr Suter says it is especially injurious to garden vegetables.

(Mr Musson thinks it possible that *L. megalodontes*, described by MM. Quoy and Gaimard in 1824, from Port Jackson, was this species.)

Amalia gagates, Draparnaud

Reported from Ohaupo and Auckland by Mr Musson, and in Hawke's Bay by Mr Colenso. Mr Suter states that this is a very variable slug and quite a number of varieties can be distinguished.

(Mr Musson thinks that *Limax maurus*, described by MM. Quoy and Gaimard in 1824, from Port Jackson, was this species.)

Amalia antipoda, Pfeiffer

Amalia fuliginosa, Gould

Amalia emarginata, Hutton

All these three species, according to Mr Suter, have to be included in the *gagates* group, and are found in New Zealand, but he gives no localities for them.

Hyalina crystallina, Müller

Mr Suter records this species as occurring in Auckland.

Hyalina alliaria, Miller

In conservatories and hot-houses; Mr Suter does not record any particular locality.

Hyalina cellaria, Müller

Mr Suter records it as occurring in gardens, meadows, etc., mostly hiding under stones, at the Bay of Islands, Auckland, Napier, and Wanganui. Mr Musson reports it from Auckland "under stones, especially about the various volcanic mounts."

Family HELICIDÆ

Helix aspersa, Müller

According to Mr Suter this species is common at most of the sea-coast towns in New Zealand, where it is a great nuisance in gardens. Writing to me in 1917, he informed me that it was much more abundant in the north than it was at Christchurch, or farther south. Mr Hutchison states that "this common garden snail is greedily devoured by rats at Napier, while the next mentioned species (*H. hortensis*?), which is also in the gardens, is not appreciated by rats for some reason." A reference to this will be found at p. 83.

Thrushes are very fond of this snail, which they carry to some hard surface, where they break the shell with the bill and eat the animal. Mr Musson, writing in 1890, says that examples from Opua in the Bay of Islands are exceptionally thin-shelled, whilst shells from Auckland are of the variety *conoidea*, thin, small and conical. He also describes this as a most voracious snail, which has been known to perforate birds' eggs for food[1].

In the Agricultural Department's report for 1897 this species is stated (in conjunction with different species of slugs) to be very destructive to orange- and lemon-trees in the Hokianga district. Captain Broun records that in another district near Auckland he examined a lemon-tree about three years old that had been nearly killed by snails; large pieces of the bark had been eaten away, and even the green wood had been injured.

Mr Huddlestone of Nelson states that English snails, introduced along with the plants from Britain, were first seen in the Nelson district in numbers in 1861. These were in all probability either *Helix aspersa* or *H. caperata*.

Helix hortensis, Müller

This snail is apparently widely spread in the North Island. It was recorded from Auckland by Captain Hutton, and from Taranaki by Mr W. W. Smith. I am told that it is abundant at Napier.

Snails of the genus *Helix* seem to be very tenacious of life, a fact which favours their distribution in hay, straw, etc. Kew records cases of *H. hortensis* which lived for 14 months without food, and of *H. aspersa*, which survived after being in a closed pot of earth for about ten and a half months, and subsequently produced fertile eggs[2].

Helicella caperata, Montagu

Mr Suter recorded this snail from Nelson in the South Island, and Paekakariki in the North Island. It is probably much more widely spread than these two isolated localities indicated. As showing how readily snails are distributed, it may be mentioned that Kew (p. 161) records a case in England where "thirteen wrinkled-snails (*H. caperata*), together with a quantity of tares were taken from the stomach of a wood-pigeon which had been shot three days previously. Most of the snails were alive, and began creeping about on being placed in a dish containing a little water."

Helicostyla tricolor, Pfeiffer

Mr Suter reports this from a garden in the Bay of Islands.

[1] Dr Binney in *Terrestrial Air-breathing Mollusks* (1851) states that the larger European snails, and particularly *Helix aspersa*, are sometimes imported into the United States, for use as food by foreign residents.

[2] Mr A. Nicols in *Acclimatisation of the Salmonidæ at the Antipodes* says (p. 46) that in the 'Mindora' shipment of salmon ova in 1869—"a living snail was found among the moss" in which the eggs were packed, and was "acclimated." By which I suppose he means that it was set free.

Helicostyla daphnis, Broderip

Found at Picton by Mr Kinsey.

Helicostyla fulgetrum, Broderip

Mr Suter states that "the specimen which was collected by Dr Dieffenbach, and was in the British Museum, is lost."

Vallonia excentrica, Sterki

Mr Suter gives this species as from Auckland, Mr W. W. Smith reports it from Taranaki. In Mr Musson's list of introduced Mollusca (1890) this species was referred to as *Helix pulchella*, Müller. His specimens were collected by Mr Cheeseman in Albert Park, Auckland.

Family Arionidæ

Arion empiricorum, Ferussac

Mr Suter reports this as found near Auckland, and at Dunedin. In the former locality Mr Musson states that they were found crawling over the roads after rain.

Arion subfuscus, Draparnaud

Found at Dunedin by Captain Hutton, who described it as a new species, which he named *A. incommodus*, but a specimen in the British Museum (also) from Dunedin, showed it to be the *cinereo-fuscus* form of *A. subfuscus*.

Arion hortensis, Ferussac

Reported as plentiful about Auckland (1890) by Mr Musson, who found it crawling about the roads after rain, along with *A. empiricorum*. Mr W. W. Smith informs me that it also occurs in Taranaki.

Arion minimus, Simroth

Mr Suter records this snail from Rangitoto Island. It was also found by Mrs Longstaff at Matihiwi, near Masterton.

Family Stenogyridæ

Cionella lubrica, Müller

Mr Suter reports this species from Auckland.

Cæcilianella acicula, Müller

This is also recorded from Auckland by Mr Suter.

Family Athoracophoridæ

Aneitea græffei, Humbert

"This slug, a native of Queensland and New South Wales, was found at Port Chalmers by Dr C. Chilton, and near Collingwood by Mr J. Dall[1]."

Order PSEUDOLAMELLIBRANCHIATA

Family Ostreidæ

Ostrea edulis, Linn.

In 1868 Mr W. C. Young shipped a quantity of oysters by the 'Celestial Queen' from London to the Otago Society. Only two arrived alive in Dunedin: "These have been deposited on an artificial oyster bed at Portobello, where it is to be hoped they will do well." Great was the faith of these early pioneers in acclimatisation work. According to a letter from Mr Frank Buckland these oysters were from the River Roach in Essex, and were presented by Mr F. Wiseman of Paglesham. Mr Buckland evidently had not sampled the oysters of Foveaux Strait for he adds: "Oysters are found naturally in New Zealand, but if the culture of a better class could be instituted there is a chance of an additional source of food being supplied to the colony."

[1] In Mr Musson's list of introduced Mollusca (1890) he gives *Zonites nitidus*, Müller, as found by himself at "Lake St John, Auckland; a dozen specimens under logs." Mr Suter informs me that the species was the indigenous *Fretum novaræ*, Pfeiffer.

Neritina fluviatilis, Linn., was reported from the Waikare River by Mr T. Kirk, and identified by Mr Musson in 1890. Mr Suter doubts the accuracy of the locality, and seems to think the specimens were taken—by mistake—from the collections in the Dominion Museum.

Chapter VIII

INSECTS WITH MYRIAPODA

MYRIAPODA

(Order CHILOPODA, see Appendix B, p. 558.)

Order SCHIZOTARSIA

Family CERMATIIDÆ

Cermatia smithii, Newport

ONE example of this species was received recently (1919) at the Dominion Museum, Wellington. It was taken at Wanganui.

INSECTA

(The scheme of classification generally adopted is that used by Dr D. Sharp in the *Cambridge Natural History*, but I have been obliged to depart from it in regard to several groups, especially those of parasitic insects, regarding which the classification is still in a condition of flux.)

Order I. APTERA

Sub-order THYSANURA

Family LEPISMIDÆ

Lepisma saccharina, Linn. (Silver-fish), Bristle-tail or Fish-moth

This species is very common throughout the North Island, but not so widespread in the South. (Dr Hilgendorf reports it as only found in heated linen cupboards about Christchurch.) I have found it by no means uncommon in Dunedin. Mr Philpott has only found it in one locality in Southland. Hudson considers it was introduced at a very early date.

Mr Howes reports it from houses in Wellington, Christchurch and Invercargill. "It is undoubtedly throughout N.Z. but cannot be said to be common."

This insect is essentially a vegetable feeder, but its favourite food is starch. Hence the damage it does to books and papers, muslin curtains, starched articles of clothing, and silk garments and tapestry which have been stiffened with starch. They are great pests in libraries, where they eat the glaze, paste of the labels and the surface of the paper.

A number of insects sent to Mr Jas. Drummond in May, 1913,

from Lake Brunner, where they were said to have recently appeared, proved to be this species.

Miller says this species only lives in damp places.

Thermobia furnorum, Rovelli. (Fire-brat)

Common about bakehouses, fireplaces, ovens or any warm dry places, especially in the North Island.

Sub-order COLLEMBOLA

Family ENTOMOBRYIDÆ

Entomobrya multifasciata, Tulb.

Two specimens of this cosmopolitan species were found in an ants' nest at Ashburton by Mr W. W. Smith and recorded in (or before) 1895. Its presence there may have been accidental. It is difficult to say whether this is an introduced or an indigenous species.

Family ACHORUTIDÆ

Achorutes armatus, Nic., Tulb.

Three specimens were found in an ants' nest at Ashburton by Mr W. W. Smith, and recorded at the same time as the preceding species. It is a very widely spread form, being found in all European countries, Sumatra, California, Brazil, etc. The same remarks apply to it as to the preceding.

Order II. ORTHOPTERA

Family FORFICULIDÆ

Forficula auricularia, Linn. (Earwig)

This common European insect was first reported from Ashburton by Mr W. W. Smith in 1900. It was recorded from North Canterbury in 1908, and it was stated then that it had been known for some years in the Wairarapa district in the North Island. Since then it has been steadily increasing its range, and its abundance. Mr Howes first met with it commonly under fish crates in the railway station at Palmerston (S.); from such a position it would be spread broadcast. It is extremely abundant in the north end of Dunedin, in Christchurch, at Waikari in North Canterbury, and on the railway station at Cass.

These insects are most destructive to vegetables, flowers and fruits, and they even penetrate into houses, and devour all starchy and saccharine food materials. They are voracious feeders, and are especially fond of the corollas of flowers, so that they are a great annoyance to gardeners. Occasionally they visit flowers for the nectar, and Knuth reports them as going into the flowers of *Tropæolum majus* for this purpose. They have also been taken on the flowers of ivy (*Hedera Helix*), poppy (*Papaver Rhœas*) and Millfoil (*Achillæa millefolium*), perhaps on the same quest. But they eat the flowers of species of *Brassica*, and of the thistle (*Cnicus arvensis*) very freely. Their destruction of fruit is chiefly that which has fallen on the ground, or which grows near the ground, like strawberries.

Kerner suggests that:

"it is very probable that the species of *Forficula*, which we frequently find working for days together in tubular flowers, so far interfere with the floral functions as that by their presence other insects, whose visits would be of use, are prevented from sucking the nectar." He adds: "I possess, however, no definite observations on this point."

A rove-beetle, *Philonthus æneus*, also introduced into New Zealand, is very generally mistaken for an earwig.

Mr G. Howes informs me that he has three unidentified species of earwigs in his collection, which were introduced from the South Sea Islands in fruit.

Chelisoches morio, Fabr.

This species, originally belonging to the Malay Archipelago, was first observed by Mr Hudson, as landed from a home steamer in Wellington in 1890. He found another specimen amongst some bananas in 1898. Mr W. W. Smith reported it as occurring round about Christchurch in 1906.

Family BLATTIDÆ

Blatta latipennis, Brunn. (*Phyllodromia opima*). (India?)
(? *Blatta orientalis*, the Black Beetle)

This species was recorded as taken at Auckland by the 'Novara' Expedition (1859).

Blatta germanica, Linn. (*Phyllodromia germanica*, Linn.).
(Cosmopolitan.) Cockroach

This species, which has spread very widely from Europe, is known in America, where it is very common, as the "Croton Bug." Mr Howes reports it as generally distributed by merchandise, and common on all New Zealand coastal boats. It has been found in parts of Dunedin and Invercargill.

Polyzosteria truncata, Brunn.

This Australian species is recorded among introduced insects in Hutton's list in the *Fauna Novæ-Zealandiæ*.

Periplaneta americana, Linn. Cockroach, Yellow Roach

This species, which is now cosmopolitan in its distribution, is a native of tropical America. It is a common cockroach on board ships. Mr Howes says it is occasionally taken in seaport towns, where it comes off boats, amongst goods, etc. He has picked it up in fruit auction rooms in various parts of New Zealand.

Family MANTIDÆ

Orthodera ministralis, Fabr. Mantis

Some time prior to 1860, Dr Sinclair took egg-cases of this Australian insect to England; they were found apparently in Auckland. In 1873–74, Captain Hutton observed it at Clyde in Central Otago; Mr W. Colenso recorded it from Napier in 1878; Mr Potts found it in Canterbury in 1880; Mr Hudson found it in Nelson in 1886, and in Wellington in 1891; Dr Hilgendorf in 1916 says that it is common, but overlooked on account of its protective colouring; and in 1919 Mr W. W. Smith says it is common in Taranaki, while Mr Howes reports it from Oamaru. It seems, therefore, to be pretty generally distributed. Captain Hutton writing in 1896 said: "I think that the species has been unintentionally introduced into Auckland from Sydney, and into Otago from Tasmania or Victoria, when large quantities of hay were brought to Otago from Australia."

Family LOCUSTIDÆ

Cædicia olivacea, Brunner

This Australian locust was found very commonly amongst sweet-briars in Nelson in 1886, by Mr Hudson, and was also recorded from Auckland by Captain Hutton in 1897. The latter authority states that "probably it was introduced into Nelson in the early days of the gold-diggings, and taken from there to Auckland." Mr W. W. Smith tells me that it is common in Taranaki in 1919.

Family GRYLLIDÆ

Gryllus servillei, Saussure. Field Cricket or Whistling Cricket

Also recorded as *Acheta fuliginosa*. Mr Hudson thinks that the introduction of this cricket from Australia occurred at a very early period. He first recorded it from the Nelson district in 1875, when it was extremely destructive. In 1896–97 the Agricultural Department reported it as very destructive in Auckland and Hawke's Bay, in the latter district eating the paddocks quite bare. It also entered houses, destroying wall-papers, boots, clothing, harness, provisions, etc.; and did much damage in orchards and gardens. In 1907 it was extremely abundant in the neighbourhood of Auckland. Mr Howes states that "the cheerful chirping of the Black Cricket can be heard every evening (in summer) about the North Island towns. I have also heard it in the Port Hills above Christchurch."

Gryllotalpa vulgaris, Linn. Mole Cricket

In 1888 this species was noted by Mr T. W. Kirk in a bank on the Tinakori Road, Wellington. Also found by Mr Robinson of Makara, west of Wellington. Dr Hilgendorf (1916) says: "the first specimen from near Wanganui was noted there some ten years ago; it is now common in sand-hills there."

Mr Howes states that it is common as far south as Nelson.

Order III. NEUROPTERA

Sub-order MALLOPHAGA

(In Neumann's classification the Mallophaga or Ricinidæ are treated as a family of the sub-order Rhyncota or Pediculinæ.)

Trichodectes scalaris, Nitzsch. Ox-louse

This ecto-parasite of cattle is not uncommon in New Zealand.

Trichodectes sphærocephalus, Nitzsch. Sheep-louse

This species appears to be very common. The report of the Agricultural Department for 1916 states that it is a very abundant parasite among sheep. Mr Miller recorded it from Weraroa in 1917.

Trichodectes latus, Nitzsch. Dog-louse

This species is found on dogs in New Zealand.

Trichodectes climax, Nitzsch. Goat-louse

Col. H. A. Reid informs me that this species occurs on goats in New Zealand.

Goniodes dissimilis, Nitzsch. Brown Chicken-louse

Col. H. A. Reid states that this parasite is found on poultry in New Zealand.

Menopon pallidum, Nitzsch

This most troublesome ecto-parasite of poultry is termed in American publications the shaft-louse or small body-louse of chickens. It is a species which moves with great nimbleness among feathers, and can be kept alive for months on fresh feathers, of which it particularly consumes the quill epidermis. Col. H. A. Reid states that it occurs among poultry in New Zealand.

Menopon biseriatim, Nitzsch. Body-louse of Chickens

This species also occurs in New Zealand, and has been identified by Col. Reid.

Sub-order PSEUDO-NEUROPTERA

Family ATROPIDÆ

Atropos pulsatoria, Linn. Book-louse; Lesser Death Watch; Book-tick

This is a very generally distributed insect, which feeds on the paste of books, wall-papers, etc. It has been long known, and was probably introduced over a century ago, with the first papers and books.

(See Appendix B, p. 558.)

Order IV. HYMENOPTERA

Family TENTHREDINEÆ. (Sawflies)

Eriocampa limacina, De Geer (*Eriocampa adumbrata*, Klug; *Selandria cerasi*, Curtis). Slug-worm, Leech. (Europe)

Osten-Sacken appears to have been the first to record the occurrence of this insect in New Zealand between 1870 and 1888. It is now very abundant all over New Zealand; the larvæ feeding on pear, cherry, hawthorn, plum, peach, and other trees of the Rosaceæ. In the North Island it is particularly destructive, and practically prevents the use of the hawthorn as a hedge-plant. In the south it seems to appear too late in the season to do much harm. Howes has found it on the native tutu (*Coriaria ruscifolia*) at Queenstown.

It is attacked and destroyed by two indigenous bugs, *Cermatulus nasalis*, Westwood, and *Nezara amoyti*, White; also by the introduced Australian wasp, *Polistes tasmaniensis*. Mr Holman of Whangarei informs me that the native cuckoo (*Eudynamis taitiensis*) eats a considerable number of these slugs when they are in the larval condition[1].

Family SIRICIDÆ

Sirex juvencus, Linn. Steel-blue Sawfly

This European species has been found at the Government plantations at Whakarewarewa; the larvæ attack *Pinus radiata*, boring into the timber.

[1] In the British Board of Agriculture's Leaflet No. 62 (March, 1900), it is stated that the *Eriocampa limacina* does much harm to pear and cherry trees in America. Harris, writing as early as 1797, says: "Small trees were covered with them, and their foliage entirely destroyed, and even the air, by passing through the trees, became charged with a disagreeable and sickening odour given out by these slimy creatures." The same thing has been noticed in England.

Family Ichneumonidæ

Ryssa semipunctata, Kirby (*Lissopimpla semipunctata*). Dark-winged Ichneumon

In 1883 Mr P. Cameron obtained this species from specimens received from Greymouth. Writing in 1899 he says:

Captain Hutton informs me that his belief is that the species has been introduced into New Zealand from Australia. The evidence undoubtedly is that it was rare in New Zealand thirty years ago, while now it is not at all rare. At Greymouth the late Mr Richard Helms took it commonly.

In 1882 it was taken by Mr Hudson in Nelson, where it was fairly common. Mr Howes reports it from Wanganui and New Plymouth, and Mr D. Miller (1919) states that it is common in the Auckland provincial district. The wasp is parasitic on the subterranean grass-caterpillars (*Porina umbraculata* and *P. cervinata*), and upon the army-worm (*Cyrphus unipunctata*). It also attacks the N.Z. Flax grub (*Xanthorœ præfectata*), as well as native locusts.

Bassus lactatorius, Fabr. (*B. generosus*, Cameron)

Introduced from Australia. This is not a useful species, for it is parasitic in the larvæ of the syrphidæ or hover flies, which themselves destroy great quantities of aphides. It was first reported by Hudson in 1883.

Dacnusa sonchivorus, Cameron

This ichneumon was identified from N.Z. in 1902 by Cameron. It is parasitic on *Phytomyza albiceps*, the caterpillar which bores in the leaves of sowthistle, cineraria and other composites. It appears to be a common species. (Cameron does not record this as an introduced form.)

Eulophus albitarsis, Ashmead

A European and North American species which was recorded from Chatham Island by Mr P. Cameron in 1902. I am not aware of its having been met with in either the North or the South Island.

Pleurotropus (*Entodon*) *epigonus*, Walker

Sir James Hector (in 1894) recorded this parasitic wasp as occurring at Marton in 1888. It was introduced from England by the Agricultural Department in 1893, as the natural enemy of the Hessian fly. Mr T. W. Kirk reported in (1895) as follows on this introduction:

In last year's report I mentioned having succeeded in rearing from the parasitized "puparia" of Hessian flies, received from England, a large number of the ichneumon fly, known as *Semiotellus nigripes*, which were liberated in the various districts where the crops had suffered from the attacks of the Hessian fly. One place selected as a depot was the farm of Mr J. Hessey, of Masterton. This gentleman took a great interest in the experiment, fencing in a portion of his wheat as a sanctuary for the parasites. Here they must have made good use of their time, for during the past season, in company with Mr Hessey, I captured over a dozen of this particular species in a very short time, and this, too, in a paddock fully a hundred yards away from the nursery or depot.

Mr Howes informs me (1919) that in a sample of oats sent from Balfour, Southland, more of these parasites emerged than Hessian flies. This is probably the species found on the Hessian fly at Lincoln, of which from two to five per cent. of the pupæ hatch out parasites. (See Appendix B, p. 558.)

Ichneumon sp.

Kirk writing in 1894 says that in Britain a parasite is found on *Phytomyza nigricornis*, the cineraria fly, "but this does not seem to have yet reached Wellington."

The species has certainly been detected since, for in 1906 Captain Broun reported its discovery in Auckland.

Ichneumon sp.

This species, which is parasitic on the diamond-back moth (*Plutella maculipennis* or *P. cruciferarum*), appears to be fairly common. Hilgendorf states that at Lincoln, about five per cent. of the caterpillar pupæ hatch out ichneumons.

Ichneumon sp.

A species which is parasitic on *Aphis brassicæ*, the green fly of the turnip. Hilgendorf states that in February and March, aphides on cabbage leaves are commonly found with holes in them, which have been made by the escaping parasite.

Trichogramma pretiosa, Riley

In 1900 a parasitic ichneumon was found in the pupæ of the codlin moth in Mr Parr's orchard, at Waikumete, near Auckland, and Mr T. W. Kirk hatched out a number of them. Specimens sent to Professor L. O. Howard, of Washington, were identified as above. Mr W. A. Boucher reported on this parasite in 1902, that "while the percentage of moth-infected fruit of early and mid-season varieties remains much the same, a percentage of the fruit of the later varieties will apparently be saved from the moth."

Zele sp.

During 1908, Mr Simms, one of the Inspectors of Orchards in the Department of Agriculture, found this species under "some bandages" placed there to prevent the codlin moth from getting up the stems of apple trees. The investigation of this species was never carried out by the Department.

Calliephaltes messer, Gravenhorst

This insect had been introduced into California from Spain as a natural enemy of the codlin moth. In 1906 Mr Boucher was sent to California, where he obtained a supply of these parasites. These, and other parasites, were reared in an insectary specially built at Auckland, and were spread far and wide to over 50 localities where the moth was prevalent. In the *Journal of Agriculture* for 1911 it is stated that the codlin moth parasitical flies which were liberated in Whangarei five years previous "are increasing in appreciable numbers in the orchards. One firm of orchardists state they have this season found the larvæ of the parasites in three out of five moth-cocoons examined."

Platygaster minutus, Lindemann

This small hymenopterous insect, parasitic on the Hessian fly, was introduced from Britain by the Agricultural Department in 1893.

(Family CHALCIDIDÆ, see Appendix B, p. 558.)

Family APIDÆ

Bombus terrestris, Linn.
Bombus lucorum, Linn.
Bombus ruderatus, Fab.
Bombus hortorum, Linn.
Bombus lupidarius, Linn.
} Humble-bees (Europe)

The naturalisation of humble-bees in New Zealand is due to the action of the Canterbury Acclimatisation Society, the object of introducing them being to bring about the fertilisation of the red clover (*Trifolium pratense*), which is very extensively cultivated, but which previous to the advent of these insects did not produce seed, except to a very limited extent.

In a notice of the humble-bee in New Zealand in the *N.Z. Journal of Science* (January, 1891), I stated in regard to the fertilisation of red clover that

> the pollen and stigma of this flower are accessible to all insects which are heavy enough to press down the keel, and if bees visit the flowers for pollen only they will no doubt bring about cross-fertilisation. This may account for an interesting example given me by Mr Wm. Martin, of Fairfield, near Dunedin, who informs me that as far back as 1858 he obtained a large quantity of very fine seed off a small patch of red clover which he had under cultivation.

The first attempt to introduce these insects was made by Dr Frank Buckland in 1873, but he failed to get the bees in time. In January, 1876, a consignment from Dr Buckland was brought out in the 'Otari,' by the Hon. John Hall, but all were found to be dead.

In the Society's report for 1880 it is stated: "Your Council has received intimation from Mrs Belfield of Timaru, of a shipment of humble-bees from Messrs Neighbour & Sons, England."

Mr I. Hopkins, formerly Chief Government Apiarist, and author of a bulletin on the "History of the Humble-bee in New Zealand," refers evidently to this attempt as follows:

"Until a few years ago" (he was writing in 1914) "I was under the impression that I had liberated the first humble-bees in this country, but was corrected by a resident in Timaru, who stated he liberated in 1883, some which came to the order of a lady, I think."

These apparently failed to establish themselves, for when bees were liberated subsequently by the Society, the rapidity of increase and of spread was phenomenal.

In 1882 Mr Hopkins sent an order for 100 queens to Messrs Neighbour & Sons, London.

After stating the object of importing the bees I left the selection of them to the firm, but gave instructions how they were to be packed, and to be brought out in the ship's cool room, at a temperature of about 40° F.

These arrived in May, 1883, but were all dead. Another lot of 145 arrived in February, 1884, but only two were alive. After feeding them, these were liberated, but there was no indication afterwards of their having established themselves.

Other consignments arrived by post, and in the steamships 'Ionic' and 'Doric,' in January and February, 1885. A total of nearly five hundred bees came in the several consignments, but all were dead except the two mentioned. From the difference in their size, markings, and colours we concluded at the time that queens of three or four species had been sent, but what they were we had no knowledge.

These lots came to Auckland.

The second consignment of bees to the order of the Canterbury Society arrived in the S.S. 'Doric' at Lyttelton in February, 1884, and the third, of 200 bees, in the 'Ruapeha' in April, 1884. Both lots were dead.

In January, 1885, the 'Tongariro' brought a fourth consignment of 282 bees, of which 45 were alive; and in February of the same year, the 'Aorangi' landed 48 out of 260 shipped. The first lot were liberated at Riccarton, and the next at the foot of the Port Hills. "Both lots of bees were strong and healthy when liberated, and doubtless the majority, if not all, of them succeeded in establishing themselves."

It is interesting to learn from the report issued by the Canterbury Society in February, 1886, that the cost of collecting and shipping the bees in England, was 1*s.* 7*d.* each; the cost of those landed in New Zealand was 9*s.* 5*d.* each.

Another interesting and noteworthy fact is that in their original instructions for shipment of bees the Society particularly asked that *Bombus terrestris* be sent out, a species in which, according to Müller, the trunk is too short to reach the bottom of the tube of the flowers of *Trifolium pratense*. It is certain that this species was among those sent out, and it is still the commonest species in New Zealand. But apparently also specimens of *B. ruderatus* (variously referred to also as *B. subterraneus* and *B. harrisellus*), and *B. hortorum* were introduced at the same time. For as soon as the insects began to increase, fertilisation of the clover ensued and abundance of seed began to be obtained. I noticed also that previous to 1889 all my species of *Primula* (primroses, cowslips, etc.), which were growing freely in my garden, failed to produce seed naturally. After humble-bees began to come about, they seeded quite freely. This, too, in spite of the fact that the smaller bees learned to bite the corolla-tubes, and so make holes half-way down, through which they could suck the nectar without disturbing either anthers or stigmas. Mr Hopkins refers to the perforation of the tubes of red-clover flowers as being done by both *B. terrestris* and *B. lucorum* in Europe, but could find no evidence of its being done in New Zealand. However, in April, 1892, the late Mr John Allan of Taurima (in the Taieri Plain) informed me that he had repeatedly seen small humble-bees biting the tubes of the red clover; these were, no doubt, small bees produced late in the season, and not full-grown. I have frequently observed humble-bees biting holes in the tubes of *Primula*, *Arbutus*, *Antirrhinum*, *Eccremocarpus*, *Salvia*, *Narcissus* and *Hyacinth*. It would seem also that they learn the trick from one another. Thus bulb-growers in one district have told me that all their hyacinth blossoms were destroyed by humble-bees, while in another district at a few miles distance the hyacinths were quite untouched.

After the bees became thoroughly established in the country, some doubt began to be expressed that the wrong species had been introduced, and the Canterbury Agricultural and Pastoral Association resolved to import more. On the advice of Lord Avebury, they obtained three shipments from London through the agency of Mr Sladen. Mr Hopkins in his pamphlet says:

Mr O. B. Pemberton, the secretary of the association, writes in January, 1913: "We got out in all three shipments, arriving as follows:

	Arrival		Number sent	Live Queens
(1)	24th February, 1906	...	15	10
(2)	29th November, 1906	...	165	71
(3)	27th December, 1906	...	145	62

The Queens we got out were *B. lapidarius* and *B. hortorum*. These were all liberated by me in different localities. I have not heard of any of the *B. lapidarius* being seen, so I presume they did not live."

The increase of the bees in the first few years after their introduction was phenomenally rapid. The first were liberated at Christchurch in January, 1885. In January, 1886, two were seen by Mr J. D. Enys on Castle Hill on the West Coast Road (64 miles), and others at Mount Peel 90 miles in another direction. Early in 1887 they were reported from Kaikoura, 100 miles to the north, and from Timaru, 100 miles to the south. At the end of the year they had made their way from Oamaru up the Waitaki Valley, through the Lindis Pass and on to the Hawea flats. In February, 1888, they appeared at Dunedin, and at the same time at Waihola 30 miles south-west. In November, 1889, they were first recorded from the head of Lake Wakatipu, and in the beginning of 1890 were observed in the neighbourhood of Invercargill. No doubt their spread to the west coast, and to Cook Strait was equally rapid, but there is no record. Both whole nests and queens were sent from time to time to the North Island from Canterbury from 1888 onwards, in which year they were first observed in Wellington, while the first record from Auckland was in May, 1890. They became thus fully established throughout New Zealand in less than ten years.

So rapid was the first increase of the humble-bees that apiarists began to take fright, and it was very commonly feared that soon the hive-bees would be crowded out from the flowers and that no nectar would be left for them. In some districts, as in thistle-infested areas, the bees swarmed to such an extent when the plants were in flower, as to deter timid persons from going through. This extraordinary rate of increase was not maintained. After a time the numbers became reduced, till in some districts where they were formerly abundant, they became almost rare. In most parts of the country now, though humble-bees are fairly common, they are nowhere so abundant as to constitute a pest or even a menace.

The causes of this diminution in numbers have never been investigated, though a few observers have sought to give some explanation of the facts. Hopkins considers that the rainfall is the chief factor, and that those portions of Marlborough, Canterbury and Otago where the annual rainfall is under 30 inches, are best suited for the growth and increase of the bees. Portions of Otago and Southland, a small part of Canterbury, Nelson, and a very small portion of the North Island have less than 40 inches but more than 30 inches of rain per annum; while most of the North Island ranges between 40 and 70 inches. Speaking of the diminution in numbers Hopkins says:

There seemed to be no plausible reason for the decline. I am inclined to think that the falling-off was due to a series of unfavourable seasons closely following each other, unusually heavy rainfalls will cause great destruction by flooding their nests.

This, however, is only a surmise, and I do not think any exact observations have been made in support of the statement. I have elsewhere shown that in many parts of New Zealand humble-bees do not hibernate at all, and it is just possible owing to this that they frequently succumb to the rapid falls of temperature which are so common in our insular climate. Even in Otago where the winters are often fairly severe—the mean winter temperature being 41° F., and the mean winter minimum for the same months being 35° F.—I have seen and recorded the bees in every month of the year. In Taranaki, W. W. Smith says: "Queens of the three forms naturalised in the North Island may be seen on the wing almost every day of warm sunshine in the public park at New Plymouth throughout winter months."

The males appear in Otago about November, which is somewhat earlier than occurs in the corresponding season in Britain.

Mr J. Attwood of Northern Wairoa stated in 1914 that all the varieties of humble-bees were common in the district, and that ten years ago he noticed the large jet-black bees (either *B. terrestris* or *B. ruderatus*).

Mr W. Hone of Waverley (1914) says that more than 40 years ago, Mr J. Dickie, Senr., sowed red clover on a part of his land near Waverley, from which he obtained a large crop of very fertile seed.

Mr W. W. Smith states that hive-bees occasionally fertilise red clover in the shorter flowers of the heads.

Mr Philpott informed me in 1917 that the most common species in Southland is *B. terrestris*, var. *virginalis*. *B. ruderatus* also occurs, but the other introduced species are not found in the south.

I have not found the humble-bees visiting many of the indigenous flowers either for nectar or pollen. They are very fond of the flowers of *Fuchsia excorticata*, and frequently suck out the nectar left by honey-birds or tuis, through the portions of the tube torn open by the birds. They have also been recorded on *Veronica elliptica*, many large hybrid veronicas, *Myoporum lætum* and *Muehlenbeckia australis*.

Mr A. Philpott informs me (February, 1917) that he found humble-bees not uncommon on the *Celmisia* blossoms on the Hunter Mountains, at a height of 4500 ft. To get to the upland open country from the Monowai flats the insects would have to traverse or pass through about six miles of *Nothofagus* forest.

The following species of plants which have been introduced into

New Zealand, and have become more or less wild, are visited by humble-bees in Europe:

Bombus terrestris

Ranunculus aquatilis, R. acris, R. repens, Nigella damascena.
Berberis vulgaris.
Papaver Rhœas, P. somniferum.
Cheiranthus Cheiri, Brassica oleracea, Sinapis arvensis, Cakile maritima.
Viola tricolor, V. arvensis, V. odorata.
Malva sylvestris.
Hypericum perforatum.
Geranium Robertianum.
Cytisus scoparius, Ulex europæus, Lupinus luteus, Medicago sativa, M. lupulina, Trifolium repens, T. hybridum, T. fragiferum, T. pratense, T. medium, T. arvense, Lotus corniculatus, Vicia sativa, Faba vulgaris, Phaseolus vulgaris.
Persica vulgaris, Prunus domesticus, P. avium, P. cerasus, Rosa rubiginosa, Rubus fruticosus, R. idæus, Cratægus oxyacantha, Pyrus malus, P. communis, P. aucuparia.
Œnothera biennis.
Ribes nigrum, R. rubrum, R. grossularia.
Daucus Carota.
Knautia arvensis.
Petasites officinale (vulgaris), Tanacetum vulgare, Bellis perennis, Cnicus arvensis, C. lanceolatus, Chrysanthemum leucanthemum, Carduus crispus, C. nutans, Onopordon acanthium, Centaurea nigra, Taraxacum officinale, Sonchus arvensis, Crepis virens.
Calluna vulgaris.
Vinca major.
Echium vulgare.
Verbascum thapsus, Linaria vulgaris, Digitalis purpurea.
Mentha arvensis, Nepeta cataria, N. glechoma, Lamium purpureum, L. album, Marrubium vulgare.
Primula elatior.
Plantago lanceolata, P. media.
Salix alba, S. fragilis.
Allium Cepa.

Bombus hortorum

Brassica oleracea, Sinapis arvensis.
Viola odorata, V. tricolor, V. arvense.
Malva sylvestris.
Hypericum perforatum.
Geranium Robertianum.
Tropæolum majus.
Cytisus scoparius, C. laburnum, Medicago sativa, M. lupulina, Melilotus officinalis, Trifolium repens, T. pratense, Anthyllis vulneraria, Lotus corniculatus, Vicia sativa, Faba vulgaris, Phaseolus vulgaris.
Prunus domesticus, P. avium, P. cerasus, Rubus fruticosus, R. idæus, Cratægus oxyacantha, Pyrus malus.
Daucus Carota.
Senecio jacobæa, Cnicus arvensis, C. lanceolatus, C. nutans, Onopordon acanthium, Centaurea nigra, Taraxacum officinale.
Calluna vulgaris.
Echium vulgare.
Verbascum Thapsus, Linaria vulgaris, Digitalis purpurea.
Mentha aquatica, Prunella vulgaris.
Primula elatior, P. vulgaris, P. veris.
Polygonum Persicaria.
Salix caprea.

Bombus ruderatus

Cakile maritima.
Cytisus Laburnum.
Trifolium pratense, Faba vulgaris.
Cnicus arvensis, Taraxacum officinale.
Carduus nutans.
Echium vulgare.
Linaria vulgaris.

Bombus lucorum

Cakile maritima.
Viola odorata.
Trifolium repens, T. pratense, T. arvense.
Rubus fruticosus, R. idæus, Pyrus communis.
Ribes grossularia.
Petasites vulgaris, Cnicus arvensis, Taraxacum officinale.
Calluna vulgaris.
Thymus Serpyllum, Lamium album.
Salix alba.

Enemies. Mice are probably the most serious enemies the humble-bees have in New Zealand. They are very common, especially in districts and neighbourhoods where cultivation is carried on, and no doubt they destroy numbers of nests and larvæ. But I have no direct evidence of this.

Mr W. W. Smith states that during the drought in 1891–93 in Canterbury, starlings attacked and ate humble-bees during the spring seasons, when food generally was scarce. Again in 1896 he recorded that "last nesting season we noticed starlings several times capturing and carrying the bees to their nests to feed their young." A North Canterbury farmer states (in 1910) that he has repeatedly seen starlings catching the humble-bees and carrying them off to their nests.

In the following year the *Akaroa Mail* reported that tuis were observed catching humble-bees and taking their honey-bags from them.

Mr Smith also reported in 1896 that numbers of dead humble-bees were found during two previous seasons with a small puncture either in their thorax or abdomen. On one occasion he saw a specimen of *Bombus subterraneus* seized by a large native fly—*Asilus varius*—which pulled it to the ground, pierced the forepart of the thorax, and killed it in a few seconds by sucking out the viscera.

In September, 1890, I found large numbers of dead humble-bees about Dunedin, and these were always thickly infested with mites, some parts of the body—especially the bare posterior upper surface of the thorax—being covered with them to such an extent as to have the integument completely hidden.

Mr W. W. Smith (April, 1919) states that the red mites were extraordinarily common on the humble-bees, and he thinks are or were largely responsible for their remarkable diminution in numbers. He adds, "I have not seen a humble-bee in the N. Island this month."

The importance and value of the humble-bee to New Zealand has been very considerable. Mr Smith states that within nine years of the liberation of 90 queen bees in Christchurch in 1885, "the sum of about £200,000 has been realized on red clover seed alone."

Humble-bees were carried over to the Chatham Islands in October, 1890.

They were also introduced into New South Wales from New Zealand many years ago.

Hive-Bees (*Apis mellifica*, Linn. and *A. ligustica*, Spin.). (Europe)

The history of the first introduction of hive-bees into New Zealand has been investigated by Mr Isaac Hopkins, late Chief Government Apiarist, and from his report the following facts are gleaned.

On 13th March, 1839, the Rev. J. H. Burnby and his sister arrived in the 'James' at the Mission station of Mungunga, Hokianga.

Miss Burnby brought with her the first bees introduced in two straw hives. These came from New South Wales. In 1840 Lady Hobson, wife of the first Governor of New Zealand, brought bees with her from New South Wales. In 1842 Mrs Allum arrived at Nelson in the 'Clifford' with the first shipment of bees from England; and in the same year the Rev. W. Cotton arrived in the Bay of Islands with another lot of bees from England.

It is probable that the bees referred to by Dieffenbach (in 1839) had swarmed from those brought over in the autumn by Miss Burnby, unless, indeed, they were an independent importation. He says: "Bees have been introduced into New Zealand from New South Wales; my excellent friend, the Rev. Richard Taylor at Waimate, had a hive, and they were thriving remarkably well."

As settlement proceeded throughout the country—most of it on the skirt of bushland—great numbers of swarms were lost in the forests, where they quickly established themselves in hollow trees. Lady Barker in *Station Life in New Zealand* says she ate bush-honey in Canterbury in 1866. Wild bees were very common in Southland in 1868.

The Hon. Herbert Meade writing in 1871 says:

> New Zealand is *par excellence* the land of honey, and although the bees have only been introduced for, I believe, about twenty-five years, the woods are already full of wild honey. A friend assured me that he had taken as much as 70 lbs. from a single tree, and known others to get 200 and 300 lbs. at one haul; another man collected a ton and a half in a few weeks.
>
> The greatest enemies to the bees here are the dragon-flies, which grow to an enormous size. They waylay the luckless bees when homeward bound and laden with honey, and after nipping off the part containing the sting, devour the remainder with the honey, at leisure.

Dragon-flies do occasionally eat bees, but they are not really formidable enemies; their numbers are too few.

Honey-bees were sent over to the Chatham Islands in October, 1890. They had been imported previously, though I have not got a date for their introduction, but were supposed to have been destroyed by spiders, which were particularly abundant.

Numbers of various kinds of bees have been introduced into the country from time to time, such as Italian, Syrian, Cyprian, Holy-land, Carniolian and Swiss-alpine; but only the first-named is cultivated.

The first Italian bees introduced were landed in 1879 by Mr J. H. Harrison of Coromandel (one hive) and the Canterbury Acclimatisation (one hive).

Then the Canterbury Society imported four hives of Ligurians in 1880, and the Otago Society ten hives in 1883; but most of those introduced were brought in by private individuals and bee-keepers' associations. Mr Isaac Hopkins of Auckland, formerly Chief Government Apiarist, tells me:

There have not been sufficient foreign bees other than Italians cultivated and escaped in New Zealand to make any difference in our wild or vagrant bees. It might be possible to find a pure Italian vagrant colony that had just escaped from some apiary, but not one of the second generation. There are too many "black bees" about, and in a state of nature they breed a tremendous number of drones, while we restrict their breeding. Therefore there is a large preponderance of black drones flying, and in most districts the chances are fifty to one that an Italian queen will meet a black drone. There are plenty of crosses, Black-Italians, about.

According to Sir Walter Buller, the native kingfisher (*Halcyon vagans*) is destructive to bees. A farmer at Paraekaretu found that his bees were disappearing, and on killing a kingfisher found its crop full of bees.

Mr W. W. Smith states that hive-bees occasionally fertilise red clover in the shorter flowers of its heads.

Indigenous plants visited by hive-bees for nectar or pollen

The following list is compiled partly from data supplied to me by various bee-keepers and partly from my own observation and is probably far from complete.

Clematis indivisa (Puawhananga), pollen only, and *Clematis fœtida*.

Ranunculus rivularis, the honey from this flower is more or less poisonous.

Melicytus ramiflorus (mahoe), *Pittosporum tenuifolium* (kohuhu or black mapau) and *P. eugenioides* (tarata or white mapau), *Gaya Lyallii* (lace-bark), *Aristotelia racemosa* (mako-mako), *Discaria Toumatou* (tumatukuru), *Sophora tetraptera* (kowhai), *Rubus australis* (bush-lawyer or tataramoa), *Acæna Sanguisorbæ* (piri-piri), *Carpodetus serratus* (piripiriwhata or putaputawheta), *Leptospermum scoparium* (manuka or tea-tree) and *L. ericoides*: the former is a particular favourite with bees, and produces highly aromatic honey of a rich pinky-brown colour. *Metrosideros lucida* (rata); probably all the species are visited by bees, for all are nectar-producing, but this is the only one I have received a record of. *Myrtus bullata* (ramarama), *M. pedunculata*, *M. obcordata* (rohutu), *Fuchsia excorticata* (kotukutuku); all bee-keepers who live near the bush report this species. *Aciphylla squarrosa* (spear-grass, taramea or kurikuri), *Panax arboreum* (whauwhaupaku), especially on the male flowers, *Griselinia lucida* (puka) and *G. littoralis* (broad-leaf, kapuka or papaumu). *Celmisia coriacea*, and on many other species bees have been recorded; also several species of *Raoulia*, which have not, however, been strictly identified.

Senecio lagopus, *S. bellidioides* and probably most of the shrubby species.

Brachyglottis repanda (pukapuka or wharangitawhito) and *B. Rangiora* (rangiora). Both these species are especially blamed for producing poisonous honey. In October, 1906, six Maoris at Rewiti, near Helensville, were poisoned (though all ultimately recovered) by eating honey which contained large quantities of pollen grains of two species, viz., manuka and pukapuka. There was no doubt whatever in the minds of those who investigated it that the poison was due to the *Brachyglottis*. The plants are known to be poisonous. A settler near Rotorua died from eating this honey in 1917.

Parsonsia heterophylla (kaiku or kaiwhiria) and *P. capsularis* (aka-kiore), *Convolvulus tuguriorum*, probably for pollen; *Veronica salicifolia*, *V. Traversii*, *V. sali-*

cornioides and probably all the shrubby species; *Muehlenbeckia australis*, *Knightia excelsa* (rewarewa or native honeysuckle), *Nothofagus* sp. (native birches or beeches), probably all the species are visited but whether for nectar or pollen is not recorded; *Podocarpus* (miro and black and white pines) several species, all for pollen; *Rhipogonum scandens* (supple-jack, kareao or pirita); *Cordyline australis* (cabbage-tree or ti). This species is blamed by some apiarists for producing very thick honey, which is difficult to extract. *Astelia nervosa*, *Phormium tenax* (New Zealand flax or harakeke), and *Bulbinella Hookeri*.

Of introduced plants probably the best honey producer is the white clover (*Trifolium repens*) which gives a beautiful pale honey. In districts where it abounds the ragwort (*Senecio jacobæa*) produces somewhat late in the season a dark, strongly-flavoured honey which is not always saleable.

In April, 1919, I noticed bees in great numbers feeding on fallen (rotten) pears, at Whangarei.

In Europe, *Apis mellifica* fertilises the following flowering plants which have been introduced into New Zealand:

Ranunculus aquatilis, *R. acris*, *R. repens*, *R. bulbosus*, *Nigella damascena*.
Berberis vulgaris.
Papaver Rhœas, *P. somniferum*, *Chelidonium majus*.
Fumaria officinalis.
Nasturtium officinale, *Sisymbrium officinale*, *Brassica Rapa*, *B. oleracea*, *Cheiranthus Cheiri*, *Sinapis arvensis*, *S. alba*, *Capsella Bursa-pastoris*, *Cakile maritima*, *Crambe maritima*, *Raphanus Raphanistrum*, *R. sativus*.
Reseda luteola, *R. lutea*.
Spergula arvensis, *Stellaria media*.
Hypericum perforatum.
Malva sylvestris.
Geranium molle, *Erodium cicutarium*, *Tropæolum majus*.
Cytisus scoparius (for pollen only), *Medicago sativa*, *Trifolium repens*, *T. fragiferum*, *T. pratense* (after the flowers have been punctured by short-trunked humble-bees), *Lotus corniculatus*, *Robinia pseudo-acacia*, *Vicia tetrasperma*.
Rosa canina, *R. rubiginosa*, *Rubus fruticosus*, *R. idæus*, *Fragaria vesca*, *Cratægus oxyacantha*, *Pyrus communis*, *Pyrus malus*.
Ribes nigrum, *R. rubrum*, *R. Grossularia*.
Œnothera biennis.
Daucus carota.
Petasites vulgaris, *Centaurea cyanus*, *Cnicus arvensis*, *C. lanceolatus*, *Cichorium Intybus*, *Hypochœris radicata*, *Tanacetum vulgare*, *Senecio jacobæa*, *Taraxacum officinale*.

(See Appendix B, p. 559.)

Family Vespidæ

Polistes tasmaniensis, Sauss. Australian Wasp

This wasp has been established about Hokianga and Whangarei, probably, indeed, all over North Auckland Peninsula for a great number of years, and was first recorded from Rawene in 1893. It was then common over the Hokianga district. In 1911 Colonel Boscawen reported of an orchard in Opitonui, that it was full of their nests. It is extremely common in the north, but has spread very slowly to the south of Auckland. In 1918 Mr Howes met with it both at Dunedin and Waipori[1].

[1] In Nicol's book, *Acclimatisation of Salmonidæ at the Antipodes*, we are told (p. 46): "a living wasp was found among the moss" in which the eggs were packed, in the shipment of salmon ova sent to Otago in 1869, in the 'Mindora.'

It attacks many species of insects, and is especially destructive to the pear-leech or sawfly (*Eriocampa limacina*).

Family FORMICIDÆ

Prenolepis longicornis, Fabr. (Europe and Asia)

Mr W. W. Smith states that this so-called "Sugar-Ant" is a great nuisance in some houses in Nelson and New Plymouth, and is common in both localities. It is also common in the Auckland Province, and is probably very widespread, especially in the North Island[1].

Order V. COLEOPTERA

LAMELLICORNIA

Family SCARABÆIDÆ

In the *Index Faunæ Novæ-Zealandiæ*, pp. 349–54, a list of introduced species of insects is given. Among those named, especially among the Coleoptera, are many regarding which no information is now available, for Captain Hutton, who compiled the list, died several years ago, and Major Broun, who made a great number of the identifications, died recently (1919). He was unable, during the last few years of his life, to supply me with the information which I sought. Hence several species are recorded, regarding which no details can now be given. Mr W. W. Smith informs me that he used to send up numerous specimens of introduced insects to Major Broun, but it is not possible to say whether any records of these were kept, or if so, where they are now.

Aphodius granarius, Linn.

Mr Wakefield found this European beetle in Canterbury in 1872. He then stated it to be scarce, but it was plentiful in the following year. In 1887 Mr G. V. Hudson reported it as taken in horse-dung in Wellington. Mr W. W. Smith informed me (in April, 1919) that the species was common in Taranaki. A favourite location for it is on the stems of tree-ferns, among the scales and hairs[2].

Trox sp.

In the *Index* it is stated: "Reported by Dr Swale."

[1] Mr G. Howes collected an Australian (?) ant at Titahi Bay near Wellington. In connection with the introduction of a lace-wing fly from the Cook Island to help to suppress aphides in New Zealand (date?), it was stated that one breeding-cage was found to contain a number of ants, which had got in and destroyed all the flies. I have no means of ascertaining what species of ant was thus introduced.

[2] Pascoe in 1875 recorded an *Aphodius* "like *A. pusillus*," and an *Onthophagus* "apparently identical with the Australian *O. fulvolineatus*, Bl." from New Zealand, but, he adds, "there could have been no pabulum for such insects formerly."

Onthophagus granulatus, Boh. Dung Beetle

In 1872 Mr C. M. Wakefield stated that specimens of this beetle had been taken by Mr Fereday in Nelson provincial district and he considered that they had been introduced with cattle from Australia. Mr Hudson met with it at Wakapuaka in Nelson, in 1882. Mr W. W. Smith reported it as common in Taranaki in 1919. He thought it was probably common throughout New Zealand, but that there were no collectors of introduced species.

Onthophagus posticus, Erichson

Mr W. W. Smith reports this Australian species as common in Taranaki, where it usually occurs under the dry bark of species of *Eucalyptus*.

Proctophanes sculptus, Hope

Taken by Hudson at Palmerston North in September, 1883, in debris.

Calathus zealandicus, Redtenbacher

Hudson found this species under stones at Karori, Wellington, in November, 1882; and later at Nelson and Palmerston North. Writing in 1917 he says: "I doubt if this insect is imported as other European Carabidæ have not arrived here." Miller thinks the species is indigenous.

ADEPHAGA

Family CARABIDÆ

Rhytisternus puella, Chaudoir

Taken by Mr Hudson at Karori, Wellington, in 1882, under stones.

Hypharpax australasiæ, Dejean

This Australian beetle was considered by Captain Hutton, on the authority of Mr W. Bates, to be an introduced species. It was recorded in 1874.

Hypharpax australis, Dejean

Common among grass, in vegetation, etc., in Taranaki, according to Mr W. W. Smith (in 1919). An Australian species first recorded in 1874.

Agonochila binotata, White

Mr Hudson reports this Australian species as occurring under bark, in the Tinakori Range, Wellington, in September and October, 1887; also as occurring at French Pass. Mr Philpott states that it was common near Invercargill in 1892, and adds (1917): "It is not so commonly met with now, owing probably to so much bush-land having been brought under cultivation. I do not remember having seen it far from the coast."

Læmosthenes complanatus, Dejean

Mr Hudson recorded this South European species in March, 1888, as occurring under packing-cases. This insect was identified by Commander J. J. Walker as *Pristonychus terricola*, Herbst.

POLYMORPHA

Family HYDROPHILIDÆ

Hydrobius assimilis, Hope

Taken by Mr Hudson in a brackish pool at Ocean Beach, Happy Valley, Wellington, in November, 1886. Mr Smith states that it is not uncommon near the sea at New Plymouth and elsewhere, in 1919.

Paracymus nitidiusculus, Broun

Mr Hudson says: "Not known to me; but it seems hardly likely that Broun would describe an *introduced* species, unless he made a synonym!" It is given in the *Index Faunæ Novæ-Zealandiæ* as an Australian beetle.

Cyclonotum marginale, Sharp

This Australian beetle is common in New Plymouth according to Mr W. W. Smith (1919). It was recorded in the *Index* in 1903.

Cercyon flavipes, Fabr.

Mr Philpott reports (1917): "I knew this species at least twenty years ago, finding it both near the coast and fifty or sixty miles inland (in Southland). It is not abundant." As far back as 1872, Mr C. M. Wakefield sent specimens of this beetle to Britain from Canterbury, and stated that it was abundant.

Family STAPHYLINIDÆ

Homalota sordida, Marsham

Taken by Mr W. W. Smith at Ashburton, about 1884.

Oxytelus rugosus, Fabr.

This Australian species is recorded in the *Index* in 1903. Taken at New Plymouth by Mr W. W. Smith.

Quedius fulgidus, Fabr.

Introduced from Europe, and recorded in the *Index* in 1903. Mr W. W. Smith informed me in 1919 that it is common about New Plymouth.

Philonthus æneus, Rossi. Rove Beetle

Taken by Hudson among decaying vegetable refuse in a garden at Karori, Wellington, in February, 1883. This insect is commonly taken for an earwig. Found about the same time at Ashburton; especially under the bark of *Eucalyptus trees*.

Philonthus scybalarius, Nordmann

The only record I can find is in the *Index* (1903).

Philonthus affinis, Roth.

Originally taken by Mr W. W. Smith at Ashburton about 1884.

Philonthus nigritulus, Gravenhorst

Only recorded in the *Index* (1903).

Xantholinus punctulatus, Paykull

Mr Wakefield recorded this species from Canterbury in 1872. I have no record of its occurrence since.

Family Histeridæ

Platysoma bakewelli, Marseul

An Australian species, recorded in the *Index* (1903). Mr W. W. Smith informs me (1919) that it occurs at New Plymouth, on trees, especially under bark.

Carcinops 14-striata, Stephens

Recorded in *Index* (1903).

Family Nitidulidæ

Carpophilus hemipteris, Linn.

This cosmopolitan species is recorded in the *Index* (1903).

Carpophilus mutilatus, Erichson

Like the last, a cosmopolitan species, first recorded in the *Index* (1903).

Osmosita colon, Linn.

Found by Hudson in September, 1888, at Karori, Wellington, amongst the skin and bones of a dead cow. Philpott writes (1917) "I used to meet with this species occasionally about twenty years ago. It was never common, and I have not seen it for many years."

Family Trogositidæ

Tenebrioides mauritanicus, Linn.

Captain Hutton records this species under the name of *Trogosita mauritanica*, as occurring in New Zealand before 1870. Mr W. W. Smith (1919) reports it as common about New Plymouth. A sunshine-loving beetle, found on dry walls, etc.

Family Cucujidæ

Silvanus surinamensis, Linn.

This cosmopolitan beetle is recorded in the *Index* (1903). It is being continuously introduced into New Zealand. It is common in all large dry-goods stores. It has recently been found in Quaker Oats, and other grain preparations, especially those from the United States.

Silvanus unidentatus, Fabr.

An European species, recorded in the *Index* (1903).

Family Coccinellidæ

Coccinella 11-punctata, Linn. Common Ladybird

Taken by Mr Hudson on top of the Tinakori Range, Wellington, in February, 1889; also recorded by him as swarming about the cairn on the top of Mount Enys (Castle Hill, West Coast road) at an elevation of 7200 ft. above sea-level on 9th January, 1893. Writing in 1896 Captain Broun stated that in previous years he had distributed colonies of this insect throughout the Waikato district. It is common now in Taranaki (April, 1919) and appears to be pretty universally distributed throughout New Zealand.

This species is a valuable help to the farmer and gardener, as it devours the aphides which attack turnips, cabbage, grain, garden plants and fruit trees. It does not confine itself to any one species, but eats most species of aphides. Hilgendorf states that as a rule it is not sufficiently numerous to do much good, but that occasionally it is abundant.

Coccinella californica, Mannh.

According to the report of the Agricultural Department for 1899 this species of ladybird was introduced from America by A. Koebele.

Philpott says (1917): "Neither this nor any of the succeeding six species seem to have spread to Southland, or at least to the neighbourhood of Invercargill."

Coccinella sanguinea, Linn.

Introduced from America, along with the preceding species, in, or before, 1899.

Rhizobius ventralis, Erichson. Blue Ladybird

In 1900 the Agricultural Department introduced this insect from Australia, and liberated small colonies at Auckland, Whangarei and Hawke's Bay, for the purpose of combating the scale-insects which attack the various species of *Citrus* (oranges, lemons, etc.). About this same year an imported scale-insect (*Eriococcus coriaceus*) was noticed in S. Canterbury attacking the various gum trees (*Eucalyptus* sp.) which had been extensively planted there. The pest soon rose to dangerous proportions and threatened the existence of the gum trees, numbers of them dying. Quantities of *Rhizobius* were then obtained from North Auckland, and other lots from New South Wales and were widely distributed in the district. By the winter of 1907 the plantations were well stocked all through the South Island, and the following year the scale-insect was well under control. The *Rhizobius* is now very common, especially in the South Island.

Novius cardinalis, Mulsart (*Vedalia cardinalis*).
Australian Ladybird

In 1894 *Icerya* appeared at two places in the Wellington district, and a number of ladybirds were reared and liberated on the properties attacked. They appeared to have exterminated the blight in the Wairarapa.

In 1899 the Agricultural Department introduced a large number of these ladybirds in order to destroy the cottony-cushion scale (*Icerya purchasi*), and continued to rear and distribute them for several years.

Dr Hilgendorf in 1917, says of the shrub-land in the Cass district: "the yellow-spotted black ladybird, *V. cardinalis*, occurs rarely."

Mr W. W. Smith (April, 1919) reports it as common in Taranaki.

Hippodamia convergeus, Guer.

In 1899 the Agricultural Department introduced this species from America. I have no further record, and do not know whether it increased or died out.

Leis conformis, Boisd. The common Spotted Ladybird of Australia

In 1896 this species was introduced from New South Wales by the Agricultural Department, as a destroyer of aphides. Two years later it was reported from Auckland, where they were originally liberated, that they were breeding.

Cryptolæmus montrouzeri, Mulsart. Red-headed Ladybird

In 1899 the Agricultural Department introduced this insect from Australia—(it had previously been introduced into Hawaii from Australia, with good results)—and liberated it in the district north of Auckland and about Whangarei. It proved itself a very valuable introduction, attacking and destroying the mealy-bug (*Dactylopius adonidum*), the orange-scale (*Lecanium hesperidum*) and the scale (*Eriococcus araucariæ*) which infested the Norfolk Island and other pines. Considerable numbers were reared and distributed in succeeding years. The species does not appear to breed south of Auckland. Colonies were taken down to Timaru in 1906, and liberated in the gum-tree plantations to combat the scale (*Eriococcus coriaceus*), but the winter of South Canterbury proved too severe, and the field was abandoned to *Rhizobius ventralis* which soon almost cleared the pest away.

Orchus chalybeus, Boisd. The Steely-blue Ladybird

This species was also introduced from Australia by the Agricultural Department (in 1899?) and liberated north of Auckland; and in 1906 a large consignment from New South Wales was liberated near Timaru to assist in the destruction of *Eriococcus*. The winter climate of Canterbury is, however, too severe for it. It is now very abundant, north of Cambridge, Waikato, on oakscale.

Orcus australasiæ, Boisd. The Six-spot Blue Ladybird

A single specimen of this Australian insect was taken by Mr Hudson among some apples, in April, 1887, in Wellington. It is probable the apples were imported.

Family Latridiidæ

Monotoma picipes, Herbst.

Recorded in the *Index* (1903).

Monotoma sub-4-faveolata, Watson

Mr W. W. Smith met with this beetle in the eighties, among rotten grain sacks, at Albury, S. Canterbury. This is probably the cause of its appearing in the *Index*.

Monotoma spinicollis, Aube.

Another European species recorded in the *Index* (1903).

Coniuomus nodifer, Westwood

Recorded in the *Index* (1903).

Family Dermestidæ

Dermestes vulpinus, Fabr. Bacon Beetle

This most destructive beetle was first recorded by A. Purdie, who found specimens among some paper, wrapping a collection of geological specimens received in Dunedin from Australia in 1884. It appears to be very common in many parts, and Captain Broun reported to the Agricultural Department in 1895, that he had spent "three days in the Government Buildings, Auckland, destroying this pest, which threatened injury to official documents and other property stored in the cellars." Mr A. M. Wright, Chemist to the Christchurch Meat Co., records an experience they had with this insect at the works at Picton, when it caused a good deal of damage to the woodwork of the manure building, and invaded the adjoining fellmongery building. I give his own account of the trouble:

At the time we were storing in the manure-shed a quantity of partly-dried shank-bones with wool adhering, and this at one time became almost a heaving mass of larvæ which spread about the whole building, boring into bones, wood, and even the lead covering of some concrete, where they buried themselves until they appeared as the insect.

The remedy lay in the removal of the partly-dried shank-bones outside, and with this precaution, as well as a general care to prevent none but dried offal manure within the building, the hatching ground of the pest was removed, and little or no trouble has been found since. For one year we specially treated all our manure in a dryer at the close of the season, in order to kill all eggs and larvæ before sending any out to clients. The danger at one time seemed so acute that plans were actually prepared to pull down the wooden block and replace it with a brick building, but fortunately the pest was practically eliminated. We were not able to trace definitely the point of origin of this insect, but its presence first became evident after the arrival of some material from Dunedin, and also from Australia.

I find that this beetle is in pretty well every manure factory in the country, and is considered a very serious pest, on account of its destructive attacks on the woodwork. It also occurs in many places where fur, hair, skin, etc., are stored, as it eats these materials. Mr Hudson reports (on the authority of Mr Creagh O'Connor) that this bettle is very destructive to sheep-skins.

Mr A. T. Potter found this beetle in a building in Whangarei in December, 1895, it having been probably imported from Sydney, and observed its habits. He says:

the beetle propagates only in very hot weather. When about to change from the larva into the pupa state, it will burrow into sound woodwork of a building, which in some cases is reduced to a honeycomb. The eggs hatch (at 70°) in from four to seven days. The mature larva after moulting several times...formed a chamber in its food material, or in any other convenient locality at hand, when it curled itself up, loosely covered with some of its own food, and the refuse round it. There it lay for five days, then moulted again for the last time, and turned to the pupa from which the beetle developed in thirteen days.

Butler says of the species:

they are the jackals of the flesh-flies, coming round when the maggots of the latter have finished up all the soft and juicy parts of a fresh carcase, and clearing off the hard and dry remnants of the skin, tendons, ligaments, etc., which their predecessors have left untouched.

Dried meats, skins, horns, hoofs and feathers are attacked; they destroy every form of preserved animal, such as beasts, birds, fishes, crabs, insects and spiders in museum cases, and are especially fond of cork. "An account has been placed on record of the destruction of a whole ship's cargo of cork by vast numbers of them."

Mr A. H. Cockayne says it is very destructive to sheep-skins in freezing works, stores, etc. In 1918 a great deal of damage was done to rabbit- and sheep-skins in the Dunedin stores.

Dermestes lardarius, Linn. European Bacon Beetle

The first record I find of this insect was in February, 1873, when Dr van Haast found a live specimen in a box of insects received at the Christchurch Museum from Australia.

It is occasionally introduced in American bacon, but does not establish itself.

Anthrenus musæorum, Linn. Museum Beetle

This species feeds upon skins, hair, feathers, and other remains of animals, and is not deterred even by such strong odours as camphor.

Mr W. W. Smith reports it as common and very destructive in houses in Taranaki (April, 1919).

Family Bostrichidæ

Lyctus brunneus, Stephens. The Ship-borer

In the Agricultural Department's report for 1903, Captain Broun stated that this beetle had only been found by him at Auckland three times, all isolated captures.

(See Appendix B, p 559.)

Family Ptinidæ

Ptinus fur, Linn.

First recorded from Canterbury by Mr C. M. Wakefield in 1872.

Mr Hudson reports that this species was found amongst chicory plants about 1909 by Mr A. H. Cockayne. Mr Philpott writing in 1917 says:

This beetle is common about houses wherever dry dead animal matter is to be found. For this reason it is often met with in disused bee-hives and comb-frames. It is the worst enemy the entomological collection has to be guarded against. Houses badly infested with the White Pine borer (*Anobium domesticum*) are often found to be infested with *Ptinus fur* also. They probably feed on such of the borers as fail to emerge from the timber. *P. fur* does not bore into the wood, though it often scoops out a little depression for the purpose of pupating in. It is not confined to buildings, but is found in the depths of the bush and high up on the mountains. I do not know anything of its first appearance in New Zealand, but its acquaintance was forced on me as soon as I began to make a collection of Lepidoptera, about 27 years ago.

It is probable that this was one of the species, referred to by Sir Joseph Banks in his *Journal* (see p. 341), as making their bread on board ship almost uneatable. It readily attacks grain. "At one time it was considered to be so largely an animal feeder as to have been called by De Geer *vrillette carnassière*, the 'carnivorous borer.'"

Mr W. W. Smith informed me (April, 1919) that this very quick and active beetle is common among plants growing in warm borders, where it shelters under fleshy leaves and in similar situations. It is abundant in New Plymouth.

Anobium sp. (probably *tesselatum*)

This species is recorded by Dr Hilgendorf from North Canterbury, where it was found to be very destructive in the English Church, Ashley, in December, 1916. The mature beetle emerges from the wood during several months of the year. It is larger than the preceding species, and has a circular gold spot on its back.

This is the insect popularly known in Britain as the *Death Watch*, on account of the ticking noise made by striking the walls of their burrows with its head or jaws.

Mr W. W. Smith (April, 1919) says it is very common in Taranaki.

Anobium domesticum, Linn. Wood-borer; White-pine-borer

This destructive beetle was first observed by Captain Broun in 1875 at Tairua. Its ravages are extended to many kinds of wood, but white-pine or kahikatea is the timber most readily attacked. "Entire buildings have been reduced to a substance resembling sawdust." Professor Kirk, writing in 1904, after an examination of white-pine and other timbers, found that all specimens attacked by the borer contained considerable quantities of starch in the wood-cells. He recommended that the timber be felled in early spring, when the starch is turned to glucose, and then rafted to dissolve out the glucose. Timber without starch or glucose was not found by him to be attacked by the borer.

Dr Hilgendorf states that this species emerges in the mature state from the timber in which it lives only in the second and third week in December of each year. Mr Philpott states that the perfect beetles begin to emerge about the middle of November, and by the middle of December are in their greatest numbers. By the end of January they have become scarce. The species is very abundant in Southland. At Weraroa the beetles are found to emerge during October and for a period of four months.

W. Riddell of Invercargill, with some 35 years' experience in the timber trade, considers that

> the white-pine beetle undoubtedly shows preference for timber grown in certain localities. For instance, White pine from a lowland forest is readily attacked, while that from a more elevated and drier locality is safe for a much longer period. He further stated that on an average about ten years elapse before the grub makes its appearance in a new house, and that there are very few houses more than ten years old in Invercargill in which the grub is not found. There are very few kinds of timber exempt from attack.

He considers it mainly a question of age and dryness, and that certain substances in the wood must evaporate or change before the insect will attack it.

E. A. Butler states that "formerly their ravages were more considerable" (in Britain) "than at the present day, owing to the then more extensive use of timber—and especially unpainted timber—in building construction."

This species makes ticking sounds similar to those of *A. tesselatum*.

The larvæ are destroyed to a certain extent by a minute wingless ant-like hymenopteron, which goes into the tunnels and attacks its victims.

Anobium paniceum, Linn. Biscuit-Weevil; Drug-store Beetle

This omnivorous insect eats anything of a vegetable substance that it meets with. Its first name is given on account of the attacks it not unfrequently makes on ship's biscuits; its second for its penchant for such things as rhubarb root, ginger and even Cayenne pepper. It is also very destructive to books, drawings and paintings. Westwood records it as perforating tinfoil for the comestible below it.

Mr W. W. Smith states that it is as common as the preceding species in any timber, especially in white-pine. It works nearer the surface than *A. domesticum* (April, 1919).

In Europe it has been found to visit the flowers of Hawthorn (*Cratægus oxyacantha*).

Family MALACODERMIDÆ

Metriorhynchus rufipennis, Fabr.

Mr Hudson considers that this is probably *M. erraticus*, which was found in August, 1895, by Major-General Schaw, at Auckland; and which he himself found abundant on dead trees at Ohakune, in January, 1912. This is probably the species which Mr Smith records as so destructive to the foliage of young blue-gum trees (*Eucalyptus globulus*) in Taranaki.

It occurs very commonly there under the bark of trees. Mr Smith says this is the commonest location for Australian and Tasmanian beetles.

Family CLERIDÆ

Necrobia ruficollis, Fabr. (Burying Beetle)

Mr Hudson recorded this beetle in 1890, as occurring commonly in decaying animal matter. It was found by him first at Karori, Wellington, in September, 1888, amongst the dried skin and bones of a dead cow.

Mr Smith (April, 1919) reports it as very common in Taranaki, occurring under dry boards, among rotten sacks, etc. Also under dead birds and dead sheep. Very abundant in stores in Wellington, associated with *Dermestes vulpinus*, in 1918.

Necrobia rufipes, De Geer. Red-legged Ham Beetle

First recorded from Auckland in 1875 by Captain Broun. Taken by Mr Hudson at the Dee River, Kekerangu, in February, 1890. Mr Philpott says (1917):

> I first met with this species in 1890, when I found it to be very abundant in a fertiliser (bone-dust) works near Invercargill. I have never found it except in, or about, sacks of bone-dust.

Family ELATERIDÆ

Lacon variabilis, Candeze. Australian Click Beetle

Recorded by Mr Hudson as taken on the hills around Nelson by Mr A. S. Atkinson in June, 1886; and by Commander J. J. Walker at Picton in the autumn of 1902. Very common in Taranaki (April, 1919).

Monocrepidius exsul, Sharp. The Potato Elater

Stated by Mr Hudson (1917) to be very common at Karori, Wellington. It flies freely in the dusk of the evening. Mr W. W.

Smith states that the beetle is very destructive to the foliage of the Virginian creeper (*Ampelopsis virginica* or *Veitchii*). The report of the Agricultural Department for 1898 states that it often cuts the stems of growing potatoes. Quite common in Wellington Province.

Agriotes sp.

The report of the Agricultural Department for 1906 states that a species of *Agriotes*, probably *A. lineatus*, Linn., the striped click beetle, or wire worm, was very common, and had done much damage, especially in potato crops, at Mangaweka. Mr W. W. Smith says (1919) that this species attacks the roots of gooseberry bushes, frequently killing the plants. Mr A. H. Cockayne informs me that it is very generally distributed.

Family Buprestidæ

Buprestis lauta, Leconte

Common in Taranaki, especially in sunny weather (April, 1919). It was recorded in the *Index* (1903).

Stigmodera gulielmi, White

Recorded in the *Index* (1903).

HETEROMERA

Family Tenebrionidæ

Tenebrio obscurans, Fabr. Meal Worm

Sir Joseph Banks in his *Journal* says, when nearing New Zealand (23rd September, 1769):

> Our bread is but indifferent, occasioned by the quantity of vermin that are in it. I have often seen hundreds nay, thousands, shaken out of a single biscuit. We in the cabin have, however, an easy remedy for this, by baking it in an oven, not too hot, which makes them all walk off; but this cannot be allowed to the ship's people, who must find the taste of these animals very disagreeable, as they everyone taste as strong as mustard, or rather spirits of hartshorn. They are of five kinds, three *Tenebrio*, one *Ptinus*, and the *Phalangium canchroides*; this last, however, is scarce in the common bread, but vastly plentiful in white meal biscuits, as long as we had any left.

Mr C. Bills informs me (1918) that bird-dealers formerly imported considerable numbers of meal worms into New Zealand for feeding introduced birds, but that they had discontinued doing so for some years.

Mr W. W. Smith informs me that they are very common in Taranaki. The probability is that they are widely spread.

Tenebrio molitor, Linn. Meal Worm

This cosmopolitan species has long been in New Zealand. It was recorded in the *Index* in 1903. Mr W. W. Smith reports it as very common in Taranaki (1919); while Mr A. H. Cockayne says it is very abundant in imported meals and artificial foods, and is constantly being introduced.

Gnathocerus cornutus, Fabr.

Recorded in the *Index* (1903).

Tribolium ferrugineum, Fabr.

A cosmopolitan beetle, also recorded in the *Index* (1903)

Family ANTHICIDÆ

Anthicus floralis, Linn.

Recorded in the *Index* in (1903).

Mr W. W. Smith reports this as common in Taranaki, on flower stalks, etc. (April, 1919).

Family ŒDEMERIDÆ

Nacerdes melanura, Schmidt. Wharf-borer

In the report of the Agricultural Department for 1902, Captain Broun recorded this species as occurring in hardwood, imported into Auckland from Australia and Tasmania.

Family BRUCHIDÆ

Bruchus rufimanus, Boh. Pea Weevil

This beetle is commonly imported into the country from Britain with peas and beans. A bag of specially selected peas imported from a London seedsman was submitted to me in 1906; *every pea contained a weevil*. Dr Hilgendorf says that in Canterbury it lays eggs in beans as they develop in the field, but fortunately they are not very common.

In Britain it is frequently found in springtime on the blossoms of gorse and broom.

Very common in Taranaki (April, 1919).

Family CHRYSOMELIDÆ

Paropsis sp.

Mr Drummond reports (November, 1916) that an Australian species was found in the bright sunshine on granite (?) rocks on Cooper's Knob near Lyttelton at a height of 1800 ft. This beetle feeds on Eucalyptus.

Family Cerambycidæ

Hylotrupes bajalus, Linn. Borer

The first record I have found of this beetle is of two specimens taken at Auckland in 1874, and identified in London by Mr H. W. Bates.

It perforates wood, and

> Kirby states that Sir Joseph Banks once gave him a specimen of sheet-lead, which, though only measuring eight inches by four, was pierced with twelve oval holes, some of which were as much as one-fourth inch in longest diameter.

The lead had covered rafters which had been bored by the insects.

Phoracantha recurva, Newman

One specimen of this beetle was captured near Christchurch in 1873. Mr W. W. Smith says this active beetle is not uncommon in Taranaki (April, 1919).

Callirhœ allaspa, Newman

Recorded in the *Index* (1903).

Tessaromma sulcatum

Mr Hudson reports this Australian species as captured at Auckland in May, 1902, by Commander J. J. Walker.

RHYNCHOPHORA

Family Anthribidæ

Doticus pestilens, Oliff. Dried-apple Beetle; Jumping Anthribid

This apple weevil imported from Australia was recorded by Mr T. W. Kirk in 1895 as abundant in orchards about Wellington, where it was still common in 1899.

Very common in Taranaki (April, 1919).

Family Curculionidæ

Otiorhynchus sulcatus, Fabr. Black Vine-Weevil

Captain Broun found this among sorrel (*Rumex acetosella*) and grass-roots on Mount Eden in 1866. Mr Pascoe received numerous specimens from Captain F. W. Hutton from Wellington in 1875. Later on it was found abundantly in a vinery at Nelson. Mr Hudson found it under stones on the Tinakori Range, Wellington, in August, 1889 and 1890, and under boards at Kaitoke in 1902.

Mr Philpott (1917) says "I find this species commonly about the base of tufts of grass when gardening in the spring."

Mr W. W. Smith reports it as very common in Taranaki (April, 1919). It is apparently widespread, among sorrel roots especially.

Gonipterus reticulatus, Boisduval

Mr Hudson informed me in 1890 that this Australian beetle had been found by Mr W. W. Maskell amongst species of *Eucalyptus*, Sydney Street, Wellington. It was very abundant there in February, 1892. It is very fond of sunshine.

Mr Smith reports it as occurring on several species of plants in Taranaki, and says that in captivity the perfect insect (imago) is a voracious feeder. It is very common among eucalypts.

Calandra granaria, Linn. Grain-weevil; Corn-weevil

I do not know how early this beetle was introduced, but it has been known for a very long time, and is very troublesome in stored grain, especially in wheat. In 1894 and 1895 the grain sheds in Timaru were very badly infested, and it did a great deal of damage.

Dr Hilgendorf informs me that about 1900 a rejected consignment of wheat was emptied on Timaru beach, and the weevils were seen crawling towards the town in thousands. They were destroyed by fumigation with carbon disulphide. Again in 1916 it was imported in vast numbers in a consignment of barley from South Australia. Dr Hilgendorf adds that as about 40 days elapse between the laying of the egg, and the emergence of the mature beetle, the pest is never very bad, except in grain at least one season old.

Calandra oryzæ, Linn. Rice-Weevil

This cosmopolitan beetle was recorded in the *Index* (1903), it is probably common in New Zealand. Mr W. W. Smith records it as common at Ashburton and New Plymouth (April, 1919).

Anthonomus pomorum, Linn. Apple-blossom Weevil

This beetle has been a long time in the country, but is not very common. In 1890 it did a great deal of harm in orchards in Canterbury. It is common at Ashburton and at New Plymouth.

Sitones lineatus, Linn. Striped Pea-Weevil

This little beetle, which attacks peas, beans and species of *Trifolium*, is a European species. The adults are found in numbers on the flowers of gorse and broom in Britain.

Mr W. W. Smith reports it as common at Ashburton and at New Plymouth (April, 1919).

Mr A. H. Cockayne says it is common in Wellington Province.

Oxyops concreta, Pascoe

This is an Australian species, the larva of which attacks the leaves of species of *Eucalyptus*. It is found about the Manawatu district, and is probably of general distribution where gum-trees have been planted in New Zealand. Mr A. H. Cockayne states that at the Government plantations at Rotorua and Waiotapu the cultivation of blue-gums (*Eucalyptus globulus*) has had to be abandoned, on account of the ravages of this weevil.

(See Appendix B, p. 559.)

Order LEPIDOPTERA

Sub-order RHOPALOCERA

Family NYMPHALIDÆ

Danaida plexippus, Linn. (*erippus*, Cr.)

Mr Hudson says of this butterfly that it is "a comparatively recent natural immigrant which probably reached New Zealand during the first half of the last century, independent of human agency."

It appears to have been first observed in 1840–41, when Mr F. W. Sturm took a specimen at Reinga, on the Wairoa River in Hawke's Bay. In 1848 the same gentleman took a number at the Waiau, a tributary of the Wairoa. It has been frequently taken in Hawke's Bay since then (in 1861, 1873, 1874, etc.). Mr Sturm says that it kept about the Lombardy poplars and *Hoheria populnea* in his garden. He thought it was an indigenous species.

Mr Colenso recorded the same species apparently in 1877 from the same district, and stated that he had known it for some years. Dr Hector found it in great abundance in Hokitika during the summer of 1873. Mr Fereday discussed the question of its introduction in 1874 (*Trans. N.Z. Inst.* vol. VI, p. 183).

In 1879 Mr Kingsley took a specimen in Nelson, and in 1890 he got several from the same district. In 1906 Mr Hudson got specimens from Makara Beach near Wellington, and saw one in the city itself. Mr Howes tells me (1919) that it has been taken at Auckland and Wanganui; it is also reported to have been seen in Dunedin, and he himself obtained one in a dying condition at Halfmoon Bay, Stewart Island.

Danaida chrysippus, Linn.

Mr Hudson states that two specimens of this species were taken at the Thames in March, 1904. Mr Howes, writing in 1919, says: "One flew into an Auckland school a few years back. Two specimens were recently taken at Wanganui."

Vanessa atalanta, Linn. English Red Admiral Butterfly

Mr T. W. Kirk captured specimens of this species in the Wellington Botanical Gardens, during the summer of 1881, and saw others on several subsequent occasions. Mr A. Philpott saw a specimen in a collection of Lepidoptera made by Mr Dunlop of the Orepuki Shale Works, who informed him that this was the only one secured out of several seen. Mr Howes, however, is of opinion that the specimen was brought out from Britain by Mr Dunlop, as he saw it shortly after his arrival here. There is evidently some doubt about its occurrence in Southland.

Vanessa urticæ, Linn. Small Tortoiseshell Butterfly

Found by Mr T. W. Kirk at the same time and in the same locality as the preceding species. No later record of its occurrence is known.

Junonia vellida, Fabr.

Mr Hudson states that this butterfly was common in the Cook Strait region in 1886–87, but that only one or two specimens had been seen since then. He considered it to be a natural immigrant from Australia, where it is common.

Miss Castles informs me that it was taken at New Plymouth in 1893 (where Mr W. W. Smith also reports it as taken occasionally), and at Motueka in 1898. In March, 1910, it was taken at Mt Greenland in Westland by Mr H. Hamilton. Lastly Mr Howes saw (but could not capture) a specimen in Dunedin in 1918.

Sub-order HETEROCERA

Family BOMBYCIDÆ

Bombyx mori, Linn. Silkworm

I cannot find out the earliest date at which silkworms were introduced into New Zealand, but about 1863, Mr T. C. Batchelor, of Nelson, was rearing Tuscan worms with considerable success.

(Further information on the subject will be found in Appendix B, p. 560.)

Family CARADRINIDÆ

Heliothis armigera, Hubner. Tomato Caterpillar

This cosmopolitan moth, which is known in America as the cotton-ball worm and the corn worm, and in Australia as the maize moth, was reported to the Agricultural Department in 1907 from several localities. Several consignments of tomatoes from Sydney were found to be infested with this pest. Mr Howes says it is somewhat rarely seen in the south of the South Island, though he has taken occasional specimens in Dunedin, and has seen it in fair numbers at Oamaru and Queenstown. Mr W. W. Smith informs me that it is common at New Plymouth, where it is very destructive to tomatoes, the larvæ eating the fruit. It is probably common throughout the North Island. The larva is fond of burying itself in the flowers of asters, etc.; indeed its wide range of food plants makes it troublesome.

Family Sphingidæ

Sphinx convolvuli, Linn. Convolvulus Hawk Moth

Captain Hutton, and probably several other entomologists, consider this cosmopolitan species to be indigenous. It was found in Hawke's Bay, and recorded by Mr A. G. Butler in 1877. Sir W. Buller reported it as common in Ohinemutu in 1879, and it was taken in Auckland in 1882. Since then it has been taken in many parts of New Zealand. Mr Howes suggests that, as it is commonly found feeding on the Kumara (*Ipomœa batatas*), it was probably introduced either along with that plant, or that it is a comparatively recent arrival from Australia.

Sphinx ligustri, Linn. Privet Hawk Moth

Mr Howes reports this species as having been taken at Titahi Bay near Wellington.

Chærocampa celerio, Linn. Silver-stripe Moth

Mr Hudson recorded this first in 1904, where four specimens were taken by different collectors in Nelson. It is a species of very wide distribution, and Mr Hudson considered it a natural immigrant from Australia, where it is common. Mr A. P. Buller recorded it the same season from Titahi Bay, Wellington, where several specimens were found at dusk by Mr C. O'Connor, feeding on the sweet-scented Christmas lily (*Lilium longiflorum*?). He considers it probable that it was brought over to New Zealand by westerly winds, in view of the fact that the Hawk-moth family are possessed of sustained powers of flight; indeed, I might mention that I have in my collection a fine *Sphinx* that flew on board the R.M.S. 'Ruahine' when the vessel was some five hundred miles off the coast of South America.

Family Sesiidæ

Trochilium tipuliforme, Clerck (*Ægeria tipuliformis*, *Sesia tipuliformis*). Currant Clear-wing Moth

Introduced with the garden currant (*Ribes rubrum*) from Europe, and first bred from larvæ so obtained by Mr Fereday in Christchurch in the early seventies. It has since then become very generally distributed. Though so common, it does not seem to do a great deal of harm, or to very seriously affect the fruiting powers of the currant bushes.

Family Arctiidæ

Utetheisa pulchella, Linn.

This moth was first observed by Mr Hudson in 1877, in the Wainuiomata Valley near Wellington, and subsequently at Petone; it was recorded as *Deopeia pulchella*. Meyrick considered it as probably only an occasional immigrant. Later it was taken in considerable numbers on the flowers of the white rata, and others in the tussock-grass, by Mr O'Connor, at Titahi Bay. Mr Howes informs me that Mr Hamilton took it at Dunedin, and that Mr Morris recorded a swarm on the river bed of the Waitaki, near Oamaru. The moth usually appears in February.

Family Geometridæ

Phrissogonus laticostatus, Walk.

This species was first taken by Mr Hudson in 1905 at Nelson and again at Wanganui. In 1914 it was collected at Otaki, by my son Dr J. A. Thomson. Mr Howes informs me that he collected it amongst seaside scrub at Auckland. Mr Meyrick states that it is "very common in Australia, whence it has been recently introduced by artificial means."

Paragyrtis inostentata, Walk. (*Adeixis inostentata*)

Mr A. Philpott records this moth as common (in 1915–16) on Seaward moss and other coastal swamps near Invercargill, and adds: "The restricted distribution in New Zealand of this common Australian species would seem to point to its recent introduction through the medium of shipping at the port of Bluff."

Mr Howes met with this species commonly in a swamp at Waimarino in the North Island.

Ophideres maturna, Linn. Banana Moth

Probably introduced from Australia with bananas. Mr Hudson reports them as first seen in 1906 when a specimen was taken at Makara Bay, near Wellington. Another was captured by Mr Howes in Dunedin in 1907.

Ophideres fullonica, Hubn.

Mr Meyrick reports one doubtful specimen of this Australian species, which was taken at Christchurch. "In any case," he adds, "it is probably only a stray immigrant."

Ophiusa pulcherrima, Lucas

Mr Hudson states one specimen of this Queensland species was obtained in 1904, at Titahi Bay near Wellington.

Ophiusa melicerte, Drury

A single specimen of this moth was taken in Mr Travers's greenhouse in Wellington in 1870, and was described by Mr Fereday as *Catocala traversii*. Since then it has been taken at Titahi Bay, Wellington, at Waitomo, at Motueka and at Orepuki[1].

It is a common moth in Australia.

Plusia oxygramma, Hubn.

Mr Hudson calls this a well-established natural immigrant. Several specimens were taken in the Thames district in 1906. Miss Castles records it as occurring at the Waitakere in 1909, and at Fielding in 1911. Mr Howes states that it is common throughout the North Island, but that south of Christchurch only occasional specimens are seen.

Agrotis segetum, Schiff. Turnip Moth

First recorded from Wanganui in the Agricultural Department report for 1904. Mr W. W. Smith (April, 1919) says it is common at New Plymouth. The caterpillars attack especially mangolds, turnips and potatoes. They hide beneath the surface of the soil, and usually attack the plants they infest at or just below the surface, and nearly always at night. They also attack wheat and grass crops, and several garden vegetables and flowers. The moth is now common everywhere.

Family PYRALIDÆ

Meliphora grisella, Fabr. Bee Moth

Apparently introduced along with honey-bees, as the larva lives in bee-hives. Mr W. W. Smith reports it as common in Taranaki. He further states that when he arrived in South Canterbury in 1877, this moth was common in the hives on the Upper Rangitata. Mr Meyrick recorded a specimen from Nelson in 1877.

Plodia interpunctella, Hub. Indian Meal Moth

First observed in 1912 by Mr Hudson; probably introduced in figs, maize, etc. It feeds on raisins, currants, prunes and other dried food-products. It was recorded again in 1914. Mr A. H. Cockayne says it is common in the Gisborne district.

[1] In his paper on some rare species of *Lepidoptera* (*Trans. N.Z. Inst.* vol. XXXVII, p. 333), Mr A. P. Buller mixes up two papers by Mr Fereday in vol. IX of the *Trans. N.Z. Inst.* in a remarkable manner.

Pyralis farinalis, Linn. Meal Moth

Mr G. Howes reported this species as common in Dunedin in 1903–4. Mr W. W. Smith also recorded it as common in Taranaki, and later as common everywhere. It attacks all kinds of cereals "whether as corn, flour, meal, bran, and even straw." The moth appears during the summer months, wherever grain or farinaceous materials are stored.

Hymenia fascialis, Cram.

Larva found on melons, etc. It has probably been introduced with fruit from tropical or sub-tropical Australia. Hitherto it has only been recorded by Meyrick in 1912.

Diplopseustis perieralis, Walk.

Found near towns, and probably introduced from Australia or Fiji, where it is not uncommon. Mr Philpott is inclined to think the species is indigenous, as it occurs throughout both islands.

Galleria melonella, Linn. Bee Moth; Wax Moth

According to the reports of the Agricultural Department, this moth was first observed at Okaiawa near Mount Egmont, in three different apiaries, in 1904. It was probably introduced from Australia, where it is said to have been brought from Europe in 1880. There was very considerable doubt thrown for some time on the identification, but the fact of this species being established in the North Island is now well known, alike to lepidopterists and to apiarists. Mr W. W. Smith informs me (1919) that the species is common in hives in Taranaki. It also occurs at Ruakura.

In the Apiaries Act 1907, this moth is included among the diseases of bees, which come under the powers of the Act.

Family Phycitidæ

Ephestia kuehniella, Zell. Mediterranean Flour Moth

Mr A. Philpott writes (August, 1916): "This species has become established in the flour-mills in Invercargill within the last two or three years, and I also have had examples sent to me from Dunedin." In 1905 it was reported from the Waikato. Mr A. H. Cockayne says it has been found in pretty well every flour-mill in New Zealand.

Crocydopora cinigerella, Walk.

Mr Meyrick reported this moth as occurring at Whangarei and Nelson in 1885–86, and says if this is "a recent accidental introduction (from Australia) it will probably be found soon to become more common and generally distributed." Mr Philpott says (1916): "I have not met with it myself and do not think it is in any South Island collection."

Mr Howes states that "the only locality I have met with this species was on the shores of Lake Taupo on shingle; apparently it was attached to a creeping convolvulus."

Family Tortricidæ

Tortrix postvittana, Walk.

Mr Hudson says:

It is undoubtedly introduced from Australia where it is abundant. The larva feeds on *Geranium* (? *Pelargonium*). It was first observed at Wellington in 1891, and though not abundant, is steadily becoming commoner.

Mr Howes records it from Auckland and Wanganui; and Mr Meyrick from Christchurch.

Tortrix indigestana, Meyr.

Mr A. Philpott says: "Meyrick suggests that this species may be an introduction from Australia. I have seen only one specimen—from Flagstaff, Dunedin." Mr Meyrick recorded it from two specimens reared from larvæ feeding on *Pimelea lævigata*, from Makara Beach. Mr Howes states that it is very common on hills above Waitati near Dunedin, where it was attached to *Dracophyllum*, and is found apparently right through New Zealand.

Laspeyresia pomonella, Linn. (*Carpocapsa pomonella*).

Codlin Moth

This moth is one of the most dreaded orchard pests in all countries where it is now found. It probably was introduced into New Zealand from various sources, from Britain and America, or from Australia, where it first appeared about 1855. Mr T. Kirk recorded having seen grub-eaten apples some years previous to 1874, at which date it was first noticed in Auckland, the fruit having come from Tasmania. In 1882 it was recorded by Mr Meyrick as having been taken at Wellington, and he added "probably widely spread, though hitherto little noticed." There is no doubt that it was very common, for in 1894 the Agricultural Department reported it as occurring in very many parts of New Zealand, and it took drastic steps to combat the pest. In addition to publishing literature with instructions to fruit-growers, it instituted close inspection of orchards and of all fruit (apples, etc.) imported into the country. It also sought for and introduced various natural enemies, especially parasitic wasps, such as *Calliephaltes messer* and *Zele* sp. belonging to the Ichneumonidæ.

In 1886 the moth was reported from Christchurch, next year from Te Awamutu; in 1891 from Paraparaumu, 1892 from Auckland, and 1897 from New Plymouth. In 1899 it was stated to be spreading to some extent about Hawera, and was also reported from Waikouaiti, Timaru, Palmerston North, Wanganui and Hawke's Bay. It was almost certainly in all those districts some time before its presence was recorded.

Mr Philpott informs me that it does not occur much further south than the lower end of Lake Wakatipu. In all cases where the pest has been reported from Southland, the species has turned out to be one of the native moths *Tortrix excessana*, Walk., or *Ctenopseustes obliquana*, Walk. The latter moth has frequently been mistaken in other parts of New Zealand for the codlin moth, for it attacks apples in much the same manner, but the damage it does is infinitesimal when compared with that of the imported pest.

The codlin moth has numerous enemies, both native and introduced. The tiger beetle (*Cicindela tuberculata*) destroys the grubs on the ground, and also climbs up trees in search of them. The common red ant (*Aphænogaster* (*Monomorium*) *antarcticus*) and the larger *Amblyopone cephalotes* also climb trees and destroy the grubs.

The indigenous *Ichneumon insidiator*, or an allied variety, attacks the grubs.

At least one fungus (*Isaria farinosa*) which is allied to *Cordiceps*, the vegetable-caterpillar fungus, also attacks and destroys the codlin-moth grub.

Family Tineidæ

Phthorimæa operculella, Zell. (*Lita solanella*, Boisduval.)
Potato Moth

This species has been known for the last 25 or 30 years, and is fairly common and destructive in the North Island, and as far south as Canterbury; but in the south its occurrence is only periodic. A few years ago it caused some loss in the neighbourhood of Dunedin. Mr A. Philpott says it does not trouble Southland. It is reported every year in the reports of the Agricultural Department, and the larva is found on stored potatoes or those exposed before being dug. Potatoes introduced from New South Wales and Victoria are frequently infested with the caterpillar. Mr Meyrick says this species has certainly been introduced with the potato, and is probably a native of Algeria. It feeds on all species of *Solanum*, and also attacks tomatoes.

Mr T. Kirk states that the larva can be reared on poro-poro (*Solanum aviculare*); and Mr Allan Wright records it as feeding on bulrush or raupo (*Typha angustifolia*). Mr Howes says that when potatoes are cultivated, the destruction of all small tubers and waste by turning pigs on to the ground after the crops are dug, together with the burning of the haulms, will easily keep the pest in check.

The larvæ of this species are attacked and destroyed by the larvæ of two common indigenous hover flies—*Syrphus novæ-zealandiæ* and *Melanostoma fasciatum*.

Endrosis lacteella, Schiff. (*E. fenestrella*). Window Moth

This moth seems to have been introduced in the very early days of colonisation. The moth itself is common throughout the year, but particularly from October to March, and is very generally distributed, being reported as a household moth from Whangarei to Invercargill. The larva feeds on seeds, dried fruits, flour, pollard, honeycomb and many other farinaceous or saccharine materials. This species appears to have got both its technical name, and its popular designation of "Milk-moth," in consequence of its being so frequently found drowned in milk jugs. It is not a clothes moth, but is commonly destroyed as such.

Borkhausenia pseudospretella, Staint. (*Œcophora pseudospretella*, Staint.) A Clothes Moth

This is probably one of the first moths which was introduced into New Zealand, and it is now generally distributed, probably in every house. The moth occurs between November and March. The larva feeds on all sorts of dry refuse, and on cloth (woollen, cotton and linen), paper, cork, etc. It is very destructive in museums. Mr W. W. Smith says it occurs, but not commonly, in Taranaki. Mr Meyrick (1883) recorded it from Hamilton, Napier, Wanganui, Wellington, Christchurch, Castle Hill and Dunedin.

It was reported by Mr Butler in 1877 in collections received from N.Z. from Dr Hector and Mr J. D. Enys. Dr Hilgendorf (1917) states that it is very common at Lincoln, and is destructive to stored wheat. It is also met with in the Chatham Islands.

Mr Butler calls this species "a detestable pest," and adds:

This is one of the most destructive insects imaginable, and is apparently a perfectly general feeder; nothing that is in the smallest degree edible comes amiss to it. The moth is fond of concealment, and often hides amongst the substances that have suffered from its depredations. When disturbed, it runs rather than flies, and that very rapidly, at once seeking shelter again. To pursue it with one's fingers is no easy task; it is so rapid in its movements and so slippery when touched, in consequence of the glossiness of its scales. The caterpillar is of an active habit, but conceals itself most effectually by spinning together quantities of the material it happens to be feeding upon.

It has been found to destroy tons of rice in a warehouse, and one of the Local Government Boards in Liverpool lost a great number of sweeping brooms, made of heather or ling, which were kept amongst their stores. Mr W. W. Smith (April, 1919) states that it is now very common, especially among grass and other seeds.

Ocystola acroxantha, Meyr.

Mr Hudson says: "First observed in 1886, and no doubt introduced from Australia along with *Eucalyptus*, to which it is attached."

Barea confusella, Walk.

Mr Hudson says: "First observed in 1908, when it was taken both at Nelson and Wellington. Also introduced from Australia where it is common, attached to *Eucalyptus*."

Mr Philpott says: "Examples were sent to me from New Plymouth in 1909. It is now common near Dunedin, but does not occur in Southland. I have also seen specimens from the Humboldt Range."

Symmoca quadripuncta, Haw.

First observed in Nelson in 1908, where, according to Mr Hudson, it was taken by Mr Sunley. It is a British species, which is not known from Australia.

Choreutis bjerkandrella, Thunb.

This cosmopolitan moth is a well-established natural immigrant, according to Mr Hudson. It has been collected at many points in both islands: Kaeo, Whangarei, Hamilton, Taranaki, Palmerston, Napier and Nelson. Mr W. W. Smith says it is very common in Taranaki. Mr Howes reports it as very common among tussocks above the bush line on Flagstaff and other hills near Dunedin. Mr A. Philpott, writing in July, 1916, says: "If this is an introduced species it has thriven wonderfully. It is found in all open situations up to about 3000 feet."

The larvæ feed on the thistles (*Carduus*, *Cnicus*, etc.), and other composites.

Plutella maculipennis, Curt. (*P. cruciferarum*, Zeller).

Diamond-back Moth; Shot-hole Moth; Cabbage Moth

This species is abundant all over New Zealand, especially (according to Dr Hilgendorf) in all regions where the rainfall is not much over 30 inches per annum. He estimates the damage done to turnips, rape and cabbages as amounting annually to hundreds of thousands of pounds, and in some seasons as exceeding £250,000. Rape is, however, the least damaged. The species was introduced certainly more than 30 years ago.

According to Mr T. W. Kirk the moth occurs from August to May; he states that it first came into notice in the Wellington district in 1879.

Hilgendorf gives the average cycle of the life of the insect as 53 days: the adult moth lives from 10 to 15 days; the eggs, which average 18 in number, are laid on the third to the seventh day, and they hatch out in nine days; the life of the caterpillar is 22 days; and the pupation period 17 days. From January to March, in badly infested districts, the moths are seen on the turnip fields in clouds; and they destroy about 75 per cent. of the crop. Smith says they attack and destroy all species of *Brassica*.

The caterpillars are attacked and destroyed by the larvæ of the common hover flies—*Syrphus novæ-zealandiæ* and *Melanostoma fasciatum*.

The eastern districts of Britain were ravaged by the caterpillars of *Plutella maculipennis* in 1891, when it was noticed that various kinds of birds were very effective enemies, especially rooks, starlings, and sea-gulls; and it was stated that where small birds had been exterminated the damage was worse.

Bedellia somnulentella, Zeller

This moth has been known for the last 20 years or more; its larva mines large blotches on the leaves of species of *Convolvulus*, *Calystègia* and *Ipomœa*. The late Mr A. Purdie of Dunedin bred it freely from its larva some 20 years ago. The moth occurs usually from September to November. Mr W. W. Smith reports it as occurring in Taranaki commonly.

Opogona comptella, Walk.

This moth was taken at Nelson by Mr Sunley in 1910, and is considered by Mr Meyrick to be an accidental introduction from South-east Australia, where the species is common. Mr Howes obtained a specimen more recently on the Raurimu Spiral.

Tricophaga tapetiella, Linn. (*T. palæstrica*).
Clothes Moth; Tapestry Moth

I do not know when this moth was first introduced. Its larva feeds chiefly on furs and woollen stuffs. Mr Philpott states that it is not at all common in the south, but Mr W. W. Smith reports it as very destructive in Taranaki. It was reported from N.Z. in 1877 by Mr G. Butler in collections received from Dr Hector and Mr J. D. Enys. It is fairly abundant in all parts of New Zealand north of Southland, not only in houses, but in the bush.

Monopis ethelella, Newman

Mr Hudson informs me that this moth was introduced from Australia, where it is common. The larva feeds on skins. Mr Philpott says it is very common and generally distributed; it occurs throughout the whole year; but he is doubtful as to its being an introduction.

Mr W. W. Smith reports it as common in Taranaki. It has been taken up to 4000 ft.

Monopis crocicapitella, Clem.

This is a rather common species, whose larva feeds on refuse. It was originally recorded by Mr Meyrick in 1887 as *M. ferruginella*. It has been collected in all parts of New Zealand, and more or less all the year round. In Nelson it has been taken at 4000 ft. elevation.

Tinea fuscipunctella, Hawthorn

Widely distributed in New Zealand, feeding on all kinds of dry refuse, both animal and vegetable; a domestic species. According to Mr A. Philpott it is rare in Southland.

Tinea terranea, Hawthorn

Mr W. W. Smith informs me that this species occurs in Taranaki.

Tineola biselliella, Humboldt. Clothes Moth

Another common and very destructive domestic species, feeding especially on woollen goods and hair. It is occasionally found in the linings of sofas and chairs, and in mattresses. The larvæ are found in the houses in Britain from February to September inclusive, and the moths from April to November.

Mr W. W. Smith says these moths were common at Oamaru in 1884, and are now (1919) very common everywhere.

Sitotroga cerealella, Oliv. Angoumois Grain Moth

Originally reported by Mr A. H. Cockayne about 1910, in imported maize from the United States. Probably frequently introduced, but not established.

Lampronia rubiella, Bjerk. Raspberry-bud Caterpillar

Introduced from Britain. The report of the Agricultural Department for 1904 states that the caterpillar was found on raspberry plants at Wellington and Hastings; and in the following year it was observed at Kaiapoi.

Family Hepialidæ

(?) *Perissectes australasiæ*, Donovan

This Australian moth was found at Woodville in the spring of 1918.

Order DIPTERA

Family Cecidomyiidæ

Cecidomyia destructor, Say. Hessian Fly

This much-dreaded pest has been present in all wheat- and barley-growing districts in New Zealand for the last 30 or 40 years, but has generally been kept well under observation and reduced by stubble burning. The fly was first detected in wheat at Marton in 1888, and at the same time Mr Hudson stated that large numbers of a hymenopterous parasite were observed. Mr Maskell thought these latter insects belonged to the family Proctotrupidæ, and that they were indigenous to New Zealand. This first attack involved some 200 acres of wheat. In 1892 about 1300 acres in the Marton district were attacked, and the fly appeared in the beginning of 1893 at Balclutha. During that year over 7000 acres in Bruce County were in wheat, and half of it was badly damaged by the fly; also about 200 acres were attacked in Clutha County. It appeared also in Tuapeka County, and in Nelson and Blenheim. The Government then arranged for some thousands of infected puparia of the Hessian fly to be sent from England, and in this way two parasites—*Pleurotropus epigonus* and *Platygaster minutus*—were introduced into the colony.

In 1894 the fly was severely felt in Masterton and Bruce, and in the former district many farmers abandoned the cultivation of wheat. In 1896 *Semiotellus* was found somewhat freely in the Bruce district, but no Hessian fly was recorded. In the following year the fly was thinly distributed about Waimate, Timaru, Oamaru and in Bruce.

In more recent years the fly has been recorded from time to time, but it is now negligible as a pest.

Hessian fly attacks rye, and it has been stated that it never attacks oats. This, however, is incorrect, for Mr Howes informs me that badly affected oats have been sent from Balfour in Southland, in which "the introduced control insects were more numerous than the Cecidomyids." The fly also has been found on the following grasses:

Yorkshire fog, *Holcus lanatus;* Timothy, *Phleum pratense*, *Triticum repens*, *Agropyrum repens*, *Elymus americanus* and *Bromus ciliatus*[1].

Family PSYCHODIDÆ

Psychoda phalænoides, Latr.

This fly, introduced from Europe, was first recorded by Hutton as being common in Christchurch in 1901; and was also found by Suter in Auckland. Mr W. W. Smith (April, 1919) says it is common in Taranaki.

Family BIBIONIDÆ

Scatopse notata, Meig

Hutton records in 1901 that: "In a letter to Mr Skuse, Baron von Osten-Sacken says that he has received numerous specimens from New Zealand. No doubt it has been introduced from Europe."

Oligotropus alopecuri, Reuter. Meadow-foxtail Midge

This species, which is found generally amongst meadow-foxtail pastures, originated in Scandinavia, and is also found in Great Britain. It appeared first in N.Z.-grown seed during 1910, but by 1914 had so increased as to become dangerous. In 1915 ten per cent. of the Manawatu crop of meadow-foxtail was infested, and in the following year the precentage was higher, so that the local production of this seed had to be abandoned. Meadow-foxtail seed was harvested only in two districts—Manawatu and Tauranga. Up till 1917 the Tauranga crop was not seriously infected. The larvæ are frequently found in seed imported from Europe.

Contarinia tritici, Kirby. Wheat Midge

Mr A. H. Cockayne states that this species occurs at Timaru and at Gore.

Dasyneura pyri, Bouché (*Perrisia pyri*). Pear Midge

This midge was first recorded from Avondale, Auckland, in 1918. The larvæ attack the young shoots of pear trees. It spread over the whole of Auckland Province within a year of its discovery; it was probably established for some years previously. It only attacks pear trees under (about) eight years of age.

It is very destructive to young trees.

[1] Mr A. Philpott, writing on 23rd April, 1917, says: "A few years ago some pods of *Phormium tenax*, attacked by the larva of a *Cecidomyia*, were sent to me. The pods were distorted and small, and the seeds less than the usual size. I do not think that the fly was *C. destructor*, but I was not successful in rearing the mature insect. Probably it was an undescribed native species."

Family Stratiomyidæ

Exaireta spiniger, Schiner

Introduced from Australia, and first recorded in 1859 from Auckland (*Reïse der 'Novara'*). Since recorded from Auckland (where it is abundant), Whangarei and Wellington (Hutton). Abundant in summer.

Mr Hudson says it was captured fairly commonly in the Wellington Botanical Gardens in 1882, and adds "I have always assumed this to be a native." Mr A. Philpott has not met with it in the south, and considers that so conspicuous an insect could not escape detection.

Family Mydaidæ

Mydas macquarti, Schiner

First recorded from Auckland in 1859 (*Reise der 'Novara'*). Later (1901) it was omitted from Hutton's list of introduced Diptera.

Family Asilidæ

Lampria ænea, Fabr.

First recorded in New Zealand by Nowicki (1875?). Later (1901) it was omitted from Hutton's list.

Family Phoridæ

Phora omnivora, Hudson

This fly, which is now abundant all the year round, was originally recorded from Wellington, Christchurch and Dunedin by Captain Hutton in 1900. It is a common meat fly, but its larva is parasitic on several moths, as *Melanchra composita*, the New Zealand army-worm or grass-caterpillar, *M. mutans*, *M. ustistriga*, *Erana graminosa*, and other species of Noctuæ. It also occurs on some Coleoptera, e.g. *Uloma tenebrionides*. Mr Hudson was under the impression that it was very destructive in bee-hives, but both Mr D. Miller and Mr Philpott think this is an erroneous idea. The former suggests that the larva is only a scavenger in bee-hives. The latter ventures to think that the hive is not "finally ruined by the wholesale destruction of the honey when the flies emerge," but by the reduction in the strength of the colony caused by the parasitic larvæ, the hive being at last so weakened that the necessary temperatures cannot be kept up and the activities of the colony cease.

Family Syrphidæ

Syrphus viridiceps, Weid. (*Syrphus obesus*, Hutton)

An Australian fly, well established in the far north of Auckland, at Parengarenga Harbour. The larvæ destroy large numbers of caterpillars and aphides. All authorities state that the larvæ of *Syrphus* destroy great quantities of aphides. Recently (April, 1919) Mr Miller has obtained it from Otira Gorge.

Syrphus novæ-zealandiæ, Macq.

First recorded in New Zealand by Mr Bigot. The species is very common throughout New Zealand. It is perhaps a native species, but may have been introduced from Polynesia. Very abundant during the summer in Otago; but in Nelson, where the climate is more equable, it occurs practically all the year round.

Eristalis tenax, Fabr. Common Drone Fly

Capt. Hutton thought this species came from Britain, but Mr W. W. Smith believes that California was the region from which it was introduced.

Mr G. V. Hudson first noticed it in Wellington in 1888, and Mr Smith took it later in the same year at Ashburton, on the flowers of *Veronica Andersoni*. It soon became very abundant. It usually appears late in autumn (Mr Howes records it as flying in August), and lasts till the end of July. Its rat-tailed larva is abundant in ditches and dirty water.

Dr Hilgendorf reports it as occurring in cess-pools and earth closets, and even found alive floating in sheep-dip, where probably the poisonous matters had sunk to the bottom, and the upper layers were purified by addition of rain-water.

Eristalis tenax fertilises (in Europe) the following flowers of species which have been introduced into New Zealand:

Ranunculus aquatilis, R. acris, R. repens, R. bulbosus.
Berberis vulgaris.
Papaver somniferum.
Brassica Rapa, B. nigra, Sinapis arvensis, S. alba, Cakile maritima, Crambe maritima.
Spergula arvensis, Stellaria graminea.
Hypericum perforatum.
Lotus corniculatus, Trifolium arvense.
Rubus idæus, R. fruticosus, Cratægus oxyacantha, Pyrus communis, P. malus, P. aucuparia, Prunus domesticus, P. avium, P. cerasus.
Ribes Grossularia.
Œnothera biennis.
Carum Petroselinum, Daucus Carota, Scandix pecten-Veneris.
Hedera Helix.
Sambucus nigra.
Sherardia arvensis.
Bellis perennis, Achillæa millefolium, Cnicus arvensis, C. lanceolatus, Cichorium intybus, Leontodon hirtus, Hypochæris radicata, Taraxacum officinale, Crepis virens, Anthemis arvensis, Matricaria inodora, Senecio jacobæa, Calendula officinalis, Carduus crispus, C. nutans, Centaurea nigra.
Solanum tuberosum.
Mentha aquatica.
Fagopyrum esculentum, Polygonum Persicaria.
Euphorbia helioscopia, E. cyperissias, E. Peplus.
Salix caprea.
Allium cepa.

Syritta oceanica, Macq.

First recorded in New Zealand by M. Bigot (date?).

Merodon equestris, Meig. Narcissus Fly

In 1906 narcissus bulbs imported from Britain were found to be infested with this fly. Originally found in Invercargill, it has since been met with in Auckland, and is probably occasionally introduced.

Recently (1918) some bulbs were received from Japan which were infested by the larvæ of this or an allied species, but as they did not develop beyond the pupal stage the species could not be determined.

Family MUSCIDÆ ACALYPTRATÆ

Drosophila ampelophila, Loew. Fruit Fly; Pomace Fly

This European species was first noticed by Captain Broun in 1904. It has become very common, especially at seaports where fruit is introduced, and is particularly partial to bananas, oranges, and pineapples, which it attacks whether they are bruised or not. Some idea of the frequency of this and other allied species of flies may be gathered from the following statement.

In 1910–11 the following cases of fruit imported into New Zealand were condemned and destroyed for fruit-fly maggot:

	Total imported	Destroyed
Auckland	299,249 cases	848 cases
Wellington	305,050 ,,	987 ,,
Christchurch	62,332 ,,	255 ,,
Dunedin	58,633 ,,	273 ,,
Bluff	10,338 ,,	9 ,,

Mr Howes considers that the term "fruit-fly" should be confined to those flies which attack growing fruits. The Drosophilidæ are usually known as "Vinegar-flies," and attack rotting vegetable matter and over-ripe fruit. He adds:

Fruit-flies have been bred from apparently perfectly sound fruit. On the other hand many fruits with accidental bruises and perforations prove, on examination, not to be infected. I have seen fruit-flies inside a Chinese fruit-shop in Lambton Quay, Wellington, crawling up the glass. It might be possible to establish these flies in the North Island, but not in the colder parts of the south.

Dacus psidii, Froggatt. Guava Fruit Fly

According to the Agricultural Department's report for 1908, this species is frequently found in fruit imported from Northern New South Wales. It has apparently not become established in New Zealand.

Tephrites xanthodes, Broun

Introduced from Suva on pineapples; and from Tonga and Rarotonga on oranges, grenadillas and mummy apples. First observed by Capt. Broun in 1903. Constantly met with on imported fruit from all those localities. Not established.

Tephrites tryoni, Froggatt. Queensland Fruit Fly

Introduced from Australia in pineapples; first observed by Capt. Broun in 1904. For many years it was very commonly met with, but was not nearly so common in 1907 and 1908. Mr W. W. Smith (April, 1919) states that it is very common in orchards in Taranaki.

Trypeta musæ, Froggatt. New Hebrides Fruit Fly

The Agricultural Department report for 1918 states that this fly was very prevalent in Mandarin oranges imported from Sydney, and that large numbers had been bred out. It has not succeeded, however, in establishing itself.

Lonchæa splendida, Loew. Tomato Fruit Fly

First observed in 1903 in tomatoes brought from Sydney, from which large numbers were reared. In the Agricultural Department's report for 1908 it is said to be on the increase in tomatoes and oranges. It is found, but not commonly, in Wellington, and probably in other centres.

Consignments of the following fruits containing maggots have been burned on the wharves--apples, apricots, bananas, cherries, figs, grenadillas, loquats, mangoes, mummy apples, mandarins, maupi fruit, nectarines, oranges, peaches, persimmons, plums, pineapples, pears, shaddocks and tomatoes. Several regulations were enforced to check the introduction of this pest, but owing to efforts of importers these regulations were temporarily relaxed, and in October, 1907, a very large consignment of fruit containing maggots was distributed from Auckland, and reached various parts of the colony. Efforts were immediately made to trace this fruit and destroy it where possible. But later in the season the flies were found breeding in peaches in several gardens in Napier, and in both peaches and tomatoes near Blenheim. In 1908 it was found in peaches both in Napier and Auckland gardens, and in many consignments of imported fruits.

Since 1908 no specimens, other than those from imported fruit, have been found.

Halterophora capitata, Broun. Mediterranean Fruit Fly

First detected in peaches on Wellington Wharf in 1898, when flies were reared from maggots in the condemned fruit by Mr T. W. Kirk. Later they were discovered by Capt. Broun in soil accompanying imported plants. The flies deposit eggs in two days after emerging from the pupa, and 29 days later, new flies emerge.

The report of the Department of Agriculture for 1908 records a new species of fruit fly introduced with *Citrus* fruit from Rarotonga and Tonga. Apparently it has not been described.

Phytomyza albiceps, Meig.

This fly is common throughout New Zealand, and the larva burrows in the foliage of many plants, especially composites. The sowthistles (*Sonchus oleraceus*, *S. asper* and *S. arvensis*) are most commonly attacked, but the following have also been noted as furnishing food plants for this species. *Dahlia*, dandelion (*Taraxacum officinale*), common groundsel (*Senecio vulgaris*), cape weed (*Cryptostemma calendulacea*), and nettle (*Urtica ferox*).

Dr Hilgendorf informs me that the larvæ are attacked and destroyed by the larva of an introduced parasitic hymenopterous insect (*Dacnusa* sp.). Messrs Watt and Miller report two other hymenopterous parasites, minute species of the genus *Chrysocaris*, which destroy this fly.

Phytomyza nigricornis, Macquart. Cineraria Fly; Marguerite Fly

This species was first recorded in Wellington in 1893. In 1897 it was said to be widely dispersed and to have become a serious pest, and it has been reported every year since as abundant. It mines the leaves of *Cineraria*, globe artichoke, sowthistle, dandelion, chrysanthemum, peas and poro-poro (*Solanum aviculare*).

Phytomyza chrysanthemi, Kowarz. Chrysanthemum Fly

This species was reported as occurring abundantly on chrysanthemums in 1907–9.

Family ANTHOMYIIDÆ

Fannia canicularis, Linn. (*Homalomyia canicularis*, Linn.). The Little House Fly

Seems to be common throughout New Zealand, though not so abundant as *Musca domestica*. It delights to hover and sport about in rooms, and is easily recognisable on window panes by its small size, and the semi-transparent patches on its body.

Abundant in early spring, and is displaced by the common house fly in summer. Its larvæ are occasionally the cause of intestinal myiasis.

Family Tachinidæ

Comptosia bicolor, Macq.

First recorded from Auckland in 1859 (*Reise der 'Novara'*); introduced from Australia.

Capt. Hutton omits this from his list in 1903, as having been included in error in the 'Novara' list.

Comptosia fasciata, Fabr.

First recorded from Auckland in 1859 (*Reise der 'Novara'*); probably introduced from Polynesia.

This species is also omitted by Capt. Hutton.

Micropalpus brevigaster, Macq.

First recorded from Auckland in 1859 (*Reise der 'Novara'*); probably introduced from Australia or Tasmania.

Omitted by Capt. Hutton.

Lamprogaster strigipennis, Macq.

First recorded from Auckland in 1859 (*Reise der 'Novara'*); an Australian species.

Omitted by Capt. Hutton.

Lamprogaster cœrulea, Macq.

First recorded from Auckland in 1859 (*Reise der 'Novara'*); probably introduced from Australia.

Omitted by Capt. Hutton.

Captain Hutton was of opinion that the five above-named species got into the 'Novara's' lists in error.

Phorocera feredayi, Hutton

First recorded from Dunedin by Capt. Hutton in 1881. Parasitic in the larvæ of the basket moth (*Liothula omnivora*).

Phorocera marginata, Hutton

Recorded from Dunedin (1881), Christchurch and Wellington. Parasitic in the larvæ of *Liothula omnivora*, *Œceticus omnivorus*, *Melanchra composita*, *M. ustistriga*, and *M. mutans*.

Phorocera nyctemeriana, Hudson

Wellington (1883) recorded by Mr Hudson; Christchurch and Queenstown (Capt. Hutton), and Dunedin by Mr Howes. Parasitic on the caterpillars of *Nyctemera annulata* and of *Leucania purdiei*[1].

[1] Philpott recorded this as reared from Porina, but is not sure that his identification is correct.

Family Dexiidæ

Dexia rubricarinata, Macq.

First observed in Auckland, 1859 (*Reise der 'Novara'*). Collected on the Kermadecs by Mr W. L. Wallace in 1908, where it was taken from the carcase of a goat. Also at Astrolabe, Nelson, on the sea-beach in 1911 by Mr D. Miller.

This species was omitted by Capt. Hutton, but these later discoveries show the accuracy of the 'Novara' list, as far as this species is concerned

Rutilia pelluceus, Macq.

Recorded from Auckland in 1859 (*Reise der 'Novara'*).

Omitted by Capt. Hutton from his list in 1903.

Aminia leonina, Fabr.

First recorded from Auckland in 1859 (*Reise der 'Novara'*) probably introduced from Australia.

Omitted by Capt. Hutton.

Family Sarcophagidæ

Sarcophaga impatiens, Walker. Flesh Fly

This is one of the commonest flies in New Zealand, particularly in the South Island. Capt. Hutton considered that it was probably introduced from Australia. He met with it in great abundance in Whangarei in 1901. Dr Hilgendorf also reported it from Christchurch and Banks' Peninsula. Mr Philpott took it at Ashburton in the month of April.

Mr D. Miller in 1909 says:

> This common fly is found in most situations. During December 1907 at Taieri Mouth, I captured a specimen near a swamp. About the middle of June 1908, I picked up another individual which was lying on the Tomahawk (Dunedin) sea-beach; and several were obtained from the swamp behind Murdering Beach during January 1909, as well as from Long Beach during the two following months of the same year.

Ten years later, in 1919, he informed me that he had obtained the larvæ in large numbers infesting the intestine of a sheep.

Family Muscidæ

Musca domestica, Linn. Common House Fly

Very abundant throughout New Zealand; and probably introduced at a very early date.

In 1870 A. Bathgate stated that it is driving out to a great extent the native blow-fly. Hutton in 1901 says:

The statement that the introduced house-fly has displaced the native blow-flies, which have practically disappeared, is quite erroneous. I doubt whether they compete in any way.

The female lays about 120 eggs.

In Europe *Musca domestica* has been found to visit and pollinate the flowers of the following plants which have been introduced into New Zealand.

Berberis vulgaris.
Cakile maritima.
Stellaria media.
Pyrus malus, *P. communis* and *P. aucuparia.*
Conium maculatum, *Carum Petroselinum.*
Bellis perennis.
Digitalis purpurea.
Fagopyrum esculentum.

Musca taitensis, Macq.

This was reported by Captain Hutton in 1881, as having been found by Dr Sinclair. He thought, however, that probably it did not belong to the genus *Musca*.

Musca corvina, Fabr.

This was recorded in the *Index* in 1903 by Captain Hutton as occurring in New Zealand, on the authority of Dr Hilgendorf.

Muscina stabulans, Desv. The Stable Fly

First recorded from Auckland in 1859 (*Reise der 'Novara'*). Now abundant throughout New Zealand; more common in the North than in the South Island.

The larvæ probably eat all kinds of decaying vegetable matter, and have frequently been found on rotten fungi. They occasionally attack growing plants, and have been found destroying shallots.

Calliphora erythrocephala, Desv. Common European Blow-Fly

First noticed by Mr Hudson in June, 1889, at Wellington, and by Capt. Hutton 1893 in Christchurch. It is now common throughout New Zealand. The fly is particularly fond of the flowers of indigenous species of *Veronica*.

In Europe this species fertilises the following plants which have been introduced into New Zealand:

Ranunculus repens.
Brassica nigra.
Hypericum perforatum.
Pyrus communis.
Ribes grossularia.
Daucus Carota.
Hedera Helix.
Cnicus arvensis, *Senecio jacobæa*, *Taraxacum officinale*, *Calendula officinalis*, *Achillæa millefolium*, *Onopordon acanthium*, *Calluna vulgaris*, *Linaria vulgaris*, *Veronica serpyllifolia*, *Mentha aquatica*, *Salix caprea.*

This species lays from 300 to 600 eggs, but is not viviparous like some of the indigenous blow-flies[1].

Pollenia villosa, Robineau-Desvoidy (*Calliphora læmica*, Walker). The Golden-haired Blow-Fly

Probably introduced from Australia into the North Island, where it became common before 1874. In that year it was first observed in Christchurch. In 1900 it was first observed at Lake Wakatipu. It is now abundant throughout New Zealand. Mr Hudson observed it in Wellington in 1881. One of the most common meat flies in houses. It deposits both eggs and maggots. In Australia it does incalculable damage by blowing the wool of sheep, the maggots burrowing into the skin of the animal, causing the wool to rot off.

Although well-established throughout New Zealand, it has not hitherto caused any appreciable damage among sheep. The cases of sheep-blowing which occur in New Zealand are generally attributed to *Lucilia sericata*, which is an erroneous idea, as any damage done is caused by *Pollenia villosa*.

Lucilia cæsar, Linn. Green-bottle Fly; Sheep-maggot Fly

First observed by Capt. Hutton in Christchurch in 1872; but it was some years before it spread. It is particularly common in the North Island and in Marlborough, where it causes fly-blow in sheep. Dr Hilgendorf states that the eggs are laid on dogs, and the maggots bore into the ham-muscles of living sheep. The species is not so common in the South Island. Mr Philpott says: "during the summer, odd specimens are often met with, and I once saw about a dozen on a dead rabbit in March."

[1] Mr Philpott writes me as follows: "There is something peculiar about ovi-depositing habits of the large blow-flies. In some localities it is quite unsafe to leave blankets or any woollen fabrics uncovered, but in others they may be left exposed from one year's end to another without being fly-blown. On the Tuatapere-Preservation Inlet track a stock of blankets has been placed in each of the Government huts, but it has been found necessary to provide zinc cases to keep them in. These huts are all practically at sea-level, but the same trouble occurs at higher elevations. The hut on the Hump is situated at about 3000 ft., but in hot weather the flies are just as great a nuisance as down on the coast. In various parts of the Wakatipu district campers are subjected to the same annoyance. On the other hand, in a hut on the Hunter Mountains (at 3000 ft.) blankets have been left for eight or nine years. They are never covered in any way, simply lie on the bunks, or hang on the rafters, and no trouble from fly-blow has ever been experienced. I have camped in this hut for four successive seasons, always in mid-summer; the blue-bottles come into the hut freely and alight on the blankets, but do no harm. Last year I found that the furniture of the hut had been supplemented by an uncured long-woolled sheepskin, but even this failed to induce the flies to oviposit. Also, I camped for a fortnight one year on Ben Lomond, at about 2000 ft. This was late in November, and although the weather was hot, my blankets were not interfered with. Unfortunately, I cannot speak with decision as to the species of fly. It may be that the Hunter Mountain fly is *Calliphora quadrimaculata*."

Mr Howes says (1919): "From a mass of wool swarming with maggots, I bred out three different flies, none of which were *Lucilia cæsar*." Mr Miller writes me: "This fly is abundant everywhere, living often on decaying vegetable matter. It is, however, particularly a carrion fly, though it also feeds on excrement." It is a fly which increases rapidly, the female laying from 3000 to 6000 eggs.

This species is a familiar flower-visitant, and in Europe fertilises the flowers of the following species of plants introduced into New Zealand:

Ranunculus sceleratus, R. acris, R. repens, R. bulbosus.
Brassica nigra.
Stellaria media, Cerastium triviale.
Geranium molle.
Cratægus oxyacantha?
Rubus idæus, Pyrus malus, Pyrus aucuparia.
Carum Petroselinum, C. carui, Conium maculatum, Daucus Carota.
Bellis perennis, Achillæa millefolium, Cnicus arvensis, Matricaria inodora, Anthemis arvensis, Chrysanthemum Leucanthemum, Senecio jacobæa, Tanacetum vulgare, Cnicus lanceolatus, Taraxacum officinale.
Solanum tuberosum.
Mentha piperita, M. aquatica.
Euphorbia helioscopia, E. Peplus.

Lucilia sericata, Macq. Sheep-maggot Fly

Mr Hudson took this species at Karori in 1883. It appeared in North Otago in 1906, when attention was drawn to its occurrence, and steps taken to arrest its progress. Also found in Mackenzie country. It reappeared again in 1907–8, in Palmerston South. In 1909 it was abundant everywhere. It is chiefly found in South Island.

Lucilia sericata (in Europe) visits flowers of *Medicago sativa, Achillæa millefolium, Cnicus arvensis*, and *Senecio jacobæa*.

Stomoxys calcitrans, Fabr. The Biting House Fly; Stable Fly

Common in both islands. I noted it near Dunedin in 1893, and Capt. Hutton found it later near Christchurch. Mr Hudson found it in Wellington in May, 1889. Dr Hilgendorf says:

An observation in 1917 showed that the inhabitants of a stable were 50 per cent. *Stomoxys*, and 50 per cent. of *Musca domestica*; of a pig-sty, 95 per cent. of *Musca* and 5 per cent. *Stomoxys*; and of a house near by 100 per cent. *Musca*.

This fly is about the size of the common house-fly (*Musca domestica*), but is at once distinguished by its needle-like proboscis, with which it gives a fierce puncture of the skin. It is much quicker in its movements also, due, no doubt, to its inherited need of escaping rapidly from the animal it has pierced. It easily punctures through one's clothes.

The species is probably common throughout New Zealand. In Europe it visits the flowers of:

Hypericum perforatum.
Achillæa millefolium and *Cnicus arvensis*; and *Mentha aquatica.*

Stomoxys nigra, Macq.

Found in 1916 at St John's Lake, Auckland, by Mr D. Miller. This is a South African species.

Family ŒSTRIDÆ

Œstrus ovis, Linn. Sheep Nasal Bot-Fly

Capt. Hutton recorded this first in Canterbury in 1873. Dr Hilgendorf states that these flies are very common; and that in the back country of Canterbury they trouble the merinos very much:

"At the Cass on hot days," he says, "the sheep may be seen stamping their feet, tossing their heads, or standing huddled together, with noses to the ground; this is because the bot-flies have laid their eggs in their nostrils."

Dr Gilruth, reporting on the subject in 1899, stated that:

flocks in New Zealand become affected without any previous agitation of the animals having been noticed. In older countries the sheep become very excited if even only one fly be in their vicinity. They suffer mostly in autumn, the larvæ lodging in the nasal passages and the bony sinuses of the skull.

The larvæ have been known to cause rhinal myiasis (i.e. in human beings).

This fly is found in all parts of New Zealand. Dr Hilgendorf considers that nearly every sheep harbours them.

Gastrophilus equi, Meigen. Horse Bot-Fly

Captain Hutton saw this species first in 1892, and for two or three years they were very abundant and caused quite a scare. Dr Hilgendorf considers that the damage they do to horses' stomachs is probably insignificant, though the annoyance to horses when the flies are laying their eggs is very great. He says: "Every horse we kill has abundance of bots in its stomach."

Mr Philpott says (1917):

The horse bot is very common now. Some horses are extremely restless when the flies are about, but others take little notice. A few weeks ago I saw a bay mare so covered with eggs about the legs and abdomen that at a little distance she appeared to be tinged with grey. Her owner told me that she never took the least notice of the flies. In the same paddock was another horse which was continually walking to and fro, whisking its tail and snapping vigorously at the insects.

Gastrophilus hæmorrhoidalis, Meigen. Horse Bot-Fly

Supposed to have been introduced by some Mexican circus-horses from San Francisco; first noticed in the North Island in 1889, and in the South Island in 1891. Now common throughout New Zealand, but according to Dr Hilgendorf not nearly as abundant as *G. equi.*

Mr D. Miller says:

During February 1909, I observed a large swarm hovering about a young horse, attacking the animal at the knees and sides, and, as they flew about the place upon which they wished to alight, they suddenly darted in and out, each time coming into contact with the horse's flesh.

Gastrophilus nasalis, Meig.

Mr W. W. Smith records this bot-fly as first attacking horses in South Canterbury in 1890, and then spreading rapidly over the South Island. It was fairly common in New Zealand in 1894.

Hypoderma bovis, De Geer. Ox-warble Fly

According to Dr Hilgendorf hardly an ox reaches New Zealand without carrying the larvæ of this fly, but none has yet got past the Quarantine Stations, which so far have successfully resisted the establishment of this dangerous pest.

In Europe the larvæ have occasionally been found in human beings; but the parasite is more common among herdsmen in America.

Mr D. Miller (April, 1919) says these larvæ were found on cows at Lea Flat Station, about 20 years ago; and at Owaka about 30 years ago.

The larvæ of this species have been known to occur in man.

Family HIPPOBOSCIDÆ

Melophagus ovinus, Latr. European Sheep Tick; Ked

Dr Reakes informs me that this insect pest is common throughout New Zealand; and Dr Hilgendorf states that it is universal.

Sub-order APHANIPTERA

Family PULICIDÆ

Pulex irritans, Linn. Common Flea

In his narrative of Captain Cook's second voyage, when in Queen Charlotte Sound in 1773, Forster states: "We were told that the people from the 'Adventure' had found the native huts exceedingly full of vermin and particularly fleas." It is rather singular that neither Cook nor Banks in the first voyage to New Zealand makes any mention of the fleas, which were associated with the natives. In his second voyage, when the 'Adventure' came into Ship Cove in Queen Charlotte Sound in April, 1773, the old Maori pa was found to be deserted, but "the presence of immense quantities of vermin was taken by the sailors as a rough and ready indication that the huts had not been long abandoned by the Maoris."

It is quite probable that there were no fleas in this country at the date of Cook's first voyage (1769), but that they were then introduced, and very rapidly increased. It would seem indeed that this is the case from the following facts.

In the vocabulary at the end of Nicholas's *Narrative of a Voyage to New Zealand*, vol. II, p. 338 (published in 1817), occurs the following:

A White Man = Packaha.

The flea is also called by this name, as the Maoris assert it to have been *first introduced into their country by Europeans.*

Angas in *Savage Life and Scenes in Australia and New Zealand*, vol. II, p. 20 (published in 1847), says:

Here we pitched our tent, overlooking the broad surface of the Waikato, at about half a dozen yards from its brink. The fear of too many visitations from that active parasite, the flea (cleverly styled *e pakea nohinohi*, or "the little stranger," by the natives, *who say it was first introduced by the Europeans*), prevented our encamping within the enclosure of the *pah*.

Mr Elsdon Best says: "I cannot remember any mention of the flea in *old* Maori traditions, as the waeroa, Kutu and namu are mentioned, which seems to support the above statements."

After their introduction fleas increased at an astounding rate, and spread from end to end of the country. The Maori pas and villages were full of them, especially in the North Island; but they are found also in vast numbers in warm dry quarters, such as abandoned settlements, old sheds and sheepyards, where the ground often literally moves with them.

Ctenocephalus canis, Curtis. Dog and Cat Flea

This is a common species in New Zealand, and "is probably the most widely distributed member of the order." It has been recorded as *Pulex serraticeps*, Gervais. Mr Rothschild considers that there are two distinct species, one *C. canis* occurring on dogs, and *C. felis* on cats, but he states that it is impossible to separate them.

Ceratophyllus fasciatus, Curtis. Rat Flea

This flea is probably common on rats throughout the country, but no one has worked out the occurrence of this important group, except Dr Russell-Ritchie, who informs me that it is common in rats which have been examined in Wellington Hospital. This is one of the species which conveys bubonic plague, but before investigations into the possible introduction of this disease can be properly undertaken, a complete study of the fleas present in the country must be made.

This species is found on *Mus decumanus* (the brown rat), and on *M. musculus* (the mouse)[1].

Ceratophyllus gallinæ, Wagner. Bird Flea

This species occurs in dirty fowl-runs, where it causes harm and irritation to the birds. It has usually been referred to as *Pulex avium*, Tasch. The report of the Agricultural Department for 1900 states that this insect is common in New Zealand.

Ctenopsyllus musculi, Wagner. Rat Flea

This species is also probably found throughout New Zealand, as it occurs on mice, brown rats and black rats (*Mus rattus*). Dr Russell-Ritchie tells me that it is found on rats in Wellington, but not so commonly as *Ceratophyllus fasciatus*.

[1] The connection between rats (or mice) and bubonic plague was evidently known to Asiatic peoples in very far-distant ages. Like many other facts of historic and scientific interest, this knowledge was lost with the wholesale destruction of peoples and libraries which has taken place from time to time in the past. But there is an interesting record of it in the early history of the Jewish people. Somewhere about the 11th century B.C., as narrated in the First Book of Samuel (Chap. iv–vi), during the ever-recurring wars between the Philistines and the Israelites, the latter were defeated with great slaughter in a battle fought near Aphek, and the Ark of God was taken, and was conveyed to the house of Dagon in Ashdod. There is a mixture of history and of priest-lore in the succeeding narrative, but we are told "the hand of the Lord was heavy upon them of Ashdod, and he destroyed them, and smote them with emerods" (= hæmorrhoids or bubonic glands?). In other words they were attacked by some very deadly and infectious plague. The lords of the Philistines sent the ark to Gath, "and it was so, that, after they had carried it about, the hand of the Lord was against the city with a very great destruction: and he smote the men of the city, both small and great, and they had emerods in their secret parts." Then they sent it to Ekron, and the plague broke out there. In their desperation, thinking it was a visitation of vengeance of the God of Israel, they resolved to return it to the people from whom it was originally taken, and to accompany it with a trespass offering. This offering consisted of five golden emerods and *five golden mice*, which probably may be more correctly translated *rats*, though mice and rats are equally transmitters of bubonic plague. The further mixture of history and priestly superstition is recorded at the end of Chap. vi. The ark was placed on a cart to which two cows were attached, and they were set towards the land of Israel and allowed to go wherever they liked. They took the road to Bethshemesh, and the arrival of the ark was hailed by the people with joy. But they carried the plague with them, for the priestly record says: "He smote the men of Bethshemesh, *because they had looked into the ark of the Lord*, even he smote of the people fifty thousand and threescore and ten men. And the people lamented, because the Lord had smitten many of the people with a great slaughter." It is a most interesting narrative, and shows that at that early date the rulers knew that there was some relation between the plague and the rats or mice, even though they did not know that fleas were the carriers of the infection.

Order THYSANOPTERA

Family THRIPIDÆ

Thrips sp.

Several species of *Thrips* and allied genera occur in New Zealand, most of which are apparently introduced, but, as far as I am aware, none have been worked out or identified. They are found in numbers in most cultivated flowers, as well as a great many indigenous species; but here again no record has been made of them. In some cases they are found in such abundance on vegetation as to amount to a destructive pest. Mr W. W. Smith has sent me foliage of *Pittosporum*, etc., from New Plymouth, which shows the effects they produce. The undersides of many of the leaves were covered with a rust-coloured dust, others were similarly affected on the upper surface. The cuticle was punctured, and the epidermal cells largely destroyed, while in most of them the chlorophyll was wanting. Whether this had been removed by the insects puncturing the upper layers of tissue, or was due to the destruction of the epidermis, I could not say. Many of the leaves were covered with white patches, which proved to be formed of thousands of minute eggs, apparently of *Thrips*. The effect on many leaves of the loss of chlorophyll and of the damage to the epidermal tissues was to give them a silvery appearance, and to cause them to present a very sickly aspect.

Order HEMIPTERA

(See Appendix B, p. 560.)

Sub-order HETEROPTERA

Family REDUVIIDÆ

Nabis lineatus, Dahlbom

This European species is recorded by Captain Hutton in the *Index* (1903) as occurring in New Zealand. In a note on those introduced Hemiptera, Mr G. W. Kirkcaldy states that this species is probably included in error.

Family CIMICIDÆ

Cimex lactularius, Linn. Bed Bug

This most objectionable insect has been in New Zealand from early days of settlement, as many of the vessels which visited the country were infested with bugs. I first met with them on a coasting steamer in 1884; and Mr Hudson reported them in 1890. They are not particularly common in the South Island, but in the warmer parts of the North Island, many houses are infested with them.

Family Sciocoridæ

Sciocoris helferi, Fieb.

In a note written in 1896 Captain Hutton says of this species "A South European species, said to have been collected in Auckland by the 'Novara' Expedition." In 1905 Mr Kirkcaldy stated that the species ought to be expunged from New Zealand lists till further confirmation. It has not been recorded since.

Family Pentatomidæ

Nezara prasina, Linn.

Captain Hutton stated in 1897 that "specimens of this species are in the British Museum from New Zealand." It is a cosmopolitan species, but I have not heard of any collector who has found it in these islands.

Sub-order HOMOPTERA

Family Psyllidæ

Psylla acaciæ-baileyanæ, Froggatt

In the report on State Afforestation in New Zealand for 1910–11, this species is stated by Mr A. H. Cockayne to occur on the plantations of black wattle (*Acacia decurrens*), "but so far it has done little harm." It has evidently been introduced from Australia.

Rhinocola eucalypti, Maskell. Yellow Aphis

This species, which is a native of Australia, is a small dark brown psyllid, the larvæ of which cluster at the tips of the foliage of young blue gums (*Eucalyptus globulus*), and cover themselves with threads of white flocculent matter, hence apparently the popular name. Mr Maskell thought it was introduced from Tasmania. The species is common in the Manawatu district, and is probably of very general distribution.

Family Fulgoridæ

Scolypopa (*Pochasia*) *australis*, Walk. Vine Hopper; Tree Hopper

Captain Broun reported this (in 1896) as common on native shrubs in the forest about 20 years ago. Later on it began to attack passion-vines, figs, etc.; preventing the formation or maturing of the fruit. In 1898 it is recorded as occasionally attacking grape-vines.

The species was probably introduced from Australia. Mr Howes found it in Auckland in immense numbers on fig trees, and on many garden plants. He also obtained it in Nelson.

Family Aphidæ

Aphis brassicæ, Linn. Cabbage Aphis

This pest seems to have appeared at an early date after settlement. It was very abundant 50 years ago, and is particularly common in those districts of New Zealand where the annual rainfall does not exceed 35 inches. It attacks every species of *Brassica* (cabbage, turnip, swede, rape, etc.), and is found on many other crucifers (*Capsella*, *Sisymbrium*, *Cakile*, etc.). Dr Hilgendorf considers that it is the most destructive insect that has been introduced into the country, and estimates that the annual loss due to its presence cannot be put down at less than £250,000; and this not so much by the direct damage it does, as the restriction it entails in the selection of varieties which while more or less blight-proof, are not nearly as productive as others[1].

Aphis persicæ niger, Smith. Black Peach Aphis

Mr D. Miller informed me that this insect, which appears to be common in parts of Australia, has recently (1917) been recorded from New Zealand (locality?).

Chermes corticalis, Kaltenbach

This species is recorded by Captain Hutton in the *Index* (1903). None of my correspondents know it.

Chermes pini, Koch

In 1884 Mr Maskell recorded that pine trees were badly attacked by a "blight" some four or five years previously. The species which suffered were *Pinus halepensis*, *P. radiata* (*P. insignis*), and *P. sylvestris*. The latter species was especially damaged in Nelson, while the former about Wellington were greatly infested. It was also destructive in plantations at Wanganui, Christchurch, Ashburton and Peel Forest. The pest seems to be fairly common, and though it does not kill pine trees, it disfigures them and greatly interferes with their vigorous growth. Mr Maskell suggested that perhaps *C. pini* and *C. corticalis* were the same species.

Chermes laricis, Hartig. The Pine White Aphis

In the report on State Afforestation for 1901–11, this species is stated by Mr A. H. Cockayne to be fairly abundant on some of the pines, "but it appears to select trees that are of weak constitution,

[1] In the report of the Agricultural Department for 1905, two species are recorded which, in the absence of certain literature, I cannot identify. They are: (1) the turnip fly (*Aphis rosæ*), which is recorded from Kohinui, and (2) the green fly (*Rophalosiphon dianthi*), which is said to have been found on diseased tomatoes from Wellington.

and will probably more or less disappear when the trees are older." Though only reported from Rotorua and Waiotapu, the insect is probably widespread.

Lachnus strobi, Fitch

Recorded by Mr D. Miller as occurring on Douglas Spruce (*Abies Douglasii*) at Palmerston North in 1919.

Phylloxera vastatrix, Planchon. Vine Louse

This insect appeared in one locality in Auckland in 1885, and efforts were made to exterminate it. In 1889 and 1890 it was found in several places near Auckland, and at Whangarei. In 1895 it was reported from three localities near Masterton, and in 1898 at Carterton. In the Agricultural report for 1898–99 it is also stated to have been found in Eden, Waitemata, Whakatane, Tauranga, Piako and Manukau counties, and to be common from Auckland North. In 1902 it was rediscovered at Mt Eden, Auckland, and in the following year was reported from Opotiki, and was said to be bad at Whangarei.

In 1920 it was still present in a number of localities in the Auckland district.

Schizoneura lanigera, Haus. (*Eriosoma lanigera*, Haus.). Woolly Aphis; American Blight

This species has been known for a very long time in New Zealand, and is universally distributed throughout the country. It is very abundant still in old and neglected orchards, but has been greatly reduced in recent years by the use of "Northern Spy" stocks on which to graft the different varieties of apples. Dr Hilgendorf considers that it is more common in the northern half of the North Island than elsewhere.

In 1861 the Otago Provincial Council passed the "American Blight Protection Ordinance," the preamble of which reads:

> Whereas considerable injury has been done to Fruit Trees within the Province of Otago by the Blight, or Insect called the American Blight; and whereas it is expedient to prevent as much as possible the increase thereof: Be it therefore enacted, etc.

This is interesting as showing how early this insect was recognised as a dangerous and common pest in New Zealand, and also how unscientific is the definition of the pest itself.

The majority of apples in New Zealand are now grafted on Northern Spy stocks, an American apple which has a hard bark and very firm wood, thus enabling it to withstand the attacks of the woolly aphis.

Siphonophora rosæ, Reaum. Rose Fly

This pest has long been known in the country, and apparently is common everywhere.

Siphonophora granaria, Kirby

This species is also widespread throughout New Zealand.

Mysus cerasi, Fabr. Black Aphis; Black Fly

This pest has been known in New Zealand since the early fifties, and appears to be common throughout the Dominion. It particularly attacks peach, cherry and plum trees, making its appearance in the early spring, and doing most damage when the trees are in bloom, and the foliage and fruit are young.

Family COCCIDÆ

The nomenclature of this family has recently undergone a complete revision, and I am indebted to Mr Guy Brittin of Christchurch for a correct list of the species. As, however, most fruit growers only know the names which appear in Maskell's papers, and in the publications of the Agricultural Department, I have retained these in parentheses for the sake of reference.

Aulacaspis rosæ, Sandberg (*Diaspis rosæ*)

This species, which attacks rosaceous plants—roses, blackberries, raspberries, etc.—has been reported from Canterbury and Hawke's Bay. It was first recorded by Maskell in 1878.

Diaspis boisduvalii, Signoret

This scale has been found on various species of wattles (*Acacia*), and on some hot-house plants. It was also recorded by Maskell in 1878.

Chionaspis citri, Comstock

Originally imported from America, this destructive insect is found on species of *Citrus* in the north of New Zealand; it is also commonly introduced on oranges from Sydney.

Chionaspis dubia, Maskell

Originally described by Mr Maskell in 1881, who stated that it was common on many plants, *Coprosma*, *Rubus*, *Asplenium*, etc. In re-describing it in 1887, he added *Pellæa* to the plants on which it feeds, and gave Canterbury and Auckland as habitats. In 1891 a small form of it was taken at Reefton on leaves of *Leptospermum*, and at Wellington on *Asplenium* and *Cyathodes*. In 1915 it was reported by Mr E. E. Green as occurring on a species of *Adiantum* introduced from Fiji. Mr Brittin states: "I am rather doubtful as to whether this is imported or not."

Lepidosaphes pinnæformis, Bouché (*Mytilaspis citricola*, Packard).
Lemon Scale; Purple Scale

In 1889 this scale insect was found on the rind of some oranges forwarded to Mr Maskell from Inangahua; the oranges came originally from Fiji. In 1895–96 it was found on lemons imported from Portugal. In the report of the Department of Agriculture for 1909 the species is reported as occurring among lemon trees in New Zealand.

Lepidosaphes ulmi, Linn. (*Mytilaspis pomorum*, Bouché).
Apple Scale

Probably introduced very much earlier, but first recorded by Mr Maskell in 1878, when it was quite common in New Zealand. It is a most abundant, and extremely destructive species. While chiefly attacking apple trees, it is stated by Mr Maskell to occur on pears, hawthorn, walnut, plum, peach, apricot, lilac, *Cotoneaster*, sycamore, ash, and many other plants. According to Dr Hilgendorf this pest has been tending to disappear during the last ten years in North Canterbury, but from natural (unknown) causes. A very minute white acarid (mite) has been observed among the eggs of this scale insect.

Lepidosaphes nullipora, Froggart

Mr Brittin informs me that he has found this species very plentifully in North Otago, on *Eucalyptus*, wattle and walnut. It has probably been introduced from Australia. It may be easily mistaken for *L. ulmi*.

Aspidiotus aurantii, Maskell (*A. coccineus*, Gennadius).
Red Scale of Orange

In 1878 Mr Maskell reported this species as occurring on orange and lemon trees, growing at Governor's Bay, near Lyttelton. It occurs, he says, in immense numbers on the oranges and lemons sold in our shops, and which are imported from Sydney. It was also reported from Auckland.

The Agricultural Department report for 1909 speaks of "the immense quantities of scale-infested Citrus fruits which both Australia and Italy have for years poured into the Dominion." Energetic measures are taken, however, to arrest the introduction and spread of the pest.

The following table gives the amount of fruit, which was fumigated at the principal ports of arrival in order to combat the introduction of scale insects, mealy bug, etc.

	Fruit imported cases	Fumigated cases
Auckland	299,249	7751
Wellington ...	305,050	2922
Christchurch ...	62,332	280
Dunedin	58,633	97
Bluff	10,338	—

Aspidiotus buddleiæ, Signoret

Found by Mr Maskell in 1877 on the silver wattle (*Acacia dealbata*) in Nelson; also in Christchurch on the same kind of tree.

Aspidiotus camelliæ, Signoret

Originally noted by Mr Maskell in 1878 on camellias in Christchurch. In 1885 he reported it as common about Wellington on *Euonymus*, weeping willow, and other garden trees and shrubs, to which it often did much damage.

Aspidiotus perniciosus, Comstock. San José Scale

First reported by Mr Maskell in 1895, but his specimens were Australian, and the insect was probably not in New Zealand at that date.

In the Agricultural Department's report for 1909 it is stated to be firmly established in portions of the Nelson district and to be found in isolated localities in other parts of both islands. It has been noticed on apples, pears, plums, peaches, nectarines, apricots, currants and gooseberries.

Aspidiotus hederæ, Bouché (*A. nerii*)

Originally recorded by Mr Maskell in 1881, as occurring in the North Island (Wellington) on *Coprosma*, and later on *Carpodetus serratus* and *Vitex littoralis*. The favourite food-plant in Europe, from whence it was introduced, is *Nerium oleander*. In 1877 he also found it on karaka (*Corynocarpus lævigata*). It has also been met with on palms and orchids in hot-houses, on grape-vines, and, in Christchurch on wattles. In 1895 it was found on the skins of a shipment of lemons imported from Portugal, and some of this fruit was sent up to Whangarei, where lemons are somewhat extensively grown.

Chrysomphalus rossi (Crawford), Maskell (*Aspidiotus rossi*)

This species was recorded by Mr Maskell in May, 1890, as having been received by him from Australia, where it was very common in

Adelaide, Melbourne and Sydney, on almost every kind of plant; but the first announcement of its appearance in New Zealand was in 1895, when Captain Broun found it on olive trees at Whangarei. Later the Agricultural Department reported it as occurring on various species of *Citrus*, and on *Camellia*. The species, which is cosmopolitan in its distribution, is now common and very troublesome in the northern parts of New Zealand.

Coccus maculatus, Signoret (*Lecanium maculatum*)

This European species was reported by Mr Maskell in 1878, as occurring on *Bouvardia* in a greenhouse in Christchurch. I do not know that it has been recorded since.

Coccus mori, Signoret (*Lecanium mori*)

Taken by Mr Maskell on *Alsophila* and other plants in the Botanical Gardens, Wellington, in 1884; and in 1893 on *Asplenium* and other ferns. In 1895 it was found to be plentiful on gorse (*Ulex europæus*) and broom (*Spartium* or *Genista*) at Fairlie in South Canterbury. (The latter is probably *Cytisus scoparius*.)

Coccus longulus, Douglas (*Lecanium longulum*, Douglas; *L. chirimoliæ*, Maskell)

In 1889 Mr Maskell recorded this species from Fiji, where it was found on the bark and leaves of the Peruvian Cherimoyer (*Anona tripetala*). In 1896 he states:

> This insect has come to New Zealand. Captain Broun sent me specimens on *Laurus*, from Northcote, near Auckland. It has evidently been imported from Fiji, between which place and New Zealand there is a rapidly-growing trade in fruit, etc.

Coccus hesperidus, Linn. (*Lecanium hesperidum*, Blanch). Broad Scale; Holly Scale; Ivy Scale

In 1878 Mr Maskell wrote: "This insect is becoming a veritable pest in this country. Hollies, ivies, Portugal laurels, and many other trees in our gardens are every year becoming more and more infested with it." In 1887 he added *Camellia*, orange, myrtle, box, etc., and stated: "this is the commonest of the Lecanidæ in this country." It is also found on oranges, gooseberries, and occasionally on grape-vines.

In 1888–89 Dr Jas. Hudson recorded this species (under the name of *Lecanium hispidum*) as occurring on orange and lemon trees in Nelson, and badly infesting the trees. In February, 1890, he observed *Rhizobius ventralis* preying on the Coccid; and in January, 1891, he stated that both the blight and the ladybird had completely disappeared.

It is kept in check to some extent by a parasitic fly, and by an undetermined species of fungus.

Coccus persicæ, Geoffroy

In 1891 Mr Maskell recorded this species, or a variety of it, as occurring on grape-vines at Ashburton, under the name of *Lecanium rosarum*, Snell.

Coccus persicæ, var. *coryli*, Linn. (*Lecanium ribis*, Fitch)

This species was recorded in 1890 by Mr Maskell, who received specimens from Ashburton, from Mr W. W. Smith. At that date it was common in gardens on gooseberries, black and red currants. In the following year he noted its occurrence from various places in Canterbury, and from Oamaru, and added: "the pest is a new arrival in the colony within the last three or four years, and seems to be spreading rapidly."

Saissetia nigra, Neitner (*Lecanium nigrum*, *L. depressum*)

This was recorded in 1878 by Mr Maskell, as occurring on greenhouse plants in Christchurch and Wellington.

Saissetia oleæ, Bernard (*Lecanium oleæ*).
"Black Scale" of California

Reported in 1884 by Mr Maskell as becoming very common throughout New Zealand, especially in the North Island. It occurs on many plants in gardens and orchards. He found it abundant on *Cassinia leptophylla*, a native composite shrub which covers the hills near Wellington; and it was reported also to be spreading on native trees near Whangarei. A very widespread species.

Saissetia hemispherica, Targioni-Tozzetti (*Lecanium hemisphericum*, *L. hibernaculorum*, Boisd.)

This European species was stated by Mr Maskell in 1878, to be common in greenhouses in Christchurch. In 1884 he found it on *Camellia* in the Hutt Valley, Wellington.

Pulvinaria floccifera, Westwood (*P. camellicola*, Signoret)

In 1878 Mr Maskell recorded this species as occurring on *Camellia* in Christchurch and Wellington, and in greenhouses in the south.

Asterolecanium variolosum, Ratzeburg (*Planchonia quercicola*, Bouché

In 1895 Mr R. I. Kingsley sent to Mr Maskell from Nelson twigs of oak thickly covered with this coccid, and he stated that "the owner first noticed the blight about fourteen years ago."

Pseudococcus longispinus, Targioni-Tozzetti (*Dactylopius adonidum*, Signoret). Mealy Bug

What I think was certainly this species was abundant in a vinery in Auckland in 1884; it was also reported from Whangarei. It was not recorded, however, until 1889, when Mr Maskell stated that it occurred in the hot-houses and stoves of Government House, Wellington. In 1895 Mr Maskell mentioned an outbreak of this pest in the Hutt Valley, near Wellington. In the following year Captain Broun reported that it had proved a terrible nuisance in some of the northern vineries; at Tauranga it was most abundant on passion-vines (*Passiflora* sp.). It was also found commonly on many fruit trees.

The Agricultural Department introduced the black ladybird (*Rhizobius ventralis*), the steely-blue ladybird (*Orchus chalybeus*), and the red-headed ladybird (*Cryptolæmus montrouzeri*), specially to combat this dangerous pest.

Pseudococcus coriaceus, Maskell (*Eriococcus coriaceus*). Blue-Gum Scale

First noticed about Timaru in 1900, having been introduced from Australia, partly among young gum trees, and partly by hardwood logs. It has been detected within recent years as being brought in in both ways. It was originally described by Mr Maskell in 1892 from specimens sent to him from New South Wales. About 1900 plantations of gums (*Eucalyptus globulus* and *E. stuartiana* chiefly) were attacked by this scale insect, and trees 40 to 80 feet in height were completely killed. It soon spread over S. Canterbury and North Otago, and threatened to destroy all the gum trees in the country (including *E. gunnii*, *E. amygdalina*, *E. regnans*, and *E. coccifera*). Later on it was found infesting European myrtle (*Myrtus communis*).

A number of black ladybirds (*Rhizobius ventralis*), red-headed ladybirds (*Cryptolæmus montrouzeri*), and steely-blue ladybirds (*Orchus chalybeus*) were brought from North of Auckland, where they had been introduced some years previously, and were liberated near Timaru. The two last-named species could not stand the winter of S. Canterbury, but the *Rhizobius* increased rapidly and very soon cleared the trees of scale. By the winter of 1907 nearly all the affected plantations were stocked with ladybirds. Mr T. W. Kirk said in 1908:

In January of this year my assistant collected at Rolleston over 1300 on ten gum trees in a little over three hours....Three years ago the plantations were swarming with the pest and to all appearances were doomed to utter destruction. It is not too much to say that within another twelve months there will scarcely be a single living scale to be found on the southern plantations.

Eriococcus araucariæ, Maskell

Mr Maskell described this species from specimens found on Norfolk Island pines (*Araucaria excelsa*) at Governor's Bay, near Lyttelton, in 1878. It is found on the same tree and on the Moreton Bay pine (*A. Bidwillii*) in the North of Auckland district. The species has been found both in Spain and in America, and is almost certainly an introduction. It is apparently held in check in New Zealand by the introduced ladybirds.

Dactylopius coccus, Costa (*Coccus cacti*). Cochineal Insect

Apparently two attempts, both unsuccessful, have been made to naturalise the cochineal insect in New Zealand. Mr Jas. Drummond states that it was introduced by Mr Walter Brodie into Mangonui about 1847. The Canterbury Society received a number of these insects in a case of food-plants, from Sir Geo. Grey, in 1868. The climate of New Zealand is too cold for this species.

Icerya purchasi, Maskell. Cottony-cushion Scale

This species was described by Mr Maskell in 1878 from specimens sent to him by the Rev. Dr Purchas, who first found it, in Auckland, where it had nearly destroyed a hedge of the Kangaroo Acacia (*A. armata*). Writing in 1883, Mr Maskell said:

Icerya purchasi has spread greatly in the last two years. It had just reached Napier at the date of my last paper; it has now established itself in that district, not only in gardens but in the native forests. In Auckland it is attacking all sorts of plants, from apple trees and roses to pines, cypresses and gorse, and it is spreading over a large district. It has reached Nelson ...where it is devouring wattles, cypresses, gorse, and many other plants.

Mr Maskell made every effort to rouse public attention to the danger arising from this dreaded pest. Fortunately the Agricultural Department awoke to the importance of meeting the problem, and by the introduction of the Australian ladybird (*Vedalia cardinalis*), this scale is now kept in check and rapidly destroyed wherever it is met with.

Chætococcus parvus, Maskell (*Cryptococcus nudata*, Brittin)

This species was found on *Hoheria* at Cashmere Hills, Christchurch, by Mr G. Brittin, and described as a new species, in 1914. It was later found to be a species which Mr Maskell had described in 1897, as feeding on wild plums in China. In 1914 it was reported by Mr Green as occurring at St Albans, Herts, England, on cherry trees which had been imported from Japan.

Sub-order ANOPLURA

Family PEDICULIDÆ

Pediculus capitis, Nitzsch. Common Louse

Sir Joseph Banks, describing the Maoris shortly after he first met them in 1769, says: "In their hair was much oil, which had very little smell, but more lice than ever I saw before[1]."

Very common in New Zealand.

Pediculus corporis, De Geer (*P. vestimenti*, Nitzsch). Body Louse

Equally common with the preceding species. Mr Howes says, what is perfectly correct, that both species are becoming scarcer. The segregation of children in schools formerly tended to spread these offensive insects, but closer inspection in later years has very much reduced the pest.

Phthirius inguinalis (*Pediculus pubis*). Crab Louse

Mr Elsdon Best writing in June, says: "The Maori carried two forms of louse, the body louse (*Kutu*), and a form called the *Werau* that, he says, infests the *aroaro* (private parts) only. Both are said to have been pre-European."

I think it more probable that this parasite was introduced by Europeans from the earliest days when they had connection with Maori women. Banks says:

Though we were in several of their towns, where young and old crowded to see us, actuated by the same curiosity as made us desirous of seeing them, I do not remember a single instance of a person distempered in any degree that came under my inspection, and among the numbers of them that I have seen naked, I have never seen an eruption on the skin or any signs of one, scars or otherwise. Their skins, when they came off to us in their canoes, were often marked in patches with a little floury appearance, which at first deceived us, but we afterwards found that it was owing to

[1] In another passage Banks says: "the disgustful thing about them is the oil with which they daub their hair, smelling something like a Greenland dock when they are 'trying' whale blubber. This is melted from the fat either of fish or birds. The better sort indeed have it fresh, and then it is entirely void of smell."

their having been in their passage wetted with the spray of the sea, which, when it was dry, left the salt behind it in a fine white powder.

During the period of the war a considerable increase in the prevalence of this pest has been noted. It is no doubt due in part to the use of the public lavatories in trains and railway stations by infected men.

Hæmatopinus ventricosus, Denny. Rabbit Louse

This louse has probably been here since rabbits were first introduced. In 1889 Mr Coleman Phillips attributed the decrease of the rabbits in South Wairarapa in 1885–86 largely to the prevalence of this insect-pest, in association with *Sarcoptes cuniculi*.

Hæmatopinus eurysternus, Nitzsch. Ox Louse

This parasite has been reported by the Agricultural Department for many years past, as being common both on cattle, and pigs. The latter occurrence is not given by Colonel Reid, and I think, therefore, that it is doubtful.

Hæmatopinus macrocephalus, Burm. (*H. asini*). Horse Louse

This species is commonly found on horses all over the Dominion.

Hæmatopinus vituli, Linn. (*H. tenuirostris*).
Long-nosed Ox Louse

Colonel Reid informs me that this species occurs on cattle in New Zealand; and Mr Miller states that it was met with at Weraroa in 1917.

Hæmatopinus ovillus, Neumann

In 1906 this louse was found among sheep in the South Island, but was not identified with any described species. Specimens were sent to Dr Neumann who described it (Agricultural Report for 1908, p. 194) under the above name. This species is found among sheep in Scotland, from whence, no doubt, the parasite was introduced, but where it was not previously identified.

Hæmatopinus pedalis, Osborn

Colonel Reid states that this species occurs among sheep in New Zealand.

Hæmatopinus urius, Nitzsch (*H. suis*, Linn.). Pig Louse

All the authorities are agreed that this louse is very common among pigs in New Zealand.

Hæmatopinus pilferus, Burm. Dog Louse

Dr Reakes and Colonel Reid inform me that this species is common on dogs.

Chapter IX

CRUSTACEA AND ARACHNIDA

CRUSTACEA

(Order Anaspidacea, see Appendix B, p. 561.)

Order ISOPODA

Family Oniscoida

Porcellio scaber, Latreille. Wood Louse

This species, which originally belonged to the more temperate regions of Europe, but has been introduced accidentally by man to nearly all the temperate regions of the world, must have reached New Zealand at an early date, for it was recorded in White's list of New Zealand Crustacea in the British Museum in 1847 as *P. graniger*. It is now common all over the Dominion, especially in greenhouses and other places near dwellings, but also further from habitations, though not (according to Dr Chilton) in the untouched native bush.

Mr W. W. Smith states that the wood lice have largely displaced native ants. He says (1901):

> In several parts of this district (Ashburton) the wood-lice have almost displaced the native ants. Instead of finding great numbers of ants' nests, as formerly, under the half-embedded stones, we found their old homes tenanted by swarms of wood-lice.

Armadillidium vulgare, Latreille

Another European species that has been dispersed by artificial means to all temperate regions. In New Zealand it is known from Nelson, where it appears to be common in the town gardens; from Mount Egmont (exact locality not known); and from Sumner, where it is common in some gardens. The date of its introduction into New Zealand is unknown, but it was established in Nelson before 1890.

Metoponorthus pruinosus, Brandt

Dr Chilton says "a species common in Europe and neighbouring countries in rather warmer climates than the two preceding. Also widely dispersed into the warmer portions of the world. Specimens from Tasmania in the British Museum were named *Porcellio zealandicus* by White in his list, published in 1847, and a similar specimen from New Zealand was obtained in 1854 (but not described till 1876, in Miers' *Catalogue of New Zealand Crustacea*), so that it must have been introduced before that date, perhaps by whaling ships. It does not appear to have established itself, and I have specimens from one locality in Hawke's Bay only."

The species is abundant in Norfolk Island, the Kermadecs, Australia, etc.

Order DECAPODA

Sub-order MACRURA

Family NEPHROPSIDÆ

Homarus vulgaris, Linn. European Lobster

The first attempt to introduce the lobster into New Zealand waters was made by the late Mr A. M. Johnson, who left London in 1864 by the ship 'British Empire' with 26 lobsters on board which he obtained from the "Mumbles" in Wales. In a letter to me dated 15th September, 1915, he says: "they developed so pugnacious a disposition that they killed each other; the remaining one I sold to one of the first-class passengers." Mr Johnson evidently kept them altogether in one tank.

In 1885 Mr S. C. Farr, on behalf of the Canterbury Society, put 12 lobsters on board the 'Kaikoura,' but they all died in the tropics.

In 1892 Mr Clifford shipped a number from London for the Otago Acclimatisation Society, but "although the experiment was gone into on a somewhat extensive scale, it nevertheless failed."

In 1891 and again in 1892, Mr Purvis, chief engineer of the 'Ionic,' attempted to bring lobsters out to Otago, on both occasions without success. But in 1893 he was successful in landing nine (out of 12 shipped) at Dunedin. These were liberated at the Mole at the entrance to Otago Harbour, a very unsuitable place, but nothing more was ever heard of them.

The next attempt was made on behalf of the Board of the Portobello Marine Fish Hatchery, when arrangements were made with the Marine Biological Laboratory to procure lobsters at Plymouth, and ship them to New Zealand. Four shipments were made on successive trips of the S.S. 'Karamea' in 1906–8, as follows:

Date of arrival at Port Chalmers	Number shipped		Number arrived	
1906, June 29th ...	13 males	12 females	—	2 females
1907, Feb. 26th ...	13 "	12 "	3 males	4 "
Aug. 25th ...	13 "	12 "	3 "	4 "
1908, March 6th ...	17 "	16 "	17 "	14 "

In 1908 some 36,000 larvæ were hatched in the tanks; and in the following year about 100,000 were hatched out. In 1910 only 33,000 larvæ were secured in the tanks; but the majority were allowed to hatch naturally in the ponds and the larvæ to escape into the open sea. This plan has been followed in succeeding years, as it has been found impossible with the other work to be done at the station, to control the rearing of young lobsters. On 1st March, 1913, Mr Anderton, curator of the Station, arrived from Britain in the 'Waimana' with 14 male and 28 female lobsters (out of 43 shipped).

In the following year, to relieve the congestion in the ponds, 12 of the old stock of lobsters (four males and eight females) were liberated at the end of the spawning season, in what was considered to be the most suitable locality in the neighbourhood of Otago Harbour.

In 1914 and succeeding years the female lobsters, numbering about 20, have borne full crops of eggs. It is considered therefore on a moderate estimate, that during the 15 years since lobsters were first introduced at Portobello, more than 1,000,000 fry have been liberated from Otago Harbour. As the young lobsters are free-swimming for the first few weeks of their existence, and as a southerly current averaging a knot and a half per hour flows past the entrance of Otago Harbour, the probabilities are that numbers of them have been carried northwards before they reached the stage at which they sink to the bottom of the sea.

No young lobsters have yet been taken on the coast, but as they take probably seven years or more to reach sexual maturity, and during that time live mostly concealed among rocks and seaweed, the chances of their being captured are few. Any day therefore specimens may be met with.

Family PENÆIDÆ

Penæus canaliculatus, W. A. Haswell. Australian Prawn

In 1892 the Wellington Society received some prawns from Captain Wheeler, which he had brought over from Sydney. They probably belonged to this species, which is commonly caught and marketed in Sydney. They were liberated at Nelson, and were never heard of again.

In 1894 Mr Clifford brought over a number (which I identified as belonging to this species) from Sydney for the Otago Society. They were liberated from the mole at the entrance to Otago Harbour, and were not heard of again. The water of the southerly current which washes the south-east coast of Otago is too cold for this species

In July, 1921, the Otago Acclimatisation Society obtained from the Fisheries Department, Melbourne, a number of fresh-water shrimps from some inland waters in Victoria. All were dead on arrival. I do not know what species this was.

Sub-order BRACHYURA

Family CANCRIDÆ

Cancer pagurus, Linn. British Edible Crab

In 1885 Mr S. C. Farr, on behalf of the Canterbury Society, shipped 12 crabs by the 'Kaikoura,' but they all died in the tropics.

No further attempt seems to have been made till the Portobello Fish Hatchery Board decided to introduce them. In August, 1907, there were landed from the 'Karamea,' three male and five female crabs; and on her next voyage in 1908, seven males and one female. The females all bore ova, but no attempt was made to deal with the larvæ as they hatched. They were liberated from the station with the outgoing tide, and considerable numbers were carried outside the Otago Heads and set free at a distance off shore. In this way it was estimated that up to 1912, over 20,000,000 fry had been liberated.

In 1913 there were shipped on the 'Karamea' at Plymouth, 50 crabs (17 males and 33 females), and of these 43 were landed at the hatchery. They did not thrive, however, and no fewer than 16 died at the approach of the cold season. It was therefore thought advisable to liberate most of the remainder, so 19 were set free in a suitable locality, and eight were retained in the ponds. Of these two died the following season. The number of larvæ liberated in 1914–15 was estimated to be 12,000,000. It is probable therefore that some 40,000,000 larvæ have been distributed since the first experimental introduction in 1907. As the larvæ remain in a pelagic condition for a long time, and pass through several metamorphoses, the death-rate must be very high; but making allowance for this the chances are that ere long specimens of this crab will be found on the New Zealand coast.

It has been found very difficult to keep these crabs under observation at the Portobello Hatchery, as they burrow in the mud, get under stones and even under the foundations of the walls of the ponds. It is probable also that the winter temperature of the ponds in Otago Harbour is too low for them, and this may account for the high death-rate. In British seas it is known that they move into deep water at the approach of winter. While the temperature of the open sea outside Otago Heads seldom falls to 50° F. in the middle of winter, that of the Harbour itself often touches 40° F., and in the ponds has been found as low as 32° F. Lobsters can stand these low temperatures, but they appear to be very detrimental, if not always fatal to this species of crab.

ARACHNIDA

Order SCORPIONIDÆ

Family BUTHIDÆ

Isometrus thorellii. Australian Scorpion

Mr W. W. Smith has obtained specimens of this Australian scorpion among imported hardwood timber at New Plymouth. It may have been introduced at other ports also, but, fortunately it does not appear, so far, to have succeeded in establishing itself anywhere in New Zealand.

Order ARANEÆ

Family PHOLCIDÆ

Pholcus phalangioides, Fuesslin

This cosmopolitan spider was collected by Comte de Dalmas in the interior both of the North and South Islands. Mr Miller records it from Nelson and Wanganui.

Family THERIDIIDÆ

Theridion tepidariorum, Koch. House Spider

This species was taken in the collections made by the 'Novara' Expedition in Auckland in 1859, but whether in houses or in the open is not stated. Comte de Dalmas says:

> This ubiquitous species is found commonly in the open air, despite the very temperate climate, and has also been collected in the Chatham Islands, while in Europe it is not found outside (buildings) even in Central France.

Theridion rufipes, Lucas

This species was recorded by Mons. E. Simon, in 1899, from D'Urville Island. It is also a species of very wide distribution.

Family ARGIOPIDÆ

Diplocephalus cristatus, Blackwall (*Walckenæra cristata*)

Mr A. T. Urquhart reported in 1891 that specimens of this European spider were taken at Nelson by Mr A. S. Atkinson and were identified by the Rev. O. P. Cambridge. It has not been reported by any other collectors since that date[1].

[1] In 1879 the Rev. O. P. Cambridge described a spider from an imperfect male example in Mr A. S. Atkinson's collection, probably from Nelson, as *Linyphia melanopygia*. In 1886 Mr Urquhart described the female from specimens taken at Te Karaka, Auckland. Le Comte de Dalmas makes the following interesting remark on this species, under the genus *Ostearius*, belong to the Argiopidæ:

"The genus *Ostearius* was proposed by Hull (in 1910) for the form found in England and described by O. P. Cambridge (in 1907) under the name of *Tmeticus nigricauda*, the diagnosis and figures of which resemble so closely those given twenty-

Lephthyphantes tenuis, Blackwall

This species was taken by Comte de Dalmas (in 1912–13) in the interior of the Canterbury and Nelson districts, and also in the interior of the North Island (" dans la seule région encore uniquement peuplée d'indigènes Maori ").

Lycosa piratica, Clerck

This species was recorded by Mons. E. Simon in 1899 from the shores of Cook Strait.

Order ACARINA

Sub-order VERMIFORMIA

Family DEMODICIDÆ

Demodex folliculorum, Owen. Hair-follicle Mite

Abundant from the earliest days, but whether the mite was found among the Maoris, or was introduced by Europeans it is quite impossible to ascertain.

Demodex folliculorum, Owen, var. *canis*

This cosmopolitan mite is common among dogs in New Zealand and sometimes causes follicular mange.

Demodex folliculorum, Owen, var. *suis* (*D. phylloides*, Cooker)

Occurs among pigs; occasionally producing a pustular affection.

Sub-order ASTIGMATA

Family SARCOPTIDÆ

Sarcoptes scabiei, De Geer. The Itch Mite

This mite has long been known in New Zealand, and was probably introduced in the earliest days of settlement.

The following, communicated to me by Mr Elsdon Best (June, 1918) may refer to this parasite:

> The hakihaki, a form of itch, which developed into a distressing skin-disease, was pre-European. In many parts one now hears little of it among the natives, but it was among the Urewera in the nineties.

eight years previously by the same author for *Linyphia melanopygia* of the Antipodes, that they seem to apply to a single species; more especially as *L. melanopygia*, founded on a single incomplete male, has been rediscovered in abundance by Urquhart, who rectifies the difference of the ocular group, figured by Cambridge with the median anterior eyes contiguous, instead of being separated almost by the width of their diameter. The two species belong in any case to one genus, which seems to be close to *Microneta*; if they are indeed a single species, then this one, common in New Zealand, but very rare and discovered only recently in England, would appear to be, in contradistinction to the others, accidentally imported into Europe."

Sarcoptes scabiei, var. *canis*

I am informed that this mite occurs in New Zealand, particularly on house dogs. It is not included in Colonel Reid's list (furnished to me) of ecto-parasites found in this country.

Sarcoptes minor, Fürst, var. *cuniculi*

This mite, which particularly attacks the head of the rabbit, produces a kind of scabies which is fatal to severe cases. Colonel Reid does not include it in his list, but Mr Coleman Phillips, in 1889, attributed the disappearance of rabbits in the South Wairarapa district largely to its attacks, conjointly with those of the rabbit louse (*Hæmatopinus ventricosus*).

Sarcoptes mutans, Robin. Scaly-leg Mite (of fowls)

In the report of the Agricultural Department for 1900 it is stated that the mite which causes this disease is not uncommon in New Zealand.

Psoroptes communis, Fürst, var. *ovis*

This mite produces the disease known as scab in sheep. At one time it was universally spread throughout the flocks in New Zealand, where it had been introduced from Australia, and originally from Europe. Before the days of Colonial administration in New Zealand, the various Provincial legislatures attempted to cope with it. Active measures were adopted wherever it occurred, and a tax of 2*s*. per annum per 100 sheep used to be levied on sheep-owners for the inspection and control of the pest. This tax at one period realised over £20,000 per annum. About 1880 the country was declared free of the pest, but the tax was not remitted till 1906, on account of the necessity for maintaining a close inspection of the flocks. There has been no reappearance of the pest in New Zealand for over 40 years.

Psoroptes communis, Fürst, var. *cuniculi*. Rabbit Mite

This ecto-parasite occurs among rabbits in New Zealand. It produces scabies in the ear, and has been found in the pulmonary organs.

Chorioptes auricularum, Bendz.

Colonel Reid states that this mite occurs in cats in New Zealand. It produces intense irritation in the ear of its victim.

Tyroglyphus siro, Linn. Cheese Mite

This cosmopolitan species was probably introduced with the earliest importations of cheese into the country. It, however, only occurs on old dry cheeses, occasionally in flour, on dried fruits, etc., and is usually rigidly excluded from all dairy factories and all good cheese stores.

Tyroglyphus farinæ, Koch. Wheat Mite

In 1893 at a meeting of the Entomological Society of London, Mr R. W. Lloyd exhibited specimens of this "wheat mite" which was found in wheat imported from New Zealand. The species was probably introduced into New Zealand from Europe at an early date. Mr W. W. Smith informed me some years ago that it was very common in grain sheds at Ashburton, and it is probably widespread. It is occasionally found on dry cheese, and is recorded as *T. siro*; but it is considerably smaller than that species.

Tyroglyphus longior, Gervais

Recorded from Wellington; probably widespread.

Cytolichus nudus, Viz. Internal mite of Fowls

Mr D. Miller reports this as first recorded in 1920.

Glyciphagus prunorum, Hermann (*Glyciphagus domesticus*, De Geer). Common mite

This cosmopolitan species has been known in New Zealand from the earliest days of European settlement. This is the cause of "Grocer's Itch."

Rhizoglyphus sp.

Mr Howes informs me that an undetermined mite, belonging to this genus, is commonly found attacking bulbs of various species (of *Narcissus*?) in New Zealand.

Sub-order METASTIGMATA

Family IXODIDÆ

Hæmaphysalis bispinosa, Neumann

Some years ago ticks found in the north of Auckland were classified as *Ixodes ricinus*. Later on, further specimens from the same locality were sent to Prof. Nuttall of Cambridge, who advised the Agricultural Department that they belonged to the genus *Hæmaphysalis*. More recently, other specimens submitted to Cooper, a co-worker with Nuttall, were classified as *H. bispinosa*.

Dr Reakes thinks that as *Ixodes ricinus* is commonly found in most temperate climates and is normally found in Britain, it is quite possible that it occurs also in New Zealand, but it has not yet been actually recorded.

H. bispinosa, previously recorded as *H. punctata*, Koch, has been found in New Zealand on horses, cattle, and casually on sheep and dogs. It only sucks the blood of its host, and is not credited with being the carrier of any disease.

Family Gamasidæ

Dermanyssus gallinæ, De Geer

This is a common and very injurious mite which is present in all but exceptionally clean poultry houses. It lives in concealment during the daytime, in the straw of the nests, and the cracks and crevices of the roosts, and comes out at night to attack the fowls. It is widespread in New Zealand.

Family Tetranychidæ

Bryobia pratensis, Garm. Red Mite of Fruit growers

In the report of the Agricultural Department for 1901 Blackmore reports a *Bryobia* as found under the leaves of fruit trees, to which it was causing great damage. In the report for 1909 it is stated to be doing great damage to apple trees, especially in the North Island. In the South Island it was found to be causing the death of various species of *Abies*.

In 1905 it was reported from such widely-separated localities as Ormond, Palmerston North and the Lillburn Valley. It is generally distributed throughout New Zealand.

Family Eriophyidæ

Eriophyes (*Phytopus*) *pyri*, Paget. Pear Mite

Reported in 1896 and 1897 as very common in the colony. It appears in spring and early summer, and attacks foliage. In 1909 the Agricultural Department reported it as very prevalent, and doing considerable damage on unsprayed trees.

Generally distributed throughout New Zealand on pear trees.

Family Uropodidæ

Uropoda vegetans, De Geer

Mr W. W. Smith stated in 1901 that this parasitic mite had been found on the following *Lepidoptera*, viz., *Xanthorrhöe beata* and *X. rosearia* at Invercargill; and on the following beetles—*Trichosternus antarcticus*, and a large carnivorous ground beetle at Ashburton; *Uloma tenebrionoides*, *Lissotes reticulatus*, *Thoramus wakefieldi*,

Œmona hirta, *Pterostichus præcox*, *Coptomma variegata* and *Xyloteles griseus*, at Ophir, in Central Otago. It was also taken on a fly in the Wellington district, and on an introduced wood louse (*Porcellio scaber*).

As far back as 1892 Mr Maskell reported this mite on a species of *Elater* (?) from Wellington; and on *Oniscus* (*Porcellio* ?) from Christchurch.

The carabid found infested with this mite at Ashburton was in a cucumber-frame which was crowded with wood lice. Mr Smith says this species has only been detected in certain districts within recent years, and he considers it to be an introduced form which is spreading rapidly.

Sub-order PROSTIGMATA

Family TROMBIDIIDÆ

Tetranychus tellarius, Linn. Red Spider

This mite has probably been in New Zealand from early days of settlement. It was reported in 1873 as occurring commonly on apple trees, but probably this was a mistaken identification, and *Bryobia pratensis* was intended. It is a common greenhouse and vinery pest, but numerous reported cases refer to the other species. As far as I know, it is not found on apple trees at all, but I have heard of its attacking violets. The Agricultural Department reported it in 1911 as common on Cape gooseberries, violets and primroses.

Tetranychus bimaculatus, Harvey. Red Spider

Mr A. H. Cockayne states, in the Afforestation report for 1910–11, that this species threatens to become a serious menace to the successful growing of certain species of *Abies* in the Canterbury district.

Chapter X

PENTASTOMIDÆ, PLATYHELMINTHES, NEMATHELMINTHES, OLIGOCHÆTA

PLATYHELMINTHES

TURBELLARIA

Family Bipaliidæ

Bipalium kewense, Moseley

PROFESSOR DENDY, writing in 1894, says:

> This species, which has been so widely distributed by the unintentional agency of man, and whose natural habitat is still unknown, was obtained by Mr T. Steel in Albert Park, Auckland, in 1892.

Family Rhynchodemidæ

Rhynchodemus moseleyi, Fletcher and Hamilton

Four specimens of this Australian species were collected by Mr Steel in Albert Park, Auckland, in 1892.

Family Geoplanidæ

Geoplana sanguinea, Moseley

In 1879 Captain Hutton described this Australian planarian as *Rhynchodemus testaceus*, from specimens obtained by him at Dunedin and Wellington. It has since been collected at Napier, Auckland and Tarawera township, and is probably widespread throughout New Zealand.

Geoplana cœrulea, Moseley, var.

Three specimens of this common Australian species were reported in 1892, from Albert Park, Auckland.

It is quite probable that all four species of these introduced Land Planarians are widely spread in New Zealand, but very few persons notice, and fewer collect them.

TREMATODA

Family Amphistomatidæ

Amphistomum conicum, Zeder

Found in the rumen and reticulum of cattle, but apparently does little harm. It has been reported from time to time both by Drs Gilruth and Reakes.

Family DISTOMATIDÆ

(*Distomum hepaticum*, Retz.) *Fasciola hepatica*, Linn.

Liver Fluke

This parasite was undoubtedly introduced with living sheep, and is not common, nor does it appear to cause loss of stock. It occurs in the livers and bile-ducts of sheep in the Hawke's Bay, Fielding and Nelson districts. Cases were found at Te Hauke in Hawke's Bay in 1897, and again in Waipawa and Te Aute in 1907. The European host is a small snail, *Lymnæa trunculata*. This species does not occur in New Zealand, but there are half-a-dozen native species of the genus, and probably one of these is the intermediary host. The presence of fluke was recognised when sheep came to be examined in the freezing works after the passing of the Slaughtering and Inspection Act in 1900. It occasionally occurs in cattle.

CESTODA

Family TÆNIIDÆ

Moniezia expansa, Rud.

In 1902 Dr Gilruth stated that this species was the common tape-worm of the sheep in New Zealand. In 1903 it was stated to be very common, and it was reported that calves were infected with it in several districts. In 1916 Dr Reakes informed me that it was occasionally found in cattle, but was rare and did no harm economically. It was also found in sheep, but he adds "it is not common and probably does little harm. No serious mortality has been traced to it." It is impossible to say when it was first introduced into the country. Colonel Reid states that it occurs also in goats in New Zealand.

Moniezia alba, Rud.

Dr Reakes states that this is occasionally found in cattle on post-mortem examination, but that it is far from common.

Moniezia planissima, Rud.

According to Dr Reakes this parasite is occasionally found in cattle and sheep, but is far from common.

Moniezia filicollis, Rud.

Colonel Reid states that this species occurs in sheep in New Zealand.

Tænia echinococcus, V. Sieb. (Larval form: *Echinococcus polymorphus*, Dies.; *E. veterinorum*, Rud.)

This is the most common tape-worm in dogs, and Dr Reakes thinks was probably introduced with the first European dogs brought into the country. It is widely distributed all over the Dominion, and Dr Gilruth considers "that it is almost impossible to find an ox or sheep perfectly free from them." They are also found in horses and in pigs. In stock they are not so dangerous as in man, where their hydatid cysts form a destructive and offensive form of disease.

The following figures, showing the extent to which hydatids in human beings occur in New Zealand, during the years 1914 to 1917, have been kindly furnished to me by Mr Malcolm Fraser, Government Statistician:

	1914	1915	1916	1917
Total cases recorded ...	68	90	78	80
Deaths due to hydatids ...	12	11	17	12

Tænia marginata, Batsch. (larva = *Cysticercus tenuicollis*, Rud.)

This tape-worm is also common in dogs, and the larval form in cattle, sheep, and, occasionally, in pigs. It has been a very long time in New Zealand, but it does comparatively little harm; the cysts are usually found in the serous membranes, particularly those covering the intestines. The slender neck enables it to be easily removed. The larvæ frequently die when encysted in oxen or sheep, become calcified, and are then occasionally taken for tubercular growths.

The parasite does not attack human beings.

Tænia solium, Rud. (larva = *Cysticercus cellulosæ*, Rud.). The Pork Tape-worm

This cestode, which produces "pig-measles," has never to my knowledge been recorded in its tape-worm form in man in New Zealand, but Colonel Reid reports the cystic form as occurring in pigs.

Tænia cœnurus, Kürch. (larva = *Cœnurus cerebralis*, Rud.)

This tape-worm occurs in the dog, but its cystic stage develops in the brain-cavity of the sheep, producing the disease known as "Gid" in England, and "sturdy" in Scotland. It is not very common, but some cases were recorded from Christchurch and Wellington in 1897, and again in 1901–12. It also occurs in cattle.

Tænia serialis, Baillet (larva = *Cœnurus serialis*, Gervais)

Colonel Reid reports this tape-worm as occurring in dogs. It is closely related to the preceding species. The cystic form occurs in rabbits and hares, but it has not been recorded from those animals in New Zealand.

Tænia crassicollis, Rud. (larva = *Cysticercus fasciolaris*, Rud.)

Colonel Reid records this tape-worm as occurring in cats in New Zealand. The cyst-stage occurs in rats and mice.

Tænia serrata, Goeze (larva = *Cysticercus pisiformis*, Zed.)

In 1888 Professor Thomas reported the larva or bladder-worm as fairly common among rabbits in the Waikato, but as rare in the Wairarapa. The tape-worm occurs in dogs, and the cyst-stage in hares and rabbits.

The parasite does not attack man.

Tænia saginata, Goeze (larva = *Cysticercus mediocanellatæ*, Davaine; *Cysticercus bovis*, Cobbold)

A case of this was reported from Invercargill in 1908; the parasite affecting the muscular tissue of the heart of a bullock. In its adult condition this tape-worm lives exclusively in the intestinal canal of man, while the corresponding *Cysticercus* is found almost exclusively in the ox. The popular name is the unarmed beef-worm, or the fat tape-worm.

It is a cosmopolitan species, found especially in the tropics. An individual of this species has been estimated to give off in a year, 550 grammes weight of proglottids.

Tænia (*Anoplocephala*) *perfoliata*, Goeze

This tape-worm is sometimes met with in horses in New Zealand; it is usually found in the cæcum.

NEMATHELMINTHES

NEMATODA

Family Ascaridæ[1]

Ascaris megalocephala, Cloq. Common White Worm

Dr Reakes states that this worm, which is found in the intestines of the horse, is widely distributed throughout the Dominion. It apparently does little harm. It was probably introduced with the first horses which came into the country.

Ascaris suis, Goeze. Large White Worm of Pig

This worm, which is not uncommon in the small intestines of the pig, is not uncommon in New Zealand. Dr Reakes states that it does little harm.

[1] In the *N.Z. Journal of Agriculture*, Vol. xxii, No. 2 (Feb. 1921), at p. 123, it is stated that pigs at Roa were infested with worms—probably *Ascaris lumbricoides*.

Ascaris mystax, Zeder. Stomach-worm of Cat

Has been noted in several parts of New Zealand, and is probably not uncommon. It is found inhabiting the stomach and small intestines of the cat. The form of this worm found in the dog, and which Colonel Reid states occurs among these animals in New Zealand, has been recorded as *A. marginata*, Rud., but it is not specifically distinct.

Heterakis papillosa, Bloch. White Intestinal Worm of Fowls

This parasite is found among domestic fowls in New Zealand, especially where they are crowded together. It occurs chiefly in the cæcum.

Heterakis perspicillum, Rudolphi

Colonel Reid informs me that this worm also occurs among fowls in which it chiefly infests the small intestine.

Family Oxyuridæ

Oxyuris curvula, Rud. Whip worm; Maw-worm

This worm occurs in the large intestines of the horse. It is common in New Zealand, and Dr Reakes states must have been in the country for a long time.

Family Strongylidæ

Sclerostomum equinum, Müller (*Sclerostomum armatum*, Dujardin; *Strongylus armatus*, Rud.). Palisade Worm

Dr Reakes informs me that he first met with this species on the lowlands about Kaiapoi, but that it was no doubt in New Zealand at an early date. It is fortunately not very common at present. It is the most dangerous worm which infests horses. The parasite is able to bore its way from the intestines into the blood stream, causing aneurisms of the larger arteries, especially the anterior mesenteric.

Sclerostomum tetracanthum, Dies. (*Strongylus tetracanthus*, Mehlis). Round Worm of the Horse

In the report of the Agricultural Department for 1900 this parasite is recorded as occurring among horses on the west coast of the South Island.

Sclerostomum hypostomum, Dujardin

Dr Reakes informs me that this parasite occurs among sheep in New Zealand, but that it is not common. It is found in the small intestines of sheep, and often causes hæmorrhages through puncturing the mucous membrane.

Strongylus contortus, Rud.

Dr Reakes thinks this parasite has been introduced within the last 20 years. It occurs in most parts of the Dominion, especially in damp, recently-fallen bushlands. He is of opinion that in conjunction with *S. cervicornis* and *S. gracilis*, it causes more loss among stock than all the other diseases of sheep put together. It is found chiefly in the fourth stomach of sheep and lambs. Colonel Reid states that it also occurs as an endo-parasite of cattle.

It is found in the true stomach of young calves, and in some countries causes heavy mortality among them, especially when they are not properly fed.

Strongylus micrurus, Mehlis. Lung Worm, causing Husk or Hoose

Dr Reakes states that this worm causes great loss among calves, but it is thought that it is not so serious a plague now as it used to be, as farmers rear and feed their stock better.

Strongylus ovis-pulmonalis, Ercolani

This parasite was apparently first reported as occurring in 1897 among sheep at Okoroire.

In 1902 Dr Gilruth found it encysted in the lungs of sheep.

Strongylus pulmonaris, Ercolani

Colonel Reid reports this species as occurring in cattle. It inhabits the bronchi of calves.

Strongylus cervicornis, McFadyean

This species was first reported in 1897 as occurring in young sheep at Okoroire. In 1899 Dr Gilruth reported parasitic gastritis in calves at Orepuki, Southland, as due to this worm. Mr McFadyean also found it in old cattle on the west coast of the North Island. He also met with it in sheep, but in their case not burrowing into the gastric glands. In the Agricultural Department's report for 1906 it is recorded as having been met with in the fourth stomach of an Angora goat.

Strongylus convolutus, Ostertag

Dr Reakes (writing in 1916) states that this parasite has "probably been introduced within the last fifteen years." It is very destructive to young cattle, causing heavy mortality, especially among those which have not been properly fed. The worm encysts itself in the walls of the stomach.

Strongylus strigosus, Dujardin

In the report of the Agricultural Department for 1906 it is stated that this parasite has been found in the stomachs of a large number of rabbits in the neighbourhood of Roxburgh, Otago.

Strongylus gracilis, McFadyean

This destructive worm has been found in the stomachs of New Zealand cattle. Colonel Reid also reports it as occurring in sheep.

Strongylus filicollis, Rud. Thread-worm of Sheep

Dr Gilruth reported this species in 1902. Dr Reakes states that when found in the small intestines of sheep it probably does little harm; but when it occurs in the abomasum its presence is serious, and it does considerable damage. Colonel Reid reports it as infesting goats in New Zealand.

Strongylus paradoxus, Mehlis. Lung-worm of Pig

This parasite occurs among pigs in New Zealand, but Dr Reakes states that it is not very common, and that it does little harm. It was first noted in the report of the Agricultural Department in 1895.

Strongylus strigosus, Dujardin

This species infests rabbits, and Colonel Reid records it as occurring in rabbits in New Zealand.

Strongylus rufescens, Leuchart

Dr Reakes informs me that this species causes pneumonia and nodules in the lung substance of sheep[1].

Œsophagostomum inflatum, Sch.

Dr Reakes reports this parasite as occurring in the large intestines of New Zealand cattle, but that it is not common. It apparently does little harm.

Œsophagostomum venulosum, Rud.

Colonel Reid states that this is an ecto-parasite of the sheep in New Zealand.

Œsophagostomum columbianum, Curtice

Dr Gilruth reported this small round worm in 1902. Dr Reakes states that it is found in the large intestine of sheep, but apparently causes little trouble. Colonel Reid also records it as an endo-parasite of cattle.

[1] In October, 1906, the Agricultural Department reported *Strongylus capillaris*, as occurring in a sheep in the Southland district. I cannot place this species.

Œsophagostomum dentatum, Rud.

Colonel Reid reports this species as occurring among pigs in New Zealand.

Family TRICHOTRACHELIDÆ

Tricocephalus affinis, Rud. Whip-worm

This worm was first reported from Wanganui in 1896, as occurring in the large intestines of lambs. In 1902 Dr Gilruth stated that it was to be found in the cæcum of nearly every sheep examined. In 1906 it was met with in the cæcum of an Angora goat. In 1916 Dr Reakes informed me that it was sometimes found in the large intestines of cattle. He did not consider that it did much harm.

Tricocephalus crenatus, Rud.

Colonel Reid reports this species as occurring in pigs in New Zealand. It lives in the intestines of the animal.

Family FILARIIDÆ

Filaria immitis, Leidy

The larvæ and nematode of this parasitic worm have been found in imported dogs while in quarantine.

Family ANGUILLULIDÆ

Tylenchus devastatrix, Kühn. Potato Eelworm; Stem Eelworm

I do not know when this pest was first reported in New Zealand, but it was referred to in the report of the Agricultural Department for 1903, as being present, but not very common. In 1905 it caused considerable damage at Tauranga and elsewhere, especially in heavily manured fields; and was again troublesome in 1907 and 1908. In 1906 it was only reported from one locality, but several shipments of potatoes from Tasmania were found to be attacked by it.

It attacks many kinds of plants, e.g. wheat, oats, hops, clover, and onions, but it is often associated with fungi (*Sclerotinia*) and it is not easy to determine whether the insect or the fungus is the cause of the disease.

Either this, or a closely allied species, is found on *Holcus lanatus*, *Anthoxanthum odoratum*, *Poa annua*, *Bellis perennis*, *Capsella Bursa-pastoris*, *Spergula arvensis*, *Ranunculus repens*, *Sonchus oleraceus*, *Plantago lanceolata* and *Centaurea cyanus*.

Tylenchus tritici (*T. scandens*), Basstian. Ear-cockle; Peppercorns; Purplers

This eelworm has been known in New Zealand for many years, Mr T. W. Kirk having met with it first about 1893, but it has never so far assumed dangerous proportions. The larvæ are very tenacious of life, and will hatch out worms after two years' drying.

Heterodera radicicola, Greef. Cucumber and Tomato Eelworm

This pest has been known for a number of years in this country, and is especially troublesome to tomatoes grown under glass.

Heterodera schachtii, Schmidt. Beet Eelworm

This species has been in the country for a long time, but it is the sort of pest that is very seldom noted or recorded. Mr T. W. Kirk informs me recently (1919) that no specimens have been sent to the Agricultural Department for some years past, which probably only shows that it has not been giving trouble to cultivators.

ACANTHOCEPHALA

Family ECHINORHYNCHIDÆ

Echinorhynchus gigas, Goeze

Colonel Reid reports this parasitic worm as occurring in New Zealand pigs. It is usually met with in the small intestine.

Family LINGUATULIDÆ

Linguatula denticulatum, Leuckart

Colonel Reid reports this worm as occurring in sheep.

Linguatula lanceolatum, Fröhlich

This species is recorded by Colonel Reid from dogs.

CHÆTOPODA

OLIGOCHÆTA

Family MEGASCOLECIDÆ[1]

Dichogaster modigliani, Rosa

This species has been reported to me, but I have failed to record by whom, as occurring in New Zealand. Dr Benham informs me that the only record of a *Dichogaster* in this country, rests on a statement by Ude (1905), that an unnamed species, allied both to *D. modigliani*, and to *D. malayana*, Horst, was gathered "at Oripi Bush, Tauranga," by a German traveller. He records it merely as *Dicho-*

[1] A. Nichols, in *Acclimatisation of Salmonidæ at the Antipodes* at p. 46, says of the shipment of salmon ova made to Port Chalmers in the 'Mindora' in 1869: "a living worm was found among the moss in which the eggs were packed, and was 'acclimated.'" Dr Benham suggests that this was probably an Euchytrœid.

gaster sp. As these worms come from Sumatra and the Malay Archipelago, he considers it likely that the tickets of localities got mixed up.

Dichogaster sylvatica, Hebdier (?)

Dr Benham states that the occurrence of this Australian species is incidentally referred to by Ude (1893), where he mentions that in a letter to him from Dr Rosa, the latter refers to a "variety from New Zealand." The specimens Ude was examining were from Australia.

Family EUDRILIDÆ

Eudrilus eugeniæ, Kinberg

Mr W. W. Smith reports this species as occurring in Taranaki. Dr Benham, however, who is extremely doubtful of its occurrence, says:

I cannot find any statement in any writer that this occurs in New Zealand, except in a list of New Zealand worms given by Beddard (1891), which includes "*Eudrilus* sp. (fide Benham)." I have no recollection or record of how I came to make that statement to him.

He adds that it is not unlikely that the worm does occur here, as although it is a native of Africa, it has been found widely distributed, apparently by man.

It was probably on Beddard's authority that the species was included in the list of introduced worms in the *Index Faunæ Novæ-Zealandiæ*.

Family GEOSCOLECIDÆ

Pontoscolex corethrurus, Fr. Müller

Recorded in the *Index Faunæ N.Z.* in 1903. Dr Benham says:

A native of Central America and the West Indies, it has been found in several of the Pacific Islands, such as Samoa, Tonga, Fiji and the Sandwich Islands, but I cannot find any definite statement by any writer that it has occurred here, though Michaelsen in his Monograph (1900) gives New Zealand as one of the localities, but gives no authority for its occurrence.

Family LUMBRICIDÆ

Eiseniella tetrædra, Savigny

Mr W. W. Smith recorded this in 1892, under the name *Allurus tetrædrus*, as occurring in still pools in the Ashburton River. Commonly met with in cultivated land. Mr W. W. Smith states that it occurs in red masses in sluggish overgrown gutters in New Plymouth.

Helodrilus (*Eisenia*) *fœtidus* (Sav.) (*Allobophora fœtida*).
British Brandling Worm

Originally recorded from Dunedin in 1876, by Captain Hutton, who described it as a new species, under the name *Lumbricus annulatus*. Dr Benham informs me that British worms of the Fam. *Lumbricidæ* are all commonly met with in cultivated land. This was reported by him in 1898 as common in Dunedin. Mr W. W. Smith reports it as occurring in great numbers in Taranaki, in heaps of rotten manure and putrid matter. It occurs in Sunday Island, Kermadec Group, "under leaves."

Helodrilus (*Eisenia*) *roseus*, Sav.

Recorded by Captain Hutton in the *Index Faunæ Novæ-Zealandiæ* in 1903. Mr W. W. Smith records it as common in Taranaki. Also found in the Chatham Islands.

Helodrilus constrictus, Rosa.

Dr Benham has no doubt this species occurs in New Zealand, although it has not been recorded. It is found in Campbell Island where he states: "its occurrence is clearly related to the habitation and cultivation of a patch of garden by the shepherds." It is also common everywhere in Sunday Island in forest on damp ground, and under nikau palm leaves and tree-fern fronds.

Helodrilus (*Allobophora*) *caliginosus* (Sav.)

In 1898 Dr Benham reported this as a very common worm about Dunedin. Specimens labelled *Lumbricus levis* by Captain Hutton, belonged to this species. It is also found in Chatham Island; and in Sunday Island in soil, and under nikau palm leaves.

(A European species of *Allobophora* was found in 1916 by Mr D. Miller in flax swamps in the Manawatu district.)

Helodrilus (*Dendrobœna*) *rubidus*, Sav.

This European worm is recorded in the *Index* in 1903. Mr Smith states that it is common in Taranaki.

Otoclasium cyaneum, Sav.

Another European species recorded in the *Index* in 1903; and also by Mr W. W. Smith from Taranaki.

Lumbricus rubellus, Hoffmeister

This was described as an indigenous species by Captain Hutton in 1876, under the name *L. campestris*. He recorded it as common in Dunedin and Wellington. In 1892 Mr W. W. Smith reported it as common in moist soil on swampy flats, and under moist cakes of cow-manure. Later he records it from Taranaki. It is no doubt a common form.

Lumbricus castaneus, Sav.

A European species recorded in the *Index Faunæ Novæ-Zealandiæ* in 1903. Mr W. W. Smith reports it from Taranaki, as occurring commonly in newly cleared forest.

Lumbricus terrestris, Linn.

Mr W. W. Smith recorded this species in 1892 as "common everywhere." His specimens were identified by Mr Beddard and Mr J. J. Fletcher. He further states that it is common in Taranaki, where it attains to a large size in rich soils. There is evidently some confusion as to nomenclature, for Dr Benham informs me that he has never come across the species in New Zealand.

HIRUDINEA

Hirudo medicinalis, Linn. var. *officinalis*. Medicinal Leech

The Otago Society introduced 200 leeches in 1867, and handed them over to the care of Mr F. D. Rich of Palmerston, who appears to have placed them in one of the backwaters of the Shag River, from whence they were probably washed out in the first big flood. The Canterbury Society introduced 12 in 1867, apparently through the agency of Mr A. M. Johnson, but there is no record of them. In 1868 Mr Howard of the Southland Society obtained some and placed them in the ponds at Wallacetown, but they were never seen again.

In addition to these efforts, chemists imported them at all the main centres, but no further attempts seem to have been made to rear them in the country.

Professor H. B. Kirk informs me that in Auckland, about 1875 leeches were found in a small pool at the foot of a clay bank on the west side of Grafton Road; and that boys used to catch them to sell to the chemists.

Of late years, probably due to the use of other remedies, the employment of leeches for medicinal purposes has greatly diminished. The great war has also contributed to their disuse as their importation has practically stopped. In 1918 it was impossible to obtain leeches from any chemists in Dunedin.

Part III

NATURALISATION OF PLANTS

Chapter XI

DICOTYLEDONS AND CONIFERÆ

OF the plants referred to in the following pages, over *six hundred species* have become more or less truly wild, i.e., they reproduce themselves by seed, and appear at the present time to be more or less permanent denizens of the country. The great majority of them have been brought in accidentally as seeds among other seeds, or in hay, straw, and other packing materials. Some of them have been introduced purposely as food or fodder plants, timber or ornamental trees and shrubs, or ornamental flowers. Some have been introduced for sentimental reasons, as very probably the briar rose was. In addition to these I have referred to several species of plants which have resisted all efforts—often long continued and numerous—to naturalise them.

It is seldom possible to assign definite dates for the first introduction of plants into a foreign country, as can often be done with animals. In the majority of cases the most that I can do is to give the date of the first definite mention of their occurrence.

Though isolated references to many introduced plants occur in various publications ranging from Banks's *Journal* to more recent times, the first list of such introductions appears to be in Sir J. D. Hooker's *Handbook to the New Zealand Flora*, which was written in 1864, though not published till 1867. This contains the names of 165 species of plants, and includes most of the commonest weeds and grass of cultivated land. A very large number of these were recorded by Mr Thos. Kirk from the Auckland provincial district, and were marked by an 'A' in Sir Joseph Hooker's list. Most of the cereal and the commoner cultivated grasses were introduced in early whaling days and by the first missionaries, that is between 1800 and 1820. Succeeding lists by Messrs Colenso, T. Kirk, Buchanan, Petrie, Cockayne and many other observers have added to our knowledge of the alien flora, and to its spread in these islands. The most complete list is that given by Mr T. F. Cheeseman in the *Manual of the New Zealand Flora* (pp. 1062–93), published in 1906. But in none of these lists is any attempt made to give dates of introduction.

In the classification adopted by me in this work, I have mainly followed Mr Cheeseman, who based his on the system used by Sir J. D. Hooker in his various British and Colonial Floras. I am not here concerned with any discussion on the principles of classification adopted or favoured by various systematists and authors. My object is to use names by which the plants referred to can be readily recognised and distinguished, and I have seen no good reason to follow any other scheme than that adopted by the most eminent of British systematists.

No work has been done in New Zealand in recording the insects which visit and pollinate the flowering plants, either indigenous or introduced, except in a few isolated cases.

I have therefore appended to many of the introduced species of plants referred to here, the names of those insects which have also been introduced into New Zealand, and which in Europe are found to pollinate their flowers, or at least to visit them for nectar or pollen.

DICOTYLEDONS

Division POLYPETALÆ

Sub-division THALAMIFLORÆ

RANUNCULACEÆ

Some authors are of opinion that the curved or hooked beaks of the achenes of the genus *Ranunculus* are a means of attaching the fruits to the plumage of birds, and in this way of distributing the species. I should think this is a rare, or at best a doubtful method of distribution. Guppy says: "I have found the achenes of *Ranunculus* frequently in the stomachs of birds in England, in partridges frequently and in wild ducks at times."

Ranunculus aquatilis, Linn. Water Crowfoot; Water Buttercup

First recorded as occurring in various parts of South Canterbury by Dr Cockayne. In 1906 I reported to Mr Cheeseman that it appeared in Otago first in the Waikouaiti River; then in the upper portions of the Waipahi, and more recently in the Pomahaka, these two latter being tributaries of the Clutha River. In 1908 Cockayne recorded it as being plentiful in a lake near the mouth of the Rangitikei River. It has since then increased very rapidly, and has in parts nearly choked some of these streams. Almost certainly introduced from Britain. Flowers in December.

This species may be spread by means of its achenes, but it is more likely that fragments of the stems are carried on the feathers and feet of aquatic birds.

It is visited by *Apis mellifica*, *Bombus terrestris* and *Eristalis tenax*.

Ranunculus Flammula, Linn. Lesser Spear-wort

Recorded by Cheeseman in 1906 as occurring in the Waiharakeke Stream, Piako; and again in 1912 from the vicinity of Kaitaia, collected by H. B. Matthews, in wet Kahikatea forest. This species is likely to increase rapidly on account of its long creeping runners, which will grow from 20 to 24 inches in a year.

Ranunculus sceleratus, Linn. Celery-leaved Buttercup

First recorded from the Otago Goldfields in 1876 by Dr Petrie. In the *Manual of the Flora*, Cheeseman records it as occurring in "damp pastures and waste places from Mongonui to Southland, local." It is, however, very abundant in many localities, and in times of drought has proved fatal to cattle. (Fl., Nov. to March.)

Visited by *Lucilia cæsar* and *Musca corvina*.

Ranunculus acris, Linn. Field Buttercup

Probably introduced at an early date in the settlement of the country; first recorded by Kirk in 1867, among the plants of the Great Barrier Island.

Cheeseman in 1906 records it as occurring in "pastures and waste places in both islands, but not common." It is extraordinarily common in some parts, for instance near Dunedin. (Fl., Nov. to Jan.)

Ranunculus repens, Linn. Creeping Buttercup

Probably introduced as early as *R. acris*, but recorded along with it in 1867. One of the most abundant weeds in New Zealand, spreading in all directions, both by its seeds and its creeping stolons. (Fl., Nov. to Jan.)

Ranunculus bulbosus, Linn. Bulbous Buttercup

First recorded in 1871 by Armstrong from the Canterbury district. Cheeseman in the *Manual* (1906) states that it is abundant in pastures and waste places in both islands. It is not, however, at all common in Otago and Southland. (Fl., Oct. to Dec.)

These three species of buttercup, *R. acris*, *R. repens* and *R. bulbosus* are in Europe visited by *Apis mellifica*, *Eristalis tenax* and *Calliphora erythrocephala*.

Ranunculus sardous, Crantz. Hairy Buttercup

First recorded from the North Cape by Cheeseman in 1896. In his *Manual* he states that it is common in pastures and waste places in both islands. Kerner states that this species is mainly dependent for dispersion by sticking to the feet of birds in hardened soil and clay. (Fl., Nov. to Dec.)

Ranunculus parviflorus, Linn. Small-flowered Buttercup

First recorded in Hooker's list in 1864. According to Cheeseman's *Manual* (1906) it is said to be abundant in pastures and waste places in both islands. It is not common in Otago and Southland. (Fl., Oct. to Dec.)

Ranunculus arvensis, Linn. Corn Buttercup

First recorded by Armstrong as occurring in Canterbury in 1879 Cheeseman states (1906) that it occurs "in cultivated fields in both islands not common." I do not think it occurs in Otago. (Fl., Jan.

Ranunculus muricatus, Linn.

In 1877 Kirk states that "specimens supposed to have been collected near Wellington are in the herbarium of the Colonial Museum."

First recorded in 1882 by Cheeseman from the Bay of Islands and waste places about Auckland and Onehunga. In the *Manual* (1906) it is said to occur in "waste places in both islands local." (Fl., Nov. to Dec.)

Ranunculus falcatus, Linn.

First recorded by Petrie as occurring in dry localities in Northern and Central Otago. Found on bare ground in the hill-country of Maniototo and Vincent Counties[1].

Nigella damascena, Linn. Love-in-a-Mist; Devil-in-a-Bush

First recorded by Cheeseman in 1882 as a garden escape in light soils near Auckland. The *Manual* (1906) gives the same distribution.

The species has not established itself to any extent.

It is visited by *Bombus terrestris* and *B. lapidarius*.

Delphinium Ajacis, Reich. Larkspur

Recorded by W. W. Smith in 1903 as occurring in Ashburton County.

Aquilegia vulgaris, Linn. Columbine

First recorded in 1882 by Cheeseman as a garden escape, about Auckland. It only occurs as a garden escape in several parts of New Zealand, though it is fairly common in many suburban areas.

[1] Kirk in 1877 reported *Ranunculus philonotis*, Retz. as occurring in Evans' Bay (Wellington), Hutt Valley and Otaki.

Aconitum napellus, Linn. Monkshood

First recorded by W. W. Smith in 1903, as a garden escape at Ashburton. There is no report of it from any other locality. It does not seem to spread readily.

BERBERIDEÆ

Berberis vulgaris, Linn. Barberry

First recorded by the author in 1873, as occurring on the site of abandoned gardens in the Taieri Plain, and near Dunedin. Since the vast increase of blackbirds and thrushes the barberry has been spread a great deal, and is to be found in many districts at a short distance from settlement. It is becoming common near Dunedin. Kerner records that when seeds of the barberry have passed through the alimentary canal of a thrush the period of germination is hastened.

PAPAVERACEÆ

Papaver hybridum, Linn. Rough Poppy

First recorded as a garden escape at Ashburton by W. W. Smith in 1903. It has not been reported from any other locality.

Papaver Argemone, Linn. Pale Poppy

First recorded by the author in 1870 from Southland among freshly sown pastures. Also reported from Ashburton by W. W. Smith. Has not spread.

Papaver horridum, DC.

Appeared in a fowl run at Maheno in North Otago in 1917; supposed to have been introduced with Canadian Wheat. The species is S. African and Australian.

Papaver dubium, Linn. Long-headed Poppy

First noted by the author in 1885 on heaps of tailings at Coal Creek, Clutha Valley. Petrie has also reported it from cultivated fields in various parts of Otago.

Papaver Rhœas, Linn. Field Poppy

First recorded in 1869 by Kirk among naturalised plants of Auckland; and from Southland by the author in 1870. In Cheeseman's *Manual* it is stated to be found in both islands in "cornfields and waste places, not common." (Fl., Dec. to Feb.) It has not spread to any extent, however, though all the conditions for its increase seem favourable.

Papaver somniferum, Linn. Opium Poppy

First recorded in 1882 as a garden escape near Auckland by Cheeseman. Noted by the author as fairly common in 1885 in the Clutha Valley in the neighbourhood of gardens. It has not established itself anywhere as a wild species. (Fl., Jan. to Feb.)

Argemone mexicana, Linn.
Mexican Poppy; Yellow Poppy; Prickly Poppy

First reported from Taranaki in 1888; and again in Hawke's Bay and Marlborough in 1894 by T. W. Kirk.

Glaucium flavum, Crantz. Horned Poppy

First recorded—as *G. luteum*—as widely diffused on shingly beaches near Wellington by Kirk in 1877; it is supposed to have been introduced in the packing material of the patent slip machinery at Wellington. Also recorded by W. W. Smith from Ashburton in 1903. Cheeseman in the *Manual of the N.Z. Flora* (1906) states that it occurs from East Cape and Wanganui to Cook Strait. Dr Cockayne has collected it at the mouth of R. Awatere, Marlborough. It has more recently appeared at Puketeraki in Otago, where perhaps it was purposely sown.

Guppy says that the mode of dispersal is problematical:

> "Its seeds have no proper buoyancy even after prolonged drying. On account of their oiliness they will float at first on still water; but they can be made to sink at once, or in a day, by dropping water on them." (Fl., Nov. to Jan.)

Chelidonium majus, Linn. Celandine

First recorded from Ashburton in 1903 by W. W. Smith. It also occurs somewhat rarely in the neighbourhood of Dunedin.

Eschscholtzia californica, Chamb.

First recorded by Kirk in 1877, from Castle Point in Wellington provincial district. In 1879 Cheeseman reported it as covering the greater part of a field at Panmure, Auckland. In 1882 it was common in the neighbourhood of Auckland in light dry soils. Cheeseman recorded it in 1906 as a garden escape in light dry soil in both islands. It is thoroughly established as a noxious weed in dry parts of Marlborough, Otago and South Canterbury, for nothing will eat it. The fruits are jerked off the receptacle when they are ripe. Hence it spreads along made ground, such as railway embankments, but does not progress over pastures to any great extent. The seeds apparently require loose soil to germinate in. (Fl., Dec. to Feb.)

FUMARIACEÆ

Fumaria muralis, Sond. Fumitory

First recorded by Kirk in 1877 as common in cultivated land about Wellington, Wairarapa, and Wanganui; and again in 1895 as found on a ballast-heap in Wellington, among refuse from Buenos Ayres. It is now an extremely common weed in many parts of New Zealand, and is a familiar garden pest. In many districts it flowers all the year round, but in the south mostly from October to April.

Fumaria officinalis, Linn. Fumitory

First recorded by Kirk in 1869 among naturalised plants of Auckland. In the *Manual* it is stated to occur rarely in cultivated fields in both islands. (Fl., Nov. to Jan.)

In Europe the flowers are fertilised by the honey-bee (*Apis mellifica*).

Corydalis lutea, DC.

Recorded by W. W. Smith in 1903 as occurring in Ashburton County.

CRUCIFERÆ

Matthiola incana, R. Br. Common Stock

First recorded by Kirk in 1877 as occurring on cliffs at Castle Point, Wellington. (Fl., Nov. to Dec.[1])

Cheiranthus Cheiri, Linn. Wallflower

Polack speaks of this species as cultivated by Europeans in the north of the North Island in 1838, and as being "acclimated."

First recorded as a garden escape naturalised in a few rocky places near Dunedin in 1873 by the author. Though widely spread it has never established itself to any extent. (Fl., Oct. to Jan.)

Nasturtium officinale, R. Br. Water-cress

Canterbury was settled in 1850, and this plant was probably introduced very soon after, for in a few years it blocked the Avon and other streams in the vicinity of Christchurch. It is given in Hooker's list of introduced plants in 1864, and in Kirk's list of Great Barrier plants in 1867. By that time it was widely spread and strongly established. In 1872 it was abundant on the margin of the Waikato River. At the present time it is one of the commonest of introduced plants in streams and wet places. (Fl., Oct. to March.)

[1] Kirk reported *Matthiola sinuata*, Br. as growing at Castle Rock in 1887.

It is frequently found to be very much infested with the common cabbage blight—*Aphis brassicæ*.

In A. R. Wallace's *Darwinism* he mentions the fact, communicated to him by Mr John Enys, that:

a natural remedy to the water-cress has been found by planting willows on the banks. The roots of these trees penetrate the bed of the stream in every direction, and the water-cress, unable to obtain the requisite amount of nourishment, gradually disappears.

This is no doubt quite true for narrow streams, with a good flow of water. In such situations also *Elodea Canadensis* tends to displace it; and I have noticed in some parts of the Avon at Christchurch, and in tributary streams, that a species of *Nitella* can strangle both of them. But watching shallow ponds near Dunedin, I have noticed that unless kept severely in check, the water-cress can put *Elodea*, *Aponogeton* and species of *Nymphæa* right out of competition in a year or two.

A few years after their introduction into the streams in the Canterbury district, water-cress plants grew to gigantic proportions, being as much as 14 feet in length, and stout in proportion. The size is now quite normal.

In Europe the flowers are fertilised by the honey-bee (*Apis mellifica*).

Barbarea præcox, R. Br. American Cress

First recorded in Kirk's list of introduced plants on the Great Barrier Island in 1867; and in Wellington in 1877. In the *Manual* (1906) Cheeseman reports it as not uncommon in waste places and roadsides in both islands. (Fl., Oct. to Dec.)

Barbarea vulgaris, R. Br. Winter Cress

Said to be common on Otago Peninsula; I am not sure of the identification.

Arabis hirsuta, Scop. Rock-cress

Reported as occurring near Ashburton by W. W. Smith.

Alyssum calycinum, Linn.

First recorded by Armstrong in 1879 as occurring in Canterbury. Cheeseman reports it from both islands as not uncommon on roadsides and in waste places. Petrie states that he has found it in Central Otago, growing up to 3000 ft. (Fl., Nov. to Jan.)

Alyssum maritimum, Linn.

First recorded in Hooker's list in 1864. Armstrong in 1871 recorded it among plants from Canterbury. Kirk in 1877 reports it from two points near Wellington. In 1906 Cheeseman states that it is found in waste places and dry sandy soils near the sea, often abundantly in both islands. (Fl., Nov. to Feb.)

Alyssum orientale, Ard.

Recorded by W. W. Smith in 1903 as occurring in Ashburton County.

Erophila vulgaris, DC. Whitlow-grass

First reported by Petrie from Eastern and Central Otago, and on ridges near Balclutha. (Fl., Oct. to Nov.)

Erophila verna, Linn.

Recorded by W. W. Smith in 1903 (as *Draba verna*) as occurring in Ashburton County.

Cochlearia Armoracia, Linn. Horse-radish

Probably introduced at an early date in settlement; first recorded in Hooker's list in 1864. Stated in the *Manual* to occur in deserted gardens and waste places, but uncommon. It is chiefly, as Kirk puts it, an outcast from gardens.

Hesperis matronalis, Linn. Dame's Rocket

Reported as a garden escape in both islands; from Poverty Bay by Bishop Williams; near Wellington by Kirk; and in cornfields at Oamaru by Petrie. (Fl., Nov. to Dec.)

Malcolmia maritima, R. Br. Virginian Cress

First reported from waste places near Wellington by Kirk, it occurs as a garden escape in several localities. (Fl., Sept. to June.)

Cardamine hirsuta, Linn. Bitter Cress[1]

The introduced form of this species (which is also indigenous to New Zealand) is one of the most troublesome weeds of cultivation in gardens and greenhouses. The explosive fruits scatter the seeds, and the plants produce cleistogamic flowers all through the winter months. It flowers most of the year.

Sisymbrium Sophia, Linn. Flix-weed

First reported by Petrie from Central Otago, growing near Alexandra and Naseby. (Fl., Dec. to Jan.) (According to Kerner, an average plant of this species produced 730,000 seeds in a year.)

Sisymbrium officinale, Scop. Hedge-mustard

Must have been common in the early days of settlement, but first recorded from the Auckland district in 1882 by Cheeseman as being

[1] The indigenous form is now recognised as a distinct species *C. heterophylla* (=*Sisymbrium heterophyllum*, Forst. f.).

generally distributed. However, Mr Cheeseman informs me that the plant recorded in Hooker's list of 1864, and in Kirk's list of Great Barrier plants in 1867, as *Erysimum officinale*, is this species.

(Fl., Nov. to Jan.; but in cultivated land flowers all the year round.)

In Europe the honey-bee (*Apis mellifica*) occasionally visits this flower; I have not noticed it in New Zealand.

Sisymbrium pannonicum, Jacquin

According to Kirk 1869 this species was introduced into the Auckland district along with European flax. If so, it failed to establish itself.

Camelina sativa, Crantz. Gold of Pleasure

Recorded in 1882 by Cheeseman as occurring in one or two places near Auckland. Also by Petrie from Oamaru. In the *Manual* it is said to occur rarely in cultivated fields in both islands. The seeds become mucilaginous on the surface when wet. (Fl., Dec. to Jan.)

Brassica oleracea, Linn. Wild Cabbage

Captain Cook, on arriving in Dusky Sound in the 'Resolution' early in 1773, during his second voyage to New Zealand, cleared a piece of garden ground and "sowed a quantity of European garden seeds of the best kinds." Cabbage was almost certainly among them but probably none of the plants established themselves. Later in the year Captain Furneaux arrived in the 'Adventure' in Queen Charlotte Sound, and made a garden on Motuaro, in which, among other seeds, cabbage was sown. Captain Cook followed in May, 1773, and made other gardens on Long Island. On returning in November he says: "I crossed to Ship Cove next day and visited the gardens. I found *cabbages*, *carrots*, *onions and parsley* in excellent condition." The Sound was visited in October, 1774, when he says: "We likewise visited the cabbage-garden on Motu-Aro, and found the plants shot into seed, which had been for the greatest part consumed by the birds." These were probably the parakeets.

On the third voyage, Captain Cook again visited Queen Charlotte Sound, and though the gardens were "over-run with the weeds of the country, we found cabbages," etc.

From this centre the cabbage has spread round the coasts of New Zealand, but there is no doubt the Maoris helped to distribute it.

Commander Bellingshausen, who visited Motuaro in 1820, says: "we gathered such a quantity of wild cabbage that we had sufficient for one meal of cabbage soup for all the servants and the officers."

A. R. Cruise, writing in 1820, says: "the excellent plants left by Captain Cook, viz., *Cabbages*, turnips, parsnips, carrots, etc., are still numerous but very degenerated." Captain Edwardson, speaking of the natives of Foveaux Straits in 1823, says: "Potatoes, *cabbages* and other kitchen vegetables introduced by the Europeans are cultivated." The mate of the brig 'Hawes' in 1828 found near Tauranga many cultivations with *cabbages*, etc.

Polack in 1831 recorded *cabbages* as extensively cultivated by the natives in the Kaipara district.

Dieffenbach, who visited East Bay near Tory Channel in 1839, says: "the cabbage, which now abounds in Queen Charlotte Sound, and which grows wild, was in blossom, and covered the sides of the hills with a yellow carpet." Later on, speaking of Captain Cook, he says: "the cabbage, which he sowed, has spread over all the open places in Cook's Straits, and early in spring the sides of the hills are covered with its yellow flowers." In the island of Kapiti he found plantations of cabbages thriving well. Bidwill, who travelled in 1839 from Tauranga to the summit of Tongariro, found that the natives used wild cabbages, which they boiled freely.

When Wilkes visited the Auckland Islands in 1840, he found cabbages growing finely on one of the points of Sarah's Bosom.

At the present time the wild cabbage is common on sea-cliffs in both islands, but especially in the neighbourhood of former Maori settlements. (Fl., Nov. to Dec.)

In Europe the flowers are visited by two of our humble-bees—*Bombus terrestris* and *B. lapidarius*. I have seen hive-bees on them.

Brassica campestris, Linn. Swede-turnip

I cannot find when swede-turnips were first introduced into New Zealand, but it is quite probable that they were among the seeds sown by Cook in 1773. The first notice I have come across is by Bidwill when travelling in 1839; at a small native settlement between Waikato and Taupo "they roasted some Swedish turnips." He clearly distinguishes between them and common turnips, of which the natives used the leaves, and which were abundant in the wild condition. At present this species only occurs as an escape from cultivation.

There is, however, a remarkable form of wild turnip found growing in Taranaki, which W.W. Smith considers to be the "Korau" of the Maori. It is a gigantic form, growing five and six feet high, with heavy branching stems, and leaves from two to three feet long. It never forms any bulb, but has a thick stem as much as three inches or more in diameter at the base. Both Cheeseman and Williams say that the "Korau" is *Brassica campestris*, Linn., but Smith thinks this form does not belong to this species. At the same time it is not *Brassica Rapa*, which is a very different plant. I suggest that it may be a hybrid between these allied species; it certainly is an interesting form.

In the Noxious Weeds Act, 1900, wild turnip (*Brassica campestris*) is included among *noxious seeds*. As no farmers and probably very few botanists can distinguish the seeds of the species of *Brassica*, probably the identification does not matter much.

Brassica Rapa, Linn. Turnip

Cook and Furneaux sowed seeds of turnip in their various clearings in Queen Charlotte Sound in 1773, and showed them to the natives. In November of the same year they found that they had seeded. The natives spread the seed throughout both islands in all probability. Marsden found them in cultivation when he landed at the Bay of Islands in 1814; but the Maoris in Queen Charlotte Sound appear to have lost the plant, for Bellingshausen gave them seeds in 1820 and showed them how to sow them.

Nicholas (1817) says that the natives "had mussels and turnips at this feast, but the latter had very much degenerated, and become long and fibrous." He says also: "the turnip is called *packahâ* from its whiteness."

The mate of the brig 'Hawes,' in 1828, found many cultivations, in which was a small sort of turnip, near Tauranga.

Polack, who was in New Zealand from 1831 to 1837, says: "the turnip is found in a wild state over the entire country," but he only saw a small portion of the North Island, mostly in the Kaipara and Wairoa districts. Dieffenbach, in 1839, found that the natives in Kapiti were cultivating turnips.

E. J. Wakefield found wild turnips on the site of Cook's old garden at the entrance of Queen Charlotte Sound in 1839. D'Urville in 1840 visited Otago Harbour and found turnips in all the native and the whalers' cultivations. Wilkes found wild turnips on the Auckland Islands in 1840. Wohlers met with them growing wild near Lake Ellesmere in 1844.

At the present time it is still found near old Maori settlements but is most commonly found as an escape from cultivation. (Fl., Dec. and Jan.)

Brassica Napus, Linn. Rape

No doubt introduced at an early date, but first recorded as a naturalised plant by Kirk in 1870 in the vicinity of Auckland. It only seems to occur as an escape from cultivation.

In Europe the flowers are visited by the honey-bee (*Apis mellifica*) and the drone-fly (*Eristalis tenax*).

Brassica nigra, Koch. Black Mustard

First recorded by Kirk in 1870 from the neighbourhood of Auckland. Cheeseman (1906) reports it as not uncommon in cultivated fields and waste places in both islands. (Fl., Nov. to Dec.)

In Europe the flowers are visited by *Calliphora erythrocephala*, *C. vomitoria*, *Lucilia cæsar* and *Eristalis tenax*.

Brassica adpressa, Boiss.

The record of this species in Cheeseman's *Manual* (1906) is the first I have noted; it is reported as occurring not uncommonly in fields and waste places in both islands.

Brassica Sinapistrum, Boiss. (*Sinapis arvensis*, L.)
Charlock; Skillock; Wild Mustard

No doubt introduced at an early date. First recorded in Hooker's list in 1864. Cheeseman reports it as occurring, but not common, in cultivated fields and waste places in both islands. It is, however, very common in cornfields in Otago and Southland, where it seriously reduces the yield of grain in many parts. The seed retains its vitality for a long time, especially when buried. (Fl., Nov. to Jan.)

Visited in Europe by *Calliphora vomitoria*, *Eristalis tenax* and *Bombus lapidarius*.

Brassica alba, Boiss. White Mustard

Sown in Queen Charlotte Sound in 1773 by Furneaux and Cook, and noted again by Cook in 1777. It does not seem to have become established. Polack records it as cultivated by Europeans in 1831 in the North Auckland district. It is now stated by Cheeseman (1906) to be found in cultivated fields and waste places in both islands, but not common. It is only found as an escape, or a weed of cultivation. (Fl., Nov. to Jan.)

Diplotaxis muralis, DC. Rocket; Wall Mustard

First recorded as occurring near Ashburton in 1899 by W. W. Smith, and as spreading in 1903. Cheeseman (1906) reports it as occurring in waste places in both islands, but local. (Fl., Dec. to Feb.)

Eruca sativa, Lam.

Reported by Kirk from Port Fitzroy in the Great Barrier Island. (Fl., Nov. to Dec.)

Capsella Bursa-pastoris, DC. Shepherd's Purse

No doubt an early introduction, but first recorded in Hooker's list in 1864. This is a most abundant weed of cultivation, and is found in every part of New Zealand. It is very liable to be attacked by a fungoid parasite, *Cystopus candidus*, the so-called white rust which spreads to cabbages and other cultivated crucifers. It is also a carrier of club-root. (Fl., Sept. to April.)

(Kerner states that an average-sized plant produces 64,000 seeds in a year.) The seeds emit mucus when moistened, and so may adhere to feathers of birds, etc.

Dr Cockayne says of this species that it is very variable in its natural habitat, and has already given rise to certain mutants. In New Zealand it varies to an astonishing degree, especially in highly manured ground.

Senebiera didyma, Pers. Wart-cress

An early introduction, first recorded by Hooker (as *S. pinnatifida*) in 1864. A most abundant weed in all parts of New Zealand, especially common in waste ground near the sea. (Fl., Dec. to Feb.)

Coronopus procumbens, Gilib (*Senebiera Coronopus*, Poir.). Wart-cress, Hog's-cress

Another early introduction; also recorded in 1864 for the first time. Very common in waste places in both islands. (Fl., Nov. to April.)

Lepidium Draba, Linn. Hoary-cress

First noted by the author in 1895, as occurring abundantly on Morven Hills Station, Otago. Also recorded from Ashburton by W. W. Smith. I have received specimens from J. B. Armstrong (1919), who informs me that it occurs in cultivated land near Christchurch, where it is a troublesome weed on account of its underground running stems.

Lepidium campestre, R. Br. Field-cress; Pepperwort

First recorded from the Taieri Plain, Otago, in 1873, by the author. Cheeseman (1906) reports it as occurring in cultivated fields and waste places in both islands, but not common.

Lepidium hirtum, Sm.

First recorded in 1882 as *L. Smithii*, Hook., by Cheeseman, as occurring in pastures near Alexandra, in the Waikato. It was very common in Southland in 1900; and in the *Manual* (1906) is stated to occur in both islands in cultivated fields, roadsides, etc., but local. (Fl., Dec. to Jan.)

Lepidium ruderale, Linn. Narrow-leaved Cress; Sheep's Cress

First recorded in Hooker's list in 1864. Stated by Kirk to be very common in the vicinity of every township in the Waikato in 1870.

In 1877 he states that it is abundant near Wellington, and is widely diffused by sheep. Cheeseman in the *Manual* (1906) states that it is plentiful in waste places and roadsides, especially near the sea. It is very abundant in Otago.

Lepidium sativum, Linn. Garden-cress

Polack (1831–37) mentions this species as common in a wild state all over the country, but he is not a safe guide. It is recorded in Hooker's list in 1864. It occurs in several parts but only as a garden escape.

The seeds become mucilaginous when wet, and possibly adhere to the feathers of birds. I do not know whether this is common to all species of the genus *Lepidium*.

Thlaspi arvense, Linn. Penny Cress; Mithridate Mustard; Canadian Stinkweed

The Agricultural Department reported this species in 1910, as recently introduced into New Zealand. It was promptly declared a noxious weed in the Third (Optional) Schedule of the Act of 1908, by Special Gazette Notice of 16th June, 1910.

Iberis amara, Linn. Candytuft

First recorded as a garden escape at Ashburton in 1903 by W. W. Smith. Cheeseman reports it from both islands, but "far from common." Like many a garden escape, I do not think it can hold its own away from cultivated ground. (Fl., Nov. to Jan.)

Rapistrum rugosum, All.

Cheeseman (1882) says:

> In the summer of 1876 this plant appeared in great abundance on the Barrack Hill, Auckland, now known as the Albert Park. The grading and laying out of the park during the past year has nearly destroyed it, but a few specimens still linger in the adjoining streets and unoccupied allotments.

He has not recorded it in his *Manual* (1906), so presumably it has disappeared.

Cakile maritima, Linn. Sea Rocket

Recorded by Cockayne in 1908 as occurring near New Brighton, Christchurch.

Raphanus sativus, Linn. Radish

First introduced by Furneaux and Cook in 1773, and sown in clearings in Queen Charlotte Sound, where it was found by them again in 1777. It was probably re-introduced by the early missionaries, 1814–20, for Polack records it as wild in many parts of the country

20 years later. In 1882 Cheeseman says it is thoroughly established in littoral situations, on sand-hills, etc., from Mongonui down to Thames and Raglan. It does not persist, however, and in the *Manual* (1906) is only recorded as a "garden escape, but uncommon." (Fl., Dec. to Feb.) Cockayne says of it that it "is abundantly naturalised near Wellington, but the roots are no longer swollen to any extent."

RESEDACEÆ

Reseda Luteola, Linn. Dyer's Weed; Weld

First recorded from "sand-hills below the block-house, Wanganui" by Kirk in 1877, and later from the Taieri Plain, Otago, in 1880, by the author. Cheeseman in the *Manual* (1906) states that it is not uncommon in fields and waste places in both islands. In Marlborough it is common on roadsides, in company with *Madia sativa*. Cockayne records it as common in Central Otago, and in the Waitaki Basin from Omarama to Kurow. (Fl., Dec. to Jan.)

Reseda lutea, Linn. Cut-leaved Mignonette

Petrie gathered this about 30 years ago at the Sowburn in Central Otago. In the *Manual* (1906) Cheeseman records this as occurring in fields at Pukeroro, on the authority of J. P. D. Morgan. In 1912 it was found on the slopes of Mount Eden, Auckland, by F. Neve.

Reseda alba, Linn. White Mignonette

First recorded in 1873 from Otago by the author as *R. suffruticulosa*; later found by Bishop Williams at Poverty Bay, and by Kirk in Canterbury. I have met with it as a garden weed in Otago.

CISTINEÆ

Cistus sp.

This species is now (1918) growing wild at Puketeraki in Otago, and is found in some gardens in Dunedin. The seed was brought from Gallipoli by returned soldiers.

VIOLACEÆ

Viola tricolor, Linn. Pansy; Heartsease

Probably introduced early last century. First recorded as a naturalised escape in 1871 by Armstrong from Canterbury, and by Kirk in 1877 as an occasional outcast from gardens. Cheeseman in the *Manual* (1906) says it occurs in both islands in cultivated fields and waste places, local. Guppy records that the seeds of this species, after lying a little time in water, were thickly covered with mucus, and that they adhered to a feather, on drying, as firmly as if gummed.

Viola arvensis, Murray

First recorded in 1873 from cultivated land in Otago by the author and again from Wellington by Kirk in 1877. It is not common and appears only as a weed in cultivated fields.

Viola odorata, Linn. Sweet Violet

I record this species, because innumerable attempts have been made to naturalise it in the open, but they have never succeeded. The probable explanation is that neither among the indigenous nor introduced insects is there found one which can fertilise its flowers. At the same time in my own garden in Dunedin it used to seed somewhat freely from cleistogamic flowers, but only in rather dry situations. Kirk reported it in 1877 from Ohariu, but added "possibly planted."

I have recorded, in connection with guinea-pigs, how these animals running wild in a garden, and on lawns in which violets were growing, enabled the latter to increase to a great extent. The guinea-pigs ate the grass very close, but would not touch the violets. Kerner has already pointed out that cattle, when grazing among grass which contains violets, will not eat the flowers of the latter.

Ionidium filiforme, F. Muell.

Recorded by Kirk, in the *Student's Flora* (1899), as found in grassy places near Lake Takapuna, Auckland, by Miss Rolleston.

POLYGALEÆ

Polygala myrtiflora, Linn.

First recorded by Kirk from Auckland district in 1869. Cheeseman (1906) reports it as a garden escape in several localities near Auckland; and from near Napier, on the authority of Colenso. Carse records it (1915) as found sparsely among sand-dunes on the west coast of Mongonui County.

Polygala virgata, Thunb.

Recorded in 1912 by Cheeseman from several parts of the North Island: "among fern and low tea-tree scrub at Mangatete, near Awanui (H. Carse); edge of forest near Kaitaia (Mrs Foley); in several places near Kihikiki, Waikato (N. M. Lethbridge)." Mr Cheeseman adds, "it is probably a garden escape, although I cannot learn that the species has been in cultivation in any of the localities quoted above."

Polygala vulgaris, Linn. Milkwort

Collected on the open ground above Waitati, near Dunedin, by Miss Eileen Woodhead, in November, 1917. Presumably introduced from Britain.

CARYOPHYLLEÆ

Dianthus prolifer, Linn. (*Tunica prolifera*, Cheeseman)

First recorded as a garden escape at Ashburton by W. W. Smith in 1903.

Dianthus Armeria, Linn. Deptford Pink

First recorded in 1882 from the Waikato district by Cheeseman. Perhaps *Silene Armeria* recorded by Armstrong from Canterbury in 1879, is the same species. In the *Manual* (1906) it is stated to occur in both islands in pastures and waste places, but not commonly. Cockayne reports it as fairly common in montane tussock-land in Marlborough and Canterbury. (Fl., Dec. to Jan.)

Dianthus barbatus, Linn. Sweet-william

First recorded in 1871 by Armstrong from Canterbury. It is occasionally found as a garden escape in many districts. (Fl., Nov. to Jan.)

Saponaria Vaccaria, Linn.

First recorded in 1870 by Kirk as occurring in the Auckland district. Later reported in the *Manual* (1906) on Kirk's authority from "Cultivated fields near Auckland and Wellington." (Fl., Dec. to Jan.)

Silene inflata, Sm. Bladder-campion

First recorded from the neighbourhood of Auckland by Kirk in 1870. In Cheeseman's *Manual* (1906) it is stated to occur, but not commonly, in both islands in cultivated fields, roadsides, etc. I have met with it from Whangarei to Dunedin.

Silene conica, Linn.

Reported by Petrie to occur in Otago.

Silene gallica, Linn. Catchfly

First recorded from Otago in 1873 by the author. It is now common in every part of New Zealand in roadsides and waste places. (Fl., Oct. to Jan.)

Silene quinquevulnera, Linn.

First recorded in Hooker's list in 1864, but no doubt introduced much earlier. It is one of the commonest weeds in the country. One of Drummond's Wellington correspondents stated that he had known it since 1855. (Fl., Oct. to Jan.)

Dr Cockayne remarks of *Silene anglica*, Linn. (which includes both *S. gallica* and *S. quinquevulnera*) that it develops more succulent leaves when growing near the sea than inland[1]

Silene noctiflora, Linn.

Armstrong reported this species from Canterbury in 1871, Kirk from Wellington in 1877, and Cheeseman in fields at Matamata in 1880.

The *Manual* (1906) omits all notice of the species.

Silene nocturna, Linn.

Reported by Kirk as occurring at Karori, near Wellington. (Fl., Dec.)

Silene nutans, Linn. Nodding Catchfly

First recorded in 1879 by Armstrong from Canterbury. Later by Cheeseman from pastures at Matamata in the Thames Valley[2].

Lychnis Flos-cuculi, Linn. Ragged Robin

First recorded in 1870 by Kirk from the Auckland district. Cheeseman reports it from Whangarei, and W. W. Smith from Ashburton. In Britain it is usually found in moist ground. (Fl., Oct. to Nov.)

Lychnis vespertina, Sibth. White Campion

First recorded in 1875 from Dunedin by the author, and later by Petrie from the same locality. Also from Ashburton by W. W. Smith. It is found in hedgerows and by waysides. (Fl., Nov. to Dec.)

Lychnis diurna, Sibth. Red Campion

Recorded by W. W. Smith in 1903 as occurring in Ashburton County. I have also noted it as a garden escape near Dunedin, but it does not establish itself.

Lychnis coronaria, Descr. Rose Campion

Reported to be common near Wellington in 1877 by Kirk. Recorded by Cheeseman in the *Manual* (1906) as an occasional outcast from gardens, in both islands. (Fl., Nov. to Dec.)

[1] In C. J. Cornish's charming little work *Wild England of to-day and the Wild Life in it* (London, 1895), there is an extract from an article which appeared in the *Journal of Horticulture* on Brading Harbour in the Isle of Wight, written by Mr C. Orchard. The writer says: "On two distinct places I have found the very rare *Silene quinquevulnera*, which I believe *has been found only in two or three places in England*." In New Zealand it is a common weed.

[2] Armstrong in 1879 recorded *Silene italica*, Pers., and *S. orientalis*, Linn., as occurring in Canterbury.

Lychnis Githago, Scop. Corn-cockle

First recorded in 1869 by Kirk from Auckland district; and in 1871 by Armstrong in Canterbury. It is not uncommon in cultivated fields in many parts of New Zealand, but nowhere seems to have become a pest. (Fl., Nov. to Dec.)

Cerastium glomeratum, Thuill. Mouse-ear Chickweed

First recorded in Hooker's list in 1864 as *C. vulgatum*. An abundant weed in every part of New Zealand. (Fl., Sept. to April.)

Cerastium triviale, Link. Larger Mouse-ear

First recorded in 1864 in Hooker's list as *C. viscosum*. Equally abundant with the preceding species throughout the country. Both these species (as also *Stellaria media*) are characterised by a development of strong tissue in the stem immediately above the root-attachment, which often enables them to hold on to the soil, when all the branches have been torn off. They then spring up again from the root-stock. Visited (in Europe) by *Lucilia cæsar*. (Fl., Oct. to April.)

Stellaria media, Linn. Chickweed

Introduced at a very early date. First mentioned by Dieffenbach in 1839. Most abundant weed of cultivation. (Fl., nearly all the year round.)

Stellaria Holostea, Linn. Greater Stitchwort

First recorded in 1871 by Armstrong from Canterbury Province.

Stellaria graminea, Linn. Lesser Stitchwort

First recorded in 1882 by Cheeseman as occurring near Auckland. In the *Manual* (1906) it is said to occur in fields and on roadsides in both islands, but not commonly. It is extraordinarily abundant on Pine Hill near Dunedin, where, when in flower, it gives the fields a greyish-white hue; it has also spread into many other districts near Dunedin. Visited (in Europe) by *Eristalis tenax*. (Fl., Nov. to Jan.)

Stellaria uliginosa, Murr.

In the *Manual* (1906) this is reported from bogs near Westport by Townson; and from Ruapuke Island in Foveaux Strait by Chas. Traill. It only grows in moist ground in Britain.

Arenaria serpyllifolia, Linn. Sandwort

First recorded in Hooker's list in 1864. Stated in the *Manual* (1906) to be "abundant in light dry soils." (Fl., Nov. to Jan.)

Sagina procumbens, Linn. Pearlwort

First recorded from the neighbourhood of Dunedin in 1879 by the author. Now extremely abundant in damp ground. Either this or the following species are used as the basis of bowling greens in some parts of the country, Invercargill, Gore, etc. It forms a very close fine sward. (Fl., Oct. to March.)

Sagina apetala, Linn. Pearlwort

First recorded in Hooker's list in 1864. Abundant in most parts of New Zealand. (Fl., Oct. to Jan.)

Spergula arvensis, Linn. Spurrey; Yar

First recorded in Hooker's list in 1864. A most abundant weed of cultivation, especially in the south. In damp seasons it is a serious menace to turnip crops, as it is most difficult to destroy. It thrives especially well, too, in ground which has been treated with phosphatic manures, which are the principal artificial stimulants used in New Zealand farming. (Fl., Oct. to Jan.) Visited in Europe by hive-bee (*Apis mellifica*) and drone-fly (*Eristalis tenax*).

Spergula pentandra, Linn.

Stated by Kirk to be naturalised near Wellington.

Spergularia rubra, St Hilaire. Sandwort Spurrey

First recorded from the neighbourhood of Wellington by Kirk in 1877, and from Ashburton in 1903 by W. W. Smith. Cheeseman in the *Manual* (1906) reports it as abundant on roadsides and waste places in both islands. While the indigenous species *S. media* is found near the sea, this introduced one is usually found in inland localities, according to Cheeseman. (Fl., Nov. to Feb.)

Polycarpon tetraphyllum, Linn.

First recorded in Hooker's list in 1864. Very abundant on roadsides and waste places in both islands. In 1877 Kirk recorded a varietal form which produced "hemispherical masses of deep green foliage" as abundant on the sands near Cape Palliser. (Fl., Sept. to June.)

PORTULACEÆ

Portulaca oleracea, Linn. Purslane

Apparently introduced by Cook in 1773 in the gardens made in Queen Charlotte Sound. It is not included in Hooker's list of introduced plants in 1864, but Kirk records it from the Great Barrier Island in 1867. Cheeseman in the *Manual* (1906) reports it as "abundant in warm dry soils as far south as the East Cape, rare and local from thence to Cook Strait." There is no reason to believe that it is a survival from Cook's time; it has been re-introduced much more recently. (Fl., Nov. to Jan.)

Claytonia perfoliata, Donn.

Found by myself as a garden escape near Dunedin in 1886, and still occasionally met with; also recorded from Cheviot by von Haast, and as a garden weed in Invercargill by Cockayne. It is a North American species, but I have reason to believe that it was introduced with seed or bulbs from Britain, where it is not infrequently naturalised. It does not spread, nor does it seem to thrive away from cultivated ground. In 1912 Cheeseman reports it from Karori, Wellington, collected by J. S. Tennant.

Calandrinia caulescens, H. B. K.

Cheeseman states that this species appeared in 1881 in a freshly-sown grass field at Otahuhu (Auckland). In the *Manual* (1906) he reports it as growing in "cultivated fields, rare and local" in both islands. It formerly was common in one locality near Dunedin, but has since disappeared. Kirk reported it from "near Christchurch." In 1882 he stated that "a white-flowered species has become plentiful in stony places near Penrose and thence to Onehunga."

HYPERICINEÆ

Hypericum Androsæmum, Linn. Tutsan

First recorded by Kirk in 1869 from Auckland district. In the *Manual* Cheeseman (1906) reports it as not uncommon on roadsides and waste places. But it is particularly a plant found about meadows, in the shelter of hedges, thickets, etc.

Now especially abundant by sides of water-courses and on the outskirts of forest land, where it often forms a dense undergrowth. Birds feed on the fruit and thus tend to spread it over wide areas.

Included in the Second Schedule of the Noxious Weeds Act by Special Gazette Notice of 1st October, 1903; and in Third (noxious seeds) Schedule by Gazette Notice of 10th November, 1904.

Hypericum perforatum, Linn. St John's Wort

First recorded in T. Kirk's list of Great Barrier plants in 1867. Cheeseman states that in 1882 it is common round Pirongia and threatens to become a dangerous weed. At Matamata "some old pastures have been completely over-run with it." It was first noticed in Otago by the author in 1894 in the neighbourhood of Dunedin. It is stated in the *Manual* (1906) to be abundant in both islands.

Cockayne (1920) records it as spreading greatly in the vicinity of Arrowtown, and as invading the bracken heath near Lake Wanaka. (Fl., Dec. to Jan.)

In Europe its flowers are visited by *Apis mellifica*, *Bombus terrestris*, *B. hortorum*, *Eristalis tenax* and *Calliphora erythrocephala*.

Hypericum humifusum, Linn.

First recorded in Hooker's list in 1864, and in many subsequent catalogues since. In the *Manual* (1906) it is said to be common, especially on clay soils, in both islands. (Fl., Dec. to Jan.)

Both of the above species were included (with *H. Androsæmum*) in the Second Schedule of the Noxious Weeds Act by Special Gazette Notice of 1st October, 1903; and in the Third (noxious seeds) Schedule by Gazette Notice of 10th November, 1904.

Hypericum montanum, Linn.

Recorded by W. W. Smith in 1903 as occurring in Ashburton County.

MALVACEÆ

Althæa officinalis, Linn. Marsh-mallow; Guimauve

First recorded from Ashburton in 1903 by W. W. Smith.

Lavatera arborea, Linn. Tree-mallow

First recorded by Kirk in 1869 in list of plants from Auckland district, and again by him in 1877 as an occasional garden escape in Wellington district. It does not appear to spread, though it frequently persists for a long time on the site of old gardens. (Fl., Nov. to Jan.)

Malva sylvestris, Linn. Common Mallow

First recorded by Kirk in 1869 from Auckland district. In the *Manual* (1906) it is reported as occurring on "roadsides and in waste places" but not commonly, in both islands. (Fl., Nov. to Dec.)

Flowers visited (in Europe) by *Apis mellifica*, *Bombus hortorum*, and *B. lapidarius*.

Malva rotundifolia, Linn. Dwarf Mallow

First recorded in Hooker's list in 1864. Reported by Cheeseman (1906) to be not uncommon on roadsides and in waste places in both islands. (Fl., Nov. to Jan.) A very common weed.

Malva parviflora, Linn.

Recorded in 1882 by Cheeseman as occurring in waste places near Auckland, but not common. It is now an abundant weed in roadsides and waste places throughout the country. (Fl., Nov. to Jan.)

Malva verticillata, Linn.

First recorded in 1882 by Cheeseman, as in immense abundance near Auckland; also at Thames and on the Coromandel Peninsula. It is now found abundantly in many parts, particularly in waste ground. (Fl., Nov. to Jan.)

Malva crispa, Linn.

Reported by Kirk as a garden escape at Port Waikato. (Fl., Nov. to Dec.[1])

Modiola multifida, Mœnch.

First recorded in 1860 by Kirk from the Auckland district as *Malva caroliniana*. In 1882 Cheeseman writes that it "must have been an early introduction, as it was nearly as abundant and as widely distributed in 1863 as it is now." In the *Manual* (1906) he reports it as abundant in pastures and on roadsides in both islands.

Sub-division DisciflORÆ

LINEÆ

Linum marginale, A. Cunn

Dieffenbach, in 1839, says: "The vegetation" (of Motu Narara) "is scanty and confined to a species of *Linum* with blue flowers." This was almost certainly *L. marginale*. I have met with it about Dunedin sparingly during the past forty years. Cheeseman in the *Manual* (1906) says it is generally distributed in both islands, but most plentifully in the north. It is very abundant to the north of Auckland, Whangarei, etc. (Fl., Nov. to Jan.)

The seeds of most, if not all, species of *Linum* become mucilaginous when wet. This may be to enable them to adhere to the soil, but it is probable also that they adhere to the feathers of birds, just as those of *Plantago* do.

Linum usitatissimum, Linn. Common Flax

In 1814 Marsden gave "Shunghee" (? Hongi) some English flax seed. This was at the Bay of Islands. There is no record, however, as to whether it was grown and utilised by the natives or by the missionaries. It was next recorded in 1869 by Kirk from Auckland Province. In 1877 he reports it as plentiful near an old ford of the Ruamahunga (Wellington), and in the Wairarapa. Cheeseman in 1906 records it as only occasionally seen as an escape from cultivation. It has not apparently established itself anywhere. (Fl., Dec. to Jan.)

[1] In 1879 Armstrong reported *Malva campestris*, Linn. from Canterbury.

Linum gallicum, Linn.

Cheeseman says this species was first seen in localities near Auckland in 1876. In the *Manual* (1906) he reports it as occurring in "fields and waste places as far south as the East Cape." (Fl., Nov. to Feb.)

Linum catharticum, Linn. Purging Flax; Heath Flax

First noted about 1895 by the author in Dunedin; then at Ashburton in 1903 by W. W. Smith. In the *Manual* (1906) it is reported as occurring in fields and waste places in both islands, but not commonly. (Fl., Nov. to Dec.[1])

ZYGOPHYLLEÆ

Tribulus terrestris, Linn.

In 1912 Cheeseman records the occurrence of a specimen of this plant collected by Mr F. Hutchinson in pure shingle at Port Ahuriri, Hawke's Bay.

GERANIACEÆ

Geranium Robertianum, Linn. Herb Robert

First reported from near Wellington by Kirk in 1877. Cheeseman says in 1882 that a few plants were seen at Devonport (Auckland) in 1879, but it has apparently died out. However, in 1910 he records it again from near Whangarei, and also from Auckland. In the West Taieri bush (Otago) it has been wild for quite 40 years, and is to be found in several localities round Dunedin; also near Wellington. In the *Manual* (1906) Cheeseman reports it from fields and waste places in both islands, not common. Introduced from Britain. (Fl., Dec. to Jan.)

Visited in Europe by ***Bombus terrestris***, ***B. hortorum*** and ***B. lapidarius***.

Geranium pratense, Linn. Meadow Crane's-bill

Buchanan recorded this species as growing at Kawau in 1876. It is a common enough garden plant in some districts, and may very occasionally occur as a garden escape, but it has never spread.

Geranium lucidum, Linn. Shining Crane's-bill

Recorded in 1903 by W. W. Smith as occurring in Ashburton County, but it has not been observed since.

[1] Armstrong records *Linum angustifolium*, Sin. (? Linn.), from Canterbury in 1879.

Geranium molle, Linn.

This species was first collected in New Zealand by Dr Lyall, who visited these islands in 1847–49 in H.M.S. 'Acheron,' and Sir J. D. Hooker in including it in the *Handbook of the New Zealand Flora* admitted that he was much puzzled with the plant. It was also found in Hawke's Bay by Colenso. Kirk in the *Student's Flora of New Zealand* treated it purely as a naturalised species. Cheeseman in including it among the indigenous plants in his *Manual of New Zealand Flora* says:

> There can be little doubt that this is introduced, but as it has had a place given to it in previous works on New Zealand plants, *and as it is now found in all soils and situations*, and would certainly be considered indigenous by a stranger unacquainted with its history, it appears best to retain it in the Flora.

I think it should have been relegated to the list of naturalised plants in the Appendix, where, unfortunately, it does not appear. It is very common in pastures at the present time.

This species is visited in Europe by *Apis mellifica*, *Bombus terrestris* and *Lucilia cæsar*.

Erodium cicutarium, L'Herit. Stork's-bill

First recorded in Hooker's list in 1864, var. *chærophyllum*, DC. reported by Kirk from Auckland district in 1869. It is abundant in waste places and cultivated ground throughout New Zealand. In the desiccated regions of Central Otago it is especially abundant, and, according to Petrie, is readily eaten by stock[1]. It dies off by midsummer, but seldom fails to mature plenty of seed.

In Europe the flowers are visited by *Apis mellifica* and *Calliphora erythrocephala*. (Fl., Sept. to March.)

Erodium moschatum, L'Herit. Musky Stork's-bill

First recorded by Kirk in 1869 from Auckland district; and in 1871 in Canterbury by Armstrong. In 1882 Cheeseman says "an abundant weed, especially in light soils." In the *Manual* (1906) he reports it as abundant by roadsides and in waste places in both islands. (Fl., Oct. to Feb.)

Erodium malachoides, Willd.

Originally recorded by Cheeseman, 1882, as *E. maritimum*, Linn., from Mongonui, Bay of Islands and Waiwera. Kirk states that this was naturalised at the Bay of Islands in 1867, but was not observed elsewhere, till he found it on a ballast heap in Wellington

[1] Dr Cockayne tells me that it is sown for sheep feed in denuded areas of the United States of America.

in 1892. The latter lot was introduced from Buenos Ayres. It does not seem to have established itself. In the *Manual* (1906) Cheeseman states that it occurs in sandy places near the sea at Mongonui and the Bay of Islands. (Fl., Oct. to April.)

Pelargonium zonale, L'Herit. Scarlet Geranium

Cheeseman records this in the *Manual of the New Zealand Flora* (1906) as occurring in the North Island, "often persisting for some years in deserted gardens." But I think it must be classed as a true garden escape, which has taken possession of considerable areas of waste ground in the neighbourhood of gardens, and which is therefore particularly common in the suburban districts of many northern towns. (Fl., Oct. to April.)

Pelargonium quercifolium, L'Herit.

First recorded by Kirk from Auckland district in 1869, and again from Wellington in 1877. It is, like the preceding species, a garden escape in the North Island, but not nearly so common.

Tropæolum majus, Linn. Indian Cress; Garden Nasturtium

Polack (1831–37) speaks of it as cultivated in European gardens in the north of the North Island. First recorded as a wild plant by Cheeseman in 1882 and as common near Auckland. It has become a most abundant weed in many suburban areas, especially in the North Island; and in Auckland, Wellington and other towns frequently fills waste sections of land with its attractive foliage and bright blossoms. I have measured gigantic wild specimens at New Plymouth in which the leaves were as much as eight inches across. (Fl., Jan. to April.)

It is visited in Europe by *Apis mellifica*, *Bombus hortorum*, and earwigs (*Forficula auricularia*), the latter after nectar.

Oxalis cernua, Thunb.

First recorded in 1882 from Auckland by Cheeseman as a troublesome weed in gardens. In the *Manual* (1906) it is stated to be "an occasional weed in gardens and orchards in the North Island." Cheeseman says (1917) that it "is still far too plentiful."

Oxalis compressa, Thunb.

This species was recorded by Cheeseman in 1882, as "a common garden escape, especially near Auckland." But in later years it was not observed, and so was expunged from the list published in the *Manual* in 1906.

Oxalis variabilis, Jacq.

First recorded in 1882 from Auckland as a garden escape by Cheeseman. Characterised in the *Manual* as "a garden escape, not common" in the North Island. (Fl., Aug. to Oct.)

Oxalis hirta, Linn.

Recorded by Kirk from Auckland in 1899; and by Cheeseman in 1906 in the *Manual* as "a garden escape in the vicinity of Auckland, rare." (Fl., Sept.)

Oxalis rosea, Jacquin

Recorded by W. W. Smith in 1903 as occurring in Ashburton County.

LIMNANTHEÆ

Limnanthes Douglasii, R. Br.(own)

Recorded by W. W. Smith in 1903 as occurring in South Canterbury as a garden escape. It does not spread.

AURANTIACEÆ

Citrus aurantium, Risso. The Orange

An officer of the brig 'Hawes' travelling near Tauranga in December, 1828, remarks: "I met with a few orange trees which have been introduced with success." Probably he was referring to native plantations, either under cultivation or abandoned. Oranges do not appear to become naturalised in any part of New Zealand.

AMPELIDEÆ

Vitis vinifera, Linn. Grape-vine

Grapes were introduced by the missionaries early last century. In 1838 Polack says that "they are largely cultivated to the north-ward of the River Thames." Kirk reported in 1877, and Cheeseman in 1882, that they were to be found in deserted gardens and old Maori cultivations, but apparently they were not spreading and the latter author practically repeats this in the *Manual* (1906).

SAPINDACEÆ

Melianthus major, Linn.

First recorded in 1877 as a garden escape near Wellington, but able to maintain its position when not disturbed by man. Also by Cheeseman from Auckland and Thames. Cheeseman states in the *Manual* (1906) that it is not uncommon in the North Island as a garden escape. (Fl., Aug. to Oct.)

ANACARDIACEÆ

Corynocarpus lævigata, Forst. Karaka

Though this plant is treated both by Hooker and Cheeseman as an indigenous species, it seems highly probable that it was introduced into New Zealand by some of the early Polynesian immigrants. If so, it was most likely brought from Western Polynesia by way of the Kermadecs. There are three species of the genus *Corynocarpus* all closely allied, one in New Zealand, one in New Caledonia, and one in the New Hebrides.

"Tradition says that one Roau came in the Nukutere canoe, landing at Waiaua, near Opotiki, and brought with him the *Karaka*, the *ti* and the *taro*." The date of this introduction was about twenty generations ago, or five hundred years. Cheeseman says of its occurrence: "Abundant (in the North Island) chiefly in lowland situations not far from the sea; (in the South Island) Marlborough and Nelson to Banks' Peninsula and Westland, but very rare and local." Once introduced into the country it would readily be spread by fruit-eating birds. The Maoris used the fruit for food.

Similarly the plant was almost certainly conveyed from New Zealand to the Chatham Islands. W. T. L. Travers in letters from his son H. H. Travers (1871) learned that the natives stated that their Maori (not Moriori) ancestors brought the tree with them. It is found "growing abundantly in the immediate neighbourhood of the various old settlements, but not in the general bush of the islands[1]."

Sub-division Calycifloræ

LEGUMINOSÆ

Papilionaceæ

Lupinus arboreus, Sims. Tree Lupin

Freely sown as a plant for sand-binding in many parts of New Zealand, but first recorded from Ashburton in 1903 as a garden escape. Cheeseman in 1906 reports it as increasing in some localities. It has got very commonly into many river beds in Canterbury and elsewhere, and has spread very considerably away from sandy areas where it has been sown. Indeed in many localities it assumes all the characters of a "pure formation."

It was included under the name of *Lupinus luteus*, among noxious seeds in the Act of 1900; and in the Second Schedule of the Act by Special Gazette Notice of 20th June, 1901. It is difficult to understand why this was done. The plant is a nitrogen-fixer, and where it has taken possession of great areas of sand-hills, it produces valuable surface soil on which other plants afterwards grow freely. (Fl., Dec. to Jan.)

[1] Dr Cockayne says this statement is incorrect, and that Karaka is the dominant tree of the lowland forest in all parts of Chatham Island.

Dr Cockayne states that this species, normally yellow, and varying but little in its native land, has undergone many changes on the dunes near New Brighton, Canterbury, in the colour of its flowers. There are, e.g., a pure white, yellows of various tints, and a great variety of purples combined, or not, with whites and yellows. These abnormally coloured plants occur in patches here and there as a general rule, and appear to get more abundant year by year. Such variations have not been noticed by him in the North Island, nor in Central Otago.

Ulex europæus, Linn. Gorse; Whin; Furze

No doubt an early introduction. When Darwin was at the Bay of Islands in 1835 he says: "At Waimate I saw gorse for fences; five years ago nothing but the fern flourished here."

Some of the early settlers in New Zealand sowed Gorse for sheep-feed. About 1890 a Mr Williams sowed a pretty large area of it in drills, and the sheep managed to keep it down for a few years, but ultimately, of course, it got ahead of them, and so filled the ground. At the time of his sowing gorse was well established over the country. This was at Pakaraka in the Bay of Islands district.

The plant spreads by means of its elastic seed vessels which throw the seeds to a considerable distance. It seeds very freely in New Zealand, but in spite of its brilliancy and attractiveness of colour I do not think its flowers are often cross-fertilised. I have examined long stretches of gorse hedges in full bloom; on one occasion a hive-bee was seen on the flowers, apparently gathering pollen; and a few blow-flies (*Calliphora*) visited them, but I never saw a humble-bee on them. In 1893 it was a very common weed in many parts of New Zealand, and to-day is most abundant. Petrie informs me that the seeds are freely eaten and distributed by Californian quail. It would be interesting, however, to ascertain whether the seeds are ground up by the birds in the process of digestion, or whether they escape trituration and are passed undigested.

This species was declared a noxious weed in the Second Schedule of the Act of 1900.

In damp seasons in Auckland and the districts to the north of it, gorse is a much more leafy and less spinous plant than it is, for instance, in Otago; it flowers from August to April in the north; but in the south it is found to bloom right through the winter.

Dr Cockayne points out that some remarkable more or less hereditary variations have come about in this species, such as colour changes from normal yellow to white, differences in shape and size of flowers, and variation in the time of blooming. He also notes that it does not spread at above 2000 ft. altitude.

In 1901 W. W. Smith stated that "the larvæ of several species of *Elater* have destroyed enormous areas of gorse fences in New Zealand during the last ten years by consuming the roots of the plants."

Drummond (Jan. 1916) records that gorse is decreasing in the Upper Waitemata. He later states that on Gouland Downs, south of Collingwood, an early settler sowed four sacks of gorse seeds. It has not thriven, the soil being perhaps too moist and swampy, and only a few stunted plants are now to be seen.

In a letter in *Nature* of 26th September, 1918 (p. 65), it is stated that gorse seed, buried for 25 years, has sprung up freely on land in Cumberland which was cleared of gorse and heather in 1893. For the preceding ten years or more, the land was in permanent pasture, but was ploughed and recultivated in 1918. I had the same experience in the neighbourhood of Dunedin. A gorse hedge was rooted out in 1876, and the ground has since been in continuous cultivation as a garden. Whenever extra deep digging was done, as late as 1908, gorse seedlings used to appear occasionally.

As early as 1859 gorse and broom began to give trouble to farmers and others, and the Provincial Legislatures of Taranaki and Nelson passed restrictive ordinances, compelling private individuals to keep their hedges pruned, to stop planting new hedges, and fining them for any plants growing on the public roads.

Cytisus scoparius, Link. Broom

No doubt an early introduction, but first recorded specifically by Armstrong in 1871 from Canterbury. It is now widely spread throughout the country, and in some districts covers wide areas of land to the exclusion of everything else. It also forms—along with gorse—immense shelter for rabbits. (Fl., Oct. to Jan., but sporadically throughout the year.)

In the Act of 1900, it is declared to be a noxious weed in the Second Schedule.

The flowers are visited, for pollen only, by *Apis mellifica*, *Bombus terrestris*, *B. hortorum* and *B. lapidarius*.

This species, like the last, exhibits numerous variations from type, and these appear to be more or less hereditary in character.

Cytisus capensis

Recorded from Canterbury in 1871 by Armstrong.

Cytisus albus, Link. White Broom

First recorded from Ashburton (as *C. albidus*), where it was common as undergrowth to the belts of trees skirting the railway lines, in 1903 by W. W. Smith. It was also growing freely on the flats on the Ashburton River. It is now common on some Canterbury river beds. In the *Manual* (1906) it is reported as occurring in both islands as "an occasional escape from gardens." (Fl., Sept. to Oct.)

Cytisus candicans, Lam.

Apparently first recorded by Cheeseman in the *Manual* (1906). It is abundant in both islands. (Fl., Sept. to Nov.)

Ononis arvensis, Linn. Rest-harrow; Wild Liquorice

Introduced into Southland in 1870 among grass-seed, but did not succeed in establishing itself.

Medicago sativa, Linn. Lucerne; Alfalfa; Purple Medick

Probably introduced very much earlier, but first recorded in 1882 by Cheeseman as occurring in cultivated fields near Auckland, but not common. Abundantly cultivated in many parts of the country but only occurs naturalised as an escape. (Fl., Sept. to Jan.)

The flowers (in Europe) are visited by *Apis mellifica*, *Bombus terrestris*, *B. hortorum* and *B. lapidarius*.

Medicago lupulina, Linn. Black Medick

First recorded in Hooker's list in 1864, but certainly introduced at a much earlier date. Now very abundant in fields and waste places. (Fl., Sept. to Feb.)

Medicago denticulata, Willd. Toothed Medick

Also first recorded in Hooker's list in 1864. Abundant throughout all cultivated districts. (Fl., Sept. to Jan.)

This and the next species are included among the noxious seeds in Schedule 3 of the Noxious Weeds Act, 1900.

Medicago maculata, Willd. Spotted Medick

First recorded by Hooker in 1864. Cheeseman in the *Manual* (1906) says: "Abundant in the Auckland Provincial District, local elsewhere." It is a most abundant weed in many parts of Otago, especially in the coastal districts. (Fl., Sept. to Dec.)

Melilotus officinalis, Lam. Melilot

First recorded in Hooker's list in 1864; no doubt introduced very many years earlier. Cheeseman in the *Manual* (1906) reports it as sparingly naturalised in fields and waste places in both islands. (Fl., Dec. to Jan.)

Melilotus arvensis, Wallr. Field Melilot

First recorded by Hooker in 1864. Now occurs commonly in waste places and on roadsides throughout New Zealand. Cockayne says it is extremely common in sand-dunes in many parts of the North Island. (Fl., Dec. to Feb.)

Melilotus alba, Desr. White Melilot

First recorded in 1871 by Armstrong as *M. leucantha* in plants naturalised in Canterbury. Reported from Napier by Kirk and Cheeseman; and from the Canterbury Plains by W. W. Smith. (Fl., Jan. to March.)

Trifolium subterraneum, Linn.

Reported by Cheeseman in 1906 as occurring in the Auckland district, and increasing rapidly. I found it common in the Whangarei district in 1916.

Trifolium arvense, Linn. Hare's-foot Trefoil

First recorded in 1879 from Canterbury by Armstrong. Cheeseman in 1882 says it was observed in a field at Otahuhu in 1876, but has not been seen since. In the *Manual* (1906) he reports it as found on roadsides and in waste places in both islands, and increasing, especially in light soils. (Fl., Jan. to March.)

In Europe the flowers are visited by *Bombus lapidarius*.

Trifolium incarnatum, Linn. Crimson Clover

First recorded in 1877 from Porirua by H. B. Kirk; then in 1882 by Cheeseman as occurring occasionally in pastures, especially in the Waikato. It is found in pastures in both islands, but not commonly. In the *Journal* of the Department of Agriculture for 1910 it is said to be largely used for ploughing into the ground for green manuring. (Fl., Dec. to Feb.)

Trifolium ochroleucum, Linn. Sulphur Clover

First recorded in 1879 by Armstrong from Canterbury; and stated by Smith to occur about Ashburton in 1903.

Trifolium pratense, Linn. Red Clover

No doubt an early introduction into New Zealand, but first recorded by Hooker in 1864. It is a most abundant plant in all parts of New Zealand to-day, but before 1885 when humble-bees were successfully imported into the country, it was only found where it had been sown, and was not spreading to any great extent, except where carried in hay and straw, and by horses.

In 1883 J. B. Armstrong stated that a certain proportion of the flowers were self-fertile. There are certainly cases on record of small yields of red clover seed before 1885, but probably the explanation is to be found in another statement made in 1883, also in Canterbury, when R. W. Fereday recorded the fact that the flowers were often fertilised by moths of the Family Noctuidæ. Since humble-bees became common, red clover has not only become permanent in pastures, but has spread far and wide throughout the country. (Fl., Nov. to Feb.)

Mr W. Hone of Waverley (Aug. 1914) says that more than 40 years ago, Mr J. Dickie, sen., sowed red clover on a part of his land near Waverley from which he obtained a large crop of very fertile seed.

Mr W. W. Smith states that hive-bees occasionally fertilise red clover in the shorter flowers of its heads.

Darwin in a paper printed in the *Ann. and Mag. of Nat. History* for December, 1858, says:

In an old number of the *Gardener's Chronicle*, an extract is given from a New Zealand newspaper, in which much surprise is expressed that the introduced clover never seeded freely until the hive-bee was introduced.

I wrote to Mr Swale of Christchurch, in New Zealand, and asked him whether leguminous plants seeded there before the hive-bee was introduced; and he, in the most obliging manner, has sent me a list of 24 plants of this order which seeded abundantly before bees were introduced. And as he states that there is no indigenous bee (perhaps this statement applies to bees resembling hive-, or humble-bees, for some other genera are known to inhabit New Zealand) the fact that these plants seeded freely at first appears quite fatal to my doctrine. But Mr Swale adds that he believes that three species of a wasp-like insect performed the part of bees, before the introduction of the latter; unfortunately he does not expressly state that he has seen them sucking the flower. He further adds a remarkable statement, that there are two or three kinds of grasshoppers which frequent flowers; and he says he has repeatedly watched them "release the stamens from the keel-petal," so that, extraordinary as the fact is, it would appear the grasshoppers, though having a mouth so differently constructed, in New Zealand have to a certain extent the habits of bees. Mr Swale further adds that the garden varieties of the Lupine seed less freely than any other leguminous plants in New Zealand; and he says, "I have for amusement during the summer months released the stamens with a pin; and a pod of seed has always rewarded me for my trouble, and the adjoining flowers not so served have all proved blind." The case of the lupine in New Zealand not seeding freely now that bees have been introduced may be accounted for by the fact, if I dare trust my memory, that in England this plant is visited by humble-bees, and not by hive-bees.

It is impossible, at the interval of over half a century, to verify or refute the accuracy of Mr Swale's statements, but hive-bees were introduced into the country in 1839 and 1840, and probably earlier, and they were certainly very abundant in many parts of the South Island early in the fifties. I am not aware, either, of any other observer who has recorded the fertilisation of flowers in New Zealand by grasshoppers.

This statement about lupines seeding is re-quoted by Herman Müller in *The Fertilisation of Flowers*, and by Henslow in *The Origin of Floral Structures*.

My wife informed me that long before humble-bees were introduced into the country, garden lupines used to seed freely in the Taieri district near Dunedin.

Guppy quotes the following from Darwin (*More Letters of Charles Darwin*, I, p. 436), regarding the distribution of clover seed: "Out of a number of seeds left in the stomach of an eagle for eighteen hours, the majority were killed; but amongst the few that germinated afterwards was a seed of clover (*Trifolium*)."

In Europe, *Trifolium pratense* is visited by *Bombus terrestris*, *B. hortorum*, *B. lapidarius* and *B. ruderatus*, the short-trunked humble-bees perforating the tubes of the flowers, and stealing the nectar; after which the flowers are visited by hive-bees which suck the remaining nectar through the holes made by the humble-bees.

Dr Cockayne has recently pointed out that red clover and cow-grass (the perennial form of *T. pratense*) vary to an astonishing extent. Many of the forms are most distinct, and the new characters are diverse, affecting colour of flowers, stems, and foliage, form of inflorescence, degree and kind of hairiness, and general habit.

Trifolium medium, Linn. Meadow Clover; Zigzag Clover

First recorded in 1870 by Kirk from the Auckland district; then from Wellington in 1877. In the *Manual* (1906) it is reported as not uncommon in pastures and meadows in both islands. (Fl., Dec. to Feb.)

Trifolium scabrum, Linn. Rough Clover

Reported as abundant on the beach at Devonport, Auckland, in 1880, by Cheeseman. In the *Manual* (1906) it is reported as occurring locally in the North Island in pastures and waste places. (Fl., Nov. to Jan.)

Trifolium glomeratum, Linn. Clustered Clover

First recorded in 1870 from the Auckland district by Kirk, then from Wellington in 1877. Cheeseman reports it (1906) as plentiful in pastures and waste places in both islands. (Fl., Nov. to Dec.)

Trifolium repens, Linn. White Clover; Dutch Clover

No doubt introduced at an early date last century, but first recorded in Hooker's list in 1864. It spread through Otago and Southland very rapidly after the opening of the goldfields, being scattered far and wide by the teams supplying the diggings in the early sixties. (Fl., Oct. to March.)

In Europe the flowers are visited by *Apis mellifica*, *Bombus lapidarius*, *B. terrestris* and *B. hortorum*.

A. R. Wallace in *Darwinism* states the "White Clover in New Zealand is sometimes displacing native species, including even the native flax (*phormium tenax*)." This story has been repeated many times since, but, as most botanists now know, the statement made by Wallace was based on defective information. Flax has been destroyed by fire and by cattle, and clover has taken its place, but in the absence of agencies introduced by man, the latter never does and apparently never can displace the former.

Clover seeds appear to pass undigested through the stomachs of cattle, for it is not uncommon to find them germinating on old patches of cow dung.

Trifolium hybridum, Linn. Alsike Clover

Must have been introduced much earlier, but first recorded by Cheeseman in 1882 as occurring in clover fields in the Waikato. It is now common in pastures and meadows throughout the country, but it is questionable whether it has spread to any extent naturally. (Fl., Jan. to March.)

Trifolium fragiferum, Linn. Strawberry Clover

The first reference to this species appears to be in Cheeseman's *Manual* in 1906, where it is recorded as occurring in "fields and waste places in the Auckland district, rare." H. Carse records it from Mongonui in 1910.

In Europe the flowers of this species are fertilised by *Apis mellifica*, *Bombus lapidarius* and *B. terrestris*.

Trifolium resupinatum, Linn. Reversed Clover

First recorded in 1882 by Cheeseman from Mongonui, Doubtless Bay, and Helensville. In the *Manual* (1906) he reports it as occurring in "fields and waste places in the North Island; very plentiful in the North Cape district, and increasing elsewhere." (Fl., Oct. to Jan.)

Trifolium agrarium, Linn.

First recorded in Hooker's list of 1864 as *T. procumbens*. Also reported from Broken River Basin in Canterbury, by Kirk. (Fl., Dec. to Jan.)

Trifolium procumbens, Linn. Hop Trefoil

First recorded in Hooker's list in 1864. Probably introduced into the country in the early days of settlement, now abundant in all parts of the Dominion. (Fl., Nov. to Feb.)

Trifolium dubium, Sibth. Yellow Suckling

This has mostly been recorded as *T. minus*. It is abundant throughout the whole of the cultivated parts of New Zealand. (Fl., Oct. to March.)

Trifolium filiforme, Linn. Lesser Trefoil

First recorded in 1867 as *T. minus* in Kirk's list of Great Barrier plants. According to Cheeseman (1906) this species occurs, but not commonly, in "various localities in Otago and Southland." (Fl., Dec. to Jan.[1])

Anthyllis vulneraria, Linn. Kidney Vetch

First recorded as occurring in Southland in 1869 by the author. According to Kirk, it is "sparingly naturalised near Nelson and Dunedin." (Fl., Dec. to Jan.)

Lotus corniculatus, Linn. Bird's-foot Trefoil

First recorded in 1864 in Hooker's list. Certainly introduced among grass and clover seeds at an early date. It occurs not uncommonly in fields and on roadsides in all parts of the country. Flowers freely in December to January in Otago.

In Europe the flowers are fertilised by *Apis mellifica*, *Bombus lapidarius*, *B. terrestris*, *B. hortorum* and *Eristalis tenax*.

Lotus uliginosus, Schkuhr. Greater Bird's-foot Trefoil

First recorded in Cheeseman's *Manual* (1906) as occurring abundantly in fields and waste places in both islands, especially in the Auckland district. It is probable that this is the species recorded in 1864 in his list of introduced plants in the Auckland district as *L. major*. (Fl., Jan. to Feb.)

Lotus angustissimus, Linn. Slender Bird's-foot Trefoil

Cheeseman records this as first seen at Remuera in 1881. In the *Manual* (1906) he reports it as occurring in several localities near Auckland, but rare. (Fl., Jan. to Feb.)

Psoralea pinnata, Willd.

Recorded in 1870 by Kirk as occurring in the Takapuna district, Auckland. He adds: "as it seeds freely, small specimens are not

[1] Armstrong in 1879 recorded *Trifolium maritimum*, Hudson, and *T. maculatum* (?), from Canterbury. Both identifications are doubtful.

uncommon in the neighbourhood of gardens, deserted homesteads, etc., where it can scarcely expect to become fully naturalised." In 1912 Cheeseman records it (on the authority of H. Oakley) as occurring wild at Waipu, and threatening to become a nuisance.

Indigofera viscosa, Lam.

First recorded by Kirk as occurring in the Auckland district in 1870. Cheeseman says (1906), "has been noticed as a garden escape near Auckland, but is scarcely naturalised."

Galega officinalis, Linn. Goat's Rue

According to Cheeseman this was found in the Manawatu river bed by H. J. Matthews about 1900. It occurs on the silt in the river, from whence it has spread considerably, down stream and into the adjacent county.

Robinia Pseud-acacia, Linn. Locust tree, False Acacia

First recorded by Kirk in 1869 as going wild in the Auckland district. In 1870 he states that "it is abundantly naturalised at Taupiri, Rangiriri, and other places" (in the Waikato), "it already forms coppices in many localities." In the Wellington district in 1877, he calls it "a mere garden or plantation escape, increasing rapidly by suckers where undisturbed." In the *Manual* (1906) it is stated to be naturalised in various localities between Auckland and the Upper Waikato. The hive-bee fertilises this flower in Europe.

Coronilla varia, Linn.

First recorded by Cheeseman in the *Manual* (in 1906) as a garden escape in the vicinity of Nelson.

Onobrychis sativa, Lam. (*O. viciæfolia*, Scop.) Sainfoin

First recorded in 1871 by Armstrong as naturalised in Canterbury. In the *Manual* (1906) it is stated to occur in both islands as an occasional escape from cultivation, but scarcely naturalised. (Fl., Dec. to Jan.)

Ornithopus perpusillus, Linn. Bird's-foot

First recorded in 1906 by Cheeseman as occurring in newly-sown grass paddocks at Brookby, Auckland.

Vicia tetrasperma, Mœnch. (*Onobrychis gemella*, Crantz). Slender Tare

First mentioned in Hooker's list in 1864. In the *Manual* (1906) it is said to be not uncommon in roadsides, hedges, etc., in both islands. It is very abundant in many parts of Otago. (Fl., Nov. to Jan.)

In Europe the flowers are visited by *Apis mellifica.*

Vicia gracilis, Loisel

Found by me in the Taieri Plain, Otago, in 1874, and identified by Kirk. The species does not seem to have established itself, and has not been found since.

Vicia hirsuta, S. F. Gray. Common Tare; Hairy Tare

First recorded in Hooker's list in 1864. In the *Manual* (1906) it is stated to be not uncommon on roadsides and in waste places in both islands. (Fl., Nov. to Feb.)

Vicia Cracca, Linn. Tufted Vetch

Originally recorded by Kirk as occurring near the Opawa River, Marlborough. I have specimens received from Mr J. B. Armstrong (1919) gathered on roadsides near Christchurch, where, however, it is not very common. (Fl., Jan. to Feb.)

Vicia sativa, Linn. Common Vetch

First recorded in Hooker's list in 1864, but probably introduced at an early date into the country. It is now a very abundant plant in cultivated fields, waste places and roadsides. (Fl., Nov. to Jan.) In Europe the flowers are fertilised by *B. terrestris*, *B. hortorum* and *B. lapidarius.*

Vicia gemella, Crantz (see *V. tetrasperma*)

Recorded by Carse in 1910 as occurring in Mongonui County.

Vicia lutea, Linn.

Recorded by W. W. Smith in 1903 as occurring in Ashburton County.

Vicia narbonensis, Linn.

First recorded by Kirk in 1867 as occurring at Port Fitzroy in the Great Barrier Island, in considerable quantity. (Fl., Dec.)

Pisum sativum. Common Pea.

Cook sowed peas in the gardens in Queen Charlotte Sound in 1773, but on returning in November of the same year, it was found that the rats had apparently eaten the plants up. Governor King gave the natives at the Bay of Islands two bushels of peas in 1793. They were probably next introduced by the missionaries (1814 onwards); and Bellingshausen in 1820 once more gave them to the Maoris of

Queen Charlotte Sound, and showed them how to sow and grow them.

The garden pea has never gone wild in New Zealand.

A. R. Cruise, writing in 1820, says:

at one place we found a number of people collected round an object, which seemed to attract great attention, and which they told us when we entered the circle was tabooed. It proved to be a plant of the common English pea, and had been growing about two months. The seed that produced it had been found in the Coromandel; it was fenced round with little sticks, and the greatest care appeared to be taken of it.

Elsewhere he says "peas were raised while we were in the country with great success, and the people promised to save the seeds and grow them again."

Faba vulgaris, Linn. Broad-Bean

The history of the introduction of the bean into New Zealand is practically the same as that of the pea; the species is nowhere wild, and it does not even occur as a garden escape.

An interesting fact regarding this species is recorded by W. O. Focke (quoted by Guppy):

A pigeon killed by some beast of prey was found in his garden in the early winter. In the following spring he noticed numerous seedlings of *Vicia faba* sprouting up from amongst the feathers that alone remained of the bird.

Guppy considers this to be the normal method of the dispersal of the Leguminosæ by birds.

Phaseolus vulgaris, Linn. Kidney-Bean

First sown in Queen Charlotte Sound in 1773, but it was not perpetuated. Occasionally runner-beans are found near gardens, but the species never spreads.

Dolichos lignosus, Linn.

First recorded by Kirk in 1870 as forming thickets in the neighbourhood of Auckland.

Dolichos Lablab, Linn.

Kirk recorded this species in 1899 as occurring on the sites of deserted homesteads from Auckland to Wellington. (Fl., Oct. to Feb.)

Lens esculenta, Mœnch. Common Lentil

First recorded in 1882 by Cheeseman as a garden escape, abundantly naturalised in Auckland Domain. In 1906 it is recorded again from the same spot, but it "does not spread."

Lathyrus odoratus, Linn. Sweet-pea

First recorded by Kirk in 1869 as an escape in the Auckland district; no doubt introduced at a very much earlier date; and again in 1877 as a garden escape, near the Hutt. I do not think it has become permanently naturalised anywhere. Cheeseman in 1906 says: "an occasional garden escape in rich warm soils in the North Island, but soon disappears."

Lathyrus latifolius, Linn. Everlasting Pea

First recorded in 1871 in Canterbury by Armstrong. According to Cheeseman (1906) it occurs in the North Island as "an occasional garden escape." (Fl., Dec. to Jan.)

Lathyrus Nissolia, Linn.

Recorded by W. W. Smith in 1903 as occurring in Ashburton County.

Lathyrus pratensis, Linn.

Recorded by W. W. Smith in 1903 as occurring in Ashburton County[1].

Sub-order MIMOSEÆ

Acacia decurrens, Willd. Black Wattle

First recorded by Kirk in 1869 as growing wild in some parts of the Auckland district. It is the species now chiefly grown for bark in the North Island. In 1897 there were some 4500 acres planted for this purpose. Cheeseman in the *Manual* (1906) says it has established itself in several localities. (Fl., Nov.)

According to E. Maxwell (of Opunake) the typical N. S. Wales form of this species spreads somewhat freely by seed, both in Auckland and in Taranaki; it is common in the poor land of the centre of the North Island, and in the Waihi district. The cultivated form is *A. decurrens*, var. *mollis*, of Tasmania and Victoria.

Acacia dealbata, Link. Silver Wattle

The early plantations of wattles planted for bark were of this species, which, however, is comparatively valueless for this purpose. First recorded as a naturalised species in various parts of Auckland by Cheeseman in 1882, where he says (1906) it is now established in several localities. (Fl., July to Sept.)

Acacia pycnantha, Benth. Golden Wattle

This species has been commonly planted in some wattle plantations, but though its bark yields more tannin than many other species,

[1] A fine red *Lathyrus* sp. is common in several parts of the Wellington-Taranaki railway line.

the plant is too bushy in habit and hence costs more to strip. It does not seem to spread to any great extent.

Acacia melanoxylon, R. Br. Blackwood

This species spreads to some extent by means of its seeds, as well as its suckers, e.g. in Taranaki.

Acacia longifolia, Willd.

Recorded by Cheeseman in 1896 as occurring in the North Cape district.

Acacia armata, R. Br. Kangaroo Acacia

Mostly grown as a hedge plant, but in some districts has spread and become a nuisance. Consequently it has been included in the Second Schedule of the Noxious Weeds Act 1900, by Special Gazette Notice of 23rd March, 1905. In 1910 H. Carse records it from Mongonui.

Albizzia lophantha, Benth. Brush Wattle

First recorded by Kirk in 1869 among plants from the Thames Goldfields. It is naturalised in many localities, especially in the Auckland and Taranaki districts, and is common in many parts of the North Island. In the South it is only a cultivated plant. Urquhart says that it "competes successfully against, and in time destroys, almost the strongest vegetation met with in the open country." (Fl., Aug. to Sept.)

ROSACEÆ

Amygdalus Persica, Linn. Peach; Nectarine

Apparently peach trees were first introduced into the Bay of Islands by Marsden in December, 1814. Cruise, writing in 1820, says: "The missionaries have got some peach trees that bear very well."

The Maoris soon scattered the seed far and wide, so that it early established itself as a wild species, for they shifted their cultivations frequently. The mate of the brig 'Hawes' in 1828 found peaches in the native cultivations near Tauranga. Polack somewhat later writes as follows:

> On a small farm I possess in the Kororarika Bay, two peach trees had been planted on the place by an early missionary. These had been allowed to grow wild for many years, but yet produced, in 1839, thousands of fruit, almost unequalled in size and flavour. This farm contained at one period nearly one hundred small trees, growing spontaneously from seed carelessly strewed about without having been planted.

He also refers to nectarines as in European gardens at the same period.

In 1869 Kirk records it as common in the Auckland district. On the introduction of the peach curl and other fungoid pests, the wild trees became badly infected in many districts, and almost ceased to bear fruit. Cheeseman in the *Manual* (1906) says: "Copiously naturalised in the Auckland Provincial District in the early period of settlement, but at the present time rarely spreads out of cultivation." (Fl., Aug. to Sept.)

Prunus Cerasus, Linn. Cherry

Probably introduced early last century by the missionaries and early settlers. Recorded as a common wild plant in Auckland in 1869 by Kirk, and in Canterbury in 1871 by Armstrong. Cheeseman in the *Manual* (1906) says: "Maintains itself in old Maori plantations and deserted orchards, sometimes forming small groves." It tends to spread in bush districts through the agency of birds.

Prunus avium, Linn. Gean

In 1871 Armstrong recorded this plant as naturalised in Canterbury, where it was probably only an occasional garden escape. In 1903 W. W. Smith reported it as dispersing freely in old bush districts. It is being scattered here and there by birds.

Prunus communis, Hudson, var. *spinosa*, Linn. Sloe; Blackthorn

Found rarely as a garden escape in one or two localities near Dunedin, where it is distributed by birds.

Rubus idæus, Linn. Raspberry

No doubt introduced at the beginning of last century at the dawn of settlement. Polack in 1838 speaks of it as overrunning the districts where it was planted in the North Island. But Polack is not a very trustworthy guide, for Cheeseman as late as 1906 in the *Manual* records it as an occasional escape from cultivation in both islands, not common. Like all succulent fruited plants, it becomes more widely spread in districts where fruit-eating birds are abundant. (Fl., Oct. to Nov.)

Rubus fruticosus, Linn. Bramble; Blackberry

Some forms of the blackberry were no doubt introduced at an early date by the first settlers. The first definite record appears to be in 1864, when Hooker's list was drawn up, which includes two sub-species *R. discolor*, Weihe, and *R. rudis*, Weihe. According to Cheeseman in the *Manual* (1906) other sub-species which occur are *R. leucostachys*, Smith, *R. rusticanus*, Weihe, and *R. macrophyllus*, Weihe. Kirk adds *R. laciniatus*, Willd. The last-named—the Italian blackberry—is the common form about Greymouth.

In many parts, especially of the North Island—though it is abundant enough in the South also—this is considered to be the most deleterious weed in New Zealand. All sorts of methods of coping with it have been suggested, and in Taranaki especially, herds of goats have been established to eat it down. But it continues to spread everywhere, and the increase of fruit-eating birds is everywhere accompanied by an increase of this pest.

Quail are reported from several localities in the North Island to be very active agents in disseminating it. (Fl., Dec. to Jan.)

Blackberry was one of the three plants declared a noxious weed without any reservation in the original Act of 1900.

Fragaria vesca, Linn. Wild Strawberry

Probably introduced by very early settlers. Polack in 1838 says: "they over-run the soil on which they are planted." Dieffenbach, in 1839, speaks of it as spreading over the country (at Matamata). Cheeseman in the *Manual* (1906) says "an occasional garden escape, not common." But I think it is rather more than that. Smith speaks of it (1903) as "abundant in masses in many parts of the Ashburton district, and I know that near Dunedin it is somewhat freely distributed by blackbirds and thrushes. Cockayne says it is common amongst manuka near Taumaranui. (Fl., Oct. to Dec.)

Fragaria elatior, Ehr. Hautbois Strawberry

Kirk reports this species as wild in the Auckland district in 1869. Cheeseman in his *Manual* (1906) says: "Has been noticed as a garden escape, in both islands, but is much rarer than the preceding." One reason of this I believe is that the species is often physiologically (though not structurally) diœcious. (Fl., Nov. to Jan.)

Herman Müller states that in the United States cultivated species of *Fragaria* incline to diœcism. The same thing is certainly true here, and I have seen beds of fine healthy strawberry plants which were covered with bloom in the spring, but which did not set a single fruit.

A writer in the *Otago Witness* on 31st October, 1892, complained bitterly that strawberry growing was handicapped by the ravages of the goldfinches. He stated that the birds picked the seeds out of the growing fruit, and thus completely destroyed the berries.

In Europe the flowers are visited by *Apis mellifica* and *Musca corvina*.

Potentilla reptans, Linn. Cinquefoil

First recorded from Wellington in 1877 by Kirk, then from Canterbury in 1879 by Armstrong, and by Cheeseman as occurring near Hamilton in the Waikato. In the *Manual* (1906) it is stated to occur by roadsides and in pastures, but to be local. (Fl., Nov. to Jan.)

(Kerner states that it is mainly distributed by the seeds adhering along with earth to the feet of birds.)

Alchemilla arvensis, Scop. Parsley Piert

First recorded in Hooker's list in 1864. Kirk also records it from Wellington in 1877. In the *Manual* (1906) Cheeseman reports it as occurring in waste places and fields in both islands, "often abundant in light dry soils." (Fl., Oct. to Feb.)

Agrimonia Eupatoria, Linn. Agrimony

Recorded by W. W. Smith in 1903 as occurring in Ashburton County.

Acæna ovina, A. Cunn

First observed by H. B. Kirk, and by Buchanan in 1877 in the neighbourhood of Wellington. Observed by the author on the hills above the Taieri Plain in 1890; the Agricultural Department reported it as spreading freely in 1897, and six years later W. W. Smith recorded it as very abundant in the hills of S. Canterbury, and becoming a great nuisance to sheep-owners owing to the manner in which the seed-heads adhere to wool. In the *Manual* (1906) Cheeseman says it is not uncommon in fields and waste places in both islands. Cockayne reports it as one of the common roadside weeds of lowland Marlborough. (Fl., Dec. to Feb.)

Poterium Sanguisorba, Linn. Salad Burnet

First recorded in 1876 from Kawau by Buchanan, then by Kirk in 1877 from hills near Castle Point. Cheeseman in the *Manual* (1906) reports it as occurring in dry pastures in both islands, but not commonly. (Fl., Dec. to Jan.)

Poterium polygamum, Waldst. and Kit. Burnet

Reported by Kirk as occurring near Lake Ellesmere and elsewhere on the Canterbury Plains.

Poterium officinale, Hook. f. Great Burnet

Buchanan reported this from Kawau in 1876. It has not been recorded again from there. It is now growing abundantly on the sand-dunes at (North) Otago Heads.

Rosa canina, Linn. Dog Rose

Introduced much earlier, but first recorded as a naturalised plant in Hooker's list in 1864. Cheeseman in the *Manual* (1906) reports it as not uncommon on roadsides and in waste places.

The flowers are fertilised in Europe by *Apis mellifica*. (Fl., Nov. to Jan.)

Rosa rubiginosa, Linn. Sweet-briar

Darwin, writing of the Bay of Islands in 1835, says "At Pahia, it was quite pleasing to behold the English flowers in the gardens before the houses; there were roses of several kinds, honeysuckle, jasmine, stocks, and whole hedges of sweet-briar." Polack, writing in 1838, says it is "acclimated" in New Zealand, i.e. in the far north, which was the only part he saw. The early settlers everywhere planted this favourite shrub, as a hedge plant, and everywhere, at least in the North Island, it got away from cultivation and quickly established itself, as a plant most difficult to eradicate. As fruit-eating birds increased it increased more rapidly, but there is little doubt that it was largely spread by horses, which will eat the hips, but are unable to digest the hard-walled achenes. Hooker mentioned it in his list of introduced plants in 1864, and it occurs in every subsequent collector's lists. It was early recognised as a great pest, and is now most abundant in all parts of the country, though it is a worse pest in the North Island than in the South Island. Cockayne says it is generally absent from wet areas, where the blackberry is the prevailing weed.

In the Noxious Weeds Act of 1900, this was one of the three plants which was declared a noxious weed without any reservation; the others being the blackberry and the Canadian (or Californian) thistle.

It is fertilised in Europe by *Apis mellifica* and *Bombus terrestris*.

Rosa multiflora, Thunb.

First recorded as a wild species in 1869 in the Auckland district by Kirk. In the *Manual* it is stated to be found "often lingering for years in deserted gardens, etc."

Cratægus oxyacantha, Linn. Hawthorn; Whitethorn

Probably introduced at an early date, but began to be scattered as soon as thrushes and blackbirds increased. First recorded as a naturalised plant by W. W. Smith in 1903. Cheeseman (1906) says that it is "scarcely naturalised, but seedlings sometimes appear in the vicinity of planted hedges." In the West Taieri (Otago) I have seen the scarlet mistletoe (*Loranthus colensoi*) growing freely on this plant. It is being spread by blackbirds. (Fl., Oct. to Nov.)

Visited in Europe by *Apis mellifica*, *Bombus terrestris*, *Eristalis tenax*, *Lucilia cæsar* and *Anobium paniceum*; the latter perhaps for pollen.

(See Appendix B, p. 562.)

Pyrus communis, Linn. Pear Tree

Probably introduced by the missionaries into the Bay of Islands and Hokianga districts. It has not gone wild anywhere in New Zealand. In Europe the flowers are pollinated by *Apis mellifica*,

Bombus terrestris, *B. lucorum*, *Calliphora erythrocephala*, *Musca domestica*, *M. corvina* and *Eristalis tenax*.

Pyrus malus, Linn. Apple

Introduced, no doubt, along with the pear. Polack (1831–37) speaks of it as cultivated in European gardens. It is, however, nowhere naturalised. In Europe the flowers are visited by *Apis mellifica*, *Bombus terrestris*, *B. hortorum*, *B. lapidarius*, *Calliphora vomitoria*, *Musca domestica*, *Lucilia cæsar* and *Eristalis tenax*.

Cydonia vulgaris, Quince

Polack records it as in cultivation in European gardens (1831–37) to the northward of the Thames River. My son, G. Stuart Thomson, tells me that it grows wild on the river banks in some parts of the Hokianga district.

SAXIFRAGEÆ

Ribes Grossularia, Linn. Gooseberry

Probably introduced early last century by the missionaries. First recorded as an escape in Canterbury in 1871 by Armstrong. In 1877 Kirk recorded it as not unfrequent in forests in the Wellington district, and added: "probably originating from seeds carried by birds."

It is particularly common in the South Island, where it is carried about by birds. In 1899 the bush about Moeraki contained a great deal of it, and it has been since found abundantly and increasingly about Dunedin and Otago Harbour, and wherever there are blackbirds and thrushes in the neighbourhood of settlement. The period of germination of the seeds of *Ribes* is found to be hastened by passing through the alimentary canal of a thrush. It is well known also that seeds of a gooseberry are not digested by passing through the human alimentary canal, but germinate freely from the fæces.

W. W. Smith tells me that in Taranaki the roots of gooseberry plants are very greatly damaged by the larvæ (wireworms) of an indigenous beetle (*Ochosternus zealandicus*).

In Europe the flowers are visited by *Apis mellifica*, *Bombus terrestris*, *B. lapidarius*, *B. lucorum*, *Eristalis tenax* and *Calliphora erythrocephala*.

Red and white currants (*R. rubrum*) and black currants (*R. nigrum*), though introduced at as early a date—probably—as the gooseberry, do not seem ever to have occurred even as garden escapes. Cockayne reports it from Kennedy's Bush, Port Hills, Canterbury.

Ribes sanguinea, Purch

Recorded by W. W. Smith in 1903 as naturalised in Ashburton County.

CRASSULACEÆ

Tillæa trichotoma, Walp.

First recorded in 1882 by Cheeseman as "growing on the sides of the road near Penrose (Auckland), and spreading rapidly on the lava-fields around Mount Stuart." Also reported from Wanganui by E. W. Andrews. In 1915 Carse reports it as found near Kaitaia by H. B. Matthews. (Fl., Sept. to Oct.)

Sedum acre, Linn. Biting Stonecrop; Wall-pepper

Recorded by W. W. Smith in 1903 as occurring in the Ashburton County; apparently introduced in unclean seed, and spreading on farms. A. H. Cockayne states that there are miles of it in some of the Canterbury river beds. Dr Cockayne says it is increasing in dry parts of the South Island.

MYRTACEÆ

Eucalyptus amygdalina, Labill. "Stringy-bark' of New South Wales; "Peppermint-tree" of Victoria and Tasmania

Maxwell reports this as spreading in Taranaki.

It is one of the species which suffered from *Eriococcus* in South Canterbury in 1900.

Eucalyptus regnans, F. v. M.

Simmonds reports this as spreading in the North Island. Also suffered from scale in 1900.

Eucalyptus coccifera, Hook.

Badly attacked by *Eriococcus* in 1900. I do not know whether or not this species has spread naturally.

Eucalyptus radiata, Sieb. (*E. numerosa*, Maiden)

At Waitati, near Dunedin, this species seeded into ground which had been covered with manuka and burned. Numerous self-sown trees sprung up, which in 1913 were 28 to 29 inches in diameter.

Eucalyptus obliqua, Lher. "Stringy-bark" of Victoria, South Australia and Tasmania

This species spreads from plantations about Auckland (Simmonds), and Taranaki (Maxwell).

Eucalyptus piperita, Sm. Peppermint Gum; Stringy-bark

Urquhart reports this as spreading from plantations in the neighbourhood of Auckland.

Eucalyptus hæmastoma, Sm. Black-butt; Spotted Gum

Urquhart reports this as spreading from plantations in the Auckland province.

Eucalyptus globulus, Lab. Blue Gum

Seedlings of this species frequently spread from plantations in both islands. It is recorded by Simmonds from Auckland. I have noted it frequently in Otago and Canterbury. In Taranaki the foliage of young plants is very much destroyed by a beetle, a Tasmanian (introduced) species of *Metriorhynchus*, probably *M. rufipennis*.

This species suffered very badly from *Eriococcus coriaceus* in Canterbury and North Otago in 1900–3, numbers of trees from 40 to 80 feet in height being completely destroyed.

Eucalyptus Gunnii, Hook. f.

One of the species which suffered much from scale in 1900–3. Spreading to a limited extent from plantations in the South Island.

Eucalyptus Stuartiana, F. Muell.

One of the species which suffered badly from scale in South Canterbury about 1900. I have not heard of it spreading.

Eucalyptus Macarthuri, Dean and Maiden

Simmonds reports this as spreading in the North Island.

Eucalyptus viminalis, Labill

According to Simmonds this species spreads from plantations in the North Island.

Eucalyptus rostrata, Schlecht. White Gum; Red Gum

Seedlings of this species spread from plantations in the North Island.

LYTHRARIEÆ

Peplis Portula, Linn. Water-purslane

Petrie found this species in various localities in the east of Otago; at Clarke's Flat, etc. (Fl., Dec. to Feb.)

Lythrum hyssopifolia, Linn. Loosestrife

First recorded in Hooker's list in 1864. In 1870 Kirk speaks of it as abundant in every moist place in the neighbourhood of Auckland. In the *Manual* (1906) it is stated to occur in moist places, ditches, etc., in both islands, abundantly naturalised. (Fl., Dec. to Feb.)

Lythrum Græferi, Tenore

First recorded by Kirk in 1869 from the Auckland district. In 1882 Cheeseman states that it occurs at Remuera, Ngaruawahia and

Thames. It has also been reported from Greymouth by Helms. In 1915 Carse reports it from Mongonui County.

GRANATEÆ

Punica Granatum. Pomegranate

Polack (1831–37) recorded pomegranates as cultivated in European gardens to the north of the Thames River.

ONAGRARIEÆ

Œnothera biennis, Linn. Evening Primrose

First recorded in 1879 in Canterbury by Armstrong. In 1882 Cheeseman reports it as occurring, but not commonly, near Auckland and Hamilton, and in abandoned Maori cultivations near Matamata. In the *Manual* (1906) it is stated to occur, but not commonly, by roadsides and in waste places in both islands. (Fl., Jan. to Feb.) In Europe the flowers are visited by *Apis mellifica*, *Bombus terrestris*, *B. lucorum* and *B. hortorum*.

Œnothera odorata, Jacq. Evening Primrose

First recorded in Hooker's list in 1864 as *Œ. stricta*. In 1868 Kirk reports it as forming a compact turf on the sands at Matauri Bay, North of Auckland; and in 1870 as chiefly confined to volcanic hills where it is abundant. In 1877 he recorded the species from Wairarapa and Wanganui. In 1882 Cheeseman reports it as common throughout Auckland Province in light soils, especially on sandy flats near the sea. Common on some of the river beds in the Canterbury Plains. (Fl., Jan. to March.)

Œnothera tetraptera, Cav.

Cheeseman records this as first seen as a garden escape near Auckland in 1878. It does not seem to have been met with anywhere else[1].

CUCURBITACEÆ

Lagenaria vulgaris, Ser. Gourd; Hue

Banks, in his *Journal* of Cook's first voyage (1773), records in the native plantations "a plant of the cucumber kind, as we judged from the seed-leaves which just appeared above ground"; and later on he adds: "they also cultivate gourds, the fruits of which serve to make bottles, jugs, etc."

According to the traditions of the east coast natives the *hue* was

[1] In 1879 Armstrong recorded *Œnothera grandiflora* from Canterbury.

introduced before the Kumara, and was said to have been cultivated by the Toi tribes prior to the arrival of the Arawa and other canoes. This would place its introduction at a distance of from 24 to 30 generations, that is from 600 to 750 years ago. The species is cultivated everywhere throughout Polynesia at the present time, and though it is being displaced by many European varieties of pumpkins, marrows and gourds it is still found in numerous native cultivations. Several varieties of it were grown.

According to Taylor the natives used to wrap the seeds in wet rags and suspend them near the fire to hasten their germination.

Citrullus vulgaris, Schrad. Water-melon

The first mention of this plant which I can find is by R. A. Cruise in 1820, who says that "they were raised with great success while we were in the country, and the people promised to save the seeds, and sow them." They were probably introduced by the missionaries shortly before that date. In 1828 the mate of the brig 'Hawes' found water-melons in native cultivations at Tauranga. Polack reports it as occurring in gardens in 1831–37. It was recorded as a garden escape by Kirk in 1869 in the Auckland Province, but he adds: "it can hardly be said to hold its ground." In 1882 Cheeseman says: "often of spontaneous (?) origin about Maori cultivations, but has never permanently established itself." And again in the *Manual* (1906) he states that it "occasionally lingers in old Maori cultivations, but scarcely naturalised."

Cucurbita Pepo, Linn. Pumpkin

In 1820 Bellingshausen gave the natives of Motuara seeds of this species. Polack in the far north in 1837; and Dieffenbach when visiting New Plymouth in 1839, found pumpkins in native cultivations.

Cucurbita ovifera. Vegetable Marrow

This was recorded by Polack as cultivated in gardens in 1837.

Cucumis sativus, Linn. Cucumber

Also recorded by Polack (1831–37) as cultivated in gardens in the far north.

None of these three species appear to have even survived as garden escapes. All were probably introduced by the missionaries.

FICOIDEÆ

Mesembryanthemum edule, Linn. Pig-face; Hottentot Fig

First recorded in 1822 by Cheeseman, as naturalised on sandy beaches near Auckland. In the *Manual* (1906) it is stated to be "often planted to check the advance of drifting sands, and spreading in several localities, especially near New Plymouth." It is very widely spread over the country, but is only naturalised in the South Island in a few localities, e.g. in the Canterbury sand-dunes.

CACTEÆ

Opuntia vulgaris. Prickly Pear; Indian Fig

E. Jerningham Wakefield in 1840 found prickly pears in gardens at Hokianga; they had, no doubt, been planted there by the missionaries. Drummond states that it was introduced into Mongonui in 1847, by Mr Walter Brodie. It has, fortunately, shown no tendency to run wild in New Zealand, as it has done in parts of warm temperate and sub-tropical Australia.

UMBELLIFERÆ

Hydrocotyle vulgaris, Linn. Penny-wort; White-rot

Recorded by W. W. Smith in 1903 as occurring in Ashburton County. It has not been collected since.

Conium maculatum, Linn. Hemlock

According to Armstrong this species was sown by a herbalist in Christchurch in 1865. The next record is from Wellington where T. Kirk found it in 1877. Cheeseman states a few plants were observed in some waste ground at the Thames in 1880, but were not to be found by him two years later. In 1895 the Agricultural Department reported three patches of the plant in Wellington. W. W. Smith reported it from Ashburton in 1903, "but not common." It was common in places near Dunedin in 1914, but was apparently not so abundant in 1919. Its occurrence, then, seems to be somewhat erratic.

It flowers in Oct. and Nov. In Europe it is visited by *Musca domestica* and *M. corvina*.

This species was included in the Second Schedule of the Noxious Weeds Act by Special Gazette Notice of 27th March, 1902; and in the Third Schedule by Special Gazette Notice of 10th November, 1904.

Bupleurum rotundifolium, Linn. Hare's-ear; Thorough-wax

Stated in the *Manual* (1906) to occur in cultivated fields and waste places in the North Island; Cheeseman reporting it from near Auckland, and Kirk from Wellington. I have also a specimen from Christchurch, where it is very rare, sent me by J. B. Armstrong (1919). (Fl., Dec. to Jan.)

Bowlesia tenera, Spreng.

Recorded by Kirk as occurring on a ballast heap in Wellington in 1895. Does not seem to have increased or re-appeared.

Apium graveolens, Linn. Celery

Celery was probably introduced by the early missionaries about 1820 or earlier. Polack (1831–37) speaks of it as found in a wild state over the entire country, but he is not a very trustworthy authority. Kirk in 1877 reported it from Wellington as "an occasional garden escape; soon dying out." In the *Manual* (1906) Cheeseman states that "it has established itself in several districts in marshy places near the sea," in both islands. (Fl., Dec. to Jan.)

Apium leptophyllum, F. Muell.

First recorded in 1882 by Cheeseman as occurring in several localities in the Auckland district. In the *Manual* (1906) said to occur in waste places from Mongonui to Wellington, but not commonly. (Fl., Nov. to March.)

Apium Lessonii

Recorded by Carse in 1910 as occurring in Mongonui County.

Ammi majus, Linn.

First recorded in 1822 by Cheeseman as occurring rarely near Auckland, but likely to spread. Evidently this prediction was not realised, for in the *Manual* (1906) he states it occurs in "waste places near Auckland, rare."

Carum Carui, Linn. Caraway

Though introduced long ago, this plant has not spread out of cultivation. It is only recorded by A. Hamilton as occurring "near Dunedin, rare."

In Europe the fly *Lucilia cæsar* has been observed on the flowers.

Carum Petroselinum, Benth. Parsley

First introduced into Queen Charlotte Sound in 1773 by Captain Cook, but it did not establish itself then. It was probably re-introduced early last century, and has spread to a slight extent. Cheeseman in the *Manual* (1906) reports it as occurring not uncommonly on roadsides and in waste places in both islands. (Fl., Dec. to May.)

In Europe the flowers are visited by *Eristalis tenax*, *Musca domestica*, and *Lucilia cæsar*.

Pimpinella saxifraga, Linn. Burnet-saxifrage

Reported by Kirk in 1869 as occurring in the Auckland district. It has failed to establish itself, and has not been recorded again.

Scandix Pecten-Veneris, Linn. Shepherd's Needle; Venus's Comb

First recorded in Hooker's list in 1864, and frequently recorded as occurring in various parts of New Zealand. W. W. Smith stated that it was very abundant about Ashburton in the eighties, but had become rare by 1903. Cheeseman in the *Manual* (1906) records it as occurring in waste places in both islands, but far from common. (Fl., Dec. to Feb.)

In Europe the flowers are visited by *Eristalis tenax*.

Anthriscus cerefolium, Hoffm. Chervil

Recorded in Hooker's list in 1864 as *Chærophyllum cerefolium*. Has not been observed since.

Anthriscus vulgaris, Pers. Beaked Parsley

This species is found at Wyndham, Otago, and has been forwarded to me by Mr Warden (1917).

Chærophyllum temulum, Linn. Chervil

First recorded in 1908 by me in the neighbourhood of Dunedin, where it occurred as an escape from cultivation. It does not seem to increase.

Fœniculum vulgare, Mill. Fennel

First recorded in Hooker's list in 1864, and frequently since then. It is one of the commonest weeds of waste ground in New Zealand, especially in the neighbourhood of towns, but is much more abundant in the North Island than in the South. (Fl., Jan. to April.)

Included in the Second Schedule of the Noxious Weeds Act of 1900, by Special Gazette Notice of 26th May, 1904; and in the Third Schedule (Noxious Seeds) by Gazette Notice of 10th November, 1904.

Ligusticum scoticum, Linn. Lovage

Found as a garden escape by the author in Dunedin in 1905; still occurs about the Town Belt, but is not spreading.

Peucedanum sativum, Bentham. Parsnip

In 1773 Captain Cook sowed parsnips in his garden clearings in Queen Charlotte Sound. They were probably re-introduced at the Bay of Islands by the missionaries about 1820 or earlier, and were given by them to the Maoris, for an officer of the brig 'Hawes,' walking towards Tauranga in 1828, noticed them growing in the native cultivations. Cheeseman speaks of it in 1906 as "an occasional escape from cultivation, but not common." I think where it does occur in the wild state it is fairly abundant. It is found in all cultivated districts of New Zealand. (Fl., Jan. to March.)

In Europe the flowers are visited by *Apis mellifica*, *Bombus terrestris*, *B. hortorum*, *Eristalis tenax*, *Calliphora erythrocephala*, *C. vomitoria* and *Lucilia cæsar*.

Cockayne writes:

The parsnip, probably the celebrated "student," which is supposed by writers on evolution to be a fixed race (Romanes writes, "that is to say, it has come true to seed for the last forty years"), came up year by year in a neglected part of my garden, but in a much deteriorated form.

Daucus Carota, Linn. Carrot

Carrots were first introduced by Cook in 1773, in Queen Charlotte Sound. The plant apparently disappeared there, for Bellingshausen gave the natives of Motuara more seeds of it in 1820. But in 1839, E. Jerningham Wakefield found carrots growing wild in the neighbourhood of Cook's old gardens at the entrance of Queen Charlotte Sound. They were probably introduced about the same time at the Bay of Islands by the missionaries, for they were found in 1828 in native cultivations by the same narrator as is mentioned in the preceding note. Wohlers mentions the plant as growing wild near Lake Ellesmere in 1844 on the site of former Maori cultivations. W. W. Smith reported it as covering acres of waste land about Ashburton in 1903. As Cheeseman stated in 1906, it is not uncommon in fields and roadsides, but it has not spread to any great extent, and is never far from cultivated land. It is, however, very abundant in parts of South Taranaki. (Fl., Dec. to March.)

Caucalis nodosa, Scop. Hedge-parsley

First recorded by T. Kirk in introduced plants of Great Barrier Island in 1867, as *Torilis nodosa*, and from other parts of the Auckland Province in 1870. I found it common at Moeraki in 1899. It occurs in many parts of both islands. (Fl., Sept. to Nov.)

ARALIACEÆ

Hedera Helix, Linn. Ivy

I have no record of the first introduction into New Zealand, but it is first mentioned by Cheeseman in 1882 as spreading in plantations and gardens, but hardly considered by him as naturalised. It flowers very freely in the south in May, June and July, rather later than the corresponding flowering season in Britain, which is Oct. to Nov.

In Europe the flowers are visited by *Eristalis tenax*, *Calliphora erythrocephala* and *C. vomitoria*.

CORNEÆ

Benthamia fragifera, Lindl. Strawberry tree

W. W. Smith informs me (1916) that this species is becoming freely naturalised in Taranaki; the fruit being carried about by birds.

Division GAMOPETALÆ

Sub-division Epigynæ

CAPRIFOLIACEÆ

Sambucus nigra, Linn. Elderberry

First recorded by Hooker in 1864. In 1882 Cheeseman speaks of it as "often planted for hedges, etc., and sometimes spreads." Blackbirds and thrushes had been introduced more than a dozen years previously, but had not had time to make much change in the existing flora. At the present time it is a serious menace in some bush districts. The Town Belt of Dunedin is full of it, and it is getting into many of the scenic reserves of the country, and is crowding out all other undergrowth. On rocky ground in Central Otago it is a useful plant. Kerner records that the seeds of this species passed through the alimentary canal of a thrush which was fed on the fruit germinated in half an hour. It flowers in Oct. and Nov. In Europe it is visited by *Eristalis tenax* and *Lucilia cæsar*. The elderberry was included in the Second Schedule of the Noxious Weeds Act of 1900 by Special Gazette Notice of 26th May, 1904.

Sambucus racemosus, Linn. Red-berried Elder

This species is growing abundantly and spreading in a wooded gully at the back of the schoolhouse at Romahapa, south of the Clutha River. The locality is close to the site of the old Puerua Manse, founded about 1848 by the Rev. Wm. Bannermen, and apparently the plant is an escape from cultivation. At present the species is being spread freely in the neighbourhood by blackbirds and thrushes.

Viburnum vulgare, Linn.

Recorded by W. W. Smith in 1903 as occurring in Ashburton County.

Leycesteria formosa, Wall. Cape Fuchsia; Spiderwort

First recorded by Kirk in 1877 from several districts near Wellington, where he says, "if left undisturbed, it would increase rapidly." Cheeseman in the *Manual* (1906) speaks of it as an occasional garden escape. But it has been known in cultivation in New Zealand for over 60 years, and since blackbirds and thrushes have become common, has been spread very widely.

It was declared a noxious weed in the Third (optional) Schedule of the Act of 1908, by Special Gazette Notice of 23rd July, 1914.

RUBIACEÆ

Galium verum, Linn. Bed-straw

First recorded by Cheeseman in 1907 as occurring at Mahurangi. It does not seem to have spread.

Galium palustre, Linn.

Reported from swamps near Mauku, Manukau Harbour, by H. Carse (in *Manual*, 1906).

Galium Mollugo, Linn.

Reported as occurring between the Manukau Harbour and the Waikato River, by H. Carse (in *Manual*, 1906).

Galium Aparine, Linn. Goose-grass; Cleavers

First recorded from the Auckland district in 1870 by Kirk, and again in 1877 from Wellington. Now a very abundant weed in waste ground, hedges, etc., in most parts of both islands. (Fl., Jan. to March.)

Galium parisiense, Linn.

First recorded by Cheeseman in 1882 as occurring in fields at Remuera, Auckland, but rare. In the *Manual* (1906) it is reported from Whangarei by Kirk, and Motueka by Kingsley. Also again from the vicinity of Auckland by Cheeseman.

Sherardia arvensis, Linn. Field Madder

First recorded in Hooker's list in 1864. A very abundant weed in all cultivated land throughout New Zealand, especially common in pastures and waste land. (Fl., Dec. to Feb.) In Europe the flowers are visited by *Eristalis tenax*.

VALERIANEÆ

Centranthus ruber, DC. Spur-valerian

First recorded by Kirk in 1877 from several parts in the Wellington district, and by Cheeseman in 1822 as an occasional garden escape at Mongonui, Auckland and Thames. It is fairly common in many suburban areas. (Fl., Nov. to April.)

Valerianella olitoria, Pollich. Lamb's Lettuce; Corn Salad

First recorded in Hooker's list in 1864 as *Fedia olitoria.* In 1882 reported by Cheeseman as occurring in waste places and roadsides at Orakei, Auckland and Hamilton. It does not seem to have spread, and has not been recorded from the South Island. In 1915 it is recorded by Carse from Mongonui County, as "not common." (Fl., Sept. to Oct.)

CALYCEREÆ

Acicarpha tribuloides, Juss.

Recorded by Kirk as occurring on a ballast heap in Wellington in 1895. Does not seem to have spread or re-appeared.

DIPSACEÆ

Dipsacus sylvestris, Mill. Wild Teasel

Kirk, in 1877, states that it occurs in great abundance in the Porirua Valley near Wellington, while Cheeseman records it as not uncommon in Tauranga in 1880; also at Pakari. In the *Manual* (1906) it is stated to occur in waste places in the North Island, but not commonly. I found it common about Marton in 1920. (Fl., Feb. to March.)

Dipsacus Fullonum, Linn. Fuller's Teasel

This plant was introduced into the Taieri Plain, and cultivated for a time for the sake of the woollen mills. It spread somewhat from its original habitant, but failed to establish itself, and soon died out.

Scabiosa arvensis, Linn. Field Scabious

First recorded in 1870 from Southland by the author. It did not spread in the south. In 1882 Cheeseman writes: "a few years ago this appeared in abundance in a cultivated field at Remuera but has since nearly died out." In the *Manual* (1906) it is stated that it occurs (but rarely) in fields near Auckland.

Scabiosa maritima, Linn.

First recorded by Kirk in 1868 from Mongonui as *S. atro-purpurea.* In 1877 reported by Kirk as a garden escape from Wellington and Wairarapa; and in 1882 stated by Cheeseman to be a common garden escape in light soils in Auckland Province. In the *Manual* (1906) Cheeseman adds Mongonui, Bay of Islands and Tauranga to the list of localities. (Fl., Jan. to April.)

COMPOSITÆ

Arctium Lappa, Linn. Burdock; Burr-thistle

First reported by Colenso in Hawke's Bay in 1882. It rapidly spread across the North Island from Hawke's Bay to the Foxton district. In 1894 it was very common about Hunterville, and was also reported from localities in Canterbury and Otago. Kingsley reported it as common in Nelson and Marlborough.

In the *Manual* (1906) it is said to be not uncommon in waste places from the East Cape, southwards, in both islands. (Fl., Feb. to April.)

It was declared a noxious weed in the Second Schedule of the Act of 1900, and burdock (of any species) was included among noxious seeds, and honoured with a second declaration of its undesirable character by special Gazette Notice of 20th June, 1901[1].

Centaurea nigra, Linn. Knapweed

First recorded in 1869 by Kirk as occurring in Auckland. Reported in 1903 by W. W. Smith as being spread freely through the Ashburton district by goldfinches, which are very fond of the seed. In the *Manual* (1906) it is stated to occur in fields and waste places, but not commonly. (Fl., Feb. to March.)

Centaurea cyanus, Linn. Cornflower

First recorded by Armstrong in 1879 from Canterbury. Also reported from Ashburton by W. W. Smith in 1903. Elsewhere it is scarcely even a garden escape. Goldfinches are also very fond of the seeds of this species.

In Europe this flower is visited by *Apis mellifica* and *Bombus lapidarius*.

Centaurea Calcitrapa, Linn. Star-thistle

First recorded in Hooker's list in 1864. In 1895 it was reported by the Agricultural Department as occurring in Canterbury, Wellington, Hawke's Bay and Auckland. Cheeseman records it in the *Manual* (1906) as not uncommon in fields and waste places in both islands. It flowers from Dec. to Feb. In Europe the flowers are fertilised by *Bombus lapidarius*.

[1] No species of carline-thistle has been recorded, as far as I can find out, by any botanist in New Zealand, as a naturalised plant, but in the list of thistles in the Second Schedule of the Noxious Weeds Act, 1900, "*Carlina* (Stemless thistle), any species" is declared a noxious seed, and is also included in the Second Schedule by Special Gazette Notice of 20th June, 1901.

Centaurea melitensis, Linn. Cockspur Thistle; Malta Thistle

In 1895 T. W. Kirk reported this species from Canterbury. It has not been recorded since.

Centaurea Solstitialis, Linn. Yellow Star-thistle

First recorded in 1864 in Hooker's list, and frequently in subsequent lists for the North Island. Cheeseman reports it in 1822 as spreading fast, especially in the Waikato. The author noted it in 1885 in the Clutha Valley. In 1895 the Agricultural Department noted it as occurring in many parts of the colony, as on the increase in pasture lands, and sometimes occurring in cornfields. In the *Manual* (1906) it is reported as occurring in fields and waste places, not uncommonly, in both islands. (Fl., Feb. to April.)

In the Act (Noxious Weeds) of 1900, *Centaurea* (any species) is included among noxious seeds; and it is placed in the Second Schedule of the Act by Special Gazette Notice of 20th June, 1901.

Onopordon Acanthium, Linn. Cotton Thistle; Scotch Thistle

Early introduced into the country, but first recorded from Ashburton in 1903 by W. W. Smith, who stated that it was very widely spread, and its seeds were scattered chiefly by goldfinches.

Petasites vulgaris, Desf. Butter-bur

Noted as a garden escape by the author in 1882 in the neighbourhood of Dunedin. An auctioneer about that date advertised a plant sale, and as a special attraction notified a number of clumps of *Chatham Island Lily*. The author went to examine the plants before the sale, and found that they were bunches of *Petasites*. The auctioneer withdrew them from sale, and later reported that the owner had thrown them out. The species had nowhere established itself as a wild plant, but wherever it has been thrown out from a garden, it is apt to increase rapidly by its underground stems. Reported from Ashburton in 1903 by W. W. Smith. This species is fertilised in Europe by *Apis mellifica*, *Bombus terrestris* and *B. lucorum*.

Cynara Cardunculus, Linn. Cardoon

First recorded by Cheeseman in 1896 from the North Cape district. In 1906 reported as occurring in waste places in the North Island, not common. In the South Island it is only known in cultivation.

Carduus nutans, Linn. Musk-thistle

Reported from Eastern Otago (Pomahaka) by Kirk in 1899. (Fl., Dec. to Jan.)

Carduus pycnocephalus, Linn. Slender Thistle

First noted by the author in 1894 as abundant near Oamaru, and also in the West Taieri district. Cheeseman in the *Manual* (1906) records it as not uncommon in fields and waste places in both islands. (Fl., Jan. to Feb.)

Its seeds are a common impurity in certain farm seeds, especially oats. This species was included in the Second Schedule of the Noxious Weeds Act by Special Gazette Notice of 12th October, 1905.

Speaking of the grass-denuded districts of Central Otago, Mr D. Petrie says of this species:

Wherever it is plentiful it affords a very considerable bulk of highly nutritious feed. There are experienced runholders who reckon it little inferior to rape. The young plants that shoot up after the earliest autumn rains form the main and almost the sole winter feed in the desert lowlands. The old dry and withered stems are also completely eaten out at this season of scarcity. A second growth from more dormant seeds usually starts up in early spring, so that its utility is not restricted to the hardest time of the year. Though the earlier plants are so closely eaten back that they would hardly be expected to flower, they later on send out secondary shoots from the axils of the stem-leaves, and then these produce enough seed to renew the crop. In other circumstances the winged thistle might be ranked as a "noxious weed," but it is the runholders' sheet-anchor in extensive areas of Central Otago and elsewhere.

Carduus crispus, Linn. Curled Thistle

Recorded by the Agricultural Department as growing near Christchurch in 1911.

All species of *Carduus* were included in the list of noxious seeds in the Noxious Weeds Act of 1900; and they were also included in the Second Schedule by Special Gazette Notice of 20th June, 1901.

Cnicus lanceolatus, Willd. Common Thistle; Spear Thistle

This species was no doubt introduced into the country at a very early date, but is first recorded in Hooker's list in 1864. It is abundant everywhere. In the earliest days of cultivation, especially in such districts as Canterbury and Otago, when large areas were cleared of native vegetation and brought under the plough, the thistle was extraordinarily abundant, and took absolute possession of the soil at first. It assumed, temporarily, all the characteristics of a "pure formation." I passed through hundreds of acres of newly ploughed land in the Oamaru district in 1873, when the thistles covered the ground to a height of 6 ft. and it was only possible to get through where cart-tracks had been made, and the growth was not more than 3 ft. high. Sometimes this dense growth prevailed for a second season, but usually it was possible to burn off the dead thistles, destroying enormous quantities of seeds at the same time, when it was found that the roots had penetrated a considerable depth into the sub-soil, opening it up and leaving a large quantity of vegetable matter in it. On reploughing and sowing wheat on such land it was quite common to take 60 bushels or more per acre in the first season.

On 14th April, 1864, Mr Macandrew asked the Provincial Secretary of Otago (in the Provincial Council):

whether or not the Government is aware of the fact that part of the Maori Reserve at the Heads, as well as some of the public lands reserved from sale adjoining thereto, are totally overrun with thistles, and if so, whether the Government purposes to take any steps towards their eradication?

In reply it was stated that the Government Gardener would make a report. This was brought up at the Council on 13th December, 1865. (It would appear that this report was made by a special committee.)

After stating that thistles were spreading, the report says:

In the northern part of the Province, it appears that it (the thistle plague) commenced about six years ago, at the Otepopo Bush and has spread towards Moeraki and along the sea coast northwards; and is now making its way to a distance of more than nine miles inland, generally avoiding the high lands, and following the course of the rivers and all the lowlands.

To the south it is fast coming down to Waikouaiti from Mt Charles, which is at the bank of the Otepopo, and from the Peninsula it will be in the Taieri in two years if not immediately checked.

The growth of the Thistle is in some places 5 or 6 ft. high, and as much as ten feet wide, being quite impervious to animals, and densely covering large patches of ground varying from ten to thirty acres in extent. When it is considered that in one patch of ground it commenced with *three thistles*, and in the short space of three years ten acres have been densely covered, the magnitude of the evil cannot easily be exaggerated[1].

The seeds of this thistle are a constant impurity among farm seeds, especially in oats, and all species of *Cnicus* are included among noxious seeds in the Noxious Weeds Act of 1900. They are also included in the Second Schedule of the Act by Special Gazette Notice of 20th June, 1901.

The species is abundant in all parts from North Cape to Stewart Island, and ranges up to the snow line. Any bush land that is recently felled and burned is liable to be immediately overrun by this thistle.

Jas. Drummond states (Jan. 1916) that the plant has died out in the Upper Waitemata.

[1] There is an amusing story current in the far north of the first appearance of this species. A Mr McInnes, an old Scotch settler at Kauri hohore, near Whangarei, found one on his farm, and he was so pleased at seeing it, that he took great care of the plant, and prevented it from being injured by either man or beast. No doubt, this was done in many other parts of the country in the very early days of settlement, but few of the instances of such care are recorded.

(Fl., Jan. to March.) In Europe the flowers are fertilised by *Apis mellifica*, *Bombus lapidarius*, *B. terrestris*, *B. hortorum*, *Eristalis tenax*.

Cnicus eriophorus, Roth. Woolly Thistle

Reported from the Upper Wairarapa by Kirk in 1899. (Fl., Jan. to Feb.) Has not been recorded since.

Cnicus palustris, Linn. March Thistle

Recorded in 1911 from Westland and Southland by the Agricultural Department.

Cnicus arvensis, Hoffm. Creeping Thistle; Californian Thistle; Canadian Thistle

The first record of this weed is in 1893 by R. I. Kingsley, who reported it from the neighbourhood of Nelson, but it must have been introduced much earlier. It came sporadically all over the country in cultivated fields, creating great alarm in the minds of farmers. Some tried to grub it out and so spread its rhizomes all over their cultivations. Others built straw-stacks over the patches of weed, while others sowed the ground with salt, thus temporarily arresting its progress. But it rapidly spread throughout the whole country, becoming particularly abundant and aggressive in half-cleared bush areas. Though still looked on as a dangerous weed, it is readily eaten by sheep at certain stages of its growth, and can thus be kept to some extent in check. It is a diœcious species, but farmers are slow to recognise the fact.

The seeds are constantly found among agricultural seeds.

The creeping thistle (but under the name of Californian thistle) was one of the three plants declared a Noxious Weed by the Act of 1900, without any reservation.

It flowers from Feb. to April. In Europe, the flowers are fertilised by *Apis mellifica*, *Bombus lapidarius*, *B. terrestris*, *B. hortorum*, *B. ruderatus*, *B. lucorum*, *Eristalis tenax*, *Lucilia cæsar*, *L. sericata*, *Calliphora erythrocephala*, *C. vomitoria* and *Stomoxys calcitrans*.

Silybum Marianum, Gærtn. Milk Thistle; or Blessed Thistle; sometimes called Scotch Thistle

First reported in 1871 from Canterbury by Armstrong.

In 1877 Kirk described the "blessed thistle" as "perhaps the most characteristic of the naturalised plants of Wellington."

"In Auckland," he says, "single specimens of this plant have been known for the past fifteen years; but, although they seeded freely, the seeds had no opportunity of germinating," owing to the dense sward of grass, "so that the thistle did not spread. A remarkable exception to this rule occurred during the formation of the Onehunga railway, where a few seeds fell on disturbed soil, grew up and flowered. The railway works being suspended, that plant increased rapidly, and spread wherever it could find disturbed soil."

Buller in his Introduction to the 2nd edition of his *History of the Birds of New Zealand*, pp. xliv to xlvi, says:

If the sparrow is fond of ripe grain it is still fonder of the ripe seeds of the variegated Scotch thistle. This formidable weed threatened at one time to overrun the whole colony. Where it had once fairly established itself it seemed wellnigh impossible to eradicate it, and it was spreading with alarming rapidity, forming a dense growth which nothing could face. In this state of affairs the sparrows took to eating the ripe seed. In tens of thousands they lived on the thistle, always giving it the preference to wheat or barley. They have succeeded in conquering the weed. In all directions it is dying out.

In the *Manual* (1906) Cheeseman records this species as abundant in fields and waste places in both islands, especially to the north of the East Cape, rarer southwards. (Fl., Jan. to March.)

The plant makes excellent ensilage, and has been used for the purpose in many parts of Australia; it contains a relatively large proportion of salt. Sheep eat the plant readily in the young state, and its seeds are very fattening for fowls.

In the Noxious Weeds Act of 1900 *Silybum* (any species) is included in the list of Noxious Seeds, and is also included in the Second Schedule of the Act by Special Gazette Notice of 20th June, 1901[1].

Vittadinia australis, A. Rich., var. *dissecta*, Benth.

In 1873 this plant was observed in abundance by roadsides and in rocky and waste places about Nelson, but later the plant died out there. In 1877 Cheeseman found it in great profusion in a new locality on the coast of Nelson, extending towards D'Urville Island. In 1878 it was found at the shingly mouths of small streams discharging into Palliser Bay near Wellington.

In the *Manual* (1906) Cheeseman states that it is naturalised in

[1] It has been found in New South Wales that this species is occasionally dangerous to stock. In 1912 Max Henry, Government Veterinary Surgeon, drew attention to losses of stock (cattle) from eating this plant. Again in 1914 S. T. D. Symons, Chief Inspector of Stock, recorded that numbers of cattle were killed through eating the plant, in several districts. This is due, apparently, to the presence in the plant of considerable quantities of hydrocyanic acid. Horses are not affected by this plant and it only hurts cattle when they are placed on it in a hungry condition, when the plants are young or stunted.

On the other hand it is recorded that in 1916 near Inverell, New South Wales, 2700 sheep, suffering from drought, got into 300 acres of land covered with this weed, and after three months were trucked to Sydney as fat sheep.

several localities, especially about Nelson; but it does not seem to have spread beyond that district. Cockayne thinks this is the plant which is forming large colonies in parts of Central Otago.

Aster imbricatus, Linn.

First recorded by Kirk in 1895 from a ballast heap in Wellington. Introduced from Buenos Ayres in 1892. It increased to a small extent, but I do not think it has established itself. It has not been noticed in any other locality.

Aster subulatus, Michx.

Cheeseman (1919) says:

> This plant was first noticed in the vicinity of Auckland about twelve years ago, and soon became plentiful, especially in moist places on harbour reclamations, by roadsides and ditches, etc. It is a native of the United States, where it is principally found in brackish-water marshes, ranging from New Hampshire to Florida.

Reported by Carse in 1915 from Mongonui County, as "spreading rapidly in all soils and situations."

Calotis lappulacea, Benth.

In the *Manual* (1906) this is reported to have been found in three localities, viz., in Poverty Bay by Bishop Williams, Nelson by Kingsley, and Banks Peninsula by Brown and Kirk. (Fl., Feb. to April.)

Erigeron canadensis, Linn. Canadian Flea-bane

First recorded in Hooker's list in 1864. Cheeseman considers it one of the earliest introductions into New Zealand, and records it as common in 1882 throughout the Auckland district. Kirk had already noted it in 1877, from several places in Wellington Province, but stated it was not nearly so abundant as in Auckland. (Fl., Feb. to April.)

In the *Manual* (1906) Cheeseman states that it is abundant throughout both islands; which is only partially correct, however, as it is not found (e.g.) near Dunedin, nor in most parts of Otago.

Erigeron linifolius, Willd.

First recorded in Hooker's list of 1864 as *Conyza ambigua*. In 1882 it was found in several localities in the Auckland Province, Northern Wairoa, Whangarei, Matamata, etc. In the *Manual* (1906) it is stated to be abundant in the Auckland provincial district, rarer southwards to Marlborough and Westport. W. W. Smith recorded it from Ashburton in 1903. (Fl., Feb. to April.)

Erigeron annuus, Linn.

H. Carse found this species in some quantity in freshly sown grass at Otukai, Mongonui in January, 1917.

Siegesbeckia orientalis, Linn.

Recorded in 1896 by Cheeseman as occurring in the North Cape district.

Bellis perennis, Linn. Daisy; Gowan

Probably introduced at an early date, but first recorded by Hooker in 1864. It is abundant in lawns and meadows in all parts of New Zealand. (Fl., Sept. to June.) In Europe this species is fertilised by *Bombus terrestris*, *Eristalis tenax*, *Musca corvina*, *M. domestica*, and *Lucilia cæsar*[1].

Xanthium strumarium, Linn. Burweed; Small Burdock

In a leaflet (No. 5) issued by the Agricultural Department in 1893, it is stated that "on a quantity of ballast brought from Buenos Ayres by the 'Silverstream,' and discharged in heaps at Wellington Railway Station, grew a quantity of this species." The plant died out later.

Xanthium spinosum, Linn. Bathurst Burr

First reported in Hooker's list of 1864, then by Kirk in 1877 at Wellington. In 1882 Cheeseman records it as occurring in waste places about Auckland and the Waikato, but adds: "it nowhere shows signs of becoming as abundant and troublesome as in certain parts of Australia." In 1893 the Agricultural Department report it as spreading in many parts of New Zealand, in 1896 in Gisborne and other parts of the Auckland district, and in 1895 as well established in Poverty Bay, especially in native lands. (Fl., Feb. to April.)

Declared a noxious weed in the Second Schedule of the Act of 1900.

Ambrosia artemisiæfolia. American Ragweed

Recorded in 1911 from Waverley by the Agricultural Department; introduced in American red-clover seed.

Pascalia glauca, Orteg.

First recorded by Kirk in 1895 as occurring on a ballast heap in Wellington. Introduced from Buenos Ayres.

[1] In 1871 Armstrong reported *Helianthus tuberosus*, the Jerusalem artichoke, as naturalised in Canterbury. The species may occur temporarily as a garden escape, and is often thrown out with garden refuse. But as it seldom flowers and does not seed to my knowledge in New Zealand, its increase is most improbable.

In 1879 Armstrong also reported *Helianthus annuus*, Linn., the common sunflower from Canterbury. This plant does occasionally occur as a garden escape, but it never establishes itself.

Gallinsoga parviflora, Cav.

First recorded, along with the preceding, by Kirk in 1895, as occurring on a ballast heap in Wellington. Introduced from Buenos Ayres.

Bidens tripartita, Linn. Bur-marigold

In 1906 Cheeseman records the first appearance of this species from near Ohaupo, Middle Waikato district.

Madia sativa, Molina

This species was first introduced about 1871 with railway material from Chili, being brought by Messrs Brogden and Co., and it appeared on more than one line of railway under construction in Otago. It has also been found in Marlborough, but I do not think it has succeeded in establishing itself. It has certainly not been met with in Otago for many years, and I do not think it seeds so far south. Kirk found it between Balclutha and Catlins, and Petrie recorded it from Bannockburn in Central Otago. Cheeseman in the *Manual* (1906) states that it occurs in waste places and roadsides in the South Island; and particularly refers to Renwicktown in Marlborough, on the authority of Reader. In regard to this last reference, A. H. Cockayne states that a farmer in Marlborough imported seed of this species with the idea of establishing an oil industry, but abandoned the scheme after the first crop. I have not, so far, been able to find the date of this introduction. (Fl., Feb. to March.)

Eclipta alba, Hassk.

Recorded in Hooker's list of 1864 as *E. erecta*, Linn. It has apparently not been seen since.

Wedelia biflora, DC.

Also recorded in Hooker's list in 1864 as *Wollastonia biflora*, DC., but not seen since.

Helenium quadridentatum, Labill.

First recorded by Cheeseman in 1869 from the North Cape district. Again in the *Manual* (1906) as occurring in waste places at Tapotopoto Bay, North Cape district. It is evidently well established there, as Carse, in 1915, speaks of it as common, chiefly on the coast.

Anthemis arvensis, Linn. Corn Chamomile

First recorded by Kirk in 1869 in the Auckland district, and in 1877 from the east coast of Wellington Province. In the *Manual* (1906) Cheeseman reports it as not uncommon on roadsides and margins of fields in both islands. (Fl., Feb. to March.)

Anthemis Cotula, Linn. Stinking May-weed

First recorded by Kirk from Wellington in 1877, and later by Cheeseman in 1882 as occurring in waste places in Auckland district. In 1906 Cheeseman reports it in the *Manual* as occurring not uncommonly by roadsides and in waste places in both islands. (Fl., Feb. to March.)

Achillea millefolium, Linn. Yarrow; Milfoil

First recorded in Hooker's list in 1864. In most cultivated localities in Auckland Province in 1882, "but nowhere abundant." In the *Manual* (1906) it is said to be not uncommon in fields and on roadsides in both islands. It is very common in many parts of Otago. It spreads in part by underground stems. It is readily eaten by sheep. (Fl., Feb. to March.)

In Europe the flowers are fertilised by the following flies: *Eristalis tenax*, *Lucilia cæsar*, *L. sericata*, *Musca corvina*, *Calliphora erythrocephala*, *Stomoxys calcitrans*[1].

Achillea tanacetifolia, All.

In the *Manual* (1906) it is reported to occur, but not commonly, in waste places and on roadsides in both islands. The only localities specified are in the vicinity of Auckland by Cheeseman, and Lincoln, Canterbury, by Kirk. (Fl., Feb. to June.)

Matricaria discoidea, DC.

First recorded by Cheeseman in 1882 as "in immense abundance in waste places about Auckland, and along most lines of road into the interior." Practically the same distribution is given in the *Manual* in 1906. (Fl., Feb. to March.)

Matricaria Chamomilla, Linn. Wild Chamomile

First recorded by Kirk in 1867 in list of naturalised plants in Great Barrier Island. In the *Manual* (1906) it is said to be not uncommon in waste places and cultivated fields in both islands.

The fruit develops an adhesive mucilage on the surface when moistened, and this may aid in its dispersion by birds.

Matricaria inodora, Linn.

First recorded in 1869 by Kirk from the Auckland district, and in 1877 from Wellington. In the *Manual* (1906) it is said to be not uncommon in fields and waste places in both islands. It is a very

[1] Kerner has pointed out that "along roads where cattle have gone to pasture, one may notice how the *leaves of yarrow*, scabious, mullein, and similar plants, have been eaten off, while the greater part of their blossoms have been left intact."

abundant weed, especially in many parts of Otago. (Fl., Feb. to March.)

In Europe the flowers are fertilised by *Bombus lapidarius*, *Eristalis tenax* and *Lucilia cæsar*.

Chrysanthemum segetum, Linn. Corn Marigold

First recorded by Kirk from the Auckland district in 1869, and in 1877 from Wellington. In the *Manual* (1906) it is stated to occur in cultivated fields in both islands, but not commonly. It is rather singular that this weed, which is so abundant in cultivation in Britain, has not established itself at all freely here. (Fl., Feb. to March.)

Chrysanthemum Parthenium, Bernh. Feverfew

First recorded in 1873 from the neighbourhood of Dunedin by the author, as *Matricaria Parthenium*; again from Wellington in 1877 by Kirk. In the *Manual* (1906) it is noted as occurring in waste places and on roadsides in both islands. (Fl., Feb.)

Chrysanthemum leucanthemum, Linn. Ox-eye Daisy; misnamed Marguerite

No doubt an early introduction into New Zealand, but first recorded by Hooker in 1864. Nearly every later list from all parts of the colony contained its name. It is now (1916) abundant in pastures throughout New Zealand. It is particularly a weed of poor land, and is readily overcome by pasture grasses where the land is freely manured. It is easily controlled by sheep, though cattle will not eat it. (Fl., Feb. to May.)

In Europe the flowers are fertilised by *Musca corvina* and *Lucilia cæsar*.

The ox-eye daisy was included in the Second Schedule of the Noxious Weeds Act of 1900 by Special Gazette Notice of 24th April, 1902.

Cenia turbinata, Pers.

First recorded by Kirk in 1895 as occurring on a ballast heap in Wellington. Introduced from Buenos Ayres.

Soliva anthemifolia, R. Br.

First recorded by Cheeseman in 1882 as occurring on alluvial flats by the Northern Wairoa River, and later by H. Carse from the neighbourhood of Mongonui. (Fl., Jan.)

Soliva sessilis, Ruiz and Pav.

First seen in the neighbourhood of Ngaruawahia and Rangiriri in 1879. Recorded by Cheeseman as *S. pterosperma*, Less.? In 1910 he states that this "has increased of late years in light warm soils in the Auckland district, and has become a troublesome weed in some localities, particularly in certain market gardens at Onehunga." (Fl., Jan. to March.)

Lagenophora emphysopus, Hook. f.

Recorded from hills near Wellington by Buchanan and Kirk, and from Banks Peninsula by Kirk. It is now abundant on Mount Victoria, Wellington. (Fl., Jan. to April.)

Tanacetum vulgare, Linn. Tansy

First recorded by Cheeseman in 1882 from the neighbourhood of Auckland. In the *Manual* (1906) it is stated to be found in waste places in both islands, not common. I have found it in Otago mostly on roadsides in the neighbourhood of gardens and farm steadings. (Fl., Jan. to March.)

In Europe the flowers are visited by *Vanessa atalanta*, *Apis mellifica*, and *Bombus terrestris*.

Artemisia Absinthium, Linn. Wormwood

First recorded in 1871 by Armstrong from Canterbury, and from Wellington in 1877 by Kirk. In 1882 noted as an occasional garden escape by Cheeseman from Northern Wairoa, vicinity of Auckland and Matamata. In the *Manual* (1906) it is stated to be not uncommon in waste places and on roadsides in both islands. Its real abundance, however, is in sheep pens and yards, wherever sheep are gathered together for mustering, shearing, etc. The scarious margins of the bracts cause the heads to adhere loosely to the wool, and the seeds are thus dropped when the animals are crowded together. (Fl., Feb. to April.)

The fruit of wormwood when wetted becomes mucilaginous and viscous. Kerner records an instance from his own observation of a small owl (*Athene noctua*) which, in catching mice, brushed against wormwood bushes, and when it flew away was all besmeared with the fruits, which had been rendered sticky by a previous shower of rain.

Helichrysum cymosum, Less.

First recorded from railway embankments near Westport, by Townson, in 1906.

Gnaphalium purpureum, Linn.

First recorded in 1906 by Cheeseman in the *Manual of N.Z. Flora*, as "not uncommon in drained swamps, freshly cleared lands, etc., from the North Cape to the Upper Waikato, rarer southwards to Wellington."

Gnaphalium germanicum, Willdenow

Recorded by W. W. Smith in 1903 as occurring in Ashburton County. I do not know the species.

Stuartina Melleri, Sond.

In the *Manual* (1906) reported on the authority of J. H. Macmahon as occurring in sandy places near the mouth of the Awatere River, Marlborough.

Senecio vulgaris, Linn. Groundsel

This weed must have been introduced at a very early date of settlement, but is first recorded by the author from Otago in 1874, and then by Cheeseman in 1882 from the Auckland district. It is one of the most abundant weeds of cultivation in every part of New Zealand. Like most other herbaceous species of *Senecio* its foliage is frequently attacked by the larvæ of *Nyctemera annulata*, the common magpie moth. (Fl., all the year round.)

When the fruit is moistened, special hairs on the outside excrete an adhesive mucilage. This is, probably, to cement them to the soil, but it may also aid in distribution by causing them to adhere to the feathers of birds.

Senecio sylvaticus, Linn.

First recorded from Wellington in 1877 by Kirk, then by Cheeseman in 1882 from the Auckland district. In 1890 the author found it in one locality near Dunedin. In the *Manual* (1906) it is said to be abundant throughout both islands, which is hardly correct, for it is certainly rare in Otago and Southland. (Fl., Jan. to April.)

Senecio jacobæa, Linn. Ragwort

I cannot find any earlier record of this weed than 1874, when I found it growing near Dunedin, but it rapidly increased, especially in Southland, and in parts of Auckland, Wellington and Taranaki. Where sheep are regularly grazed it is unable to make headway, but it is particularly troublesome in pastures only fed on by cattle and horses. It is now to be found more or less in every provincial district in New Zealand.

It was declared a noxious weed in the Second (optional) Schedule of the Act of 1900, but was placed in the First Schedule in the Act of 1908.

Since about 1880 a disease, which began in the neighbourhood of Winton in the Oreti Valley (and consequently came to be known as "Winton disease") has proved fatal to horses, and has caused very considerable mortality. It also occasionally affects cattle and sheep.

After a time the popular belief grew that the disease was due to the affected animals being fed on ragwort which was very abundant in the district. On investigation by the officers of the Veterinary Department this was found to be the case. The disease, which is cirrhosis of the liver, is by no means confined to New Zealand, but has been known under other names in other countries, notably in parts of North America, wherever *Senecio jacobæa* abounds. Thus in Canada it is known as the Picton cattle disease.

Since the vast increase of this weed, there has been a correspondingly enormous increase in numbers of the New Zealand magpie moth—*Nyctemera annulata*—the larva of which feeds mainly on this plant. In summer the moths are frequently to be seen almost in clouds in the infested districts. The larva is hairy and distasteful to birds, and there is apparently nothing to check its increase, but it is quite unable to cope with the vast increase of the weed.

In June, 1913, ragwort was reported as common on the Volcanic Plateau, an area of 5,000 acres lying to the south of Mangatautari and west of the Waikato River, and coincidently with its spread has been an enormous increase of *Nyctemera annulata*.

In the weed-infested districts bee-keepers complain that their summer honey is dark-coloured, and so strongly flavoured with the nectar of the ragwort, which is developed in great profusion, that it is almost unsaleable.

The plant flowers from Feb. to April. In Europe it is visited by *Apis mellifica*, *Bombus lapidarius*, *B. terrestris*, *B. hortorum*, *Eristalis tenax*, *Lucilia cæsar*, *L. sericata* and *Calliphora erythrocephala*.

Senecio erucifolius, Linn.

According to the report of the Agricultural Department for 1897 this species of giant groundsel occurs along with the preceding in Southland.

Senecio aquaticus, Hill

Recorded in the *Manual* (1906) as collected by the Rev. F. H. Spencer in the Buller Valley. (Fl., Feb. to March.)

Senecio mikanioides, Otto. Climbing Groundsel

First recorded by Kirk in 1869 from the Auckland district and in 1877 from Wellington, as *S. scandens*, DC. Originally a garden escape, it soon became established in the district. It has also spread in many parts of the North Island, and is occasionally met with in the south, e.g. in the neighbourhood of Dunedin. It does not stand frost, but will grow in any sheltered locality at the edge of the bush. In Taranaki it grows on the banks of many of the streams, and climbs the trees and shrubs, festooning them densely to a great height. (Fl., June to Oct.)

It is a favourite food plant of the larva of *Nyctemera annulata*, the New Zealand magpie moth.

Erechtites valerianæfolia, DC.

Recorded by Carse in 1915 as occurring at Otukai in Mongonui County; not common.

Erechtites Atkinsoniæ, F. v. Muell.

Recorded by Carse in 1915 as a troublesome weed, spreading in Mongonui County.

Cryptostemma calendulacea, R. Br. Cape-weed

First recorded in 1869 by Kirk from the Auckland district and in 1877 from Wanganui. In 1900 the Agricultural Department stated that it was plentiful in Auckland, and becoming common in Taranaki and Hawke's Bay. Cheeseman records it in the *Manual* (1906) as abundant in pastures and waste places in the North Island. (Fl., Nov. to March.)

This plant was declared a noxious weed in the Third (optional) Section of the Act, by Special Gazette Notice of 16th June, 1910.

Calendula officinalis, Linn. Marigold

First recorded by Cheeseman in 1882 as a garden escape in the Auckland district. It is only as such that it occurs now in many parts of New Zealand. (Fl., Feb. to April.)

Osteospermum moniliferum, Linn.

First recorded in 1869 by Kirk from the Auckland district, from Titirangi to the East Cape. Cheeseman in the *Manual* (1906) speaks of it as "an occasional garden escape in the vicinity of Auckland, rare." (Fl., Oct. to Feb.)

Lapsana communis, Linn. Nipplewort

Probably introduced at an early date, but recorded for the first time by Hooker in 1864, as *Lapsana pusilla*, Willd. (*Arnoseris pusilla*, Gærtn.) Kirk next recorded it from the Great Barrier Island in 1867, and later from Wellington in 1877. Then in 1882 Cheeseman reported it as a common weed in all cultivated districts. It is now found abundantly by roadsides and in waste places throughout New Zealand. (Fl., Jan. to April.)

Tolpis umbellata, Bertol.

Cheeseman states that this species was first observed in 1868, and that in 1882 it was in abundance between Penrose and Panmure in the Auckland district, but was not observed elsewhere. In 1910 he records it as occurring in "sandy soil near Helensville, Kaipara, and apparently increasing."

Cichorium Intybus, Linn. Chicory

First recorded by Hooker in 1864. Again in 1877 from Wellington by Kirk. In the *Manual* (1906) it is said to occur in waste places and on roadsides. It is by no means uncommon, but frequently dies out of districts, as e.g. near Balclutha, where it used to be cultivated, but is now rare; and near Dunedin. In 1909 the Agricultural Department reported it as increasing very much and likely to become a troublesome weed. The seed is a frequent impurity in clover seed (red clover and cow grass).

It flowers from Jan. to March. In Europe it is fertilised by *Apis mellifica* and *Eristalis tenax*.

Cichorium endivium, Willd.

Kirk recorded this species as occurring on a ballast heap in Wellington in 1895. It does not seem to have spread or re-appeared.

Hypochœris glabra, Linn. Smooth Cat's-ear

First recorded by Kirk in plants of Great Barrier Island in 1867, and from Wellington where it was common in 1877. Throughout the Auckland district in 1882, but not plentiful. In the *Manual* (1906) Cheeseman states that it is abundant in fields, etc., throughout New Zealand. (Fl., Jan. to March.)

Hypochœris radicata, Linn. Cat's-ear; Cape-weed of South Island settlers

First recorded in Hooker's list in 1864. This is one of the most widely spread and abundant weeds in all New Zealand; some authorities think it the most abundant. Yet it disappears before sheep, and in this respect is much more easily disposed of than many other introduced plants. Hilgendorf states that even on the longest and most sunshiny days of the year the flowers are open from 8.30 a.m. to 3.30 p.m. (Fl., Nov. to June.)

In Europe the flowers are fertilised by *Apis mellifica*, *Bombus lapidarius* and *Eristalis tenax*.

Tragopogon pratensis, Linn. Goat's-beard

Recorded in Hooker's list of 1864 as *T. minus*, and by Kirk in 1869 as occurring in the Auckland district (as *T. minor*). The species has not been observed since.

Tragopogon porrifolius, Linn. Salsafy

No doubt introduced as a vegetable much earlier, but first recorded from the Auckland district as growing wild in 1869 by Kirk. In the *Manual* (1906) it is stated to occur in fields and waste places, but not commonly, throughout both islands. (Fl., Dec. to Jan.)

Picris (*Helminthia*) *echioides*, Linn. Ox-tongue

First recorded in T. Kirk's list of naturalised plants in Great Barrier Island in 1867; and later from Wellington in 1877. In 1903 T. W. Kirk reports it as spreading rapidly from the Bay of Islands to Otago. The *Manual* (1906) reports it as generally distributed in fields and waste places throughout New Zealand. (Fl., Jan. to April.)

Leontodon hirtus, Linn. Lesser Hawkbit

First recorded in Hooker's list of 1864 as *Thrincia hirta*, and from Wellington by Kirk in 1877. In 1882 Cheeseman stated that it occurred in the vicinity of Auckland, but not plentifully. In the *Manual* (1906) it is stated to occur, but not commonly, in fields and waste places in both islands. It is, however, very abundant in Otago. (Fl., Nov. to March.)

In Europe the flowers are fertilised by *Bombus lapidarius* and *Eristalis tenax*.

Leontodon hispidus, Linn. Common Hawkbit

First recorded by Cheeseman in 1882 from pastures at Remuera and Epsom, near Auckland; but rare. In the *Manual* (1906) it is reported as occurring in fields and waste places and to be plentiful in many localities. (Fl., Dec. to March.)

Leontodon autumnalis, Linn. Autumnal Hawkbit

First recorded in Hooker's list of 1864 as *Apargia autumnalis*, and by Kirk from Wellington in 1877. In 1882 found at Panmure, Otahuhu and near Alexandra in the Auckland Provincial district. In the *Manual* (1906) it is stated to occur in fields and waste places from Auckland to Wellington, but not commonly. (Fl., Feb. to April.)

Lactuca saligna, Linn. Willow Lettuce

Only reported from Petane in Hawke's Bay, by A. Hamilton (*Manual*, 1906). (Fl., Jan. to Feb.)

Lactuca scariola, Linn.

Reported by Cheeseman, on the authority of J. P. Kalaugher, as occurring in Onehunga Railway Station yard in 1912.

Lactuca muralis, E. Mey. Wall Lettuce

Only recorded from Marlborough by Macmahon, and by Kirk, as abundantly naturalised in the Kaikoura Mountains. (Fl., Dec. to Jan.)

Lactuca sativa. Garden Lettuce

This species was no doubt introduced at an early date into the county. The first reference I have come across is in D'Urville's voyage. When he visited Otago Harbour in 1840, he met with it in the native cultivations "in the forest," along with other vegetables.

It has not become naturalised anywhere.

Crepis virens, Linn. Hawk's-beard

First recorded in Hooker's list in 1864. An extremely abundant weed in fields, waste places and roadsides throughout New Zealand. (Fl., Nov. to June.)

In Europe this species is visited by *Bombus terrestris* and *Eristalis tenax*.

Crepis fœtida, Linn.

First recorded by Kirk, as *Barkhausia fœtida*, as occurring in the Auckland district in 1869. In the *Manual* (1906) it is stated to occur in fields on the Auckland Isthmus, but not commonly. (Fl., Nov. to Feb.)

Crepis taraxacifolia, Thuill.

First recorded from Auckland district in 1869 by Kirk as *Barkhausia taraxacifolia*. Recorded from the same locality in the *Manual* (1906). (Fl., Nov. to Feb.)

Crepis setosa, Hall. f.

Stated by Cheeseman in the *Manual* (1906) to occur, but not commonly, on waste places on the Auckland Isthmus. (Fl., Jan. to Feb.)

Sonchus arvensis, Linn. Corn Sowthistle

First recorded by Hooker in 1864. In the *Manual* (1906) it is stated that it occurs in "cultivated fields near Auckland, rare."

Two species of sowthistle are perhaps indigenous to New Zealand; of these *S. littoralis*, Kirk, was collected by Banks and Solander. *S. oleraceus*, Linn., is a doubtful native. Kirk says (1893): "It is certain that seeds of two forms (*S. asper* and *S. oleraceus*) must have been repeatedly introduced since that period" (1773) "and that cross-fertilisation has taken place." Both species are extremely abundant in waste ground in all parts of the country at the present day. They were largely used by the natives as a food plant. (Fl., all the year round.)

Hieraceum aurantiacum, Linn. Orange Hawkweed

Reported in 1911 by the Agricultural Department as occurring at the Waiau[1].

STYLIDIEÆ

Stylidium graminifolium, Swartz.

A solitary specimen of this species was picked up on clay hills near Auckland by Colonel Bolton in 1861, and accordingly it was placed among indigenous species in Hooker's *Handbook to the New Zealand Flora*. But the plant has not been collected since.

CAMPANULACEÆ

Campanula Trachelium, Linn. Nettle-leaved Campanula

First recorded by Kirk in 1877 as a garden escape from Ohariu, Wellington district.

Campanula hybrida, Linn.

Recorded by W. W. Smith in 1903 as occurring in Ashburton County.

Jasione montana, Linn. Sheep's-bit

Recorded by Cheeseman in 1912, on the authority of H. Guthrie Smith, from Tutira Run, near the source of the Mohaka River, Hawke's Bay.

Specularia hybrida, A.DC.

Recorded as occurring in cultivated fields at Ashburton by W. W. Smith.

Sub-division HYPOGYNÆ

EPACRIDEÆ

Epacris purpurascens, R. Br.

First discovered in the fifties by Dr Sinclair and General Bolton at Manurewa near Manukau Harbour; again recorded in 1869 by Kirk, and in 1881 by A. T. Urquhart. According to Cheeseman (1906) it occurs in the open country at the head of the Manukau Harbour, near Papakura and Drury.

[1] In 1879 Armstrong reported *Lobelia erinus* from Canterbury. This common garden species does not seem to occur even as an escape.

Epacris microphylla, R. Br.

In 1875 A. T. Urquhart discovered three plants on the southern side of Manukau Harbour. Six years later a dense mass of plants 60 yards in circumference occupied the ground. Its present occurrence is almost the same as the preceding species.

Epacris pulchella, Cav.

Found by Urquhart in the same locality as the two preceding species. Urquhart thought it probable that the seeds of all three species were carried over from Australia by high winds.

ERICACEÆ

Calluna vulgaris, Salisb. Ling; Heather

In 1909 D. Petrie reported that some two or three acres of land near Opepe Bush in the Taupo district was covered with this plant. He was informed that the seed was sown by Major Roberts (formerly Stipendiary Magistrate at Tauranga, during the latter part of the war with the natives in Te Kooti's time (about 1869). Another version of the introduction of *Calluna vulgaris* into the Taupo district is given in a letter from Captain Gilbert Mair to Mr J. D. Macfarlane of Napier. He says:

> In 1870 I received a case used for Portland Cement, full of heather sprays. Whenever I rode about the country, I sowed this seed from the saddle, but it must have fallen on stony ground, or bush fires may have destroyed it, for there were no results in this district.
>
> I then sent a native orderly named Hammond to Major Scannell at Opepe, with the bulk of the seed, and in July 1871, he told me he had entrusted Sergeant McCarthy with the sowing of it.

In later years, Mr Cullen, Head of the Department of Police, has planted large numbers of plants in Tongariro National Park, in anticipation of the grouse which he hoped to see liberated there. A very strong protest against this action has been made by Captain Ellis, the recently appointed Director of Forestry in New Zealand. The plant is establishing itself in more than one locality.

In Europe the species is fertilised by *Apis mellifica*, *Bombus hortorum*, *B. lucorum*, *B. lapidarius*, *B. terrestris*, *Eristalis tenax* and *Calliphora erythrocephala*[1].

[1] In 1882 the Canterbury Acclimatisation Society imported seeds of the American cranberry (*Oxycoccos macrocarpus*, Pers.) and distributed them. I am not aware that they ever grew or increased.

PRIMULACEÆ

Repeated attempts have been made to naturalise various species of *Primula*, especially the primrose (*P. vulgaris*), the cowslip (*P. veris*), and numerous hybrid polyanthus primroses, but none of these have succeeded anywhere. Before humble-bees were introduced, the plants would not even seed naturally in gardens; but since they were brought in they seed freely. Yet where planted and sown in the bush or in the open, they have failed utterly to become established. The reason is difficult to explain. Cockayne attributes it chiefly to the rank growth of grasses and other plants.

In Europe, *Primula vulgaris*, *P. elatior* and *P. veris* are chiefly pollinated by long-tongued Hymenoptera (*Bombus hortorum* and *B. lapidarius*), less commonly by *Apis mellifica*, while *Bombus terrestris* bites a hole in the corolla-tube and steals the nectar. These insects are very common now in New Zealand.

Lysimachia Nummularia, Linn. Creeping Jenny; Money-wort

Recorded by Townson in 1906 as occurring in the Westport district.

Anagallis arvensis, Linn. Pimpernel; Poor Man's Weather-glass

Probably introduced at a very early date in the settlement of New Zealand; first mentioned by Dieffenbach in 1839. It was stated to Drummond by a Wellington resident that he knew it since 1855. Nearly every succeeding list of introduced plants from 1864 mentions it. At the present time it is a most abundant weed in fields and waste places throughout both islands.

In the Agricultural Department's report for 1910 there is a report of mortality among sheep in Hawke's Bay, and it was suspected to be due to poisoning by this species. I do not know that the matter was ever thoroughly investigated, or the suspicion refuted.

OLEINEÆ

Olea europæa, Linn. Olive

The oldest olive tree in Auckland was planted about 1848 by Col. Matson at Brookside, Parnell[1].

[1] In 1871 Armstrong recorded the privet (*Ligustrum vulgare*, Linn.) as naturalised in Canterbury. The plant is common in hedges, but can hardly be characterised even as an escape. The succulent fruit is no doubt eaten by birds, and the seeds so distributed, but I have never heard of a self-sown specimen of the plant being found away from a hedge.

APOCYNEÆ

Vinca major, Linn. Periwinkle

Probably introduced at a much earlier date but first recorded by Kirk in 1877 from several localities in the Wellington district. Then as a garden escape but plentifully established in waste ground about Auckland, and increasing, by Cheeseman in 1882. In the *Manual* (1906) it is stated to occur as a plentiful garden escape by roadsides and waste places in both islands. It was included in the Second Schedule of the Noxious Weeds Act of 1900, by Special Gazette Notice of 26th May, 1904.

In Europe this species is visited by the two humble-bees, *Bombus hortorum* and *B. terrestris*.

Vinca minor, Linn. Lesser Periwinkle

Recorded by W. W. Smith in 1903 as occurring in Ashburton County.

ASCLEPIADEÆ

Gomphocarpus fruticosus, R. Br.

First recorded by Kirk in 1869 from the Auckland district as *Asclepias nivea*, Linn. In the *Manual* (1906) it is stated to occur, but rarely, as "an occasional garden escape near Auckland and Napier."

GENTIANEÆ

Erythræa Centaurium, Pers. Centaury

First recorded in Hooker's list in 1864. Later by Kirk in 1877, as found in various localities near Wellington. By 1882 it was generally distributed in the north. It is now found abundantly in pastures throughout the whole of New Zealand, especially where cattle are pastured, as they will not eat it on account of its bitter taste.

Erythræa australis, R. Br.

Colenso reported this species in 1894 as having been found by H. Hill in woods and highlands near East Cape. I am not aware of its having been collected since.

Chlora perfoliata, Linn.

In 1919 Cheeseman reports this as found among Manuka scrub at Parengarenga, North Cape district.

POLEMONIACEÆ

Collomia coccinea, Lehm.

Recorded by Cheeseman in 1882 as a garden escape in one or two localities near Auckland. Later reported from Ashburton by

W. W. Smith; and from localities in Central Otago, Roxburgh and Cardrona, by Petrie and Kirk.

Gilia squarrosa, Hook. and Arn. Californian Stink-weed; Digger's Weed

First recorded by Kirk in 1870 (as *Navarettia squarrosa*) as abundant at Ngaruawahia in the Waikato. In 1885 it was abundant in the valley of the Clutha River up to Lake Wanaka; and was also reported as common in Marlborough. It is borne about by carriers, and by sheep all over the country, especially in dry districts, and neither stock nor rabbits will eat it. Cheeseman in the *Manual* (1906) reports it as occurring not uncommonly in dry pastures. The seeds of species of *Gilia* become mucilaginous when wetted.

Gilia cærulea (Auct.?)

Recorded by W. W. Smith in 1903 from Ashburton County.

BORAGINEÆ

Amsinckia angustifolia, Lehm.

Recorded from Alexandra and Black's in Central Otago, by Petrie.

Symphytum officinale, Linn. Comfrey

J. B. Armstrong sent me a specimen from Mairehau near Christchurch, where it is growing wild (1918). The species appears to have been introduced somewhat recently into the country for fodder purposes. L. D. Ayson of Rotorua, who has investigated the distribution of this species, informs me (in 1919) that it was introduced into the Rotorua district over 40 years ago as a fodder plant. It now grows abundantly throughout all that district, and is now met with in Wanganui and throughout Taranaki, at Opotiki, Matamata, Te Aroha and in many parts round Auckland and up to Northern Wairoa. It does not seem to be known in the South Island except near Christchurch.

Borago officinalis, Linn. Borage

First recorded in 1877 from Otago Peninsula by the author. Kirk reported it in the same year from Johnsonville, near Wellington. In 1882 Cheeseman reported it as occurring in waste places on the Auckland Isthmus, "but rare." In 1896 the Agricultural Department reported it as spreading in Marlborough. In the *Manual* (1906), Cheeseman only reports it as "not uncommon in waste places from Auckland to Wellington," but it also occurs locally in the South Island.

In Europe this species is visited by *Apis mellifica* and *Bombus terrestris*.

Anchusa italica, Linn.

This species was recorded in 1879 from Canterbury, but it is not known there outside of gardens.

Anchusa officinalis, Linn. Alkanet

Recorded by W. W. Smith in 1903 from Ashburton County.

Anchusa arvensis, Bieb.

Recorded by W. W. Smith in 1903 as *Lycopsis arvensis*, Linn., from Ashburton County.

Lithospermum arvense, Linn. Corn Gromwell

First recorded in Hooker's list in 1864. Stated by Cheeseman in 1882 to occur at various places in the Auckland district. In the *Manual* (1906) it is stated to be plentiful in roadsides and waste places in both islands.

This species is fertilised in Europe by *Apis mellifica* and *Bombus lapidarius*.

Myosotis palustris, Linn. Forget-me-not

First recorded by Kirk from several parts of Wellington Province in 1877 as *M. strigulosa*, Rehf.; then by Cheeseman in 1882 from the vicinity of Auckland, and Motuihi Island. Cheeseman in the *Manual* (1906) states that it is not uncommon in wet places in both islands.

Myosotis cæspitosa, Schultz.

Stated by Cheeseman in the *Manual* (1906) to be not uncommon in wet places in both islands.

Myosotis sylvatica, Hoffm.

First recorded from Ashburton in 1903 by W. W. Smith. In the *Manual* (1906) stated to be found in waste places in both islands, but not commonly. It is common (1919) in Dean's Bush, Christchurch.

This species in Europe is fertilised by *Apis mellifica*, *Musca domestica* and *M. corvina*.

Myosotis arvensis, Lam.

First recorded by the author in 1875 as occurring in the Leith Valley, Dunedin, and by Cheeseman in 1882 in one locality at Whangarei. In the *Manual* (1906) it is said to occur locally in fields and waste places in both islands.

Myosotis collina, Hoffm.

First reported by Cheeseman in the *Manual* (1906) as occurring in fields and waste places in both islands, not common.

Echium vulgare, Linn. Viper's Bugloss

First recorded in 1869 by Kirk as occurring on the west side of the Firth of Thames; and in 1882 by Cheeseman at Matamata, and near Hamilton. In 1896 it was reported by the Agricultural Department to be spreading in Marlborough. W. W. Smith reported it as spreading greatly in the Ashburton district, and as being fertilised by humble-bees. In the *Manual* (1906) said to occur locally on roadsides and waste places in both islands.

Included in the Second Schedule of the Noxious Weeds Act of 1900 by Special Gazette Notice of 23rd March, 1905.

Echium plantagineum, Linn.

First recorded by Armstrong in Canterbury in 1879 (as *E. violaceum*). In 1882 Cheeseman stated: "it has recently appeared in one or two localities near Auckland." In the *Manual* (1906) it is said to occur only in the North Island somewhat rarely, Cheeseman reporting it from the vicinity of Auckland, and Kirk from a ballast heap in Wellington. Yet in a Special Gazette Notice of 23rd March, 1905, wild borage (*Echium violaceum*) is included in the Second Schedule of the Noxious Weeds Act. This species is not a borage, but a bugloss. It is a serious pest in some parts of Australia, where it is known as "Paterson's Curse." But it has not assumed dangerous proportions in New Zealand.

In more recent years the plant has become quite abundant in some localities. Dr Colquhoun sent me specimens in 1919 from the bed of the Otaio River in South Canterbury, which he informed me was full of it, and was quite blue with its flowers in January. My son, Dr J. Allan Thomson, also collected it about the same time from Awapiri in Marlborough, where he makes the same statement, viz., that the country is blue with it in January.

Cynoglossum micranthum, Desf.

This species was recorded by Hooker in 1864 from the Auckland district. It has not been collected since.

Cynoglossum furcatum, Willd.

Recorded by W. W. Smith in 1903 as occurring in Ashburton County. I do not know the species.

CONVOLVULACEÆ

Ipomæa batatas, Poir. Kumara; Sweet Potato

When Captain Cook first came to New Zealand the natives were found to have excellent plantations of at least four kinds of plants. Banks in his *Journal* says:

Their plantations were now hardly finished, but so well was the ground tilled that I have seldom seen land better broken up. In them were planted *sweet potatoes*, cocos, and a plant of the cucumber kind.

Later on he says:

nor does their cultivated ground produce many species of esculent plants, three only have I seen, yams, *sweet potatoes* and cocos. They also cultivate gourds.

It is difficult to assign a date for the introduction of the kumara into New Zealand. The plant was certainly not introduced by the first immigrants, who appear to have come from Western Polynesia, and who had a strong mixture of Melanesian blood in their veins. At a later date a purer Polynesian invasion took place from Eastern Polynesia, and it was these people who brought in the plant.

Mr S. Percy Smith tells me:

There are several accounts of the introduction of the kumara, in fact most tribes have their own account and they differ a good deal. It was such a very important article of food, and everything connected with its growth and harvesting so sacred, and accompanied with so many ceremonies, that each tribe sought to accredit its own ancestors with the honour of its introduction. As a matter of very strong probability, most, if not all of the canoes of the last migration, which took place in the middle of the 14th century, brought the kumara with them.

The first migration of the pure Polynesians that came from Tahiti, under the celebrated ancestor Toi-te-huatahi, arrived here long before the 14th century; indeed it has been long settled that the period of that celebrated chief was about the year 1150. He did not come direct, but first made the Chathams, after calling at various islands of the mid-Pacific, and that is the reason probably that he did not introduce the kumara, viz., that the sea-stores were all exhausted on the voyage; for it is quite certain that the kumara was in cultivation in Tahiti when Toi left his home, and doubtless he would have provisioned his vessel with the usual class of stores.

The fact of Toi not having introduced the kumara is generally considered to be proved by the well-known story of Taukata and Hoake, who arrived at Whakatane in the Bay of Plenty some seven or eight generations after him. On their arrival they gave Toi the name of Toi-kai-rakau, or the wood-eater, because he or rather his descendants had only the native plants of New Zealand to live on.

Another account of the introduction of the kumara by a very well-known Maori accredits the 'Horouta' canoe with introducing it. And probably this is also true, for most of the later migrations brought the kumara with them. 'Horouta' came here at the time of the great migration; i.e. about 1350 A.D.

Mr Elsdon Best writes me:

The Kumara was introduced by Aotea, Arawa, Tainui and other immigrants twenty generations ago, but traditions say that the first to reach New Zealand were fetched from Polynesia by voyagers who left Whakatane for that purpose in a vessel called 'Te Aratawhao.' This was about twenty-four generations ago, or say six hundred years. (1300 A.D.)

The kumara is still largely cultivated by the Maoris, and they have a good many varieties. It is not a wild species anywhere.

Convolvulus arvensis, Linn. Small Bindweed

First recorded by Armstrong in Canterbury in 1879. In the *Manual* (1906) it is stated to occur, not uncommonly, in fields and waste places in both islands.

Cuscuta racemosa, Mart.

Kirk reported this species as *C. Hassiaca*, Pfeiff., in 1884, as occurring in Canterbury, parasitic on lucerne, knot-grass, etc. It does not appear to have been reported from anywhere else, nor has it been found recently. This is a Chilian species, but was introduced from California.

Cuscuta Epilinum, Weihe. Flax Dodder

Apparently only recorded from Canterbury.

Cuscuta Epithymum, Murr. Lesser Dodder;
Clover Dodder

First recorded in 1870 in Southland by the author on clover; and by Kirk the same year in the Waikato district. When the cultivation of red clover was greatly extended in the Canterbury Plains after the introduction of humble-bees, dodder overran many parts. In 1901 A. Wilson reported it as common at the head of Lake Wakatipu, parasitic on piri-piri (*Acæna*). T. W. Kirk reports it in 1909 as found on the following introduced species: red clover (*Trifolium pratense*), Alsike clover (*T. hybridum*), white clover (*T. repens*), Lucerne (*Medicago sativa*), broom (*Cytisus scoparius*), gorse (*Ulex europæus*), knotweed (*Polygonum aviculare*), dandelion (*Taraxacum dens-leonis*), Canadian flea-bane (*Erigeron canadensis*), and ox-eye daisy (*Chrysanthemum leucanthemum*).

Clover dodder (*Cuscuta Epithymum*, var. *trifolii*) is included among *noxious seeds* in the Third Schedule of the Noxious Weeds Act of 1900.

SOLANACEÆ

Lycopersicum esculentum, Mill. Tomato

I do not know how early this was first introduced, but it is recorded as a naturalised species by Hooker in 1864, and again in Auckland Province by Kirk in 1869. The latter speaks of it in 1870 as abundantly naturalised in many localities in the Waikato. Cheeseman (1882) calls it a garden escape of short duration.

Solanum marginatum, Linn.

First recorded by Cheeseman in 1882 who writes: "A garden outcast near Auckland. A large clump existed for many years in Alten Road, but is now nearly destroyed." Apparently it has not spread.

Solanum sodomæum, Linn. Dead-sea Apple

First recorded by Cheeseman in 1882 as common on the volcanic hills of the Auckland Isthmus, etc., and also noticed at Mongonui, Bay of Islands, and in the Waikato. In the *Manual* (1906) it is reported to occur from the North Cape to the Waikato, as not uncommon on warm dry soils and on sand-dunes.

Solanum auriculatum, Ait.

First recorded in 1882 by Cheeseman from the neighbourhood of Auckland, and from Mahurangi. In the *Manual* (1906) he states that it is increasing in waste places in the vicinity of Auckland. Carse in 1915 reports it from near Kaitaia in Mongonui County.

Solanum tuberosum, Linn. Potato

It is probable that potatoes were first planted at Dusky Sound when Captain Cook anchored there early in 1773, for a piece of ground was cleared and left as a garden. Geo. Forster, in his *Journal* of the voyage of the 'Resolution,' says:

> We re-embarked all our instruments and utensils, and left no other vestiges of our residence than a piece of ground, from whence we had cleared the weeds. We sowed indeed a quantity of European garden seeds of the best kinds; but it is obvious that the shoots of the surrounding weeds will shortly stifle every salutary and useful plant, and that in a few years our abode, no longer discernible, must return to its original chaotic state.

When the 'Adventure' arrived in Queen Charlotte Sound early in the same year, Captain Furneaux cleared a spot at the pah (or "Hippah," as he calls it), at the S.W. point of Motuara and made a garden, planting potatoes among other things. The natives appreciated this attempt at cultivation, for writing in his *Journal* on 29th May, Captain Cook says:

One of these people I took over to Motuara, and showed him some potatoes planted there by Mr Fanner, Master of the 'Adventure.' There seemed to be no doubt of their succeeding; and the man was so well pleased with them, that he, of his own accord, began to hoe the earth about the plants. We next took him to the other gardens and showed him the turnips, carrots and parsnips; roots which, together with the potatoes, will be of more real use to them than all the other articles we had planted.

These other gardens referred to here, included those made early in May, 1773. Forster writes:

On May 22nd we went over to an island in the Sound to which Captain Cook had given the name of Long Island. Captain Cook, who was determined to omit nothing which might tend to the preservation of European garden plants in this country, prepared the soil, sowed seeds, and transplanted the young plants in four or five different parts of this island. He had cultivated a spot of ground on the beach of Long Island, another on the Hippah rock, two more on the Motu-Aro, and one of considerable extent at the bottom of Ship's Cove, where our vessels lay at anchor. He chiefly desired to raise such vegetables as have useful and nutritive roots, and among them particularly potatoes, of which we had been able to preserve but few in a state of preservation.

On his third voyage Captain Cook visited these gardens and in his *Journal*, under date 15th February, 1777, he says:

When the 'Adventure' arrived first at Queen Charlotte Sound in 1773, Mr Bayly fixed upon this place (the Hippah or fortified village at the S.W. point of Motuara) for making his observations; and he and the people with him, at their leisure hours, planted several spots with English garden seeds. Not the least vestige of these now remained. It is probable that they had been all rooted out to make room for buildings, when the village was re-inhabited, for, at all the other gardens then planted by Captain Furneaux, although now wholly over-run with the weeds of the country, we found cabbages, onions, leeks, purslain, radishes, mustard, etc., and a few potatoes. These potatoes, which were first brought from the Cape of Good Hope, had been greatly improved by change of soil; and, with proper cultivation would be superior to those produced in most other countries. Though the New Zealanders are fond of this root, it was evident that they had not taken the trouble to plant a single one (much less any other of the articles which we had introduced); and if it were not for the difficulty of clearing ground where potatoes had once been planted, there would not have been any now remaining.

Though Captain Cook did not look hopefully on the cultivation of the potato by the Maoris—those natives he met with at Queen Charlotte Sound being rather a poor lot—there is no doubt that the tubers were carried about by them and scattered throughout the various tribes north and south. In 1813 Captain Williams reported that the natives at the Bluff had a field of considerably more than one hundred acres

of potatoes which presented one well-cultivated bed, filled with rising crops of various ages, some of which were ready for digging, while others had been but newly planted. Captain Fowler of the 'Matilda' was for eleven days in Otago Harbour in 1813, and got potatoes from the natives.

The Rev. Samuel Marsden who landed in the Bay of Islands in 1814 says of the native cultivations:

Their potato plantations are all very neatly fenced in, and were in as high condition as the gardens in and near London, as they do not suffer a single weed to remain that would injure the growing crops. The flat where the natives were encamped might contain somewhat about a hundred acres or more, part of which was enclosed and planted with potatoes. We were furnished with a good supply of potatoes and pork.

A. Hamilton, writing of a massacre which took place at Otago Heads in 1817, says:

De Surville was, with Cook, supposed to have been the introducer of the potato to the Maoris of the North Island and the northern part of the South Island. Many old Maoris contend that *tiwas* were known and largely cultivated before the advent of Europeans. The Maoris certainly had a number of named varieties as early as 1820, and here we find them in Otago in 1817 able to supply large quantities to whalers as a recognised article of trade.

Commander Bellingshausen, who visited New Zealand in 1820, says:

At present the New Zealander also grows potatoes which are as good as the English species. They learned to grow this vegetable from Captain Cook, and although after forty-seven years they grow sufficient quantities, they only use the potatoes for themselves and do not part with any.

Captain Edwardson found the natives about Foveaux Straits in 1823 cultivated potatoes, which he says they preserved during the winter "by the same process as that employed by the Irish."

The mate of the brig 'Hawes' found potatoes in the native cultivations near Tauranga in 1828.

The records of the various whaling stations, which were so industriously collected by the late Dr R. McNab, show that cultivation of the potato had become so common, that considerable quantities were exported to Sydney. The vessels trading between New South Wales and New Zealand very frequently took potatoes back with them, and their cargoes contained potatoes from Kapiti, from Chatham Island, "eight tons from Otago," "four tons from Preservation Inlet," and so on. Most of the cultivation was carried on by native women. This was between 1813 and 1820, and later the whaling settlements at Port Pegasus (in Stewart Island), Tautuku Bay in S.E. Otago, Waikouaiti and elsewhere on the east coast, supplied large quantities of potatoes.

Darwin who travelled from the Bay of Islands to Waimate in 1835 says:

> After travelling some miles we came to a little country village, where a few hovels were collected, and some patches of ground cultivated with potatoes. The introduction of the potato has been the most essential benefit to the island; it is now much more used than any native vegetable.

Bidwill travelling in 1839 between Tauranga and Tongariro says:

> the potato might be taken for an indigenous plant, as it is impossible to go anywhere without finding it growing wild. Maoris only grow potatoes in land which is just cleared, and after about three crops abandon it, and clear another portion of forest.

This mode of cultivation was also noticed by Dieffenbach in the same year in the island of Ararapa. He remarks: "Half-burned stems of trees were lying in confusion over each other, and in the places between were patches of potatoes." He says in one place in his account of this trip:

> as we passed through the woods we found two plantations of potatoes. As my natives never seemed to consider that these kind of plantations belonged to anybody, we always used to help ourselves when we came to any of them, without compunction. In fact I suppose that these patches must have been planted by some of the mission-natives, on purpose to save trouble when they went their journeys between the two stations.

Wilkes, who visited the Auckland Islands in 1840, says: "Some attempts at forming a garden were observed at one of the points of Sarah's Bosom, and turnips, cabbages and *potatoes* were growing finely, which if left undisturbed, will soon cover this portion of the island."

The "Maori potato," as it used to be called, has been largely displaced by modern varieties, but it still persists in the neighbourhood of old Maori cultivations. Thus it was found comparatively recently in waste ground near the sea on Otago Peninsula and on the southern side of Blueskin Bay. The Maoris have practically disappeared from there, and only a few half-castes remain in the neighbourhood, where formerly there was a large native population, but the wild cabbage and the wild potato still persist. The latter had a very firm tuber, rather bluish in colour, and very solid when cooked.

Mr Elsdon Best tells me that some varieties of potatoes still linger in scrub and fern in the North Island. In the Uriwera country he has found them growing in land now covered by blackberries, but which were Maori clearings 50 years ago. The potato haulms grew up through the tangle of blackberries to a height of five or six feet, but bore abundance of excellent firm tubers.

In Europe the flowers of this species are visited by *Apis mellifica*, *Bombus hortorum*, *B. lapidarius*, *B. terrestris* and *Eristalis tenax*[1].

Atropa belladonna, Linn. Deadly Nightshade

Recorded by W. W. Smith in 1903 as naturalised in the County of Ashburton, but as dying out.

Capsicum annuum, Chili

Recorded by Kirk in 1869 in his naturalised plants of the Auckland district, but it practically never spreads out of cultivation in New Zealand. Polack records it as cultivated by the Europeans in 1831.

Physalis alkekengi, Linn. Winter-cherry

Also recorded by Kirk in 1869 from the Auckland district. Only occurs as a rare garden escape, and does not appear to have established itself anywhere as a naturalised plant.

Physalis peruviana, Linn. Cape Gooseberry; Tipari

Probably introduced very much earlier, but recorded first as a naturalised plant by Hooker in 1864; again by Kirk in the Auckland district in 1869 and from Wellington in 1877. Cheeseman says in 1882: "warm sheltered localities throughout the district, but not so common now as fifteen or twenty years back." In the *Manual* (1906) it is recorded as an escape from cultivation in the North Island only, but it occurs as far south as Moeraki in Otago, where it grew in sheltered localities quite freely a few years ago.

Polack (1831–37) mentions it as in cultivation in European gardens in the north of the North Island.

Nicandra physaloides, Gærtn.

First recorded from Ashburton district in 1903 by W. W. Smith. In the *Manual* (1906) it is stated to occur very rarely in waste places in the vicinity of Auckland.

Lycium chinense, Mill.

First recorded in Hooker's list in 1864, and again by Kirk in 1869 as occurring in the Auckland provincial district; both as *L. bar-*

[1] In 1868 Kirk recorded the occurrence of *Solanum virginianum* at Mongonui. I do not know what species he refers to. On submitting my doubt to Mr Cheeseman he writes (Sept. 1916): "I have no idea what plant Kirk had in mind, but it was not *S. virginianum*."

In 1869 he recorded *S. indicum*, Linn., from Auckland, and in 1877 the same species from Wellington. Cheeseman writes: "I think this was a mistake of Kirk's for *S. mammosum* of later lists."

barum. Kirk also reported it from Wellington in 1877. In the *Manual* (1906) it is stated to be not uncommon in waste places and roadsides in both islands.

Lycium horridum, Thunb. Box-thorn

Introduced 40 or 50 years ago as a hedge-plant. Mr Cheeseman states that in the Auckland district it does not maintain itself outside of cultivation. But in many parts it has spread very considerably, and has become a serious nuisance. It was declared a noxious weed in the Third (optional) Schedule of the Act of 1908, by Special Gazette Notice of 10th September, 1908.

Datura Stramonium, Linn. Thorn-apple

First recorded in Hooker's list in 1864. Stated by Kirk in 1870 to be occasionally met with in the Auckland Province. In 1895 the Agricultural Department reported it from Auckland, Rangitikei, Wairarapa and Picton. W. W. Smith states that it was very abundant in the Ashburton district in 1895–97, but after the long drought of later years it became rare. In the *Manual* (1906) Cheeseman reports it as not uncommon in waste places in rich warm soils as far south as Canterbury.

Hyoscyamus niger, Linn. Henbane

Reported by Kirk as occurring among ballast at Wellington. Cheeseman records it in 1912 from near Pakuranga, Auckland, collected by R. Green[1].

In Europe *Bombus terrestris* and *B. lapidarius* visit the flowers.

Nicotiana Tabacum, Linn. Tobacco

I have no record of when the tobacco plant was first introduced into New Zealand, but it must have been in the very early days of settlement, the seed being brought by the missionaries and by whalers in the earlier years of the 19th century. The natives were very desirous of obtaining the plant, and numerous tales were current of the sale of dockseed to them by unscrupulous traders (see pp. 466–467).

The first record of its occurrence as a naturalised plant is in Bidwill's account of his visit to Rotorua in 1839, where he says: "there was plenty of very find tobacco growing near, although I never at any other place met with any that was worth gathering." The natives had an extraordinary fondness for tobacco, but they took little trouble either to cultivate the plant or cure the leaf. Kirk in 1869, in his list of plants of the Auckland Province, notes its occurrence; and again in 1877 from Wellington. Cheeseman in the *Manual* (1906) says it occurs as "an occasional escape from cultivation in rich warm soils."

[1] Kerner records that an average-sized plant produces 10,000 seeds in a year.

Nicotiana acutiflora, A. St Hil.

First recorded by Kirk in 1895 as occurring on a ballast heap in Wellington. Introduced from Buenos Ayres.

Nicotiana suaveolens, Lehm. Native Tobacco of Australia

The Agricultural Department's report for 1899 states that a single specimen of this plant was collected in (Napier?) Hawke's Bay, evidently introduced in ballast.

Petunia parviflora, Juss.

First recorded by Kirk in 1895 as occurring—along with the preceding species—on a ballast heap in Wellington. Introduced from Buenos Ayres in 1892. I do not think it has succeeded in establishing itself.

OLEINEÆ

Syringa vulgaris. Lilac

This species only occurs in New Zealand as a cultivated shrub, but I refer to it here because a common native longicorn beetle (*Prionoplus reticularis*) has taken to it as a convenient dwelling place. W. W. Smith of New Plymouth finds that the larvæ bore extensively into the stems of old plants, and he has reared the mature insect from them.

SCROPHULARINEÆ

Verbascum Thapsus, Linn. Mullein

First recorded in Hooker's list in 1864, and then by Kirk in 1877 from Wellington. Cheeseman records it from volcanic hills near Auckland; and as plentiful at Matamata in 1879. In the *Manual* (1906) it is said to be "abundantly naturalised in dry places." It is particularly in evidence in rabbit-infested areas such as Central Otago, as it is one of the few plants rabbits will not eat, on account of its densely woolly character.

In Europe the flowers are visited by *Apis mellifica*, *Bombus hortorum* and *B. terrestris*.

Verbascum Blattaria, Linn. Moth Mullein

First recorded by Kirk in 1870 as occurring at Mt Eden, Auckland and on the North Head, Waitemata; and again from Wellington in 1877. Cheeseman in 1882 states that it occurs in waste places and pastures, from Auckland to Waikato. It is not uncommon in many parts of both islands at the present time.

Verbascum phœniceum, Linn.

First recorded in Hooker's list in 1864, and again in Kirk's list of Great Barrier plants in 1867, but has not proved permanent. Mr Cheeseman, who knows the flora of the country, and especially of the Auckland district thoroughly, has never met with it.

Celsia cretica, Linn.

Recorded by Cheeseman in the *Manual* (1906) as occurring not uncommonly in fields in the Auckland district.

Linaria vulgaris, Mill. Toad-flax

First recorded as a garden escape from Ashburton in 1903 by W. W. Smith. Also recorded from near Lake Brunner by J. W. Brame. It occurs near Dunedin in several localities as a garden escape, and is reported by Cockayne from the Taieri Plain, and from Kinloch, Lake Wakatipu.

In Europe the flowers are visited by *Apis mellifica*, *Bombus hortorum*, *B. lapidarius*, *B. terrestris* and *Calliphora erythrocephala*.

Linaria purpurea, Mill.

First recorded in 1873 from the neighbourhood of Dunedin by the author. In the *Manual* (1906) said to be a garden escape in some parts of Otago and Canterbury.

Visited by *Apis mellifica* in Europe.

Linaria latifolia, Desf.

Reported by Kirk as a garden escape in the vicinity of Wellington.

Linaria Elatine, Mill.

First recorded in Hooker's list in 1864. Cheeseman reports it from the vicinity of Auckland, Otahuhu, Ngaruawahia, etc. In the *Manual* (1906) it is stated to be not uncommon on roadsides and in waste places in both islands.

Linaria Cymbalaria, Mill. Ivy-leaved Toad-flax

First recorded as a garden escape from Ashburton in 1903 by W. W. Smith. In the *Manual* (1906) reported as an occasional garden escape throughout both islands.

Antirrhinum Orontium, Linn.

In the *Manual* (1906) reported from waste places near Auckland by Cheeseman and from Napier by A. Hamilton.

Scrophularia aquatica, Linn.

Recorded by W. W. Smith in 1903 from Ashburton County.

Mimulus luteus, Linn.

First recorded in 1876 by the author as common on streams near Dunedin. In 1877 Kirk reported it as abundant in swampy or moist places in several parts of Wellington Province. In the *Manual* (1906) it is stated to occur in damp places, sides of streams, etc., from Wellington southwards. Mr J. Belton tells me that about Lake Ellesmere it fills the ditches for miles.

Mimulus moschatus, Dougl. Musk

First recorded by the author as a naturalised weed in Otago in 1875, and again in 1877 by Kirk from two points in the Wellington provincial district. It does not appear to have established itself further north, as Cheeseman in the *Manual* (1906) reports it as occurring in damp places, sides of streams, etc., from Wellington southwards. It is abundant in Otago.

This species tends in many localities, especially in the southern parts of New Zealand, to monopolise the ground where it grows, and thus to appear as a "pure formation."

Digitalis purpurea, Linn. Foxglove

First recorded by Hooker in 1864, then in 1877 in Wellington by Kirk. Cheeseman in recording it in 1882 from Auckland Isthmus, Thames and Whangarei, states that it is by no means common. Even in the *Manual* (1906) it is said to be "an occasional garden escape, not common," in both islands. It is, however, very common in certain localities, and in the far north, e.g. near Hokianga, has taken complete possession of many old lava flows.

A northern correspondent writes:

> at the present time purple and white foxgloves, growing in profusion on the Wangamoa hills (north-east of Nelson), present a glorious blaze of colour. This noxious pest has apparently got completely out of hand, and threatens to ruin the countryside.

At Tokaanu at the south end of Lake Taupo, in 1906, it formed a "pure formation" over very large areas of country, having nearly crowded out all other vegetation.

It used to be very common in places near Dunedin, but its comparative disappearance is explained by this peculiarity, that it tends to die out of pasture land when it is not pulled out of the ground. Wherever it is pulled out of the ground and the surface soil is thus disturbed, fresh seedlings spring up. I am informed that in the Wairarapa district some farmers have expended as much as £150 a year in their endeavours to clear the land of foxglove, by pulling it out. Others who have elected to leave it, have been fined *fifty shillings* for a breach of the Noxious Weeds Act, and their land has become nearly clean by the plants dying out of the pastures. It is abundant in the Taieri gorge.

In Europe the flowers are visited by *Bombus hortorum* and *B. terrestris*.

This species was included in the Second Schedule of the Noxious Weeds Act of 1900 by Special Gazette Notice of 21st December, 1905.

Veronica agrestis, Linn.

First recorded in Hooker's list in 1864, then by Kirk in 1877 as "not unfrequently in cultivated land" in Wellington Province; common in the Auckland district in 1882. In the *Manual* (1906) it is stated to be abundant in fields and waste places in both islands.

In Europe it is visited by *Apis mellifica*.

Veronica Buxbaumii, Ten.

First recorded in Hooker's list in 1864, and in most subsequent lists of introduced plants, as plentiful in cultivated districts. First noted in Otago by the author at Port Chalmers in 1885. In this instance the plant came up where a quantity of immigrants' bedding had been condemned and burned. Now found abundantly in cultivated land throughout all parts of New Zealand. It is a remarkably strongly-rooted species.

Veronica Anagallis, Linn.

When crossing the Ruahine Ranges in 1845, Colenso gathered specimens of this species. In writing the account of it later he states that:

> I noticed a *Veronica* with blue flowers which grew in the water and was not unlike our English *V. beccabunga* or *V. anagallis*. (I mention this particularly, as I fear it has of late years quite disappeared from this district, not having seen a plant anywhere for more than twenty years.)

Kirk (in 1869) doubted the accuracy of the identification, but Cheeseman saw the specimens, which undoubtedly belong to *V. Anagallis*, and he has included the species in his Flora as doubtfully indigenous. I think there can be little doubt that it is an introduction. It is remarkable, as Cheeseman states in his Flora, that the plant should have apparently disappeared. But I may record a somewhat similar instance. In 1870 in a field of newly broken up land in Southland, sown with grass-seed imported from Lawson and Sons, of Edinburgh, there came up and flowered several plants of *Veronica Chamædrys*, Linn. The field was not grazed the first year, but a light crop of hay was cut off it. The plant never re-appeared. I am inclined to think that no insects capable of fertilising the flowers being available in those days—when hive-bees were rare and humble-bees had not been introduced—many flowering plants failed to reproduce themselves by seed.

Veronica arvensis, Linn.

First recorded in Hooker's list in 1864. Now found abundantly in fields and waste places in all parts of New Zealand.

Veronica saxatilis, Linn.

In 1871 Armstrong recorded this species as occurring in Canterbury. It has not been observed since, and I think the identification is doubtful.

Veronica serpyllifolia, Linn.

First recorded in Hooker's list in 1864. One of the most widely spread and common weeds in all parts of the country.

Hermann Müller records *Calliphora erythrocephala* as sucking the flowers of these plants, which were kept in a room.

Veronica officinalis, Linn.

First recorded in Hooker's list in 1864. Kirk, in 1869, expressed the opinion that this was an error in identification, and that perhaps *V. serpyllifolia* had been mistaken for it. It was later recorded from the Canterbury district by both Armstrong and W. W. Smith. Cheeseman states that the species has not been seen in the North Island, in the locality where it was originally supposed to have been found.

Veronica Chamædrys, Linn.

First recorded in 1870, then in 1917. Found growing among lawn grass in Dunedin by Mrs G. S. Thomson. (The previous occurrence is referred to under *V. Anagallis*.)

Bartsia viscosa, Linn. Tar-weed

First recorded in 1869 as *Rhinanthus crista-galli* in Southland by the author; common in Otago in 1875. Reported by Cheeseman in 1882 from Helensville; neighbourhood of Auckland, and in great abundance between Pukekohe and Tuakau.

OROBANCHEÆ

Orobanche minor, Sutt. Broom-rape

According to the late Mr T. Kirk, var. *picridis*, F. Schultz., was first observed at Whangarei in 1867, growing on the roots of cat's-ear (*Hypochœris radicata*). The typical form was first observed on red clover in the Waikato near Cambridge in 1885. It is now well established in both islands as far south as Canterbury (but not in Otago and Southland so far), growing on *Trifolium*, *Medicago*, *Lotus*, *Crepis*, *Lathyrus*, *Hypochœris*, etc. It is particularly abundant in the Auckland district, where it does considerable damage. Grazing animals never eat parasitic or saprophytic plants like *Orobanche*.

LABIATÆ

Mentha viridis, Linn. (*Mentha spicata*, L.) Spear-mint

No doubt introduced at a very early date after settlement began. First recorded as an escape from cultivation by Hooker in the Auckland district in 1864. Found occurring abundantly in the extreme south of Stewart Island in Wilson Bay by the author in 1874. In 1877 Kirk reported it from several places in the Wellington district. By 1880 it was common in ditches and waste ground in most districts of the colony. In the wetter parts of the Canterbury Plains it chokes many ditches and drainage channels. Recorded by Poppelwell in 1911 as occurring in old clearings in Codfish Island and Rugged Island. This species spreads freely by means of its underground stems.

Mentha piperita, Linn. Pepper-mint

First recorded by Kirk in 1869 from the Auckland district, where it was reported as spreading in 1882 by Cheeseman. Noted by the author in several localities in the valley of the Clutha in 1885. In the *Manual* (1906) it is stated to be an occasional garden escape in damp places in both islands. The flowers are visited in Europe by *Lucilia cæsar*.

Mentha aquatica, Linn. Water-mint

First recorded by Hooker in his list of 1864. Reported as spreading in Auckland district in 1882. Cheeseman reports it in the *Manual* (1906) as occurring, but not commonly, in wet places in the Auckland district.

In Europe this species is visited by *Apis mellifica*, *Bombus hortorum*, *Lucilia cæsar*, *Musca corvina*, *Eristalis tenax* and *Calliphora erythrocephala*.

Mentha arvensis, Linn. Corn or Field Mint

First recorded by Armstrong in Canterbury in 1879, and by Cheeseman from Auckland in 1882. In the *Manual* (1906) said to be not uncommon in fields and waste places in the North Island. This is one of the worst weeds of arable land in Britain, but fortunately it has not spread much in New Zealand. In Europe it is visited by *Apis mellifica*, *Bombus terrestris* and *Lucilia cæsar*.

Mentha Pulegium, Linn. Penny-royal

First recorded by Cheeseman in 1882 from Whangarei, and several places about Auckland. Its subsequent history in the North Island is one of continued and rapid increase and aggression. The Agricultural Department reported on it continually. In 1896 it was spreading about Gisborne in the river beds and in old Maori plantations. In 1899 it was proving a very troublesome weed in grassland; very abundant in the Auckland district, less frequent in Hawke's Bay, Taranaki, Nelson and as far south as Canterbury. I found in April, 1919, that in the north of Auckland the flowers were visited by immense numbers of the common butterfly *Lycæna labradus* (formerly known as *L. phœbe*). In 1901 it was included in the Second Schedule of the Noxious Weeds Act by Special Gazette Notice of 20th June, and in 1904 in the Third (Noxious Seeds) Section by Gazette Notice of 10th November.

Mentha australis, R. Br.

Reported by Cheeseman to be plentiful in 1877 on roadsides between Raglan and Ruapuke, Auckland district. Imported from Australia. It was also recorded from the Wairarapa by Kirk[1].

Thymus serpyllum, Linn. Thyme

Recorded by Cheeseman in the *Manual* (1906) as an occasional garden escape, in both islands; not common.

Calamintha Acinos, Clairv. Basil Thyme

This species was recorded by Kirk in 1869 as occurring in the Auckland provincial district, but Cheeseman in 1882 stated that it did not appear to have been noticed of late years. It had not apparently been observed again.

Satureia hortensis, Linn. Summer Savory

In 1876 Buchanan recorded this species from Kawau. It is not uncommon as a garden pot-herb, but I do not think it is naturalised anywhere in New Zealand.

Melissa officinalis, Linn. Balm

First recorded from Ashburton as a garden escape in 1903 by W. W. Smith. In the *Manual* (1906) it is reported as a garden escape in a few localities throughout New Zealand.

[1] In 1870 Kirk recorded *Mentha dentata* from the neighbourhood of Auckland; and in 1877 he named it in his list of the introduced plants of Wellington. The name is of doubtful application. It is possible that Kirk saw one of the forms of *M. sativa* or *M. gentilis*, which has not perpetuated itself in New Zealand. I do not know the species.

Salvia Verbenaca, Linn. Wild Sage

Cheeseman reports of this species in 1882 that it "appeared by a roadside in the suburbs of Auckland some years ago, but seems to have become extinct." In the *Manual* (1906) he reports it from "waste places near Auckland," on the authority of Kirk.

Nepeta Cataria, Linn. Cat-mint

First recorded by T. Kirk in his list of Great Barrier Island plants in 1867. Reported from vicinity of Auckland and the Waikato by Cheeseman in 1882. In the *Manual* (1906) it is stated to occur, but not commonly, in waste places and on roadsides in the Auckland provincial district. Specimens have also been received from Nydia Bay, Marlborough, from Mr E. F. Paton.

In Europe the flowers are visited by *Bombus terrestris*.

Nepeta Glechoma, Benth. Ground Ivy

Recorded from the vicinity of Wanganui by Kirk. The seeds of this species become mucilaginous when wet, according to Guppy.

Cedronella triphylla, Mœnch.

Reported by Cheeseman in 1882 as a garden escape: "This has become very abundant on the lava streams around Mount Eden (Auckland), forming dense clumps many feet in diameter and 3–4 feet high." In the *Manual* (1906) it is reported by Cheeseman as occurring in waste places near Auckland, and by Kirk from near Wellington. Cockayne says it is fairly common in damaged semi-coastal forest near Wellington.

Prunella vulgaris, Linn. Self-heal

First recorded in Hooker's list in 1864. One of the most abundant and widespread naturalised plants in New Zealand. It is particularly common on poor, rather sterile land, and is best coped with by manuring the ground, and so improving the herbage.

Guppy states that the nutlets of this species emit mucus when wetted, and adhere firmly to feathers on drying. Darwin also found that the seeds were often found in hardened earth taken from the feet of birds, and he thinks the plant depends mainly on this mode of dispersion.

In Europe the flowers are visited by *Apis mellifica*, *Bombus lapidarius*, *B. terrestris*, *B. hortorum* and *B. subterraneus*.

Marrubium vulgare, Linn. Horehound

First recorded in Hooker's list in 1864, and by Kirk in 1877 from Foxton, Wairarapa, etc. In the Auckland district reported as not uncommon on roadsides and waste places from Auckland to the Waikato in 1882. In the *Manual* (1906) it is reported abundant throughout both islands. It is extremely common where sheep camp.

In Europe the flowers are visited by *Apis mellifica* and *Bombus terrestris*.

Stachys germanica, Linn. Woundwort

First recorded from Ashburton by W. W. Smith in 1903. Not found since.

Stachys palustris, Linn.

First recorded by Kirk from Wanganui in 1877. There is no later report.

Stachys arvensis, Linn.

First recorded in Hooker's list in 1864. In 1877 Kirk reported it as common in cultivated ground at Wellington. Reported by Cheeseman in 1882 as a troublesome weed in the Auckland district in cultivated ground. It appears to be confined to the North Island where it is abundant in cultivations.

Stachys annua, Linn.

First recorded from Ashburton in 1903 by W. W. Smith. Not collected since.

Teucrium Scorodonia, Linn. Wood Sage

Recorded by Armstrong in 1879 as occurring in South Canterbury and by W. W. Smith in 1903 from Ashburton County.

Galeopsis Tetrahit, Linn. Hemp-nettle

Reported by Cheeseman as occurring in waste places near Otahuhu, Auckland, in 1881.

Lamium purpureum, Linn. Purple Dead-nettle

First recorded by Kirk as occurring in cultivated ground at Wanganui in 1877; then by Armstrong in Canterbury in 1879. W. W. Smith also records it from Ashburton.

Lamium album, Linn. White Dead-nettle

First recorded by Armstrong in 1871 from Canterbury. In 1906 the Agricultural Department report it from Tarata, Taranaki, and wrongly call it cat-mint. It grows somewhat freely in Dean's Bush, Christchurch[1].

[1] In 1879 Armstrong records two Labiates as occurring among the naturalised plants of Canterbury, viz., *Lamium amplexicaule*, Linn., the Henbit dead-nettle; and *Lamium maculatum*, Linn. They have not been reported by any later collector.

VERBENACEÆ

Verbena officinalis, Linn. Vervain

First recorded in Hooker's list in 1864. Then from several districts in the Wellington provincial district in 1877 by Kirk. In the *Manual* (1906) it is stated to be not uncommon in fields and on roadsides in both islands. In many districts, especially in Taranaki, it has become a troublesome pest. In Britain it is an erect, usually slender plant from one to two feet in height. In Taranaki it is gigantic, and Mr W. W. Smith has sent me specimens over 10 ft. high and with stout straight stems. It covers acres of ground, and is harsh and almost scabrid.

In Europe the flowers are visited by *Apis mellifica*.

Verbena bonariensis, Linn.

First recorded by Kirk in 1870 as occurring in the Auckland district. Cheeseman in the *Manual* (1906) says it occurs rarely in waste places in the Auckland district. In 1912 it was reported by Carse as occurring near Kaitaia.

Verbena hastata, Linn.

Recorded from Waverley in 1911 by the Agricultural Department; introduced in American red-clover seed.

Lantana Camara, Linn.

This weed, which is one of the commonest and most troublesome pests in North-eastern Australia, is found wild in the neighbourhood of Kohu-Kohu on the Hokianga River where it was first recorded about 1895. It probably occurs in other parts in the far north as a garden escape. Mr Maiden, Government Botanist for New South Wales, informs me that in Queensland, where all attempts to check the pest have proved unsuccessful, it has been found that land which has been completely overrun by it has become permanently enriched. In Hawaii special insects have been introduced to combat the increase of the plant.

PLANTAGINEÆ

Plantago major, Linn. Greater Plantain

No doubt introduced by the earlier settlers; first recorded by Dieffenbach in 1839. Bidwill in his journey from Tauranga to Rotorua and Taupo in 1839 says: "the common plantain is everywhere quite as general as in England; not being an article of food, the natives can tell nothing about how or when it came." He found the species very abundant at Rotoiti, and he was the first European to visit that locality. Hooker recorded it in his list in 1864, and in all succeeding lists of introduced plants it has appeared. At the present time it is most abundant in fields and waste places throughout the country.

Guppy (and others) have pointed out that the seeds of *Plantago major*, *P. lanceolata*, etc., become coated with a mucilaginous material when wetted. He says:

> In 1892, when experimenting on these plants, I found that the wetted seeds adhere firmly to a feather, so that it could be blown about without their becoming detached. My readers can readily ascertain by a simple experiment that a bird pecking the fruit-spikes in wet weather would often carry away some of the sticky seeds in its plumage. Several years ago, when I was endeavouring to examine the condition of these seeds in the droppings of a canary, my efforts were defeated by the bird itself, since, in spite of all my care, some seeds and capsules were always carried by the bird on its feathers into the clean cage reserved for the experiment.

For some reason or other which is not understood *Plantago major* is not eaten by cattle or other grazing animals, and hence it spreads more or less undisturbed. Kerner suggests that it is probable that to eat them would be injurious to grazing animals, because though it contains no alkaloids, and is not poisonous to men, yet they carefully avoid it.

Kerner states that an average-sized plant of this species produces 14,000 seeds in a year.

Plantago media, Linn.

First recorded by Kirk in 1869 in his list of Auckland plants. In 1877 Kirk, on the authority of J. Buchanan, reports a single specimen from Wellington. In the *Manual* (1906) it is stated to occur in fields and waste places in both islands, but not commonly.

Plantago lanceolata, Linn. Ribwort; Rib-grass

First recorded in Hooker's list in 1864. It has appeared in every succeeding list of naturalised plants, and at the present time is one of the commonest introduced plants in the country.

In Europe the flowers are visited by hive-bees (*Apis mellifica*) for pollen; and less commonly by *Bombus terrestris*.

Plantago varia, R. Br.

Kirk in 1877 states: "this plant has for several years maintained a struggling existence in Boulcott Street, Wellington, but appears doomed to speedy extinction from the progress of street improvements." Recorded by Cheeseman in the *Manual* (1906) as "sparingly naturalised in several localities between the East Cape and Banks Peninsula." Introduced from Australia.

Plantago hirtella, H. B. and K.

Recorded by Cheeseman as *P. virginica*, L. (?) in 1882, from Rangiriri, Ngaruawahia and other places in the Waikato. In the *Manual* (1906) it is said to occur in "moist shaded places in the North Island; not uncommon." Apparently introduced from America.

Plantago Coronopus, Linn.

First recorded by Kirk from Wellington in 1877, then by Cheeseman in 1882 as occurring in waste places and sandy soil near the sea from a number of localities in the Auckland Province, from Bay of Islands down to Tauranga and Poverty Bay. In the *Manual* it is reported as not uncommon in sandy and gravelly places in both islands. Cockayne says it is so common in many salt-meadows that it might easily be mistaken for an indigenous species.

Division INCOMPLETÆ

POLYGONEÆ

Polygonum lapathifolium, Linn.

First recorded from Ashburton in 1899 by W. W. Smith, and stated in 1903 to have spread up the bed of the Ashburton River. He also reports it (in 1919) as having gone wild at Whangamomona, Taranaki Province, in great abundance along the roadsides and in waste ground.

Polygonum Persicaria, Linn.

First recorded by Kirk, on the authority of J. Buchanan, from Wellington in 1877, then by Cheeseman in 1882 in fields near Panmure, Auckland. In the *Manual* (1906) it is stated to occur, but not commonly, in ditches and on roadsides in both islands. It is extremely common in flax-areas after draining.

In Europe the flowers are visited by *Eristalis tenax*.

Polygonum Hydropiper, Linn. Water-pepper

Recorded from East Cape district in the North Island by Bishop Williams.

Polygonum aviculare, Linn. Knot-grass

First recorded by Hooker in 1864. Cheeseman in the *Manual* (1906) states that this species, which is a most abundant weed in roadsides, waste places and edges of fields, is most probably an immigrant. The late Mr Kirk always considered it to be indigenous, and its position was the subject of much dispute between him and Mr W. T. L. Travers (see *N.Z. Inst. Trans.* vols. IV and V). The consensus of opinion is, however, strongly in favour of its being an introduced plant.

Guppy says:

In England I have found the nutlets of *Polygonum convolvulus*, *P. persicaria* and *P. aviculare* in the stomachs of a wild duck, and a curlew, and they came frequently under my notice in the crops and intestines of different kinds of partridges and of wood-pigeons. Though most of the fruits were generally injured, a few of them were not uncommonly obtained in a sound condition.

Polygonum Convolvulus, Linn. Black Bindweed

Armstrong reported this species from Canterbury in 1871. It was common in gardens in Dunedin in 1874. Kirk found it in cultivated land in Wellington Province in 1877. Cheeseman recorded it in 1882 from roadsides and waste places, but not common. It is now a common weed of cultivation in many parts of both islands. In New Plymouth I saw it (April, 1919) climbing 15 to 20 ft. high, and bearing bunches of fruit over six inches in diameter.

In Europe this species is visited by *Apis mellifica*[1].

Fagopyrum esculentum, Mœnch. Buckwheat

I do not know when or where this was first introduced, but Kirk first records it in 1869 from Auckland district, and remarks: "This may become a weed of cultivated land, but at present can scarcely be called naturalised." W. W. Smith reported in 1903 that it used to be common on railways and cuttings near Ashburton, but disappeared after drought seasons. The *Manual* (1906) considers it only an escape from cultivation in both islands. My own opinion is that it is nowhere naturalised. It is largely grown by bee-keepers.

In Europe it is visited by *Apis mellifica* (most abundantly), *Bombus lapidarius*, *Musca corvina* and *Eristalis tenax*.

Rumex obtusifolius, Linn. Common Dock

This obnoxious weed was no doubt introduced into the country at an early period of settlement, as few samples of grass or other agricultural seeds brought from Britain were quite free from it. Earl mentions that the dock was a great nuisance in Maori plantations at Hokianga in 1834. But it increased so freely that legends of its introduction sprung up, which may or may not be true. Thus Darwin, who visited the Bay of Islands in 1835, says: "The common dock is also widely disseminated, and will, I fear, for ever remain a proof

[1] In 1879 Armstrong reported *Polygonum Dryandri* as occurring in Canterbury; the identification is doubtful.

of the rascality of an Englishman, who sold the seeds for those of the tobacco plant."

Colenso crossed the Ruahine Range in 1845, and speaking later of the village of Te Awarua on the east side of the Rangitikei River, which was then the centre of a great potato cultivation, says:

In visiting these localities in after years, I was surprised to find such an extensive and formidable growth of English docks (*Rumex obtusifolius*) 4–5 feet high and densely thick, so that in some places I could scarcely make my way through them. On enquiry I found, when some of these people had visited Whanganui to sell their pigs they had purchased from a white man there some seed, which they were told was tobacco seed. In their ignorance they took their treasure back with them, and carefully sowed it in some of their soil, which they also had prepared by digging; and lo, the crop proved to be this horrid Dock; which, seeding largely, was carried down the rivers and filled the country.

The same iniquitous trick had also been played with the natives of Poverty Bay, as early as 1837, when, at their request, I visited some young plants they had raised from seed, fenced in and tabooed, believing them to be tobacco.

This is the commonest of all the docks, and is much the most abundant species in the South Island.

Dock (*Rumex*, any species) was included among noxious seeds in the Act of 1900, and was declared a noxious weed and included in the Second Schedule of the Act by Special Gazette Notice of 20th June, 1901.

Rumex palustris, Smith

First recorded by Kirk, on the authority of J. Buchanan, as having been found at Wellington in 1872, but it was extinct in 1877. Then it was reported by Armstrong in Canterbury in 1879. There are no other records of its occurrence in New Zealand.

Rumex crispus, Linn. Curled Dock

No doubt introduced by the settlers at a very early date, but first recorded by Dieffenbach in 1839. His identification, however, is doubtful. It is recorded in Hooker's list in 1864, and was generally distributed in the north in 1880. Cheeseman in the *Manual* (1906) reports it as abundant in fields and waste places in all parts of New Zealand.

Rumex sanguineus, Linn.

First recorded by T. Kirk in his list of naturalised plants in Great Barrier Island in 1867, as *R. viridis*, Sibth., and again from Wellington in 1877. Cheeseman in 1882 reports it as generally distributed in the Auckland district.

In the *Manual* (1906) it is stated to occur in fields and waste

30—2

places and to be abundant throughout both islands. "The form with the veins of the leaves green, not red (*R. viridis*, Sibthorp), is the one most abundant in New Zealand." I have not met with it in Otago.

Rumex conglomeratus, Murr.

First recorded in Hooker's list in 1839. In 1877 Kirk says it is reported from Wellington anonymously in the *Education Gazette* of 1874, and adds: "It is probable that starved specimens of *R. viridis* have been mistaken for this." Cheeseman in the *Manual* (1906) says it occurs, but not commonly, on roadsides and in waste places near Auckland and Wellington.

Rumex acetosa, Linn. Sorrel

First recorded in Hooker's list in 1864. In the *Manual* it is stated to be not uncommon in fields and waste places in both islands.

In Europe the honey-bee (*Apis mellifica*) visits the flowers for nectar.

Rumex acetosella, Linn. Sheep's Sorrel

First recorded in Hooker's list in 1864. No doubt an early introduction into New Zealand, and at the present time one of the most abundant weeds in the country. It occurs particularly in cultivated ground, but is mostly kept down by sheep wherever they graze freely. It has a certain food-value too in many parts. Liming the land reduces the weed.

Speaking of the grass-denuded areas of Central Otago, Mr D. Petrie, says:

> Sheep's sorrel (*Rumex acetosella*), widely spread over all the dry denuded flats and many of the barren hill-slopes, is another important food-plant. It grows fairly well all through the dry season, and being perennial and spreading freely by long creeping rootstocks, it is in little danger of being eaten out[1].

Emex australis, Steinh. Three-cornered Jack; Spiny Dock; Cat's Head

Cheeseman reports that "this has appeared twice in Auckland, but does not seem to increase." This was in 1882. In 1892 it appeared on a heap of ballast in Wellington taken out of the ship 'Silverstream' from Buenos Ayres; but it did not come up again the second year. Cheeseman in the *Manual* (1906) reports it from near Auckland and in the Bay of Plenty; and from Westport on the authority of Townson.

[1] "The Golden Dock (*Rumex maritimus*, Linn.) is stated, on anonymous authority in the *Educational Gazette*, vol. I, p. 46 (1874), to occur at Pipitea Point, I cannot but think erroneously, as so conspicuous a plant would, of necessity, have attracted the attention of local botanists." (T. Kirk.)

In 1919 he records it from Kaipara, and adds: "this species appears to be of uncertain occurrence in New Zealand, and never lingers long in any one locality."

NYCTAGINEÆ

Mirabilis Jalapa, Linn. Marvel of Peru

First recorded by Cheeseman in 1882 as a garden escape near Auckland. It does not seem to spread.

ILLECEBRACEÆ

Herniaria hirsuta, Linn. Rupture-wort

Reported by Cheeseman as occurring on sandy flats north of the Manukau Heads.

AMARANTACEÆ

Amarantus caudatus, Linn.

First recorded by Kirk in 1869 in the Auckland district. Cheeseman in 1882 states that it is occasionally seen about gardens, but is hardly naturalised.

Amarantus retroflexus, Linn.

First recorded by Kirk in 1869 in the Auckland district. Cheeseman in 1882 reports it from the streets of Auckland, and waste places and gardens in the suburbs, not common. In the *Manual* (1906) it is said to be not uncommon in waste places and gardens in the North Island, and also to occur in Nelson.

The only flower visitor in Europe was *Musca domestica.*

Amarantus hybridus, Linn.

First recorded by Cheeseman in 1882 as abundant in waste places about Auckland; also at the Thames and in most of the country townships in the Auckland district, and as becoming a troublesome weed in gardens in rich or highly-manured soils. In the *Manual* (1906) it is stated to be "common to the north of the East Cape."

Amarantus Blitum, Linn.

First recorded in 1869 by Kirk as occurring in the Auckland district, in 1877, as a rare fugitive weed in Wellington gardens, and again by Cheeseman in 1882 from waste places and streets of Auckland, but not nearly so common as the preceding. In the *Manual* (1906) it is reported as not uncommon on roadsides and in waste places as far south as Nelson and Westport.

Amarantus viridis, Linn.

Recorded in Hooker's list of 1864 as occurring at the Thames. In 1869 Kirk found it in a solitary locality at the Thames. Later recorded by Kirk as occurring in the Auckland district as *Euxolus viridis*, Moq. In 1882 Cheeseman reports it as in waste places and streets of Auckland, and adds that it was "gathered many years ago at the Bay of Islands, by Allan Cunningham." In the *Manual* (1906) it is reported to be not uncommon as far south as Wellington.

Amarantus lividus, Linn.

Recorded by Kirk in 1869 from the Auckland district.

Amarantus oleraceus, Linn.

Also recorded by Kirk in 1869 in Auckland, and stated (in 1870) to be common on volcanic hills, roadsides, etc.

As neither of these species has been found since, it looks like a case of mistaken identification.

Amarantus deflexus, Linn.

Recorded by Kirk as occurring on a ballast heap in Wellington in 1895. Does not seem to have re-appeared.

Teleanthera sp.

Cheeseman records a species of this genus found on "ballast at Aratapu, by the Northern Wairoa River," and adds "I have failed to precisely identify this, which is probably an introduction from South America."

CHENOPODIACEÆ

Chenopodium album, Linn. Fat-hen; Goose-foot

First recorded in Hooker's list in 1864, again by Kirk in the Auckland district in 1869 as *C. viride*, L., and in Wellington in 1877. A most abundant weed in cultivated land throughout New Zealand. Horses refuse to eat chaff which contains much of this plant.

Included among noxious seeds in Schedule 3 of the Noxious Weeds Act of 1900.

Chenopodium murale, Linn.

First recorded by Kirk in 1869 in plants from the Thames Goldfields, and in 1877 from Wellington. Cheeseman in 1882 reports it as plentiful in Auckland district in waste places, roadsides, etc. The *Manual* (1906) states that it is abundant in similar localities in the North Island.

Chenopodium Bonus-henricus, Linn.
Good King Henry; All-good

First recorded in 1871 from Canterbury by Armstrong. Cheeseman states that it was "noticed at Onehunga in 1878, but perhaps only an escape from cultivation." This is repeated in the *Manual* 1906. It does not appear to have spread.

Chenopodium urbicum, Linn.

Though this species appears both in Hooker's *Handbook* and in Cheeseman's *Manual of the New Zealand Flora* as an indigenous plant, it is most certainly an introduction, as indeed both recognised. It is first recorded by Kirk in 1869 as a naturalised plant in Auckland. It occurs in a few localities in the North Island, but is fairly common in the South Island, especially in Canterbury and Otago.

Roubieva multifida, Moq.

First recorded by Kirk in 1895 as occurring on a ballast heap in Wellington, introduced from Buenos Ayres three years previously. It did not ripen seeds, and did not appear the second year. I am not aware that it has been found elsewhere.

Beta vulgaris, Linn. Beet

This species was no doubt introduced in the early days of settlement, but it does not seem to have become wild anywhere in the South Island. Polack, who visited the Kaipara district in 1831, records it as cultivated by the Maoris. In the North Island Cheeseman records it in the *Manual* (1906) as "an occasional escape from cultivation[1]."

PHYTOLACCACEÆ

Phytolacca octandra, Linn. Ink-weed; Poke-weed

First recorded in Hooker's list in 1864 as *P. decandra*, but as Kirk points out this is evidently a mere clerical error. In 1870 Kirk speaks of it as common on roadsides or wherever the soil is disturbed on volcanic hills. In 1882 becoming increasingly abundant, especially on lava streams, and by sides of bush tracks. Cheeseman in the *Manual* (1906) records it as abundant in the Auckland district. Its seeds are freely distributed by fruit-eating birds. My son, G. Stuart Thomson, writing from Whangarei in 1916, tells me that the fruit is largely eaten by pheasants, and in consequence their flesh becomes very dark-coloured.

A. Kerner states that thrushes are made ill by the berries of *Phytolacca*, but this seems to have been an individual case with a cage-bird. In New Zealand thrushes feed freely on the fruit

[1] In 1877, among introduced plants of Wellington, Kirk records *Beta cycla*, as occurring at Hutton Road. I have no idea what species this is.

PROTEACEÆ

Hakea acicularis, R. Br.

First reported by Cheeseman in 1882, who says of it that it "has established itself over several miles of open Manuka country at the foot of the Waitakerei Range, and is increasing fast. Its origin can be easily traced to a planted hedge in the neighbourhood." In 1893 the Agricultural Department recorded it as not uncommon in most of the gumfields, Bay of Islands district. In the *Manual* (1906) Cheeseman records it as "often planted for hedges in the Auckland District, and frequently spreads."

This species was declared a noxious weed in the Second Schedule of the Act of 1900.

Hakea saligna, Knight

Occasionally planted for garden hedges. In 1907 Cheeseman reports it as having established itself near Waihi, and as spreading rapidly.

EUPHORBIACEÆ

Euphorbia Helioscopia, Linn. Sun Spurge

First recorded in Hooker's list in 1864 from Auckland Province. In 1877 Kirk reported it from Wellington. In 1882 Cheeseman stated it to be plentiful—in light rich soils—in the Bay of Islands and Whangarei, but scarcer to the south. The author found it in the neighbourhood of Dunedin in one locality in 1889. Cheeseman in the *Manual* (1906) says it is not uncommon on roadsides and in waste places as far south as Canterbury.

In Europe the flowers are visited by *Eristalis tenax* and *Lucilia cæsar*.

Euphorbia Peplus, Linn. Milk-weed

First recorded in Hooker's list in 1864. Nearly every succeeding list contained it; and at present it is one of the most abundant weeds—especially in gardens—in New Zealand.

In Europe the flowers are visited by *Eristalis tenax* and *Lucilia cæsar*.

Euphorbia Lathyris, Linn. Caper Spurge

First recorded by Kirk in 1869 in the Auckland district and from Wellington in 1877. In 1882 reported from several localities in the Auckland district, but not common. At present it is found in waste places and gardens in most parts of New Zealand, but is very local in its distribution.

Euphorbia hypericifolia, Linn.

First recorded by Cheeseman in 1882 from the streets of Auckland. In the *Manual* (1906) he adds: "once well established, now nearly extinct."

Euphorbia ovalifolia, Engelm.

First recorded by Kirk in 1895 as occurring on a heap of ballast (from Buenos Ayres) in Wellington. It apparently did not spread.

Euphorbia segetalis, Linn.

In 1912 Cheeseman reports this as found on sand-dunes at Tauroa, near Ahipara, by R. H. Matthews and H. Carse.

Euphorbia cyparissias, Linn.

Also recorded in 1912 by Cheeseman; collected by J. W. Murphy near Culverden, North Canterbury.

Mercurialis annua, Linn. Dog's Mercury

Recorded by W. W. Smith in 1903 as occurring in Ashburton County. Has apparently not been collected since.

Jatropha Curcas, Linn. Physic-nut

Recorded in Hooker's list in 1864 as occurring in the Auckland provincial district. Cheeseman writes me in 1916: "I believe an attempt to cultivate this at the Bay of Islands was made in the whaling days, when it was probably seen by Hooker." It has certainly not been met with since.

Poranthera ericæfolia, Rudge

Also recorded in Hooker's list in 1864. Said to have been collected near Auckland by Dr Sinclair. Has not been seen since. It is a New South Wales species.

Ricinus communis, Linn. Castor-oil Plant

First recorded in Hooker's list of 1864 as *R. Palma-Christi*. Cheeseman reports it as not uncommon in warm and dry localities near Auckland. In the *Manual* (1906) he reports it from "waste places on warm rich soils from Mongonui to the Waikato River, not common."

URTICEÆ

Humulus Lupulus, Linn. Hop

First recorded by Kirk in 1877 from various spots in the neighbourhood of Wellington, and then by Cheeseman in 1882 as an occasional escape from cultivation in the Auckland district. In the

Manual (1906) it is stated to occur in "waste places, hedges, etc." in both islands; an occasional escape from cultivation. I have seen it growing wild, but as an escape only, about New Plymouth.

Ficus Carica, Linn. Fig

Probably introduced early last century by the missionaries. Polack found it in cultivation in European gardens in the north of the North Island in 1831. Recorded by Jerningham Wakefield in 1840 as growing in Hokianga; and by Kirk in 1870 as abundantly naturalised in many localities in the Waikato district. In 1877 he reported it as occasionally found on the sites of abandoned gardens in Wellington Province. Cheeseman in 1882 reported it "as wonderfully tenacious of life, and not easily killed when once planted. It is thus frequently seen in abandoned gardens, etc., but can hardly be considered naturalised." That is practically the position to-day.

Urtica urens, Linn. Small Nettle

First recorded by Hooker in 1864 in his list of introduced plants, and by Kirk in 1877 from Paikakariki, Wellington. Cheeseman in 1882 says this species has made its appearance in waste places about Auckland, but it does not seem to spread. In the *Manual* (1906) it is stated to occur in waste places in both islands, but not commonly. It occurs occasionally, mixed with horehound, in and around sheep-pens and sheep camping ground. I am not aware of any device by which the fruit can adhere to wool or hair.

Urtica dioica, Linn. Common Nettle

First recorded by Dieffenbach in 1839, but his identifications are often doubtful, as Hooker does not record it in 1864. However in 1877 Kirk reported it from Wellington, and in 1882 Cheeseman records it as occurring in a few places about Auckland, but not spreading. This species occurs sparingly in waste places in both islands.

MOREÆ

Broussonetia papyrifera, Lam. (?)
Paper-mulberry; Aute

Sir Joseph Banks, who accompanied Captain Cook on his first voyage to New Zealand, records the occurrence of this species:

"After this," he says in his *Journal*, "they showed us a great rarity, six plants of what they called *aouta*, from whence they make cloth like that of Otahite. The plant proved exactly the same, as the name is the same, *Morus papyrifera*, Linn. (the Paper Mulberry). The same plant is used by the Chinese to make paper. Whether the climate does not agree well with it,

I do not know, but they seemed to value it very much; that it was very scarce among them I am inclined to believe, as we have not yet seen among them species large enough for any use, but only bits sticking into the holes of their ears."

According to a Maori tradition the *aute* shrub was brought to New Zealand by the Oturereao canoe which made the land at Ohiwa, where her crew settled. This was at the time of the great emigration, about twenty generations or five centuries ago.

The plant is quite extinct in New Zealand now, nor do the Maoris know it; but *aute* sails and awnings are mentioned in Maori lore. It is still abundant in many tropical and subtropical regions.

JUGLANDEÆ

Juglans regia, Linn. Walnut

Walnuts were probably introduced early last century by missionaries, but the first of which I can find any record, was a tree planted in Newmarket (Manukau Road), Auckland, about 1842. The species has nowhere become naturalised.

SALICINEÆ

Salix fragilis, Linn. Crack Willow; Withy

First reported as a naturalised plant by Armstrong in Canterbury in 1879. In the *Manual* (1906) it is stated to be "abundantly naturalised on the banks of the larger rivers of both islands." It would perhaps be more correct to say that it has been freely planted on the banks of many rivers, and has been spread widely by floods, etc.

Visited in Europe by *Bombus lapidarius*, *B. terrestris* and *B. lucorum*.

Salix babylonica, Linn. Weeping Willow

First reported by Cheeseman in 1882 as follows:

> The "weeping-willow" was planted many years ago at the Mission Station, at Tangiterora on the Northern Wairoa River, and from branches and twigs floated down the river has established itself in profusion on the banks, often fringing them for miles and in some places impeding the navigation. It is also naturalised on the banks of the Waikato, but not nearly to the same extent.

Quite recently (1916) five weeping-willows were cut down in Eden Crescent, Auckland, which were brought out by Major Mavis, of the Commissariat Department, about 1861. The vessels coming out touched at St Helena, and the cuttings were taken from the trees growing at Napoleon's grave; they were planted in a box and so transferred to Auckland. The species has been very extensively planted in all parts of New Zealand.

Salix alba, Linn. White Willow

Cheeseman reports in 1882 that this species is naturalised on the banks of the Northern Wairoa and the Waikato. Writing in 1912 he suggests that the willows (of various species) which now form a continuous fringe along the banks of the Waikato, doubtless originated from seeds, etc., floated from Taupiri, where they were planted by missionaries prior to 1850.

In Europe the catkins are visited by *Bombus lapidarius*, *B. lucorum* and *B. terrestris*.

CUPULIFERÆ

Quercus Robur, Linn. Oak

Apparently the first oak trees planted in New Zealand were those at Hokianga and the Bay of Islands, introduced nearly a century ago.

The first planted in Auckland grew from acorns sent from Sydney; Mr Cleghorn, Superintendent of Public Works, sowed them in 1841 or 1842 in the Government Gardens. The young trees were planted at Government House in 1844 or 1845. Those in the Domain were planted by Mr Chalmers in 1863.

They come up freely in plantations (Whangarei, Katikati, etc.) and are spreading over fairly wide areas. Both varieties, *sessiliflora* and *pedunculata*, are common.

The larva of the native longicorn beetle (*Prionoplus reticularis*) attacks oaks in northern districts of New Zealand. W. W. Smith says it is more severe on *Q. pedunculata* than on *Q. sessiliflora*.

BETULACEÆ

Alnus glutinosa, Linn. Alder

Cheeseman in 1912 reports that:

> old trees of the alder have spontaneously appeared in not a few stations along the banks of the lower Waikato, from Huntly to within a few miles of the mouth of the river. Probably they have originated from seeds floated from Taupiri, where I understand it was planted by the missionaries prior to 1850.

CONIFERÆ

Pinus Pinaster, Sol. (*P. maritima*, Lamarck)

This species spreads very freely and to a great distance. Maxwell thinks "the seeds are carried by imported birds, which are very fond of most of the pine seeds." It is said to be growing wild at Hokianga, in many parts of the Auckland Province, and in Taranaki. An interesting occurrence of it is on the Balmoral estate in stony sour clay land, lying between the Balmoral Hills and the Hurunui

River in Canterbury. There, some twenty trees are growing quite irregularly in the poorer parts of the land where it is covered with manuka scrub, and they were certainly not planted. Mr T. H. Foote of Christchurch tells me that similar trees grow at the glens of Tekoa homestead ten or twelve miles to the north-west, and while he thinks the seeds are too heavy to be wind-blown, yet the prevalent wind is from the north-west, and it occasionally blows with great violence.

W. W. Smith reported it in 1903 as spreading in the Ashburton County.

T. W. Adams of Greendale, Canterbury, a leading authority on arboriculture, says:

> *Pinus pinaster* has so well established itself in some parts of the country as to have been considered by some enthusiast or other as a native, and many years ago seeds were sent to Europe as such, and named *Pinus Nova-Zealandica*. I have seen it myself growing quite in a natural manner in all sorts of unlikely places but I look upon it as a very useless pine to cultivate.

Mr J. Attwood (March, 1916) says this species is plentiful at Riverhead in the Waitemata district. A few miles out of Silverdale on the Auckland Road, hundreds of acres of scrub-covered land are sprinkled with this tree.

Pinus radiata, Don. (*P. insignis*)

Cockayne says of this species that it is spreading naturally in pumice country about Taupo, and is becoming common. He considers that it is distributed by wind, as is the case on the Hurunui Plains, and at Glyn Wye. Similar evidence is given in the report of the Royal Commission on Forestry in 1913 (probably from the same source), where it is stated to be seeding freely and to be spreading, especially on manuka country. Maxwell also reports it as spreading to a distance from plantations, but attributes its dispersal to birds. He states that the seedlings occur mostly on clay banks or bare clay places.

A correspondent of Mr Jas. Drummond's says it grows strongly at Manganuiteao, west of Ruapehu, and reproduces itself among the scrub.

Sequoia sempervirens, Endl. Californian Red-wood

Self-sown seedlings of this species were found in the plantations at Hora Hora (Whangarei) in April, 1919.

Chapter XII

MONOCOTYLEDONS AND FERNS

MONOCOTYLEDONS

EPIGYNÆ

HYDROCHARIDEÆ

Elodea canadensis (*Anacharis canadensis*), Planch.
Canadian Water-thyme

According to Armstrong (who wrote in 1871) this species was introduced in 1868 by the Canterbury Society, and put into their ponds, from whence it escaped into the Avon. It was brought over from Tasmania at various times in the water containing perch, tench and goldfish, and was also sent over to the Society by Mr Morton Allport of Hobart "as a valuable pond-weed for fish." Only female plants were introduced, but it spreads by portions of the stem, and is now abundant throughout New Zealand. In the Avon at the present time (1916) it is being strangled in many parts by a species of *Nitella*.

Vallisneria spiralis, Linn.

This was first introduced into Canterbury by Mr A. M. Johnson in 1864. In 1885, Mr E. Bartley introduced it from Melbourne to Auckland, and placed some in Lake Takapuna, where it soon increased, and became a very serious pest. Fortunately it does not seed, as only one sex—the female—was introduced. This is the case also in Australia.

Ottelia ovalifolia, L. Rich.

This species was first observed by Cheeseman in 1897 in ponds near Manakau Harbour, but whether brought purposely from Australia or introduced by accident, is not known. It is now found in lakes, ponds and streams in the Auckland district, and I have met with it at Hawera. It is becoming a very serious pest in the Waikato.

SCITAMINEÆ

Canna indica, Linn. Indian Shot

First recorded by Cheeseman in 1882 as a garden escape of moderately frequent occurrence in Auckland Province. The same report appears in the *Manual*, 1906.

IRIDEÆ

Iris germanica, Linn. Flag-lily

First recorded in 1869 by Kirk as occurring in the Thames Goldfield, and in Wellington in 1877. Cheeseman in 1882 reports that "this species, originally a garden escape, has now firmly established itself in most districts." It is abundant in many parts of the North Island, but has scarcely spread in the south, where it never ripens seeds.

Visited in Europe by *Bombus hortorum*.

Iris susiana, Linn.

In 1877 Kirk recorded this species among introduced plants of Wellington, as "not unfrequent on the sites of abandoned homesteads, etc." It has probably died out.

Iris pseud-acorus, Linn. Yellow Flag

First recorded in 1877 by Kirk as occurring in the Hutt Valley near Wellington. Kirk also recorded it later from near Nelson, and S. Percy Smith from Mount Egmont Ranges. It occurs in many localities in water holes, ponds and lagoons, but has in most places been purposely introduced.

It is visited in Europe by *Apis mellifica* and *Bombus hortorum*.

Sisyrinchium chilense, Hook.

Probably this was the species collected at Matamata and recorded by Kirk in 1869 as *S. anceps*, Linn. It was reported again by him in 1877 from the hills about Wellington. In 1906 Cheeseman records it as occurring in "fields and waste places from Auckland to Otago, but often local." Cockayne records it from the neighbourhood of Queenstown in 1908. Not uncommon about Wellington in 1918.

In November, 1918, I saw sparrows working among wild plants of this species, near the Botanical Gardens, Wellington. They jumped on the plants so as to bend the flowering stems, and appeared to be sucking the nectar. I could not see or find anything else they could get.

Sisyrinchium Bermudianum, Linn. Blue-eyed Grass

Buchanan recorded this from Kawau in 1876, and Cheeseman from fields in Auckland Isthmus in 1882. As he does not refer to it in the *Manual*, it may be that it is a synonym of the preceding species.

Sisyrinchium striatum, Cavan.

Recorded in 1903 by W. W. Smith as occurring in Ashburton County.

Sisyrinchium micranthum, Cav.

First recorded by Cheeseman in 1896 from the North Cape district. In the *Manual* (1906) he defines the locality more exactly as "sandy shores of Spirits Bay, North Cape district."

Gladiolus byzantinus, Linn.

Recorded by Kirk in 1869 from the Auckland district, but apparently only a fugitive garden escape, which does not succeed in naturalising itself. In 1882 Cheeseman records *Gladiolus*, sp. as a frequent garden escape.

Antholyza æthiopica, Linn.

First recorded in the Auckland district in 1869 by T. Kirk. Cheeseman in 1882 states that it "has established itself in several localities near Auckland." In the *Manual* (1906) it is reported as occurring in "fields and waste places from Auckland to Otago, but often local."

Sparaxis tricolor, Ker.

Cheeseman records this in 1882 as a garden escape near Auckland, not common. Apparently it has not established itself.

AMARYLLIDEÆ

Agave americana, Linn. American Aloe

First recorded as a garden escape in the Auckland district by Kirk, and in 1877 near deserted homesteads in Wellington. Cheeseman in 1882 states that old plants throw up a multitude of suckers. The author noticed it in great profusion at Russell in the Bay of Islands in 1884.

Narcissus biflorus, Curt.

Recorded among introduced plants on Kapiti in 1906 by L. Cockayne.

DIOSCORIDEÆ

Dioscorea alata. Yam

This species was cultivated for a short time at the Bay of Islands, but by whom introduced, I am not sure, and was recorded by Hooker in 1864. Kirk could find no wild specimens in 1869. Polack says the Kaipara natives cultivated the yam, which he calls Kaipakeha, in the Kaipara district (1831); but his identifications are doubtful.

NUDIFLORÆ

BUTOMEÆ

Hydrocleis nymphæoides, Buchen.

Reported by Cheeseman in 1898 as occurring in the Thames Valley at Te Aroha. In 1912 he records the history of its introduction as follows:

> I am indebted to Mr Neve for numerous specimens of this handsome water-plant, which he informs me is now plentiful in several lagoons or backwaters near the Thames River, at Te Aroha, and is apparently rapidly increasing. According to inquiries kindly made for me by Mr Neve, it was planted nearly twenty years ago by a Mr Wood in a lagoon on his property about a mile and a half from Te Aroha. In this locality it now covers an area of more than an acre in extent, and has become a considerable nuisance, blocking up drains and water-channels. The beauty of the flowers has induced several settlers to transfer it to other localities near Te Aroha. In all of these it is rapidly increasing, and there is every probability of its spread in suitable places in the Thames Valley.

ALISMACEÆ

Alisma Plantago, Linn. Water Plantain

First recorded by Colenso as collected by H. Hill in water-courses in Hawke's Bay (R. Tukituki) in 1892. Later from the interior of Otago, Tokomairiro River, E. of Milton, by Petrie, and by Cheeseman from near Marton in 1906.

NAIADEÆ

Aponogeton distachyon, Thunb. Cape Water-lily

Kirk first records this in the Auckland district in 1869 as abundant in streams and ponds in the Bay of Islands and Whangarei, but says he was informed by Hutton that it was introduced by the missionaries at Waimate. Cheeseman says (1921) that it is common in shallow water throughout the North Island.

CORONARIEÆ

LILIACEÆ

Asparagus officinalis, Linn. Asparagus

Introduced early last century. First recorded as a garden escape by Kirk in 1869 in the Auckland district. He also reports it from near Wellington in 1877, and adds: "Solitary plants are sometimes found originating from seed carried by birds. It can scarcely expect to maintain its position, except perchance in maritime localities." Cheeseman in 1882 states that solitary plants are frequently seen, doubtless originating from seeds conveyed by birds from gardens.

Aloe latifolia, Haw.

First recorded by Cheeseman in 1882 as an escape from gardens near Auckland. Same record in 1906.

Asphodelus fistulosus, Linn.

First recorded in 1868 by Kirk as growing at Mongonui; where it was also reported as plentiful by Cheeseman in 1882. In the *Manual* (1906) it is reported as not uncommon on roadsides and in waste places from Mongonui to Napier. Townson also reports it from Westport.

Allium vineale, Linn. Crow Garlic; Wild Onion

Both Taylor (in *Te Ika a Maui*) and Polack state that "Marion sowed garlick, which has taken possession of the Bay of Islands." Captain Marion visited New Zealand in 1772. Darwin in 1835 says of the Bay of Islands district: "a leek has overrun whole districts, and will prove very troublesome; but it was imported as a favour by a French vessel." Cheeseman reports it in 1882 as "not uncommon, especially in abandoned Maori cultivations, and sandy flats near the sea." It seems to be confined to the Auckland provincial district. Polack states that the natives cultivated the garlic in the Kaipara district in 1831, but his identification is doubtful.

In Europe the species is visited by *Bombus lapidarius*.

Allium Cepa, Linn. Onion

Captain Cook sowed onions in his newly-cleared gardens in Queen Charlotte Sound in 1773, and they were found again when the spot was revisited in 1777.

Dieffenbach found them in native cultivations in New Plymouth in 1839.

Polack in 1831 speaks of shallots as cultivated in the Kaipara districts, and he distinguishes them from onions, which he also specifically names.

Allium Porrum, Linn. Leek

This species also was among the seeds sown by Furneaux in Queen Charlotte Sound in 1773. According to Buick: "it is said that to-day in remote parts of the Sound, leeks are still to be seen growing wild, the perpetuated progeny of those sown by Captain Furneaux in 1773." Dieffenbach found them cultivated in native gardens in New Plymouth in 1839.

Polack found them cultivated by Maoris in Kaipara in 1831.

Allium fragrans, Linn. Wild Onion; Sweet-scented Garlic

The Agricultural Department reported this in 1899 as "spreading in warmer portions of the Colony." The identification is doubtful; Cheeseman does not mention the species.

Cordyline terminalis, Kunth. Ti pore

In 1895 Mr T. Kirk described a species of palm lily which was found at Ahipara in a garden there, as *Cordyline Cheesemanii.* Two specimens were growing which had been found in different localities in the same neighbourhood. In 1900 Archdeacon Walsh described two specimens which were in the garden of a Miss M. A. Clarke in Waimate North, and which that lady found in a long-deserted native village in the neighbourhood. Mr J. B. Clarke told Canon Walsh that in the fifties the same plant was to be found in many native settlements in the north, and that he had seen it about Lake Omapere. The natives formerly used to eat the root, which was very large, long and succulent; they called the plant "*Ti pore.*" Cheeseman identified this with *Cordyline terminalis*, a species found very abundantly in Polynesia and westwards to India. The Ahipara specimens were found to belong to the same. In the *Manual of the New Zealand Flora*, Cheeseman says this species was

"formerly cultivated by the Maoris in the Bay of Islands and other northern districts; now nearly extinct." "Walsh mentions other instances of *C. terminalis* having been found in old Maori cultivations, and argues with much probability that the plant was originally introduced by the Maoris on their first colonisation of New Zealand."

A tradition, mentioned by Mr Best, and already referred to here (p. 12) states that the *ti* was brought by the Nukutere canoe which landed at Waiaua near Opitiki; that is about five hundred years ago.

In a recent letter Mr Elsdon Best writes (Aug. 1916) as follows regarding *Cordyline terminalis*:

I do not know and have not seen this. Walsh calls it *Ti pore*. But the *ti para*, which T. F. Cheeseman says is quite distinct from *C. terminalis*, was a cultivated species or variety on both coasts. The *ti tawhiti* of Taranaki seems to be the same thing. I have never seen the *ti para* growing wild, that is away from places where natives have lived. This is curious. Apparently it has no specific name. I am told that it is still found up the Whanganui River, where it is known as *ti Kowhiti*. It was cultivated for food purposes. The leaves of one I grew were an inch and a half wide; this came from the last (I think) surviving plant in the Bay of Plenty. The natives do not seem to have preserved traditions of the introduction of *Cordyline*; it was possibly brought, with the Karaka, from Sunday Island.

In the list of Maori names of plants in the Appendix to Cheeseman's Flora the name *ti para* is given on the authority of Williams as "*Cordyline* sp. cultivated for the sugary root."

COMMELYNEÆ

Tradescantia fluminensis, Vell.

Reported by Carse in 1915 from Mongonui County as growing on creek-banks and in lowland woods, spreading rapidly. In 1919 Mr Cheeseman states that it is a garden escape in many localities in the vicinity of Auckland, where it has received the local name of "wandering jew." It has become specially abundant on portions of the Mount Eden lava-fields. Mr Aston states that it is spreading fast in the vicinity of Wellington.

PONTEDERIACEÆ

Eichhornia crassipes, Solms. The Water-hyacinth

Introduced as an ornamental water-plant. Has become naturalised at Te Aroha in the Thames Valley; first recorded by Cheeseman in 1898. It is commonly cultivated in ponds and tanks, and tends to spread where not subjected to frost. In 1912 he reports it again on the authority of F. Neve, as established in a lagoon at Te Aroha, and issues a warning against its being allowed to spread. This is a most dangerous pest in many Australian rivers, causing floods by its completely blocking their channels. Fortunately most New Zealand streams are too cold for it.

JUNCEÆ

Juncus glaucus, Sibth.

Recorded by Kirk as occurring between Hokitika and Ross.

Juncus Gerhardi, Loisel

Reported from near Dunedin by Petrie, near Anderson's Bay, Otago Harbour. It has not been met with for over twenty years.

Juncus obtusiflorus, Ehr.

Found at the southern end of Lake Waihola in Otago, by Petrie. Not collected since its first discovery.

ARACEÆ

Richardia africana, Kunth (*Calla æthiopica*)
White Arum; Arum-lily

Probably introduced at an early date last century. First recorded as a garden escape in 1869 in the Auckland district by Kirk, and in 1877 from the Hutt, Wellington, "probably planted." Stated by Cheeseman in 1882 to be plentiful about Auckland and many country townships, in ditches and waste places. The author noticed it in abundance in 1884 in the Bay of Islands. It is truly naturalised in many places in Taranaki and Wellington as far south as Cook Strait; in the South Island it is mostly a garden escape.

Colocasia antiquorum, Schott. Taro

This plant was cultivated by the Maoris at the time of Captain Cook's visit to New Zealand, and Banks refers to it under the name of Cocos. According to Cheeseman it still lingers in many deserted plantations. It is extensively cultivated in several tropical regions, and there is no doubt the Maoris brought it with them from Polynesia.

Mr Elsdon Best informs me that "tradition says that one Roau came in the Nukutere canoe, landing at Waiaua near Opotiki; and brought with him the Karaka, the Ti and the *Taro*. The two latter are known as Te Huri a Roau." This was about twenty generations or five hundred years ago.

Mr S. Percy Smith says: "the strong probability is that it was introduced after the times of Toi" (circa 1150 A.D.) "and probably by most of the canoes subsequent to his time."

Polack, writing of 1831–37, says:

Various species of Taro are planted in the island, especially southward. The *taro oia* or soldier taro has a blue cast, within a thin atramentous skin. Another species (espèce) has a remarkable, lotus-like leaf, and thrives best in a swampy soil.

Canon Walsh states that Maori whalers in the early part of the 19th century introduced the *taro hoia*, a large coarse variety; evidently the one referred to by Polack.

According to Mr Best, the taro is now (1917) found growing in water courses in the Bay of Islands district, near old native cultivations.

Mr A. E. Pickmere of Te Aroha describes the flowering of the Taro, as does Miss Coutts of Onehunga, in June, 1910. The former thinks the plant seldom flowers in New Zealand. Mr Cheeseman says it occasionally flowers, but the natives regularly multiplied it by root-division[1].

[1] In Hooker's list of introduced plants in the Auckland district published in 1864, he gives *Alocasia indica* ((?) *A. macrorhiza*, Schott, and (?) *Calocasia macrorhiza*, Schott) as recently introduced by the natives. Kirk in 1869 reported that he had not seen specimens.

In Buchanan's list of naturalised plants occurring on Kawau in 1876, he gives *Caladium esculentum*, Willd., but, apparently by a slip records it under Cucurbitaceæ.

GLUMIFLORÆ

CYPERACEÆ

Cyperus rotundus, Linn. Nut-grass

Cheeseman writes in 1882: "the well-known 'nut-grass' has found its way into several gardens in the vicinity of Auckland, and is likely to prove a serious pest, as its numerous tubers make it difficult to eradicate." In 1906 he states that it is a troublesome weed in these localities.

Cyperus lucidus, R. Br.

First recorded (but not named) by Cheeseman in 1882, from Mongonui. Later R. H. Matthews and H. Carse report it from Rangaunu Harbour and Kaitaia, and Cheeseman records it from Nelson.

Cyperus vegetus, Willd.

Recorded by Kirk in 1895 as occurring on a ballast heap in Wellington. Since found in several localities in the Auckland and Wellington districts.

Carex divisa, Huds.

Stated in the *Manual* (1906) to occur rarely in waste places near Auckland.

Carex muricata, Linn.

Reported from Pelorus Valley, Marlborough, by J. Rutland.

Carex flacca, Schreb.

In the *Manual* (1906) this is recorded as occurring in fields and waste places in various localities in the North Island; found at Whangarei by H. Carse, in the vicinity of Auckland by Cheeseman, and near Wellington by Kirk.

Carex panicea, Linn.

First recorded by Cheeseman in 1882, from the vicinity of Auckland, and Mahurangi.

Carex longifolia, R. Br.

Recorded in the *Manual* (1906) as occurring in fields near Auckland, rare[1].

[1] In 1903 W. W. Smith recorded *Sclavia verticillata*, Palisot, from Ashburton County. There does not appear to be any grass with this generic name, and the nearest approaches to it appear to be *Sclaria*, a labiate allied to *Salvia*, and *Scleria* a genus belonging to the Cyperaceæ.

GRAMINEÆ

Andropogon annulatus, Forsk.

Cheeseman reports this in the *Manual* in 1906 as occurring at Mongonui, where Carse also found it in 1910.

Anthistiria imberbis, Retz. Kangaroo-grass

First recorded in 1864 in Hooker's list as *A. australis*. Again referred to in Kirk's list of Auckland plants in 1869 as having been found by Dr Sinclair, and there named *A. ciliata*, Linn.

Apparently this is the species recorded by Kirk as *A. australis*, Br., in 1877, which was collected on the Lower Rangitata. In the *Manual* (1906) it is reported to occur in fields and waste places at the Bay of Islands, Whangaparaoa near Auckland, and in the vicinity of Wellington.

Paspalum dilatatum, Poir

First recorded by Kirk in 1895 as occurring on a heap of ballast (from Buenos Ayres) in Wellington. In 1896 recorded by Cheeseman from the North Cape district. Messrs Yates and Co. (of Auckland and Sydney) claim that they introduced it as a fodder plant from Australia in 1895, but it is a South American species. Cheeseman speaks of it in 1906 as "increasing in several localities." My son, G. Stuart Thomson, writing from Whangarei in 1916, says:

> it is now one of the most aggressive plants we have in the north, and is looked on by some farmers as a pest, as bad as tall fescue. When it gets a hold in cultivated land, ploughing is almost an impossibility.

It is now used as a pasture grass in Westland.

Panicum miliaceum, Linn. Millet

In 1772 Crozet sowed seeds of this grain in the garden he formed on Moutouaro Island, and apparently, it sprouted and appeared above ground before the expedition sailed from New Zealand. It has not, however, been cultivated to any extent in this country, and has never been found in the wild state.

Panicum sanguinale, Linn. Crab-grass

First recorded in Hooker's list in 1864. Kirk reports it from near Castle Point in 1877, as *Digitaria sanguinalis*, Scop. Recorded in 1882 by Cheeseman as a common and troublesome weed in light rich soils throughout the Auckland district. In the *Manual* (1906) it is characterised as a common weed in waste places and cultivated ground.

Panicum glabrum, Gaud.

First recorded from the Auckland district by Kirk in 1869 as *Digitaria humifusa*, Pers. In 1882 Cheeseman recorded it from the vicinity of Auckland, but not common. In the *Manual* (1906) it is stated to be an occasional weed in cultivated ground.

Panicum Crus-galli, Linn. Cockspur-grass

First recorded from Auckland district by Kirk in 1870 as *Echinochloa crus-galli*. Stated by Cheeseman in 1882 to be not common. In the *Manual* (1906) it is reported as not uncommon in waste places as far south as Canterbury and Westport.

Panicum colonum, Linn.

First recorded in Hooker's list in 1864. Cheeseman reports it in 1882 only at Onehunga. In the *Manual* (1906) it is said to occur rarely in waste places from Auckland to Wellington.

Panicum Linderheimeri, Nash

Recorded from near Kaitaia, Mongonui County, in 1915, as rare; found by H. B. Matthews. Cheeseman states (1919) that it was originally found on the summit of a hill by Kerikeri Pa, near Kaitaia; but it has since been observed in several localities in the district.

Setaria glauca, Beauv.

First recorded in Hooker's list in 1864 as *Panicum glaucum*. Reported in 1882 by Cheeseman as a weed in a few gardens at Onehunga. In the *Manual* (1906) it is reported as not uncommon in waste places and cultivated fields.

Setaria verticillata, Beauv.

In the *Manual* (1906) stated to occur at Napier, collected by Colenso, and at Ashburton by W. W. Smith.

Setaria viridis, Beauv.

First recorded from the Auckland district by Kirk in 1869. Cheeseman reports it in 1882 from the vicinity of Auckland, rare. Kirk refers to *S. italica*, P. de Beauv., as having been first observed by W. T. Bassett at Papatoitoi in 1863, and "now (1869) being found for miles by the roadsides." Cheeseman thinks (1916) that Kirk's plant is *S. viridis*. In the *Manual* (1906) it is stated to be not uncommon in waste places and cultivated fields in the North Island.

Setaria imberbis, Roem. and Schult.

Collected by Kirk on a ballast heap at Wellington in 1895. It did not increase, and has not been found since.

Stenotaphrum glabrum, Trin. Buffalo-grass

First recorded in 1871 by Kirk among introduced grasses in Auckland Province. Cheeseman says of it in 1882: "Has been planted in many localities, and in some is spreading; but, as it seldom ripens perfect seed, its increase is necessarily slow." In the *Manual* (1906) it is said to be a common escape from cultivation in the North Island. It is now (1921) abundant in some of the Auckland sand-dunes.

Zizania aquatica, Linn. Canadian Wild Rice

Recorded by Cheeseman in the *Manual* (1906) as being "naturalised by the Northern Wairoa River, near Aratapu." Various acclimatisation societies are constantly trying to introduce this and other allied grasses as food for wild ducks.

Phalaris canariensis, Linn. Canary-grass

First recorded in Hooker's list in 1864; then by Kirk from Wellington in 1877. It is now (1917) an abundant weed in fields and waste places in both islands.

Phalaris arundinacea, Linn. Reed-grass

First recorded by Kirk in 1877 as occurring "by a tributary of the Waiwetu; probably planted." It does not seem to have been found anywhere else, or recorded since.

Zea Mays, Linn. Maize; Indian Corn

In 1793 Governor King gave the natives of the Bay of Islands district two bushels of Maize, and they appear to have made good use of the seed. Nicholas, who accompanied Marsden to New Zealand, recorded Indian corn as being cultivated freely in 1914; and Earle reports it as very abundant in and about Russell in 1827.

The first recorded introduction of maize is in 1772, when Crozet formed a garden on Moutouaro Island, and sowed seeds of this among other grains. He says: "everything succeeded admirably, several of the grains sprouted and appeared above ground," before the expedition sailed away from New Zealand.

I am not aware of its occurrence anywhere in the country as a naturalised form.

Anthoxanthum odoratum, Linn. Sweet Vernal-grass

Probably introduced early last century; first recorded in Hooker's list in 1864. It occurs in every subsequent list of introduced grasses, and is remarkably abundant in all parts of the country. It flowers from September to November.

Stipa verticillata, Nees

In the *Manual* (1906) reported by Kirk from near Wellington, and by Travers from Nelson.

Phleum pratense, Linn. Timothy

Certainly an early introduction; first recorded in Hooker's list in 1864. Cheeseman remarks (1882) that it is not nearly so abundant as it should be, considering the extent to which it is sown. The author's experience is that it dies out of most pastures in two or three years. It is, however, abundant in grass lands throughout New Zealand.

Phleum arenarium, Linn.

Recorded by W. W. Smith in 1903 from Ashburton County. Not observed since.

Alopecurus pratensis, Linn. Meadow Foxtail

Introduced at an early date into New Zealand; first recorded by the author as occurring in Southland in 1870. Cheeseman reported it in 1882 in fields from Auckland to Waikato, but not common, and in the *Manual* (1906) as not uncommon in meadows and pastures in both islands. One of our best introduced grasses.

Alopecurus agrestis, Linn. Slender Foxtail; Field Foxtail

First recorded in Hooker's list in 1864. Cheeseman reported it in 1882 as occurring in most districts of the north in fields and by road-sides, but nowhere common. In 1906 the same comment holds good.

Milium effusum, Linn. Millet-grass

Recorded by W. W. Smith in 1903 as occurring in Ashburton County. Not collected since.

Polypogon littoralis, Sm.

Recorded by W. W. Smith in 1903 from Ashburton County, but not collected since.

Polypogon monspeliensis, Desf. Beard-grass

First recorded by Kirk in 1877 from Wellington, then by the author in 1878, as occurring on heaps of tailings, etc., in Strath-Taieri. Cheeseman in 1882 reported it from muddy places on the shores of the Manukau and Waitemata harbours, increasing rapidly. In the *Manual* (1906) it is reported as abundant on roadsides and in waste places in both islands.

Polypogon fugax, Nees.

First recorded by Cheeseman in 1882 as occurring and increasing rapidly in waste places and ditches on the Auckland Isthmus, and at Thames. In 1912 he states: "this has spread very rapidly in the Auckland Provincial District, and is now common in most districts in brackish-water swamps."

Agrostis vulgaris, With. Red-top;
Fine Bent-grass; Black Couch

First mentioned in Hooker's list in 1864. Now (1917) very abundant in fields and waste places throughout both islands. A useless weed, very troublesome in arable land.

Agrostis alba, Linn. Fiorin

No doubt introduced at an early date, but recorded for the first time by Kirk in 1877 from Wellington and then by Cheeseman in 1882, as generally distributed in the Auckland district, but affecting somewhat stiff and damp soils. An abundant grass in pastures and waste ground in 1917, in all cultivated parts of New Zealand. Frequently crowds out other and better grasses.

Gastridium australe, Beauv. Nit-grass

First recorded by T. Kirk in 1867, in introduced plants of Great Barrier Island, as *G. lendigerum*, Gaud., and in 1869 in list of Auckland plants as *G. lanigerum*. In 1877 it was reported from Wellington on the authority of Buchanan. Cheeseman in 1882 reported it from Auckland Isthmus, Waitakerei and Otahuhu, and in the *Manual* (1906) stated that it was not uncommon in waste places and on roadsides in the North Island.

Ammophila arundinacea, Host. Marram-grass

First recorded in 1872 by Buchanan, from Miramar, Wellington. In 1882 Cheeseman stated that it had been planted on the west coast of the Auckland Province to check the progress of sand-dunes and that "it may be expected to increase, as it has done at Taranaki and Nelson." It was planted freely about Dunedin before 1894, and at New Brighton near Christchurch in 1899; and has since been distributed somewhat freely wherever efforts are made to bind drifting sand. I do not think it has spread to any very great extent naturally.

Lagurus ovatus, Linn. Hare's-tail Grass

First recorded in 1872 by Buchanan from Miramar, Wellington. According to Cheeseman, extremely plentiful on Motuihi Island in 1882, but rare in Auckland. It is not uncommon in waste places, especially near the sea, in both islands and is common on wet ground in parts of Central Otago.

Holcus lanatus, Linn. Yorkshire-fog; Soft-grass

Probably introduced very early; first recorded by Hooker in 1864. A most abundant grass in fields and on roadsides, in both islands. In damp land this is readily eaten by cattle, but as a general rule it is a poor grass, and takes the place of better pasture grasses. According to Cheeseman sheep eat this more readily than any of the indigenous grasses on runs.

Holcus mollis, Linn. Fog; Soft-grass

With the last an early introduction; first recorded by Hooker in 1864. Also a common grass in similar localities to those occupied by the preceding species, but by no means so abundant.

Aira Caryophyllea, Linn. Hair-grass

First recorded by Kirk in 1870 from the neighbourhood of Auckland, where it was common in most localities in 1882. Also in 1877 from Wellington. In 1917 an abundant grass in meadows, waste places and sides of roads.

Aira præcox, Linn. Hair-grass

Cheeseman reports that a few plants of this species were observed near Waiuku (Auckland) in 1877. In the *Manual* (1906) it is reported as not uncommon in fields and waste places in both islands.

Deschampsia flexuosa, Trin. Wavy Hair-grass

First recorded by Cheeseman in 1882 as occurring in fields on the Auckland Isthmus. In the *Manual* (1906) said to occur locally in heathy places in both islands.

Eleusine indica, Gærtn.

This species is indigenous in the Kermadec Islands, but has been introduced into New Zealand, and was first recorded in 1864, in Hooker's list, as occurring near Auckland. It has also been met with at Westport.

Eleusine coracana, Gærtn.

Recorded by Kirk in 1895 as occurring on a ballast heap at Wellington. It evidently did not become established.

Avena sativa, Linn. Common Oat

Apparently first introduced by Captain Cook in 1773. Forster tells us that in the various gardens made in Queen Charlotte Sound "he had likewise sown corn of several sorts." In 1882 Cheeseman made the interesting statement that this species "has become extensively naturalised on sea-cliffs in the northern and central portions of the Auckland district, in addition to frequently occurring in fields as an escape from cultivation." It is chiefly as an escape that it is met with now (1921) in all parts of New Zealand.

Avena fatua, Linn. Wild Oat Grass; Havers

First recorded in 1871 by Armstrong, from the Canterbury district. In the *Manual* (1906) stated to be abundant in waste places and cultivated fields throughout New Zealand.

Avena strigosa, Schreb. Hairy Oat; Bristle-pointed Oat

I do not find this recorded earlier than 1906, in Cheeseman's *Manual*, where it is reported to occur in cultivated fields in both islands, not common. It has probably been mostly overlooked.

Avena pubescens, Huds. Downy Oat-grass

Recorded by Cheeseman in the *Manual* (1906) as occurring not uncommonly in fields in Canterbury and Otago.

Avena flavescens, Linn. Yellow Oat-grass

Introduced as a pasture grass into Southland in 1870 by the author's father, but does not seem to have established itself.

Arrhenatherum avenaceum, Beauv. Tall Oat Grass

First recorded in 1870 by Kirk from the Auckland district, and in the following year by Armstrong from the Canterbury district as *A. bulbosum*. Again by Kirk in 1877 from near Wellington. In the *Manual* (1906) it is stated to occur, but not commonly, in fields and waste places in both islands.

Cynodon dactylon, Pers. Doab-grass

First recorded in Hooker's list in 1864. In 1877 Kirk stated that it occurred at Castle Point, and also formerly at Te Aro Beach, Wellington. Stated by Cheeseman in 1882 to be plentiful throughout the Auckland district. In the *Manual* (1906) reported as abundant in fields and waste places in both islands. It is, however, chiefly a tropical or sub-tropical species, and is not common in the southern half of the South Island.

Triodia decumbens, Beauv.

First recorded by Cheeseman in 1882 from Kaipara. In the *Manual* (1906) he records it as not uncommon in fields and waste places in the North Island.

Eragrostis major, Host.

In the *Manual* (1906) this is reported, on the authority of S. J. Vining, to occur at Mangatangi, near Mercer.

Eragrostis minor, Host.

First recorded by Kirk in 1895, as occurring on a heap of ballast (from Buenos Ayres) in Wellington. Apparently not seen since.

Eragrostis Brownii, Nees

According to Kirk (1869) this grass was abundant over a large district at Keri Keri in the Bay of Islands, where it was first observed by H. T. Kemp in 1865. It was then spreading rapidly. In 1882 Cheeseman further reported it from Northern Wairoa, Whangarei, and near Auckland. In the *Manual* (1906) it is recorded as abundant in fields and waste places in the North Island.

Briza maxima, Linn. Trembling-grass

First recorded in Hooker's list in 1864. Plentiful about Wellington in 1877 according to Kirk. Cheeseman in 1882 reports it from two or three localities in the neighbourhood of Auckland. In the *Manual* (1906) reported as an occasional garden escape.

Briza minor, Linn. Lesser Trembling-grass

First recorded in Hooker's list in 1864, then by Kirk in 1877 from Wellington. Very generally distributed in the north by 1882. In the *Manual* (1906) stated to be not uncommon in fields and waste places in both islands.

Dactylis glomerata, Linn. Cock's-foot Grass

Probably introduced early last century; first recorded in Hooker's list of 1864. Now (1917) one of the most abundant grasses in pastures and waste places throughout both islands. In many districts (e.g. in Banks' Peninsula), the harvesting of cock's-foot for seed, is one of the important events of the year. In New Zealand this grass has become a most agressive plant, and in light bush tends to crowd out all other undergrowth.

Cynosurus cristatus, Linn. Dog's-tail Grass

First recorded in 1864 in Hooker's list. In 1877 Kirk states that it is more generally naturalised in the Wellington district than any other grass. In 1882 Cheeseman reported it from the Auckland district as not uncommon in various parts on stiff soils. Now (1921) one of the most abundant grasses in the Dominion, especially in parts of the South Island.

Cynosurus echinatus, Linn.

In the *Manual* (1906) recorded, on the authority of Bishop Williams, as occurring in waste places near Gisborne. In 1919 Cheeseman recorded it from roadsides at Waihi.

Arundinaria macrosperma, Michx.

Reported by Cheeseman in 1882 as lingering in several old Maori cultivations, but cannot be looked upon as truly naturalised. In 1917 he states it still lingers on in one or two abandoned cultivations, but no instance has been observed of its reproducing itself.

Brachypodium pinnatum, Beauv.

I collected this grass in 1897 near the Ross Creek Reservoir, Dunedin, but have not found it since. It has not succeeded in establishing itself.

Poa annua, Linn. Annual Meadow-grass

Probably one of the earliest introduced grasses; first recorded in 1864 in Hooker's list. The commonest of all grasses on roadsides and in waste places, especially where regularly trodden on.

Poa compressa, Linn.

First recorded by Cheeseman in 1882 as occurring on the Auckland Isthmus, but not commonly. In the *Manual* (1906) reported as occurring, somewhat sparsely, in fields and waste places in the North Island.

Poa pratensis, Linn. Meadow-grass

No doubt a very early introduction; first recorded in Kirk's list of introduced plants in Great Barrier Island in 1867. One of the most abundant grasses in fields and waste places in all parts of New Zealand. In some districts, it takes exclusive possession of the pastures, and is most difficult to eradicate. It is in such parts looked on as a dangerous "couch" grass, and is a pest in cultivation.

Poa trivialis, Linn.

First recorded in Hooker's list in 1864. In 1877 stated by Kirk to be frequent about Wellington and other places in the district. Reported by Cheeseman in 1882 from waste places about Auckland, and occasionally in pastures. In the *Manual* (1906) said to be not uncommon in fields and waste places throughout both islands.

Poa nemoralis, Linn.

First recorded from Southland in 1870 by the author; then by Cheeseman in 1882 as occurring in the Auckland Domain. In the *Manual* (1906) reported as occurring sparsely in shaded places in both islands.

Glyceria aquatica, Wahlenb. Reed-grass

First recorded from the Ashburton district in 1903 by W. W. Smith. Later by Petrie from the Taieri Plains and Catlins River. It is not uncommon in Otago and Southland.

Glyceria fluitans, R. Br.

First recorded by Buchanan in 1876 as occurring in Kawau. In 1877 Kirk reported it as common in many localities in Wellington district. In 1882 Cheeseman wrote that this grass "made its appearance in some wet places on the Auckland harbour reclamations, about two years ago, but has lately been destroyed." Reported by the author from the Taieri Plain in 1894. In the *Manual* (1906) reported to be not uncommon in wet places in both islands.

Atropis distans, Griseb.

Recorded by Cheeseman in the *Manual* (1906) as occurring not uncommonly in salt marshes in both islands.

Festuca elatior, Linn. Meadow Fescue

First recorded from Southland in 1870 by the author. In the *Manual* (1906) reported as not uncommon in meadows and pastures in both islands. A bad weed in wet pastures and flax areas of the Wellington district.

Festuca ovina, Linn. Sheep's Fescue

First recorded from Southland in 1870 by the author. In the *Manual* (1906) reported as not uncommon in pastures and waste places in both islands. This species is also indigenous in New Zealand, but there is no doubt that much, if not most, of the sheep's fescue is the introduced species. It ranges up to 4500 feet on the southern mountains.

Festuca rubra, Linn.

First recorded from Southland in 1870 by the author. The distribution is the same as that of the preceding species, and like it, the introduced form is probably more abundant than the indigenous. (See Appendix B, p. 563.)

Festuca Myurus, Linn.

First recorded in 1870 by Kirk from the Auckland district as *F. sciuroides*, and from Wellington in 1877. Writing of Auckland plants in 1882, Cheeseman stated: "the true plant is by no means abundant, but is increasing. The variety *sciuroides* = *F. bromoides*, Sm., is plentiful through the district." In the *Manual* (1906) it is stated to occur, not uncommonly in dry places, in both islands.

Festuca bromoides, Linn.

First recorded in 1864 in Hooker's list. In the *Manual* (1906) stated to be abundant in waste places and pastures in both islands.

Festuca pratensis, Hudson

First recorded from Southland in 1870 by the author. Does not seem to have established itself, and has not been met with since.

Sclerochloa rigida, Griseb. = (*Glyceria rigida*, Sm.)

First recorded by Kirk in 1895, as occurring on a heap of ballast (from Buenos Ayres), but stated by him to occur in Hawke's Bay, Otago, etc. In the *Manual* (1906) reported as occurring, but not commonly, in roadsides and waste places in both islands.

Bromus erectus, Huds.

First recorded from the Auckland district by Kirk in 1869. In the *Manual* (1906) reported as occurring, not uncommonly, in fields and waste places in both islands.

Bromus sterilis, Linn. Sterile Brome-grass

First recorded in 1864 in Hooker's list, and in 1877 by Kirk from Wellington. In 1882 Cheeseman reported it from Auckland as plentiful, especially in waste or sandy places near the sea. In the *Manual* (1906) reported as abundant in fields and waste places in both islands.

Bromus madritensis, Linn.

First recorded in 1870 by Kirk from the Auckland district. In the *Manual* (1906) stated to occur in "waste places and roadsides, apparently not common," in both islands.

Bromus tectorum, Linn.

First recorded from the Auckland district by Kirk in 1869. Cheeseman in 1906 states: "I have not seen New Zealand specimens."

Bromus mollis, Linn. Soft Brome-grass

First recorded in 1864 in Hooker's list. An abundant weed in fields and waste places in most parts of New Zealand; useless for stock.

Bromus racemosus, Linn. Smooth Brome-grass

First recorded in 1864 in Hooker's list. Equally abundant as a weed, along with the preceding species, in fields and waste places.

Bromus commutatus, Schrad.

First recorded in 1864 in Hooker's list. Fairly abundant in most parts of New Zealand in fields and waste places.

Bromus patulus, Mert. and Koch.

First recorded from the Auckland district in 1869 by Kirk. Cheeseman, in the *Manual* (1906), has apparently overlooked this occurrence, and reports it as found on a ballast heap in Wellington by Kirk in 1895.

Bromus arvensis, Linn. Field Brome-grass

First recorded from the Auckland district by Kirk in 1869, and by Buchanan from Wellington in 1877. Stated by Cheeseman to occur in waste places near Auckland, but not commonly. In the *Manual* (1906) it is reported to occur in waste places in both islands.

Bromus vestitus, Thunb.

Recorded by Kirk in 1895 as occurring on a ballast heap in Wellington. Not observed again.

Bromus secalinus, Linn.

Recorded by W. W. Smith in 1903 from Ashburton County; not collected since.

Bromus unioloides, H. B. K.

First recorded in Kirk's list of Great Barrier plants in 1867 as *B. Schrœderi*, and again from the Auckland district in 1869 as *Ceratochloa unioloides*. It was also collected in Wellington in 1877. In 1882 it was stated by Cheeseman to be the prevailing grass in many of the streets and waste places about Auckland, but as not so common in the country districts, as it will not bear close cropping. In the *Manual* (1906) it is stated to be abundant in fields and waste places in both islands. This does not hold for Otago and Southland, where it is uncommon.

Lolium perenne, Linn. Rye-grass

Probably introduced early last century. John Balleny, who visited Chalky Inlet in 1838, found it growing there "admirably." Recorded later by Hooker in 1864. The most generally used pasture grass in New Zealand, abundant throughout the country. Carse has an interesting note on this species in his notes on the flora of Mongonui County (1915). He says:

> As a rule, this grass is not very enduring, being subject to "rust," but near Kaitaia is a large area of river-flat, sown with this grass more than sixty years ago, where it has held its place without deterioration.

Lolium italicum, A. Br. Italian Rye-grass

First recorded in 1870 by Kirk from the Auckland district. In the *Manual* (1906) stated to be not uncommon in fields and pastures in both islands. It is, however, less permanent than the preceding species, except in the far north, where it seems to be the dominant form.

Lolium temulentum, Linn. Darnel

First recorded in 1864 in Hooker's list. In the *Manual* (1906) it is reported as not uncommon in cultivated fields and waste places in both islands.

Lepturus incurvatus, Trin.

First recorded by Kirk as occurring on the sands at Waitemata, in 1864 in Hooker's list. He also reports it as "common on shingly beaches" in many parts round Wellington Harbour, in 1877. Cheeseman in 1882 stated that it was common in brackish-water swamps in the Auckland district. It had apparently not greatly extended its range in 1906. It is a grass specially of salt meadows.

Agropyrum repens, Beauv. Couch-grass; Twitch

Probably introduced long previously, but first recorded by W. W. Smith from Ashburton in 1903, as *Triticum repens*. In the *Manual* (1906) reported as not uncommon in fields and waste places in both islands. This grass, considered one of the worst weeds of arable land in Britain, has fortunately shown very little tendency to become a pest in New Zealand. This is somewhat remarkable, as the plant in ordinary soil puts out underground stems from 25 to 30 cm. (10–12 inches) in length.

Agropyrum pectinatum, Beauv.

First recorded from Central Otago by Petrie in 1899. Later from Hawke's Bay by A. Hamilton.

Brachypodium sylvaticum, R. and S.

Recorded by W. W. Smith from Ashburton County in 1903.

Triticum sativum, Lam. Wheat

Apparently wheat was first introduced by Crozet in 1772, who formed a garden on Moutouaru Island, in which he sowed, among other seeds, grains of wheat from the Cape of Good Hope. He writes: "everything succeeded admirably, and the wheat especially grew with surpassing vigour." In the following year Captain Cook sowed wheat in the gardens made by him on Long Island in Queen Charlotte Sound. It did not, however, become established. In 1793 Governor King gave the natives of the Bay of Islands a bushel of wheat. It is doubtful, however, whether they succeeded in retaining it in cultivation, for Marsden re-introduced it in the same locality in 1810, and sent some more over from Sydney in 1911. Nicholas, who accompanied Marsden to the Bay of Islands on his first visit there, saw it growing well in native cultivations in January, 1815. Bidwill, when

32—2

travelling on the shores of Lake Taupo in 1839, "found a fine plant of wheat. The natives could not say how it got there, and Mr Chapman, the Missionary, was the only European who had ever been there, and that only three weeks before."

At the present time wheat only occurs wild as a very occasional escape from cultivation.

Triticum caninum, Huds.

Recorded by W. W. Smith in 1903 as occurring in Ashburton County; not found since.

Triticum junceum, Linn.

Recorded by W. W. Smith in 1903 from Ashburton County; not gathered since.

Elymus arenarius, Linn. Sea-lyme Grass

The Agricultural Department reports this as well distributed in New Zealand, but it is everywhere a planted species, which has been utilised in several parts for binding sand-dunes. It does not appear to have spread anywhere.

Hordeum vulgare, Linn. Barley

Probably included in the corn sown by Captain Cook in 1773 in the gardens in Queen Charlotte Sound. Only occurs as an escape from cultivation, and nowhere grows as a wild species.

Nicholas, who accompanied Marsden to New Zealand in 1814–15, speaks of a plantation of Siberian barley at the Bay of Islands in February of the latter year.

A correspondent of the *Farmers' Union Advocate* gives the following interesting extract from the diary of the late Rev. J. G. Butler:

> On the morning of Wednesday, May 3, 1820, the agricultural plough was for the first time put into the land of New Zealand at Kiddikiddi (Kerikeri), and I felt much pleasure in holding it after a team of six bullocks, brought down by the "Dromedary." I trust that this auspicious day will be remembered with gratitude and its anniversary kept by ages yet unborn. Every heart seemed to rejoice on the occasion. I hope it will still continue to increase, and in a short time produce an abundant harvest.

The results of the ploughing were reaped from 12th December, 1820, to 8th January, 1821, regarding which Mr Butler wrote: "While I am writing down these lines, I have seven natives in sight reaping down a field of barley on land which 12 months ago was overrun with ferns." The Rev. Mr Butler left New Zealand in 1823 and returned about 1840, and settled at Petone, where he died.

Hordeum murinum, Linn. Barley Grass; Way-bent

First recorded from Auckland Province in 1869 by Kirk. It is a most abundant weed now (1917) in waste ground, especially near the sea. The seed gets into the wool of sheep, and frequently causes a certain amount of deterioration.

Hordeum maritimum, With. Squirrel-tail Grass

First recorded in 1871 from Canterbury by Armstrong. In the *Manual* (1906) it is reported from Akaroa, on the authority of Kirk[1].

Order FILICES

Tribe POLYPODIEÆ

Pteris cretica, Linn.

This species was collected near Tapuacharuru, Taupo, and described by Colenso as *P. lomarioides*, Col., in 1880. A specimen was also found in Oxford Forest, Canterbury, by Mr Vincent Pyke in 1883, and sent down to the author for identification. This was gathered in the bush, away from the immediate vicinity of settlement, and Mr Pyke was certain that it was indigenous.

Pteris longifolia, Linn.

Stated by Buchanan (1882) to have been gathered at Tarawera, between Napier and Taupo, by Mr Lascelles of Napier. Buchanan says it was collected "under circumstances which preclude the possibility of its having been introduced." No one has yet recorded how far fern-spores have travelled on air-currents.

Tribe OSMUNDEÆ

Osmunda regalis, Linn. Fern-royal

This fern was found by Dr Curll at one spot in Rangitikei, from whence it disappeared, and was rediscovered at the same place by H. C. Field of Wanganui in 1885. It does not seem to have spread. The introduction of all three species of foreign ferns was probably quite accidental in each case.

Order LYCOPODIACEÆ

Selaginella denticulata, Link.

Cheeseman (1919) reports, on the authority of T. H. Trevor, that this species has been known for many years as a garden escape at Pakaraka, Bay of Islands, and has lately appeared in great abundance on the banks of several swampy creeks in the neighbourhood. B.C. Aston also reports it as not uncommon in several localities near Wellington.

[1] In 1871 Armstrong recorded *Hordeum distichum*, from Canterbury. The species has not been observed since.

Part IV

Chapter XIII

INTERACTION OF ENDEMIC AND INTRODUCED FAUNAS

THE enormous impetus given to Natural History by the publication of Darwin's *Origin of Species* in 1859 led many of those who were interested in biological studies in New Zealand to give much consideration to the relation of his theoretical views to the problems which faced them here. I was among those who came early under the spell. I read with care and avidity every work of Darwin's which bore on the subject of evolution and natural selection, and followed this up by a careful study of contemporary writers, Hooker, Lyell, Huxley, Wallace, Asa Grey, Haeckel and numerous others. As a lad I had some slight knowledge of the British fauna, and had begun to collect the flora under my old teacher of botany, Professor John Hutton Balfour of Edinburgh. I made extensive botanical and zoological collections in many parts of New Zealand from the Bay of Islands to the south of Stewart Island between 1868 and 1892, collaborating first with that prince of field naturalists, Captain F. W. Hutton, and later with Mr D. Petrie and Professor C. Chilton. At the end of 1882, an injury which effectually lamed me, prevented the prosecution of further field work, but I continued to collect and work on the invertebrate fauna, especially the Crustacea.

The conviction early grew upon me that here in New Zealand was a field in which the accuracy of Darwin's views in certain directions could be put to the test. The way in which certain species of introduced animals and plants seemed to "run away," as it were, from their recognised specific characters, led to the expectation that new forms would spring up in this country under the altered conditions, and that we should here observe the "origin" of new species. I certainly was not alone in this half-expectation. It was somewhat generally, though vaguely, held. Examples were apparently numerous. Rabbits increased at an appalling rate, and appeared to be developing many coloured breeds; small birds—especially common sparrows, greenfinches, skylarks, etc.—multiplied prodigiously, and we were on the lookout for all sorts of changes in colour, food, nesting habits, colours of eggs, and so on; trout grew so rapidly in the streams and lakes into which they were introduced as to belie all previously recorded experience. The same thing occurred among plants. Water-cress—a plant of two to four feet in length in European waters—grew in some streams to a length of from twelve to fourteen feet, and with stems as thick as one's wrist; the common spear-thistle, which is from two to five feet high in Britain, formed in some districts vast impenetrable thickets six to seven feet in height; brambles, briars and other weeds took possession of whole districts, and threatened to choke out all other vegetation. It seemed indeed as if the laws of natural selection and the principle of the survival of the fittest had been temporarily suspended, and nature was running riot.

It is no wonder that all the younger naturalists in the country were almost inclined to think that perhaps we would here see the rise of new varieties, which would become "fixed," and would soon rank as "species," distinct from those from which they were descended. It may be that our ideas as to what constituted varietal and specific distinction were somewhat vague—in that respect we were not very different from the majority of those who used these terms in a somewhat loose manner—but there they were. Some such feeling was still in my mind as late as 1891, when I read a short paper before the Biological Section of the Australasian Association for the Advancement of Science at its Christchurch meeting, "On Some Aspects of Acclimatisation in New Zealand." The following sentences show the trend of my views at the time:

> One of the most interesting points connected with the successful naturalisation of foreign species is the observation of the changes which they undergo in their altered conditions. Nearly all our introduced animals have been brought from lands where the struggle for existence is very keen, and where natural enemies abound. In their new home they have been set free from these old trammels, and the enemies have been left behind. Under such circumstances it is not surprising to find that sports in colour, which in Europe would be strictly eliminated as soon as they appeared, owing to their rendering their possessors too conspicuous to their enemies, are here preserved and tend to be reproduced.

I then went on to instance what occurs among hares, rabbits, sparrows and other birds, humble-bees, etc., both with regard to change of colour and of habits. My mind was evidently quite prepared to find such changes, though I had to admit that the evidence sought for was not forthcoming to any extent.

The subject continued to occupy my thoughts from time to time, but I was not able to devote much consecutive attention to it until 1915, when it seemed advisable to me to resume the thread of my ideas as propounded twenty years previously, and see how far these views were correct. It soon, however, became apparent to me that it would be a much more important work, and one preliminary to any exhaustive examination of the subject, to ascertain—as far as possible—what species of animals and plants had become naturalised in the country; what species had failed to establish themselves; to seek for the reasons of their success or failure; and to ascertain what effects had been produced on the native fauna and flora.

Changes in Indigenous Fauna due to Naturalisation of Foreign Species

The question is sometimes asked: "What effects have been produced on the native fauna by the introduction of foreign animals and plants into the islands of the New Zealand group?" Any answer which can be given can only at best suggest some of the changes which have taken place. The effects have been so far-reaching and so complex that it is impossible to present any summary of them, and all that can be done is to show various aspects of the problem, and to consider facts in detail.

When knowledge of the native fauna first began to be acquired considerable changes had already commenced to take place, and others were in progress. It was, however, long before any systematic knowledge of the indigenous animals was accumulated and published, and it must be borne in mind that as far as the terrestrial invertebrates are concerned, this knowledge is still very fragmentary. Indeed for many groups, as for example, that of the Insects, it is probably the case that the fullest catalogue which could be made to-day would not include many species which were in existence at the time of Cook's first visit to these islands. Probably very many species have either ceased to exist, or have become very rare. There is no actual knowledge of the fact, it is only inferred from what we know to have taken place with regard to native birds and lizards. But that profound changes have taken place is familiar to all who have observed the development of the last fifty years. The changes have, necessarily, been most rapid during the latter half of last century and since, and especially from the time that the various acclimatisation societies sprang into activity.

An examination of the published papers and books on the zoology of New Zealand shows that very much of the present knowledge of the subject is quite recent. Since the days of Captain Cook isolated papers on the zoology appeared in many and various publications, especially during last century, and a good deal of general information was accumulated. A good summary of this is to be found in Hutton's Introduction to the *Index Faunæ Novæ Zealandiæ*. Apparently the first full lists, as far as they go, are contained in the appendix to Dieffenbach's *Travels*, published in 1843. In this work, J. E. Gray gives a catalogue of 84 species of birds and six species of lizards.

In 1862 Gray in *The Ibis* gave a synopsis of the birds collected in the "Voyage of M.M.S.S. Erebus and Terror," and enumerated 122 species as occurring in New Zealand and the Chatham Islands. Buller's *Birds of New Zealand*, published in 1873, contained descriptions of 147 species; his *Manual* in 1882, gave 176 species; and the second edition of the *Birds of New Zealand* in 1888, recorded 195 species. The *Index Faunæ Novæ Zealandiæ* (1904) gives 194 species of permanent birds, 222 of wandering species, and 15 species of lizards.

Fishes are first recorded in 1843 in Richardson's list (Dieffenbach), where 92 species are named. In Hector and Hutton's Catalogue in 1872, 141 species are recorded; the Index (1903) gives 254 names. Yet the fishes are very imperfectly known even yet.

J. E. Gray in 1843 gives a list of 222 species of Mollusca and 3 species of Brachiopoda; Hutton's *Manual of Mollusca*, published in 1880, gives 598 species of Mollusca, 8 of Brachiopoda, and 191 of Polyzoa; while Suter's *Manual* in 1913 gives 1187 species of Mollusca.

White and Doubleday wrote the account of Insecta for Dieffenbach's work in 1843, and referred to the following numbers of insects, namely: Orthoptera, 4 species; Hymenoptera, 5; Coleoptera, 50; Lepidoptera, 23; Diptera, 6; Neuroptera, 2; Homoptera, 4; and Hemiptera, 3; a total of 97 species. In 1873 Hutton published a list of New Zealand insects of the above eight orders containing 742 species, of which 265 were Coleoptera. Captain Broun brought out his first *Manual of Coleoptera* in 1880, and it contained 1321 species. In 1881 Hutton's Catalogue of Diptera, Orthoptera and Hymenoptera brought up the number of species in these three orders from 151 (1873) to 227. In 1898 Hudson's *Moths and Butterflies* described 238 species of Macro-lepidoptera alone. Finally, the following numbers are given in the *Index Faunæ Novæ Zealandiæ* (1903): Hymenoptera, 155; Lepidoptera, 608; Diptera, 343; Coleoptera, 2787; Hemiptera, 166; Neuroptera, 60; Orthoptera, 17; and Aptera, 3; making a total of 4139 species, a number which has been considerably added to since. Many of these orders of insects are still very imperfectly known.

White and Doubleday in 1843 recorded 11 spiders and 2 Myriapods; the Index (1903) gives 251 species of Arachnida and 30 of Myriapods. Similarly the same authors name 30 species of Crustacea; Miers' Catalogue of Stalk- and Sessile-eyed Crustacea (1876) describes 140 species; while the Index records 532 species—a number which has since been increased to over 600 species.

I quote this rather formidable list of figures to show how greatly our systematic knowledge of the fauna has been extended within the last seventy years.

Mr Hedley suggests that as the European or cosmopolitan weed-florula is able, with human assistance, to oust the indigenous flora, so there may be a weed-faunula (mouse, sparrow, snail, etc.) which may act similarly on the indigenous fauna.

At the date of my arrival in the country (1868) the following nine species of birds, which have either altogether or to a great extent disappeared since, were to be met with, in some cases fairly commonly: native crows (*Glaucopis wilsoni* and *G. cinerea*), huia (*Heteralocha acutirostris*), native thrushes (*Turnagra tanagri* and *T. crassirostris*), stitch-bird (*Pogonornis cincta*), kakapo (*Stringops habroptilus*), native quail (*Coturnix novæ-zealandiæ*), and the white heron (*Herodias timoriensis*). In addition, several species of originally very limited distribution (e.g. Stephens Island Wren and Chatham Island Fern-Bird) are now quite extinct.

Others which were abundant have been driven back into areas where settlement has not yet penetrated to any great extent[1].

It must not be supposed that it is the introduced animals alone which have produced this effect, even though rats, cats, rabbits, pigs, cattle, stoats and weasels, as well perhaps as some kinds of introduced birds, have penetrated beyond the settled districts. It is largely the direct disturbance of their haunts and breeding places, and the interference with their food supply, which has caused this destruction and diminution of the native fauna.

What is true of birds is equally true of other groups of animals, though it is more difficult to arrive at the facts. Many insects which were common in the bush fifty years ago must have been displaced and have largely disappeared. I cannot appeal to figures, but the surface burning of open land which prevailed, especially in the South Island, and the wanton destruction and burning of forest which has marked so much of the North Island clearing, must have destroyed an astonishing amount of native insect life, and made room for introduced forms. The clearing of the surface for cultivation and grazing, the draining of swamps, and the sowing down of wide areas in European

[1] They include the following sixteen species: native robins (*Miro albifrons* and *M. australis*), saddle-back (*Creadion carunculatus*), fern-bird or grass-bird (*Sphenœacus punctatus*), native canary (*Mohua ochrocephala*), tui (*Prosthemadera novæ-zealandiæ*), kaka (*Nestor meridionalis*), parakeets (*Cyanorhamphus novæ-zealandiæ* and *C. auriceps*), kiwis (*Apteryx mantelli, A. australis, A. oweni* and *A. haastii*), laughing owl (*Sceloglaux albifacies*), Mo-pork (*Ninox novæ-zealandiæ*), and native pigeon (*Hemiphaga novæ-zealandiæ*). In some settled localities, such as the neighbourhood of Wellington, Mo-porks are still met with.

pasture plants, have all contributed to this wholesale destruction and displacement of indigenous species. The disappearance of mosquitoes and sand-flies in settled districts where they were formerly very common is a case in point;—where they disappeared, it is tolerably certain other species not so well known became scarce at the same time. Unfortunately accurate data are not obtainable, for at the time that settlement began, very little was known of the insect fauna as I have shown, and even now, as already stated, whole groups are imperfectly known.

The introduction of many insectivorous birds, including pheasants, quail, starlings, minahs, species of Fringillidæ, and others, must also have accounted for the destruction of immense numbers of insects. At one time the wide tussock-covered hills and plains of both islands were just alive with grasshoppers, and in the summer months they sprang up before the pedestrian literally in thousands. To-day they are rare over wide areas where formerly they abounded. Fereday wrote in 1872:

> In the early days of the Canterbury Settlement quails, larks and other birds that fed upon insects and their larvæ abounded on the plains, but the quails have been exterminated, the larks have become comparatively scarce, and the other birds have almost disappeared. So long as the plains remained open and uncultivated, extensive grass fires, sweeping over the land consumed an enormous amount of insect life, and took the place of the counter-check which was being removed by the decrease of the birds; but within the last few years inclosures and cultivation have been rapidly extending around Christchurch, and forming a nursery for the preservation and increase of the insect race. A luxuriant and abundant vegetation has sprung up for its food and shelter, and it is comparatively freed from the ravages of fire and the attacks of its feathered foes.

Lizards (especially *Naultinus elegans*, *N. grayi* and *Lygosoma moco*) were also extremely common in the open country fifty years ago, but are now comparatively rare. Burning the surface growth is largely responsible for their destruction, but the abundance of wild cats, which in some parts are encouraged as counteracting to some extent the rabbit pest, is also accountable for many.

The introduction of species of Salmonidæ, of perch, carp, etc., into the lakes and rivers of most parts of the country has produced equally important changes. Some species of fish have been exterminated wherever the introduced forms have established themselves, e.g. native grayling (*Prototroctes oxyrhynchus*), kokopuru, minnow, etc. (species of *Galaxias*), and smelt (*Retropinna richardsoni*). Similarly the native species of crayfish (*Paranephrops planifrons*, *P. zealandicus* and *P. setosus*), shrimp (*Xiphocaris curvirostris*), and species of fresh-water Amphipoda, have been eaten out in streams stocked with trout.

Immense reduction has also taken place among aquatic species of Diptera, Neuroptera, etc.[1]

Among naturalists of the last half century there has been a strong belief that the introduced fauna and flora have been directly responsible for the diminution of many, and the disappearance of some indigenous animals and plants[2].

Thus Darwin in the *Origin of Species* says:

> From the extraordinary manner in which European productions have recently spread over New Zealand, and have seized on places which must have been previously occupied, we may believe, if all the animals and plants of Great Britain were set free in New Zealand, that in the course of time a multitude of British forms would become thoroughly naturalised there, *and would exterminate many of the natives*. On the other hand, from what we see now occurring in New Zealand, and from hardly a single inhabitant of the southern hemisphere having become wild in any part of Europe, we may doubt, if all the productions of New Zealand were set free in Great Britain, whether any considerable number would be enabled to seize on places now occupied by our native plants and animals. (The italics are mine.)

In examining this subject in connection with the introduced plant life I think it will be shown that where man does not interfere with the vegetation, the indigenous species can hold their own against the imported forms. It is human intervention—either direct or indirect—which completely alters the conditions. The same probably holds good to some extent of the animal life, only the problem is more difficult to follow out.

Against the wholesale destruction of native animals which has taken place within the last half century, there have to be recorded some cases of increase in the native fauna, apparently due to some adaptation to the altered conditions.

One of the most interesting cases is that of the bell-bird or mako-mako (*Anthornis melanura*). This species has very largely disappeared from the North Island, though I cannot learn from any one why it

[1] Mr Elsdon Best reports the disappearance of the grayling from the rivers on the east coast of the North Island. Mr O'Regan also records the same for the rivers on the west coast of the South Island. The latter says the popular belief is that the trout have exterminated them, but he says these fish disappeared from the Inangahua before trout were introduced, and he never heard of trout eating graylings. Though formerly they were very common on the west coast, he does not now know of any river where they may be found.

Mr Murray Campbell says the graylings come up the Cascade River in South Westland in great shoals for the greater part of the year. (April, 1910.)

[2] A. R. Wallace (in *Darwinism*, 3rd edition, p. 34) says: "A native fly is being supplanted by the European house-fly." This kind of statement is the sort of terse sentence which is apt to be quoted by writers, but it is both utterly indefinite and quite misleading. No one can say what native fly is meant, nor is it certain what European house-fly is referred. It may however be stated *definitely that no species of European fly has supplanted a native fly*.

should have done so. It was common in Whangarei in 1860, and then began to get scarce. It was, however, common about the watershed of the Wanganui and Mangawhero rivers in 1909, and is also very abundant on some outlying islands, e.g. Mayor Island. But in the South Island it has certainly become more abundant in recent years, even though there has been very wholesale destruction of bush. It is many years since Mr W. W. Smith reported its increase in the forest belt of South Canterbury. I have myself noted it year by year in the neighbourhood of Dunedin, where it is found in the Town Belt and in suburban gardens during many months of the year, especially when certain trees are in flower. In Southland, Mr J. Crosby Smith and Mr Philpott also report the species as increasing. The case is certainly peculiar, for the tui (*Prosthemadera*) persists in the North Island, but has become comparatively rare in the settled districts in the South Island. The habits and food of the two birds are somewhat similar, only the tui is a much more conspicuous bird.

The harrier (*Circus gouldii*) is another bird which has certainly increased to a great extent.

Mr Richard Reynolds of Cambridge, Waikato, an enthusiastic sportsman and naturalist, writes me as follows (June 23rd, 1916):

> The morning after £50 worth of partridges had been liberated here, my son found eight of them killed, with a hawk on each; he shot a rabbit, poisoned it with strychnine and secured eighteen hawks with it. In driving from here to Tepapa (24 miles) a few days ago I passed (within sight) 162 hawks, 17 of them in one bunch.

This gives some idea of the enormous abundance of this bird in the North Island. These hawks visit Stephens Island in Cook Strait apparently in pursuit of young petrels. But as the Island is a preserve for the Tuatara lizard, the lighthouse keepers are instructed to destroy all hawks met with. I am informed by Mr Newton of the Public Works Department that between 23rd January, 1917, and 28th February, 1919, no fewer than 1582 hawks were killed. They all seemed to come from the North Island *via* Kapiti. I have no corresponding figures for the South, but in open country the hawk is the bird most in evidence everywhere. They are protected birds on account of the damage they are supposed to do to rabbits, but even where they are very abundant the rabbit thrives without much hindrance from the hawk. In the North they are more destructive to hares than to rabbits.

Some of the introduced small birds attack the harrier, as their ancestors in Britain attacked kites and sparrow-hawks. Some larger birds also "go for" this hawk. Mr Holman of Whangarei tells me he has often seen guinea-fowls attacking and beating off the hawks from the chickens.

The increase of this bird has no doubt been due to the vast increase of rabbits, whose young are often caught by it, and of small birds; and also to the protection accorded to the species. A humourist could find excellent material on which to exercise his talents were he to summarise the literature and the oratory which has been expended on this hawk in New Zealand. The sheep-farmer protects it because it is more or less destructive to rabbits which are his great pest, though it is really too slow in its movements to keep down such an active and wary creature as the rabbit. Formerly it lived mainly on lizards and insects, but it has now a much larger menu. Wherever game fanciers prevailed, the bird was mercilessly destroyed, and rewards were offered for heads, as it certainly is a most active agent in keeping down pheasants, quail, wild ducks and other imported game. Where farmers grew grain crops the hawk was protected as an antidote to the small bird pest; but where poultry were largely kept, it was ruthlessly trapped and destroyed on account of its predilection for chickens. Altogether the record is a very mixed one, but on the whole the bird seems to have more than held its own, and is very common throughout New Zealand.

The grey warbler (*Pseudogerygone igata*), yellow-breasted tit (*Petrœca macrocephala*), the fan-tailed flycatchers (*Rhipidura flabellifera* and *R. fuliginosa*), and the pipit or ground lark (*Anthus novæ-zealandiæ*) appear to have more than held their own. This may be due to a certain measure of protection accorded them by settlers, but they have been mostly left alone, and their most active enemy is the cat. Yet it is quite surprising how common the ground larks are in all open country, and especially in country districts, and how the others named penetrate into gardens, even in the larger towns. Their food supply consists mostly of small insects and other invertebrates, many of them introduced species, and these are very abundant.

The wax-eye or blight-bird (*Zosterops cærulescens*) has apparently increased very much since it was first recorded in 1832; the facts which are known as to its spread and increase are recorded at p. 161. These little birds are very fond of meat, and especially of fat, and come about houses and stock-yards for the sake of the animal food to be obtained. Dr Fulton says that many of his correspondents consider that the long-tailed cuckoo (*Urodynamis taitensis*) has become increasingly numerous during the past thirty years, and they attribute this to the increase of small European birds, on whose eggs and young they feed. In and about trout-hatcheries they do a great deal of mischief. They also destroy numbers of young birds, such as sparrows, wax-eyes (*Zosterops*), goldfinches, etc.

Recorded cases of increase among other groups of indigenous animals other than birds, are very few.

The two moths, whose larvæ are known as "flax-grubs"—*Xanthorhoë præfectata* and *Melanchra steropastis*—are both considered to be much more abundant now than they were formerly. It may be that by the clearing away of other native vegetation on which they formerly fed, their attacks are now made more persistently on flax-covered areas, but the pest has assumed such serious proportions in recent years, as to lead to combined efforts to cope with it.

Another native moth—*Œceticus omnivorus*—appears to have increased also to a considerable extent. Within the last few years also very many complaints have been received of the steady increase of a moth *Venusia verriculata*, whose larva feeds on and occasionally nearly destroys the cabbage-tree (*Cordyline australis*).

The common magpie-moth (*Nyctemera annulata*) has certainly become extremely abundant wherever the introduced ragwort (*Senecio jacobæa*) has become a common pest. Formerly *Nyctemera* appears to have fed chiefly on the indigenous *Senecio bellidioides*, *S. Lautus* and glabrous species of *Erechtites*, but it has transferred its attention now chiefly to the introduced species of *Senecio* (including *S. vulgaris* and *S. mikanoides*). Wherever ragwort has spread and become an abundant weed the *Nyctemera* has also increased enormously, and may be seen rising in vast swarms from the plants during the adult moth stage. I have never seen a bird catching the moths, and the hairy caterpillars appear to be very distasteful to them. I know chickens will not touch them. Hence they seem to increase almost without check from enemies. The only bird I have heard of as eating them is the shining cuckoo (*Chalcococcyx lucidus*); my informant is Mr Holman, Curator of the Whangarei Acclimatisation Society, who is a careful observer.

It is difficult to get any information as to changes in the abundance of native insects due to the introduction of foreign species of plants. One species of beetle, *Odontria zealandica*, the common "grass-grub," has apparently greatly increased with the introduction and increase of European grasses. It is extraordinarily abundant in some pastures, and is most destructive to lawns. Two allied species, *O. puncticollis* and *O. striata*, have been found to be very destructive to seedling trees in the State nurseries. The former is most active in the North Island, destroying sometimes as much as 30 to 40 per cent. of the seedling larch (*Larix Europæa*). *O. striata* is most in evidence in the South Island, where it destroys both larch and *Pinus laricio*. The common longicorn beetle—*Prionoplus reticularis*—cannot be at all so common as it was before the wholesale destruction of forest trees in which its larva used to bore. But it has now learned to bore into certain introduced plants, and I have received from Mr W. W. Smith of New Plymouth, specimens of oak trees (*Quercus robur* var. *pedunculata*), and of lilac (*Syringa vulgaris*) very freely tunnelled by these large grubs.

Changes in introduced Fauna since Naturalisation

In the introduction to this chapter I stated that the observations of naturalists thirty or forty years ago led to the prevalent belief that in this new country variation would proceed very rapidly among introduced species of animals and plants owing to the removal or rather absence of those checks which acted upon the species in their native habitats. After nearly fifty years of fairly close observation I have to state very definitely that such a belief has been absolutely dissipated. *I am aware of no definite permanent change in any introduced species.*

For a long time I collected all the information I could lay my hands on as to albinism in birds, expecting to find that it was greatly on the increase. Buller referring to this matter says:

> A remarkable feature in the New Zealand avifauna is the inherent tendency to albinism. The condition itself is no doubt due to the absence of colouring pigment in the feathers; but the difficulty is to find any sufficient cause for this in a temperate climate like that of New Zealand. In India, as is well known, the tendency is in the opposite direction, melanism being of very frequent occurrence. Strange to say, there is the same tendency to albinism in the imported birds. Albino sparrows are *far more common than they are in their native country*, and even the skylark not unfrequently changes its sober dress for a yellowish-white one.

The italics are mine, but the statement shows that Buller—writing nearly fifty years ago—was of opinion that colour changes were rapidly taking place.

Writing in 1891 I said:

> Among house sparrows in particular, variation in colour and especially development of white feathers is extremely common, and is certainly on the increase. From evidence I have collected in Otago, I find that the development of white plumage, usually in the wings and tail, is a very common feature, and by no means confined to sparrows. I have numerous recorded cases of the occurrence of more or less complete albinism, as well as of the development of bright colours in thrushes, blackbirds, linnets, skylarks and starlings. Such variations are, of course, not uncommon in the original habitats of all these birds, but it never tends to increase. Here, on the other hand, it is of frequent occurrence, and seems to me to be very decidedly on the increase. My own observations lead me to think that birds with any bizarre or distinctly abnormal colouring are wilder and more shy than their normally coloured fellows; but as

I have only watched them in the neighbourhood of towns, where they become objects of interest to passers-by, and especially to those great enemies of birds—the small boys,—they are in such localities subjected to an amount of attention and persecution which they do not receive in more sequestered parts.

I have attempted to find whether the colour changes recorded are permanent, or merely transient, but have not been able to obtain evidence on the subject. I think it will be found on examination that very frequently abnormal colouring at one period of the bird's life may disappear at succeeding moults. Mr A. Binnie tells me that a dealer got a pure white goldfinch in a catch near Dunedin one day. He refused 10*s*. for it. At next moult it reverted to the natural colour. He also stated that another bird was nearly black, but it retained its colour at successive moults.

Writing now (1917), nearly thirty years later, I consider that the inferences I drew from my own observations were wrong. I was so busy looking for anomalous characters that I met with a good many, and so came to the conclusion that they were on the increase, but I cannot find now any more white, coloured, or white-feathered birds than were to be found in 1876. There are different varieties of birds in different parts, but no variation seems to be now taking place more than occurred before. The race of house-sparrows common in Wellington appears to be of a darker brown colour than those common in Canterbury and Otago, but this would merely show that a darker strain was originally imported into the Wellington district.

Similarly my observations on rabbits do not show that there is any tendency to the production and perpetuation of multi-coloured varieties. I thought differently in 1891, when I wrote:

> In the neighbourhood of towns and villages, where cats and dogs, and sportsmen abound, the sober greys and browns of the wild rabbit are the colours commonly seen; but in the country districts, away from all enemies except professional rabbiters and phosphorised oats, neither of which are likely to exert much selective action on their colour, it is as common to see black, white, yellow, and piebald rabbits as the ordinary greys.

It is clear that I did not attach much importance to the native harrier as an agency for keeping down the rabbit pest, nor do I now. The subject could almost be treated numerically, for skins are exported by the million every year, but I have been unable to obtain exact figures. However, rabbit-skin dealers and exporters tell me that black, white, buff, yellow and piebald skins do not tend to increase at all in numbers. One large exporter informed me that greys amount to about 97 per cent. of all the skins which come into the market.

Mr B. C. Aston, in an account of a trip to Marlborough makes an interesting remark on the colour changes in introduced animals, to this effect:

> The strong insolation of the area, is, I think, affecting the colour of the wild animals, goats, pigs and rabbits. Where the normal colour is khaki, as in rabbits, melanism is very frequent, but in the Clarence valley, bicoloration is common. This in a pig makes for better concealment. I noticed this particularly. A huge spotted boar running on a hillside about a mile away was conspicuous owing to his movement, but he disappeared as soon as he stopped.

Seasonal changes common to certain species in the Northern Hemisphere tend to be reproduced in New Zealand, wherever the climatal changes correspond. Thus in the colder parts of Canterbury and Otago, where severe winter frosts are often experienced, hares, stoats and weasels tend to grow white winter coats. But I have no record of such changes occurring in the warmer districts.

The introduction of stoats and weasels, which was undertaken primarily as an antidote to the rabbit pest, has had other far-reaching consequences on the introduced fauna. It has certainly helped to prevent the establishment of such species of birds as partridges, ducks and other game birds; and it has probably also greatly reduced the reproduction and spread of well-established birds like pheasants. On the other hand it has enormously reduced the numbers of rats and mice in the country, and in this way may have indirectly helped in the preservation of the native weka (*Ocydromus*), and the Californian quail, both of which appear to have increased of late years, in districts where stoats and weasels are quite common. Probably also the humble-bees have benefited by the destruction of mice.

It might be thought that the reversal of the breeding season between the northern and southern hemispheres might affect introduced species, but no such changes have been recorded. With introduced individuals it might be noticeable. I am informed by bird fanciers and dealers that some introduced birds suffer very seriously at the moulting season. This is particularly the case with imported canaries; but these are highly artificial and domesticated birds.

Scarcely any attention has been paid to the relations between introduced animals and plants and indigenous species. It is, however, probable that native ichneumon flies have found suitable hosts in several species of introduced insects, but no one seems to have kept any definite record of the facts, if they have occurred.

One definite example has been frequently reported. A native lady-bird—*Coccinella tasmanii*, White—which is common in both islands, has been found to be very destructive to melons, cucumbers, etc., and also to *Schizoneura langigera*, Haus., the woolly aphis.

The larva of a native moth—*Cacœcia excessana*, Walker—forms a tough netting and covers it mimetically with dead leaves, under which it lives. Mr W. W. Smith has noted it as common on the surface of the leaves of the laurel (*Prunus lauro-cerasus*), and on many of the cultivated fruit trees. Other examples of interaction between native insects and introduced plants are given in the next chapter.

Chapter XIV

ALTERATION IN FLORA SINCE EUROPEAN OCCUPATION OF NEW ZEALAND

THE changes produced on the indigenous vegetation by the introduction of new types of animal life have long been under observation by many competent naturalists, and are more easily followed than those which have affected the native fauna. The subject has been dealt with by many writers. I have dealt with it in some detail when referring to individual species of animals, so I will only give a general summary here.

Cattle, sheep, goats, horses, deer, pigs, rats, and especially rabbits, have done a great deal of damage in many localities by eating down, and in some cases by eating out the native vegetation. Some mammals and several kinds of birds have actually brought about an increase of the indigenous flora by distributing the seeds of a few species of plants. Introduced insects have produced very considerable changes, but they are somewhat difficult to follow, and have only been partially recorded.

No instance can be recorded of any species of native plant which has been exterminated owing to this or any other cause dependent on European occupation of the islands. But local extermination has taken place, and the following cases can be cited.

Lepidium oleraceum, Forst., was formerly abundant round the coasts of all parts of New Zealand, but is now becoming very scarce. Cheeseman says of it:

> It was originally discovered by Banks and Solander during Cook's first voyage, and at that time must have been abundant, for Dr Solander speaks of it as "*copiose in littoribus marinis*," and Cook states that boat-loads of it were collected and used as an antiscorbutic by his crew. It is now quite extinct in several of the localities he visited, and is fast becoming rare in others. Its disappearance is due to *cattle* and *sheep*, which greedily eat it down in any locality they can reach.

Aston remarks that it "has been eaten out along the Wellington coast, and is now generally only to be found growing on inaccessible rock-faces." The same is to be said of it in Otago, where it is now extremely rare, though abundant forty years ago.

Speaking of *Hibiscus Trionum*, Linn., and *H. diversifolius*, Jacq., Cheeseman says: "Both are being rapidly destroyed by *cattle*, fires, etc., and are now rare or almost extinct in localities where they were plentiful twenty or thirty years ago."

Entelea arborescens, R. Br., is "greedily eaten by *cattle* and *horses*, and consequently is fast becoming rare on the mainland, except in comparatively inaccessible situations. It is still plentiful on most of the small outlying islands on the north-east coast of the Auckland district."

Pomaderris apetala, Lab. This species was formerly abundant at Kawhia, but is now extinct there, having been completely exterminated apparently by *goats*—according to Kirk. It is now restricted to the South Head of the Mokau River, and to the Chatham Islands. It is clear, however, that it must formerly have had a wider distribution over New Zealand, for it is a common Australian species, and therefore it would seem probable that its extinction in this country was proceeding apace even before goats were introduced.

Clianthus puniceus, Banks and Sol. This plant, according to Cheeseman, is "exceedingly rare and local in the wild state, and is fast becoming extinct. It was formerly cultivated by the Maoris in many localities on the shores of the North Island." I mention this species here, not as an example of a plant undergoing extermination owing to the direct action of man or of the animals introduced by him, but as one which was evidently dying out when Captain Cook first landed in the country. Why it should have been in this decadent state I cannot suggest, but it must have been widely spread in former times, for it and an Australian species *C. Dampieri*, A. Cunn., are the only two species known. It is now common as a cultivated shrub, but does not seem able to re-establish itself naturally.

There is a popular idea that *Clianthus puniceus* was introduced by the Maoris. This is expressed by Taylor in *Te Ika a Maui* as follows:

> A better gift, was, I believe, the Kowai-ngutu-kaka (*Clianthus puniceus*), which was most probably introduced by his ship. The Taranaki slaves, when released by the Ngapuhi on their embracing Christianity, took the seed with them as a remembrance of the land of their captivity.

The botanist would naturally ask, from whence did they bring the plant? There is no trace of a *Clianthus* in any Melanesian or Polynesian island, as far as is known.

Angelica Gingidium, Hook. f. This plant was formerly extremely common throughout both islands, but "as it is everywhere greedily eaten by stock, it has become scarce in many districts."

Myostidium nobile, Hook., is now found in many gardens in both islands, especially near the sea, where it thrives. But in its native habitat in the Chatham Islands it is now quite rare as a wild plant, though formerly it was very abundant. *Cattle* and *pigs* have eaten it out. This is another species which was on the verge of extinction when discovered. It must formerly have occurred over a considerable portion of the New Zealand area, for its relationships are more with those of the Australian species of *Cynoglossum* than with *Myosotis*, which is still well represented in this country.

Gastrodia Cunninghamii, Hook. f. This orchid has large thick tuberous roots which contain erythro-dextrin in place of starch. These are eaten greedily by *pigs* and by both common species of *rats* (*M. rattus* and *M. decumanus*). Kirk remarks that it

has become very rare in districts where the *black rat* is plentiful. On one occasion, in 1874, I found three remarkably fine specimens, quite two feet in height, with tubers 6 in. or 7 in. in length, and placed them in what seemed a safe place in a hut at Omaha, but during the night they were carried off by the rodents.

I have noticed this disappearance of this species from the neighbourhood of Dunedin, where it used to be common during the last thirty years, but am more inclined to put it down to another cause than the fondness of rats for its roots. It is commonly given as a character of this genus that its roots are usually parasitic on the roots of other plants. My own observations would tend to make me question the accuracy of this statement, at least I have never been able to verify it. A great number of bush plants appear to be dependent on symbiotic fungi for their nourishment, and this, I am inclined to think is one of them. It certainly will not grow on any garden soil, but the same difficulty is experienced with other bush plants which are not root parasites. I should term it a saprophyte.

Marattia fraxinea, Smith. This fine fern is also fast disappearing. Cheeseman says: "The large starchy rhizome was formerly eaten by the Maoris, and hence the plant was occasionally cultivated near their villages." Aston blames the *pigs* for eating it out, as they are also very fond of the rhizomes.

Local destruction of the native flora is quite a different matter from extermination, and frequently is not due so much to introduced animals as to the direct action of man. Thus by bush clearing, surface burning, draining, and breaking up the surface for cultivation, immense areas have had their native vegetation nearly quite destroyed, and introduced animals have assisted in this work.

A remarkable example of this is to be seen in Central Otago where an area of many thousands of acres of what was once good grazing land, covered with heavy tussock grasses, has had its surface nearly denuded of vegetation and has been reduced to the condition of a desert. In this area the three active agencies in desiccating the country have been fire, overstocking *with sheep*, and *rabbits*. By enclosing areas with rabbit-proof fences, and keeping grass-feeding animals completely off it, it has been found possible to re-grass certain areas and if this policy is persisted in, the country will be again redeemed and vegetation will largely reassert itself.

The Town Belt of Dunedin is an interesting tract of ground, of limited extent, on which it has been possible to study the changes in vegetation which have taken place during the last half century. I have had it under close observation since 1871. The Belt is a strip of land only a few chains wide which surrounds the city, and is of very diversified character. Its total area is about 500 acres. In 1871 considerable changes from its primitive condition had already taken place, for settlement began in 1848, the gold rushes to Otago began after 1863, and the heavy bush was everywhere cut out, not only outside the limits of the town, but even within the Belt itself. Most of the timber that would burn had likewise been destroyed. But there gradually sprang up a feeling that the Belt must in future be preserved intact as much as possible, and it is now many years since anything in the way of plants or timber has been carried out of the area. It has also been strictly preserved from cattle, sheep, etc., and the few rabbits which have occasionally managed to exist, have been too few to do any damage. Thus it comes about that the changes which have more recently taken place have been largely due to other causes than fire and animals. One of these is no doubt the firming of the ground in many parts by people moving about on it, the opening up of the undergrowth and consequent drying of the surface, and the consequent disappearance of the masses of symbiotic fungi which are so characteristic of unbroken forest. Another cause has been the great increase of introduced fruit-eating birds, which have spread plants with succulent fruits very extensively.

The following native species which were not uncommon in various parts of the Belt in 1871, have not been found for some years past:

Plagianthus divaricatus, Forst.
Elæocarpus dentatus, Vahl.
,, *Hookerianus*, Raoul.
Pennantia corymbosa, Forst.
Nertera setulosa, Hook. f.
Galium tenuicaule, A. Cunn.
,, *umbrosum*, Sol.
Asperula perpusilla, Hook. f.
Senecio sciadophilus, Raoul.
Microseris Forsteri, Hook. f.
Teucridium parvifolium, Hook. f.
Mentha Cunninghamii, Benth.
Korthalsella (*Viscum*) *Lindsayi*, Engl.
Podocarpus ferrugineus, D. Don.(?)
,, *spicatus*, R. Br.
Dacrydium cupressinum, Soland.
Bulbophyllum pygmæum, Lindl.
Corysanthes rotundifolia, Hook. f.
Corysanthes macrantha, Hook. f.
Gastrodia Cunninghamii, Hook. f.
Arthropodium candidum, Raoul.
Echinopogon ovatus, Beauv.
Poa imbecilla, Forst.
Hymenophyllum dilatatum, Swartz.
,, *scabrum*, A. Rich.
,, *flabellatum*, Lab.
,, *Tunbridgense*, Smith.
,, *unilaterale*, Willd.
,, *multifidum*, Swartz.
,, *bivalve*, Swartz.
Trichomanes venosum, R. Br.
Asplenium flabellifolium, Cav.
Todea hymenophylloides, A. Rich.
Ophioglossum vulgatum, Linn.
Botrychium ternatum, Swartz.
Tmesipteris tannensis, Bernh.

The following notes regarding the partial extermination of certain species are taken from Dr Petrie's paper "On the Flowering Plants of Otago" (1895):

Aciphylla squarrosa, Forst. Seedlings are readily cropped by sheep and cattle, so that when the old plants die off there are few young ones to take their place.

Celmisia densiflora, Hook. f., will soon be all but exterminated through burning and the attacks of stock and rabbits.

C. coriacea[1], Hook. f., rapidly disappearing before the attacks of rabbits and stock.

C. discolor, Hook. f. Once abundant in Central and Western Otago, but now getting rare.

Raoulia Hectori, Hook. f. Formerly common on all high mountains of Central Otago, now rapidly dying off from the drying of the ground through burning and close cropping.

Further examples are furnished by B. C. Aston in his paper on "Some effects of Imported Animals on the Indigenous Vegetation."

Senecio Greyii, Hook. f., at Mukumuku, Palliser Bay, is restricted—possibly chiefly owing to goats—to the cliffs, where it is with great difficulty that specimens can be secured.

Lepidium tenuicaule, T. Kirk. It is feared that this species, common in Titahi Bay in 1907, has been entirely eaten out.

Ligusticum aromaticum, Hook. f., is being exterminated high up on Ngauruhoe by rabbits.

Gymnogramme leptophylla, Desv. This fern, once abundant on Wellington coasts, is believed to have been exterminated by sheep and rabbits. Being an annual it would be eaten before the spores were shed.

Stilbocarpa polaris, A. Gray. At Campbell Island, which is inhabited and farmed as a sheep-run, this plant is being eaten out by sheep.

Pleurophyllum speciosum, Hook. f. At Port Ross, Auckland Island, in January, 1909, at 1100 ft., numerous pig-tracks were observed, and this plant appeared to have been eaten out on all stations but inaccessible rock-faces.

Pleurophyllum Hookeri, Buch. On Auckland Island, in November, 1907, at Flat-topped Mountain, Carnley Harbour, and above the scrub line, pigs had eaten freely of this species, having grubbed up the plants to get at the root-stock.

Poa littorosa, Cheesm. This grass was evidently being exterminated at Enderby Island by the cattle.

One effect of the introduction of animals has been to cause an increase in the abundance of certain indigenous species of plants.

Very few native plants are so adapted as to have their seeds spread by passing animals. There being apparently no animals in

[1] Cheeseman states that in presence of rabbits *C. coriacea* has increased enormously, e.g. on the Takitimo Mts., and in the Hanmer Plains area. I have observed the same on Maungatua in Otago. It has succumbed to fire in many parts.

New Zealand which could do this in the past except, perhaps, a few ground birds, it was not to be expected that the flora would contain plants which were so distributed. There are only two genera which are well provided in this respect, viz., *Acæna* and *Uncinia*.

In the piri-piri (*Acæna* sp.) the lobes of the calyx are produced into spines, and in several species these are barbed, and hook on to hair, wool and feathers very effectively. The plants grow not only out in the open, but also especially along the edges of scrub and bush, so that animals living in the shelter are almost certain to catch the dry fruit as they pass out and in. Sheep and dogs have spread these plants far and wide, and they have thus increased greatly.

The sedges of the genus *Uncinia* are also provided with a very efficient hook whereby they catch on to passing animals. From the inside of the base of the utricle a long bristle, which is evidently the produced axis of the spikelet, projects to the length of an inch or more, and this is very strongly hooked at the tip. I cannot say whether the species of *Uncinia* have spread as those of *Acæna* have done, though it is highly probable that they have. The latter are very abundant and appeal to every one, for they not only get into wool to a very serious extent, but they catch on to dogs, and also very freely hook on to clothing. *Uncinia* is not nearly so abundant in individuals, not being so efficiently provided with grappling appliances, and it has not been observed to any great extent.

It is very noteworthy that in both genera the species with the best apparatus for distributing the seeds have a very wide range outside of New Zealand, and almost certainly have developed this character in regions where mammalia are to be found. Thus the species (or group of species) known as *Acæna sanguisorbæ*, Vahl, is found in Australia, Tasmania and Tristan d'Acunha and the plants are all strongly barbed. On the other hand *A. Buchanani*, Hook. f., has the barbs frequently reduced to hairs; in *A. microphylla*, Hook. f., the spines themselves are frequently wanting; while in *A. glabra*, Buch., the fruiting calyx is quite unarmed. These three species are confined to New Zealand, and probably originated in these islands.

The evidence is not so clear in the genus *Uncinia*. *U. tenella*, R. Br., *U. compacta*, R. Br., and *U. riparia*, R. Br., occur in the Australian region; *U. Sinclairii*, Boott, in Fuegia; and *U. australis*, Pers., is said to occur in the Sandwich Islands. The other seven species are endemic, but are in most cases very nearly allied to some of the first-named species. In *U. filiformis*, Bott., which is endemic, and is the same as or closely allied to *U. debilior*, F. v. Muell., of Lord Howe Island, the bristle is very feeble and the hook has nearly lost its catching power.

Daucus brachiatus, Sieb., the native carrot, has hooked bristles on its fruit, and is no doubt spread by hairy and woolly animals. On the other hand, the plant is so readily eaten by sheep and rabbits, that it is nearly exterminated in those districts where it used to be common.

Fruit-eating birds, especially blackbirds and thrushes, have increased to an enormous extent, and while spreading introduced succulent fruit plants very freely (e.g. blackberry, briar-rose, elderberry, etc.) have also caused a very great increase in certain indigenous species. For instance—in the Town Belt of Dunedin, to which reference has just been made, there has been a great increase within the last thirteen years of the following species: *Melicytus ramiflorus*, Forst.; *Rubus australis*, Forst.; *Fuchsia excorticata*, Linn. f., species of *Coprosma*; *Loranthus micranthus*, Hook. f.; *Tupeia antarctica*, Cam. and Schl.; and *Muehlenbeckia australis*, Meissn. In other districts other succulent fruit plants have been spread. No doubt the Parapara, *Pisonia Brunoniana*, Endl., which has a large viscid and succulent perianth, is also distributed by birds, but the species has a considerable range through Norfolk and Lord Howe Islands into North-Eastern Australia, and I cannot say by what birds it was distributed. But it is remarkable that certain small birds, e.g. the (recently introduced?) wax-eye, *Zosterops cærulescens*, which is a great eater of fruit such as pears, peaches, etc., frequently get caught on the fruit of *Pisonia*, and are unable to liberate themselves.

It is difficult to obtain any information upon or evidence of effects produced on the native vegetation by introduced insects. At least two species of *Thrips* attack many native trees, as recorded by W. W Smith of New Plymouth, and as I have seen from the specimens submitted to me. It is probable also that several Coccidæ and Aphides have taken to living on indigenous trees and plants, but there is practically no evidence on the subject.

On the other hand certain indigenous insects have found introduced plants to their liking.

Among Lepidoptera the fine green *Hepialus virescens*, Doubleday, does a great deal of damage to introduced trees. Large ash trees (*Fraxinus excelsior*) in Taranaki have been killed by the larvæ boring and tunnelling into the solid wood. Similarly, large oaks (*Quercus robur*) and elms (*Ulmus europæus*) are destroyed by these larvæ. Whereas in Puriri (*Vitex lucens*) the tunnels remain open, those in oak, which are generally very moist, are closed. In Puriri the opening is covered by the larva with a tough leathery screen, but in oak the bark and woody tissue form a thick callus, which effectually closes up the larva in an air-tight chamber. *Benthamia fragifera* is also attacked by this larva, as are occasionally walnut (*Juglans regia*) and a species of *Eucalyptus*. The larva of *Melanchra* (*Mamestra*) *mutans*, Walk., frequently does very serious damage to the leaves of young wheat, and considerable portions of the crops have been destroyed at various times, particularly in the Canterbury district. *M. composita*, Guenée, known as the "grass-moth" or "grass-caterpillar," does great damage and causes heavy losses to bush farmers especially. Fortunately these animals are kept in check to a considerable extent by certain ichneumons as *Probolus sollicitorius*, Fabr., *Mesostenus albopictus*, Smith, as well as the introduced (?) *Ryssa semipunctata*.

Liothula (*Œceticus*) *omnivora*, Fereday, has occasionally done much damage among raspberry plantations.

The caterpillar of *Ctenopseustes obliquana*, Walk., which usually feeds on manuka and other native shrubs, has developed a fondness for the ripe fruit of plums and apricots.

The larvæ of *Cacœcia excessana*, Walk., attacks the outsides of apples, and these have frequently been sent in to the Agricultural Department under the impression that they were attacked by the larvæ of the introduced codlin moth (*Carpocapsa pomonella*). The latter, however, attacks the core of the fruit, while that of *Cacœcia* only attacks the outside. This insect also attacks and disfigures the cherry-laurel (*Prunus lauro-cerasus*).

Two native species belonging to the Caradrinidæ may be mentioned in this connection. *Agrotis ypsilon*, Bott., is known in New Zealand as the tussock moth, the larvæ doing considerable damage in tussock country; it is also a very common pest in vegetable gardens. In Hawaii, where it is known as the greasy cutworm, it is very prevalent, and attacks cotton and tobacco.

Cyrphus unipunctatus, Hawthorne, which is very abundant in the North Island, occasionally attacks grain crops, but is apparently kept in check to a considerable extent by birds. In North America it is one of the most serious of the cereal crop pests.

Among Coleoptera only a few instances are recorded.

Œmona humilis, Newman, attacks lemon trees in the northern orchards, and eats into the wood.

Xyloteles griseus, Westwood, has been reported several times as attacking fig trees, while *X. lætus*, White, sometimes called the "apple-borer," occasionally destroys the apples on the trees.

All three species of beetles belong to the Cerambycidæ, one of the few families reported to contain destructive pests, but none of them have become serious.

The larva of *Odontria striata*, White, commonly known as the "grass-grub," is a very destructive insect, especially in pastures and lawns. It cuts the grass at the roots, and in this manner destroys the plants, so that the surface growth dies right out. The larvæ are sufficiently deep in the soil to be beyond the range of starlings and other insectivorous birds, and the chief means of destruction are either flooding the land, or rolling with a very heavy roller.

Two indigenous scale insects have been recorded as attacking introduced plants. In the Agricultural Department's Report for 1909, *Diaspis santali*, Maskell, is reported as becoming troublesome in orchards, where it attacks the plum trees; while *Dactylopius aurilanatus*, Maskell, has established itself on Norfolk Island Pines and other species of *Araucaria* in the North Island.

Though these few are the only examples recorded of indigenous species of insects attacking introduced plants, it must be noted that the small number is probably due to the fact that observations on the point have not been made. Anyone taking up a research on this subject would find that the list might be multiplied to a very great extent.

Inter-relation of Native and Introduced Flora

The introduction of foreign plants into New Zealand has wrought a very radical change in the facies of the vegetation. The distinctive character of the native flora has disappeared from nearly all closely settled portions of the country, and what may be called a cosmopolitan type of vegetation has taken the room formerly occupied by the displaced species. This fact led many naturalists to the conclusion that the indigenous fauna was doomed to destruction, and would in time be exterminated by the alien introductions. Sir J. D. Hooker, who visited New Zealand in 1841, and published an account of its plant life in the *Flora Novæ-Zealandiæ*, which forms the second part of the *Botany of the Antarctic Expedition* of Sir J. Ross, published in 1854–5, was the first who voiced this conclusion, and the pre-eminent position he occupied as the leading British botanist, deservedly and naturally caused his views to be widely accepted. He came to the conclusion that the northern or Arctic element in all the south temperate floras, including that of New Zealand, was due to the wonderful aggressive and colonising power of what he termed "the Scandinavian Flora." He says:

> When I take a comprehensive view of the vegetation of the Old World, I am struck with the appearance it presents of there being a continuous current of vegetation (if I may so fancifully express myself) from Scandinavia to Tasmania.... Scandinavian genera, and even species, reappear everywhere from Lapland and Iceland to the tops of the Tasmanian Alps in rapidly diminishing numbers it is true, but in vigorous development

throughout;...and in New Zealand and the Antarctic Islands, many of the species remaining unchanged throughout.

Darwin in the *Origin of Species* states that:

from the extraordinary manner in which European productions have recently spread over New Zealand, and have seized on places which must have been previously occupied, we may believe, if all the animals and plants of Great Britain were set free in New Zealand, that in the course of time a multitude of British forms would become thoroughly naturalised there, and would exterminate many of the natives. On the other hand, from what we see now occurring in New Zealand, and from hardly a single inhabitant of the southern hemisphere having become wild in any part of Europe, we may doubt, if all the productions of New Zealand were set free in Great Britain, whether any considerable number would be enabled to seize on places now occupied by our native plants and animals.

Arguing from the facts advanced by Sir J. D. Hooker, A. R. Wallace in *Island Life* (1880) has discussed the question at some length, and I quote that portion of his remarks which bears directly on the New Zealand flora. He says (p. 479):

The first important fact bearing upon this question is the wonderful aggressive and colonising power of the Scandinavian flora, as shown by the way in which it established itself in any temperate country to which it may gain access. About 150 species have thus established themselves in New Zealand, often taking possession of large tracts of country; about the same number are found in Australia, and nearly as many in the Atlantic states of America, where they form the commonest weeds. Whether or not we accept Mr Darwin's explanation of this power as due to development in the most extensive land area of the globe where competition has been most severe and long-continued, the fact of the existence of this power remains, and we can see how important an agent it must be in the formation of the floras of any lands to which these aggressive plants have been able to gain access.

Wallace was very strongly impressed by several instances of the rapid spread of European plants in New Zealand which had been communicated to him by Enys, Travers, and other good observers. Some of the examples given by Wallace in *Darwinism* may be referred to here. Thus he gives instances of the spread of water-cress in the streams of the Canterbury Plains; of white clover displacing native species and even exterminating the native flax (*Phormium tenax*); of *Hypochœris radicata* (the so-called Cape-weed of the South Island):

which destroyed excellent pasture in three years, and absolutely displaced every other plant on the ground. It grows on every kind of soil, and is even said to drive out the white clover, which is usually so powerful in taking possession of the soil;

of knot-grass (*Polygonum aviculare*), common dock (*Rumex obtusifolius*), sowthistle (*Sonchus oleraceus*) and sorrel (*Rumex acetosella*), spreading in all directions, and in many places absolutely occupying the ground they invaded.

Many of these statements are unintentionally misleading. Thus *Hypochœris radicata*, the most abundant and most widespread introduced plant into New Zealand, can only assert itself in land grazed over by cattle (it asserts itself in lawns), which eat out all the good grasses and the clover, and leave the others. When sheep are introduced the Cape-weed is itself quickly eaten out. Dr Cockayne is very emphatic on the subject. He says:

The often quoted stories of white clover being able to wipe out *Phormium tenax*, of *Salix Babylonica* overcoming the water-cress, of *Hypochœris radicata* displacing every other plant of excellent pastures in Nelson, are without foundation. *P. tenax* has certainly been eradicated in many places, and perhaps, in a sense, replaced by white clover, *but not until fire and feeding of stock had killed the plant.*

Dr Cockayne's reference is to a well-known passage in *Darwinism*:

A curious example of the struggle between plants has been communicated to me by Mr John Enys, a resident of New Zealand. The English water-cress grows so luxuriantly in that country as to completely choke up the rivers, sometimes leading to disastrous floods, and necessitating great outlay to keep the streams open. But a natural remedy has now been found by planting willows on the banks. The roots of these trees penetrate the bed of the stream in every direction, and the water-cress, unable to obtain the requisite amount of nourishment, gradually disappears.

Enys is partly right in this statement. I have recently noticed that a species of *Nitella*, whether native or introduced I do not know, is killing out the water-cress in some of the streams about Christchurch.

W. T. L. Travers, addressing the Wellington Philosophical Society in 1871, in referring to the destruction of the native alpine and subalpine flora of Nelson province said:

Indeed I have no doubt, from the comparative rarity of many plants which were formerly found in abundance in such districts, that in a few years our only knowledge of them will be derived from the dried specimens in our herbaria.

In another place, quoted by Cheeseman, he stated:

Such, in effect, is the activity with which the introduced plants are doing their work, that I believe if every human being were at once removed from the islands for even a limited number of years, looking at the matter from a geological point of view, the introduced would succeed in displacing the indigenous fauna and flora.

I have no doubt that had Travers been living to-day he would have completely reversed this judgment. The opinion of all botanists in New Zealand to-day is that when the direct, or—to a large extent—

the indirect influence of man is eliminated, the native vegetation can always hold its own against the introduced. Those plants which have thriven abnormally in this new country, and have impressed visitors by their abundance, are found in settled and cultivated districts, and belong chiefly to what are known as *weeds of cultivation*, that is, plants which have become adapted to conditions caused by the direct and indirect action of human beings, and which only thrive where those conditions are maintained.

In a paper on plant acclimatisation in New Zealand which was published in 1900 I said:

Seeds of such plants as violets, primroses, cowslips, bluebells, heaths, etc., and of fruits like the bilberry (or blæberry) and cranberry, have been sown by numbers of persons during the past 50 or 60 years in all sorts of situations, but they have not established themselves. They cannot always succeed even when growing in open competition against the indigenous vegetation, and they never make the slightest headway against many of the vigorous introduced forms. Even where individual plants become established, they nearly always fail to produce seed, and this is the chief reason why such species do not become naturalised. In their native countries their flowers are visited and fertilised by certain species of insects, and these are totally wanting here. Our indigenous insects are unable to fertilise them, and so they do not produce seed. There are no doubt other differences which affect their success in the struggle for existence. The rapidity of germination of their seeds, the subsequent rapidity of growth of the young plants, and many other factors, which have not been sufficiently looked into in this connection, all bear on this question. I have in past years sown quantities of the seeds of many flowering plants of Great Britain along the wayside in one of the suburban roads leading through our Town Belt, but from none of them have plants appeared except from those of foxgloves, whose strong coarse foliage enables it to hold its own against most of its neighbours. If the others have germinated they have nearly always been smothered by cocksfoot or other coarse grass. In gardens many of our European flowers seed now on account of the general prevalence of humble-bees, but many others remain unfertilised.

Most of the common naturalised plants are capable of self-fertilisation if they are not habitually self-fertilised. For instance the following are found producing seeds in midwinter from flowers which never open and which are more or less imperfect in structure: Shepherd's Purse (*Capsella Bursa-pastoris*), winter-cress (*Barbarea vulgaris*), bitter-cress (*Cardamine hirsuta*), hedge-mustard (*Sisymbrium officinale*), wart-cress (*Senebiera didyma*), chickweed (*Stellaria media*), mouse-ear chickweed (*Cerastium glomeratum* and *C. triviale*), groundsel (*Senecio vulgaris*), sowthistle (*Sonchus oleraceus*), spurge (*Euphorbia Peplus*), and perhaps others. This faculty of producing more or less imperfect self-fertilised flowers is almost an essential feature in all such plants, many of which are thus enabled to produce fruit at all seasons of the year, and almost independent of the weather. Another point is that most of them produce very small and very numerous seeds; and still another, that a large proportion of them come to maturity very rapidly, and that their seeds germinate quickly. These characters are all retrogressive from one point of view—that is to say—the plants exhibiting them have tended to become less instead of more specialised in their development; but by this degradation of their reproductive organs they have really become better adapted for the peculiar conditions which are imposed upon them in their struggle with the gardener and agriculturist.

Commenting on this paper D. Petrie wrote as follows:

The spread of weeds is mainly due to useful plants—their competitors—being regularly checked and eaten down, while the weeds are mostly allowed to grow without check of any important nature. Almost all weeds found in our northern pastures owe their spread to this; e.g., several buttercups, numerous docks, pennyroyal, *Holcus mollis* and *H. lanatus*, and many weedy grasses, various spurges, mallows, mulleins, and so forth. In many cases their spread is facilitated by the ready germination of their seeds, by the long time that the seeds retain their vitality in the soil, and by the readiness with which their earliest roots strike deep down into the soil, which allows the plants to establish themselves in hot, dry weather. Black Medick (*Medicago lupulina*), meadow plantain (*Plantago lanceolata*), the docks and spurges, all start and thrive in hot, dry weather, when more superficial-rooting seedlings die off. The introduced speedwells and poor man's weather-glass (*Anagallis arvensis*) are much in the same case. The decline of plants that have taken possession of a district for some years is no doubt due to temporary exhaustion of some element of plant-food needful for their vigorous growth. This principle lies at the base of the theory of rotation of crops. In Central Otago, when I first knew it, *Carduus lanceolatus* was the prevailing weed on open downs and dry hill-slopes. Some years after *C. pauciflorus*[1] completely replaced it, and this will, no doubt, be now giving way to something else. *The doctrine that the Scandinavian plants possess extraordinary vigour, which is the cause of their aggressive character, seems to me very doubtful.* In each single species particular advantages can generally be assigned that will readily explain their rapid spread. In the peninsula north of Auckland there are very large areas of land on which European weeds have but slightly established themselves, though the ground is frequently cleared of all native vegetation by fires. In these areas native plants mostly grow up with great readiness, especially species of *Leptospermum* and *Pomaderris*, *Haloragis tetragyna* and *H. minuta*, besides various cyperaceous plants. The pre-eminence in aggressive characters of North European plants is decided enough, but many non-European plants are now widely spread here and are indeed very aggressive. I may instance *Modiola multifida*, a North American malvaceous plant, an Australian *Plantago*, two species of *Erigeron*, and *Kyllinga*. The rat-tail grass, too, is no doubt introduced, and has been most aggressive, while the South African *Cyperus* (*minimus*?) is nearly as ubiquitous as sorrel.

Here, as in the South, a few native plants are spreading: *Aristotelia racemosa*, *Fuchsia excorticata*, *Pomaderris phylicifolia*, *Erechtites*, etc.; but

[1] ?*C. pycnocephalus*.

the most aggressive plant of all is *Pteris aquilina*, which is rapidly over-running much of the land that has been cleared of bush, and which permanently establishes itself before its roots are sufficiently decayed to admit of ploughing.

The most abundant and most wide-spread introduced weed is *Hypochœris radicata*.

Cheeseman writes as follows on the naturalised plants of the Auckland district:

387 species are catalogued. Of these 280 are natives of Europe, many of them also ranging into temperate Asia and North America, and some into North Africa. Ten species, not European, are from the eastern portion of N. America, and four are from the western. Total, 294 species from the north temperate zone. From Australia only 19; from Chili and the cool portions of S. America 9; from the Cape of Good Hope 21. Total 49 species from the south temperate zone. Finally 53 species subtropical and tropical.

31 are trees or shrubs, and 356 herbaceous. Of this latter number, 176 are annual, 28 biennial, 152 perennial. The large proportion of annual species is noteworthy, as in the indigenous flora nearly all the herbaceous plants are of perennial growth.

With the above facts before us, we are better able to enquire into the general subject of the naturalisation of plants in New Zealand and to attempt an answer to the question why the native vegetation should apparently be unable to hold its own against the numerous intruders streaming in on every side. In considering the subject, it appears to me most important to bear in mind constantly that the conditions of plant-life now prevailing in New Zealand are in great measure different to those that existed when European voyagers first visited its shores. When Cook landed here the whole country was covered with a dense native vegetation, hardly interfered with by man. The cultivations of the Maoris were small in areas, and as they rarely tilled the same plot of ground for many years in succession, preferring to abandon it when the soil showed signs of exhaustion and to make new clearings elsewhere, there is little chance of the establishment of a race of indigenous weeds. In fact, it can be roundly said that the New Zealand Flora contained no such class. At that time there were no herbivorous animals of any kind, either wild or domesticated, to graze upon the vegetation, or to interfere with it in any way. Thus no check existed to the growth of many species which can now hardly live in a district where our introduced cattle are abundant. And the repeated burning off, year by year, of large tracts of open country, was then a circumstance almost unknown. The Maori rarely wantonly destroyed the vegetation, and if he used fire in making his new clearings, generally took precautions that it should not spread further than was absolutely required. It is hardly necessary to dwell longer on this point; for all must admit that the advent of European settlers and the colonisation of the country have brought into operation a set of conditions injurious to both the indigenous fauna and flora. The chief of these conditions may be conveniently grouped under three heads: first, the actual destruction of the vegetation by the settlers to make room for their cultivations, or in the construction of roads, or in the cutting down of the forests for timber, etc.; second, the introduction of sheep, cattle and horses, and their spread over the greater part of the country; third, the practice, now very generally followed, of burning off the vegetation in the open districts at regular intervals.

If the above facts are duly considered there will not be so much cause for wonder in the introduction and rapid spread of so many foreign plants. For instance, it might be expected that the weeds of our corn-fields and pasture—which now form such an important and conspicuous element in the naturalised Flora—would be almost wholly composed of introductions from abroad. The native Flora possessed few plants suitable for the places they have taken, and these few could hardly compete with a chance of success against species that have from time immemorial occupied the cultivations of man, and whose best adapted varieties have been rigorously selected. The introduced weeds flourish and multiply because they have an environment suited to them and to which they have been modified; the native ones fail because the conditions have become altogether different to those they had been accustomed to.

Similarly it was to be expected that foreign plants would in some degree displace the indigenous ones in districts grazed over but not actually cultivated. Many native species will not bear repeated cropping, and soon decrease in numbers when cattle and sheep are brought in. Their places will, therefore, be taken by plants that are indifferent to this, or escape by reason of being unpalatable....At the same time it must be remembered that any native plants possessing similar advantages would also increase; and in many cases this has actually taken place. The spread of such indigenous plants as *Poa australis* and *Discaria* in the river valleys in the interior of Nelson and Canterbury, of *Cassinia* on the shores of Cook Straits; and of some grasses (as *Danthonia semiannularis* and *Microlæna stipoides*) in Auckland, are well-known examples, and it would be easy to enumerate more.

But although we may safely credit the changed conditions of plant life with being a powerful reason for the spread of naturalised plants in New Zealand, it is impossible to consider it as the sole explanation. For we find that not a few species have penetrated into localities where cultivation and cattle are alike unknown, and where man himself is a rare visitant; where in fact, the conditions are still unchanged. This is the most interesting part of the subject, for it proves conclusively, as remarked by Mr Darwin, that the indigenous plants of any district are not necessarily those best suited for it. In most cases it is impossible to assign any obvious reason for the fact that these intruders should be able to thrust on one side the native vegetation; but it is significant that all, or nearly all, are common and widely distributed in their native countries, in short, are predominant species; and that they have followed almost everywhere the footsteps of man, being as extensively naturalised in many other countries as in New Zealand. We may, therefore, suppose that by long-continued competition with other species, in different localities and in different climates, they have gained a vigour of constitution and a faculty of adapting themselves to a great variety of conditions which enable them to readily overcome plants that have not been so advantageously modified.

This supposition will also throw some light on the curious fact that the vast majority of our plants are of northern origin. It is now generally admitted by geologists that the present continents are of immense antiquity, and that there has been no great alteration in the relative proportions of land and water during vast geological epochs. Mr Darwin therefore argues that as the northern hemisphere has probably always possessed the most extensive continuous land area, so the wonderfully aggressive and colonizing power of its plants at the present time is due to development where the competition of species has been the most severe and long continued, owing to the presence of facilities for natural migration. The plants of the comparatively isolated countries of the southern hemisphere have not been subjected to the same degree of competition, and consequently could not be so advantageously modified.

Mr Kirk, who paid special attention to the naturalisation of plants in New Zealand, in a paper on the naturalised plants of Port Nicholson, says:

> At length a turning-point is reached, the invaders lose a portion of their vigour and become less encroaching, while the indigenous plants find the struggle less severe and gradually recover a portion of their lost ground, the result being the gradual amalgamation of those kinds best adapted to hold their own in the struggle for existence with the introduced forms, and the restriction of those less favourably adapted to habitats which afford them special advantages.

Further on in the same article Mr Kirk combats the view that the majority of our native plants will become extinct, stating that the particular species for which this danger is to be feared might almost be counted upon one's fingers.

My own views on this difficult question are much nearer to Mr Kirk's than to those of Mr Travers. I can certainly find little evidence in support of the opinion that a considerable proportion of the native flora will become extinct. Even in isolated localities of limited areas, like Madeira and St Helena, where there is little variety of climate and physical conditions, and where the native plants have been subjected to far more disadvantageous influences, and to a keener competition with introduced forms than in New Zealand, the process of naturalisation has not gone so far as to stamp out the whole of the indigenous vegetation, although great and remarkable changes have been effected, and many species have become extinct. I fail to see why it is assumed that a greater effect will be produced in New Zealand, with its diversified physical features and many varieties of soil, situation, and climate. Surely its far-stretching coast-line, bold cliffs, and extensive sand-dunes, its swamps and moorlands, its lofty mountains and wide-spreading forests, will afford numerous places of refuge for its plants until sufficient time has been allowed for the gradual development of varieties better suited to the changed conditions. No doubt some few species will become extinct; but these will be mostly plants whose distribution was local and confined even when Europeans first arrived here; and probably all will be species that have for some time been tending towards extinction, and whose final exit has thus been hastened. I cannot call to mind a single case of a plant known to be widely distributed when settlement commenced that is at present in any danger of extinction. Species have been banished from cultivated districts, of course, but they are still abundant in other situations, and probably there will always be a sufficient area of unoccupied and uncultivated lands to afford them a secure home.

Speaking generally, I am inclined to believe that the struggle between the naturalised and the native floras will result in a limitation of the range of the native species rather than in their actual extermination. We must be prepared to see many plants once common become comparatively rare, and possibly a limited number—I should not estimate it at more than a score or two—may altogether disappear, to be only known to us in the future by the dried specimens preserved in our museums.

Perhaps the most emphatic testimony as to the staying power of the indigenous vegetation is that borne by Dr Cockayne in his *Ecological Studies in Evolution* (p. 32), where he says:

There have been recorded for New Zealand up to the present time some 555 species of introduced plants, but less than 180 can be considered common, while others are local, rare, or even not truly established as wild plants. Many at first sight appear better suited to the soil and climate than are the indigenous species, and over much of the land they give the characteristic stamp to the vegetation; *but this is only the case where draining, cultivation, constant burning of forest, scrub, and tussock, and the grazing of a multitude of domestic animals have made absolutely new edaphic conditions which approximate to those of Europe*, and where it is no wonder that the European invader can replace the aboriginal. On the other hand, although this foreign host is present in its millions, and notwithstanding abundant winds and land-birds (introduced, not native birds), *the indigenous vegetation is still virgin and the introduced plants altogether absent where grazing animals have no access and where fires have never been.* On certain sub-alpine herb-fields the indigenous form of the dandelion (*Taraxacum officinale*, Wigg.) is abundant, and yet the introduced form, with its readily wind-borne fruit, has not gained a foothold, nor even the abundant *Hypochœris radicata*, L., though it may be in thousands on the neighbouring tussock pasture, less than one mile away. On Auckland Island, introduced plants occur only in the neighbourhood of the depots for castaways, but on Enderby Island, where there are cattle, they are much more widely spread. Even where the rain forest has been felled or burnt, and cattle etc., are kept away, it is gradually replaced by indigenous trees and shrubs —i.e., in localities where the rain-fall is sufficient.

Some of the indigenous species are quite as aggressive, or even more so, than any of the introduced. In primeval New Zealand each would have its place in the association to which it belonged—there would be no aggression; but when the balance of nature was upset by the fire or cultivation of Maori or European, then the plants best equipped for occupying the new ground become dominant, their "adaptations" for that purpose fortuitously present. The miles on miles of *Leptospermum scoparium* and *Pteridium esculentum* were absent in primitive New Zealand. So, too, the pastures of *Danthonia semiannularis* in Marlborough, and the many acres of *Chrysobactron Hookeri* in the lower mountain region of Canterbury. *Celmisia spectabilis*, an apparently highly specialized herb for alpine fell-field or tussock-steppe conditions, is now on the increase in many montane parts of the Ashburton-Rakaia mountains and valleys, owing to its being able to withstand fire, the buds being protected by a close investment of wet decayed leaf-sheaths.

Nor are all the introduced species aggressive, by any means. Some can barely hold their own; others are limited to certain edaphic conditions. Thus, *Glaucium flavum*, Crantz, occurs, as yet, only on the coast of Wellington, chiefly in the neighbourhood of Cook Strait. It is confined to gravelly or stony shores, and appears unable to grow on the clay hillside. And yet where the latter is, in one place near Lyall Bay, covered with gravel there is a large colony of the plant, whence none have found their way on to the adjacent hillside. *Lupinus arboreus*, now so common on New Zealand dunes, appears unable to spread beyond the sandy ground[1].

The fortuitous introduction of foreign species of animals and plants into the country has been going on continuously since the European settlement of New Zealand began. This was first brought to my notice when we settled in Southland, and took up a farm there.

In 1870 we imported from Messrs Lawson and Sons, Edinburgh, a quantity of grass-seed for permanent pasture invoiced to contain the following species:

Alopecurus pratensis	*Phleum pratense*
Anthoxanthum odoratum	*Poa nemoralis*
Trisetum flavescens	,, *pratensis*
Dactylis glomerata	,, *trivialis*
Festuca elatior	*Lotus corniculatus*
,, *duriuscula*	,, *major*
,, *heterophylla*	*Medicago lupulina*
,, *pratensis*	*Trifolium pratense*
,, *rubra*	,, *hybridum*
Lolium perenne	,, *repens*

In addition to these the following appeared, the ground being newly broken up—and having never previously been cultivated:

Chrysanthemum leucanthemum	*Ononis arvensis*
,, *segetum*	*Anthyllis vulneraria*
Lychnis Githago	*Vicia sativa*
Geranium molle	*Knautia arvensis*

[1] The latest expression of views on this interesting subject will be found in Dr Cockayne's *New Zealand Plants and their story*, Chap. X (Wellington, 1919).

Leontodon Taraxacum	*Sinapis arvensis*
Prunella vulgaris	*Papaver argemone*
Veronica Chamædrys	,, *Rhœas*

Cuscuta Trifolii or *Trifolium repens* and *T. pratense.*

Similar experiences have occurred, but have not been recorded, many thousand of times in the intervening years.

In all agricultural seeds imported into the country a certain proportion of seeds of foreign plants is introduced, amounting in the aggregate to many millions. The number of those which do not grow, or if they germinate, do not succeed in establishing themselves, is very remarkable.

In recent years the Agricultural Department has set up a seed-testing bureau at its experimental farm of Weraroa, near Levin, and has drawn up a "List of Extraneous Seeds found in Commercial Samples of Seed, showing relative frequency of occurrence of each, and whether harmful or otherwise." The list for 1916 contains the names of 224 species of plants, all except three being introduced species. It included the following twenty species which have never yet been observed growing in the country, although evidently the seeds are introduced from time to time:

Alyssum alyssoides	*Lolium westwolticum*
Anthemis austriaca	*Matricaria discoidea*
Camelina microcarpa	*Odontites rubra*
Carex cephalophora	*Ornithopus sativus*
Centaurea picris	*Plantago aristata*
Crepis capillaris	,, *Rugeli*
Erysimum cheiranthoides	*Rudbeckia hirta*
Geranium pusillum	*Setaria italica*
Lepidium intermedium	*Triodia Thompsoni*
,, *virginicum*	*Verbena urticifolia*

In this connection it is very remarkable to notice the number of foreign seeds introduced into this country with agricultural seeds, *but which do not grow*[1].

[1] The following newly introduced grasses were placed in the trial beds at Ruakura Farm in 1909:

Agropyrum tenerum (slender wheat-grass, U.S.A.)	*Ecrucaria cuspidata*
	Panicum brazilianum
A. Smithii (western wheat-grass, *U.S.A.*)	*P. paranensis*
	Panicum sp. Giant-couch
Bromus pacificus	*P. altissimum*
Bouteloua curtipendula	*Paspalum stoloniferum*
Briza geniculata	*Phalaris cærulescens*
Chloris abyssinica	*Piptatherum Thomasii*
C. barbata	*Sporobolus Wrightii*
Elymus submuticus	*Tristegis glutinosa*
E. canadensis	*Uniola speciosa*
E. giganteus	*Vilfa arguti*
Erianthes ravennæ	

Chapter XV

ACCLIMATISATION WORK; GENERAL CONSIDERATIONS

In the various experiments made in the attempt to acclimatise, or, as I prefer always to call it, to naturalise, certain animals, very little forethought was given as to the possibilities of any particular case. We are wise now after the event, and we blame the acclimatisation societies, the Government, or various private individuals for mistakes made. But it is doubtful whether we would have done any better ourselves. There is one curious fact which never seems to have entered into the minds of most if not all of the acclimatisation enthusiasts, viz., the migratory character of the species which it was sought to establish in this new land. This applies especially, of course, to the birds. It probably explains the failure of certain species to become established here. Thus the blackcap (*Sylvia atricapilla*) and nightingale (*Daulias luscinia*) are summer visitors only in the British Isles, the former wintering in Southern Europe, Northern and Tropical Africa, and the latter in Tropical Africa. Only a very few examples of these two species were introduced, but it would not have mattered how many had been successfully conveyed to New Zealand, and liberated, the result must have been the same. After a time the birds would almost certainly have become possessed by the desire to migrate to a warmer climate, but having no hereditary line of migration to follow, they would probably have proceeded northwards and perished at sea. There is no information at all as to the effect of the migratory instinct on birds which have been taken from their own country to a totally different hemisphere. We can only assume that the desire to migrate would come on them strongly, and if so, it would inevitably prove fatal to them in an island country like New Zealand. Seebohm (in *Siberia in Asia*, p. 196) states that

the migration of birds follows ancient coast-lines. The migration from the south of Denmark over Heligoland to the coast of Lincolnshire seems to correspond so exactly with what geologists tell us must have been the old coast-line, that it is difficult to believe it to be only a coincidence.

The following species which have failed to become established in New Zealand are purely winter visitors in Britain: brambling (*Fringilla montifringilla*), white-fronted goose (*Anser albifrons*), brent goose (*Branta nigricans*), and grey plover (*Squatorola helvetica*). These birds are in the same category as the previously named group.

The following species are partial migrants, being resident in Britain at all seasons, but having migratory representatives which visit Britain in summer, and which winter somewhere to the south: linnet (*Linota cannabina*), twite (*Linota flavirostris*), redbreast (*Erithacus rubecula*), and teal (*Nettion crecca*). The following, which have also failed to establish themselves, are partial residents, but are also largely winter visitors in Britain: gadwall (*Chaulelasmus streperus*), pintail (*Dafila acuta*), wigeon (*Mareca penelope*), pochard (*Nyroca ferina*), and golden plover (*Charadrius pluvialis*).

As several writers on the subject of migration have pointed out, racial forms of certain species appear regularly in Britain, either as seasonal visitors or as occasional guests, and these include species of which resident representatives occur all the year round. Again in catching birds for exportation, it is impossible for the professional bird-catcher to ascertain whether the specimens captured belong to residents or to migrant races. The probabilities seem to me to be that they mostly belong to the latter, because the bird-catcher goes where he knows or expects to find birds gathered together in flocks, and these mostly come together preparatory to migration. It is quite possible that the specimens of linnets and twites brought to this country were migratory representatives of these species; this may explain their failure to remain here. The following birds, all of which have become naturalised in New Zealand, are also resident in Britain all the year round, but it is well known that there are numerous migratory races of them: rook (*Corvus frugilegus*), starling (*Sturnus vulgaris*), goldfinch (*Carduelis elegans*), greenfinch (*Ligurinus chloris*), skylark (*Alauda arvensis*), song thrush (*Turdus musicus*), hedge-sparrow (*Accentor modularis*), Mallard (*Anas boscas*). In a note to his list of partial migrants, Clarke states that: "Starlings marked in Britain have been recovered in France, Greenfinches in France, Linnets in France, Song Thrushes in France and Portugal, and Mallard and Teal in Germany." Seebohm also (in *Siberia in Europe*, p. 245) says:

> Many birds, such as the Robin, the Blackbird, the Song Thrush, etc., which are resident in England, are migratory in Germany. There is every probability that it is only within comparatively recent times that these birds have ceased to migrate in England, and we may fairly conjecture that should the English climate remain long enough favourable to the winter residence of these birds, they will develop into local races, which will eventually have rounder and shorter wings than their Continental allies.

In connection with the naturalisation of the rabbit, it is interesting to note that the earlier introductions all failed to establish themselves

There were rabbits in Otago at three or more different localities in the very early days of settlement. None of these colonies succeeded, but we cannot tell now to what particular breeds they belonged. It is just possible that they were of races which had been long under domestication, which would certainly be the kind that would carry best, and would be the most likely to be brought out in immigrant ships as pets. It was not till after the introduction of grey rabbits in the sixties, rabbits of a particularly aggressive and vigorous type, that these animals became a nuisance.

Very little is known about the movements and migrations of fishes in the sea, and in New Zealand practically nothing definite has been recorded. The date of the movements of the eels from the Waikato to the sea; the invasion of the southern rivers by lampreys in October; the northern drift of vast shoals of *Clupea* (pilchards and sprats) along the east coast of Otago in January and February—such general facts as these have been recorded, but no detailed observations have been made. It is to be regretted that steps were not taken in the past to mark specimens of Atlantic salmon (*Salmo salar*) when liberating them, in order to try, if possible, to gain some knowledge of their movements. The fact is that for over fifty years in this country, and for even longer in Tasmania, the young of this species have been turned out into the rivers by the million. They have found their way to the sea, and yet none have ever returned. What comes of them is a mystery regarding which no solution has ever been offered.

Another interesting case is the failure, so far, of the European lobster (*Homarus vulgaris*) to appear again in our seas. In this instance numerous larvæ—averaging at a low estimate 100,000 per annum, and in many cases at a fairly advanced stage of development—have been liberated from the Portobello Fish Hatchery into Otago Harbour, for twelve years past. In addition, several adults were liberated within the harbour. Allowing even ten years for a lobster to come to sexual maturity, when all the larger crustaceans usually begin to move about in large shoals, there has been ample time for adult specimens to show themselves, if they are in New Zealand waters, either near the Otago coast, or further north, whither the northerly current would carry them while they were still in the free-swimming stage. It is hoped, and with good reason, that they will appear some day in quantity, but there are no indications so far. The same remarks apply to the European edible crab (*Cancer pagurus*) of which between thirty and forty million larvæ, and some adults, have been liberated. Periodical migration of Crustacea—large and small—does take place in the seas, the causes of which are quite unknown, so both the lobster and the crab may yet be met with as permanent residents of New Zealand waters.

Other interesting problems in connection with introduced bird-life in New Zealand are the partial migrations which take place, and the extent to which certain species have succeeded or failed to establish themselves in various districts. These problems may be best illustrated by taking specific examples.

The Australian magpie (*Gymnorhina leuconota*) was introduced in considerable numbers into Canterbury and Otago between the years 1865 and 1871. In the latter district it seemed at first as if it were going to become strongly established, for the birds began to build in the neighbourhood of Dunedin, and as far south as Inch Clutha, but for some unexplained reason they gradually disappeared completely. Meanwhile they throve in the neighbourhood of Christchurch, especially to the south, and have since been gradually working their way southwards. Within the last year or two (1918–19) they have been reported as far south as Hampden and Moeraki. In the North Island they are fairly common in certain parts, but Mr W. W. Smith informs me that they are not so abundant now in Taranaki as they were some years ago. No doubt the food supply, which varies to a considerable extent according to the prevalence or scarcity of other insect-eating birds, is a main cause of this partial migration, but it is most difficult to arrive at the facts of the case.

The minah (*Acridotheres tristis*) is another species which has largely changed its location since the early days of its introduction, only in this case the cause of its gradual disappearance from certain districts is more directly traceable to its competition with the ubiquitous starling. Minahs were introduced into all the main centres—Auckland, Wellington, Nelson, Christchurch and Dunedin—in the early seventies. They increased in all these localities for a time, but as the starlings multiplied much more rapidly, the minahs either all died out, or were driven away in the south. By 1890 there were none left in Dunedin or its neighbourhood, only a few about Christchurch, and none in Nelson. They have all but disappeared from Wellington, and from the towns of Wanganui and New Plymouth, though they are still common in the country districts of Taranaki, Manawatu, Wairarapa and Hawke's Bay.

The chaffinch (*Fringilla cœlebs*) was introduced into all the main centres in the early seventies. These birds are most abundant in many parts now, but their occurrence in Otago has been most erratic. A few years after their introduction they became fairly common and then nearly disappeared. Probably the eating of poisoned grain was the cause of the partial disappearance of these and other small birds, but while the latter have re-established themselves in great numbers, the chaffinches still remain rare birds in most of the lower levels of Otago. In the higher country, however, wherever bush abounds, up to 3000 feet, they are reported as abundant. Their comparative absence in the southern half of the South Island, and their abundance in the north, especially from Lake Taupo northwards, seem to bear out the idea that poisoned grain has been the disturbing factor in the first-named districts.

Something of the same kind perhaps explains the erratic occurrence of the cirl bunting (*Emberiza cirlus*). These birds were introduced into Otago about 1871, and became fairly common. Then for a time they nearly quite disappeared, at least from districts where cultivation of the land was carried on, but recently they have been more in evidence. In the north, for example along the coast from Manawatu to Taranaki, they are quite common. These birds, as well as yellow-hammers (*Emberiza citrinella*) and red-polls (*Linota rufescens*) gather in large flocks at certain seasons of the year, just as migratory species do in Europe, but there is no evidence that they leave the country.

Chapter XVI

LEGISLATION

VERY early in the history of the colony of New Zealand legislation was passed to cope with animal and vegetable pests, which had been introduced along with live stock, or in seeds and plants, and which began to threaten the productivity of the new settlements. The pests increased as rapidly, and in some cases much more rapidly than the organisms they attacked or the plants they displaced, so that it became necessary to take concerted action to eradicate them if possible, or, at any rate, to keep them under control. In many cases the early legislation was an utter failure, and this is shown not only by the spread of the organisms themselves, but also by continual recurrence of Acts of Parliament to amend the legislation previously passed.

In the early days of settlement the various local legislatures did not pass Acts, but Ordinances, and the first of these was a "Dog Nuisance Ordinance" of New Ulster, passed on 17th July, 1844, by the Legislative Council of New Zealand, sitting at Auckland. The provincial districts of New Ulster, New Munster and New Leinster (Stewart Island), were apparently not defined till 1848. On 28th January of that year Sir George Grey by proclamation defined New Ulster as that portion of New Zealand north of the parallel of latitude running through the mouth of the Patea River, or about 39° 46′ S. The Southern Province was called New Munster. This Act was entitled "An Ordinance to provide a summary mode of abating the nuisance of Dogs wandering at large in Towns." The method was quite simple. The dog was to be seized, kept for a night and a day; if claimed it was to be delivered to the owner on payment of a fine of five shillings, and if not claimed it was to be destroyed. In succeeding years most of the Provincial Legislatures also passed "Dog Nuisance Acts." This would seem to show that stray dogs were early recognised as a nuisance and a danger to the community, and it helps to explain the extraordinary abundance of wild dogs in later days, for it shows that great numbers of these animals were allowed to breed and to learn to find food for themselves.

On 6th November, 1846, there was passed a "Duties of Customs Ordinance," by which horses, mules, asses, sheep, cattle, *and all other live stock and animals*, as well as seeds, bulbs and plants were admitted duty free into New Zealand. (The italics are mine.) Not only was there no inspection at this period to guard against various pests being introduced along with the stock, but some cranks would have brought some strange "cattle" into the country. Thus one proposal was to introduce foxes from England, so that the noble sport of fox-hunting might be indulged in. This was actually done in Victoria, and many Australians rued it afterwards. At a later date it was suggested to introduce the Arctic fox to cope with the rabbit, and at the same time to furnish valuable furs. Some one actually introduced bears into Canterbury, but fortunately they were not liberated.

One effect of the unrestricted importation of live stock was the prevalence of scab and other serious diseases among the sheep flocks of the country, and the seriousness and widespread nature of the trouble is shown by the passing of preventive and protective laws in every provincial district. The first of these appears to be the "Scab Ordinance" of 1849, passed by the Province of New Munster. This was "An Act to prevent the extension of the infectious disease called the Scab, as well as the disease called the Influenza or Catarrh, in Sheep or Lambs."

The history of this legislation may be outlined concisely as follows. The Wellington Provincial Council repealed the above Act in 1854, and passed another "Scab and Catarrh Act." In 1856 a third Act was passed; in 1862 there was a "Scab, Catarrh and Sheep Inspectors Act," followed by an Amendment Act in 1864. In the Session of Parliament of 1868 it was stated that all flocks in the Wellington district were free from Scab, except one at Castle Point, and another at Waikaraka, which were only slightly affected.

The Taranaki Provincial Council passed "An Ordinance to prevent the extension of the contagious disease in Sheep called Scab," re-enacted it in 1863, again in 1864 and 1866, and an Amending Ordinance in 1875.

The Auckland Provincial Council in 1854 passed "An Act to prevent the Extension among Sheep of the Disease called the Scab," a second Act in 1856, and a third in 1868.

The Otago Provincial Council passed a "Scab and Catarrh Bill" in 1854, an amending Act in 1856, and a "Sheep Ordinance Amendment" in 1861.

In the case of scab—caused by the presence of a mite (*Psoroptes communis*, var. *ovis*)—the precautionary measures taken were successful in stamping out the disease, and by 1880 it had quite disappeared from the country. But preventive legislation was continued until 1906,—and a "Sheep Act" was passed by the Colonial Parliament in 1890 "to provide for eradication and prevention of parasitic and other Diseases in Sheep."

In nearly all the preceding Ordinances and Acts provision was made that all sheep dying of catarrh were to be burned or buried at least three feet under ground, no carcasses were to be thrown into any stream, and no infected sheep were to be slaughtered for sale.

In 1876 there was passed "An Act to restrict the importation of Cattle and other animals into the Colony of New Zealand in certain cases," and Section 2 prohibits "all cattle, sheep, horses, swine, goats or other animals," etc., which are likely to propagate any infectious or contagious disease amongst men or animals.

The next subject to attract political attention was the increase of thistles. Wellington was first in the field in 1854 with an Act, the wording of which is interesting. It was entitled "An Act *to prevent the propagation of certain plants* known as Thistles" (the italics are mine), and penalties were imposed for allowing thistles to run to seed. I do not know whether the fines were actually enforced, but it is highly improbable that they were, otherwise a large proportion of the population would have been heavily penalised. The Act was amended in 1856, and again in 1857.

The Taranaki Provincial Council in 1856 passed "An Ordinance to prevent the spread of the Scotch Thistle," and amended this six years later to provide "better prevention of the growth, etc." of the Scotch thistle. This was further amended in 1863. In the previous year, also, "A Thistle Ordinance" was passed for the prevention of *noxious* thistles. I do not know whether Scotch thistles were not considered noxious, or whether it was that they were specially so, but they certainly had a Bill all to themselves.

The Auckland Provincial Council passed a Thistle Act in 1857, and an Amending Act in 1859; the Nelson Council passed one in 1859, and an Amending Act in 1861; while the Otago Council passed "A Thistle Prevention Act" in 1862.

Gorse and broom gave trouble even in the early days of settlement. Thus in 1859 the Provincial Council of Taranaki passed a "Furze Ordinance,"—"an Ordinance to impose a Penalty on the growth of Furze within the Town of New Plymouth." In 1868 another "Furze Ordinance" was enacted,—"An Ordinance to provide for the eradication of Furze growing on Public Roads," and this was amended and re-enacted in 1875.

In 1861 the Provincial Council of Nelson passed "An Act to prevent the planting of Gorse Hedges in the City of Nelson," and imposed a penalty not exceeding five pounds on any one disobeying this law. Any one also who did not keep an existing gorse or furze hedge pruned was liable to the same penalty.

The early settlers were great law-makers, but also great lawbreakers, for it is of no avail to make laws which cannot be kept or at least enforced, and in a great many of these restrictive ordinances Nature was too strong for the settlers and beat them very frequently.

In view of the keen desire which arose early in the history of the colony to enter on an extensive introduction of foreign species of animals and plants, protective legislation was passed as early as 1861. In that year the Provincial Council of Nelson passed "An Act to provide for the protection of certain Animals, Birds, and Fishes imported into the Province of Nelson." There were practically no specially imported animals at the time, but the law was for the protection of such species as "*may at any time* be imported into the said province." And to make it drastic enough a penalty up to fifty pounds was imposed for killing, taking or destroying, or selling or offering for sale any such animal, bird, or fish; with a penalty not exceeding ten pounds "for taking or destroying any egg of any bird or spawn of any fish, proclaimed to come under the operation of this Act."

In the same year a "Protection of Certain Animals Act 1861" was passed by the Colonial Parliament, in which Section 2 provided that: "No Deer of any kind, Hare, Swan, Partridge, English Plover, Rook, Starling, Thrush or Blackbird, shall be hunted, taken or killed at any time whatever before the first day of March, which shall be in the year 1870, and after that day only during the months of April, May, June and July in any year." Section 3 adds pheasant and quail to this list; and 4 says: "No Wild Duck or wild goose of any imported species shall be hunted, etc." Later sections prohibit the poisoning or trapping of any of these animals; no one is to have them in possession without lawful excuse, nor to sell or offer to sell, or buy or offer to buy such animals or birds." It is all rather curious and interesting legislation, for it was almost entirely anticipatory, as the animals referred to were mostly not introduced till a later date. Four years later this Act was re-enacted. In 1866 it was amended, and the ideas dominating the minds of the legislators of that time appear chiefly to have been the preservation of "game" for such as could afford to shoot. It defines the word "game" to include deer of any kind, swan, wild goose or *wild duck* of any imported species whatever, *hares*, pheasants, partridges, *grouse*, *English plover*, quail, *heath-* or *moor-game*, *black grouse* or *bustards*. Yet at the time of the passing of this Act none of the animals whose names are in italics had been introduced into the colony, and some, such as bustards, have not even been attempted. Provision was made for proprietors or tenants destroying hares on their own enclosed lands, for coursing hares by greyhounds or hunting them with beagles, for killing deer on enclosed lands or hunting them with hounds. Lastly no game was to be sold without a licence, the fee for which was fixed at five pounds. It is very interesting to notice how strong was the conservative instinct to preserve game so that the privileged few should have sport. Any one who within three years of the passing of this Act was convicted of exposing a dead hen pheasant for sale was to be fined not less than *five* nor more than *twenty* pounds.

Next year, 1867, the Acts of the two preceding sessions were repealed and a new one was passed: "An Act to provide for the Protection of Certain Animals and for the Encouragement of Acclimatisation Societies in New Zealand." Section 22 reads as follows: "Any person who shall wilfully take or wilfully destroy the eggs of any game birds shall be liable to a penalty not exceeding Five Pounds." Limitation of power to introduce certain animals is covered by Section 29, which reads: "It shall not be lawful for any person to introduce any fox, venomous reptile, hawk, vulture, or other bird of prey into the colony," or to allow such if already in possession to go at large. In the Schedule which defines "game," the *Snipe* and the *Antelope* are included.

This Act was amended in 1868 so as to enable game to be taken for the purposes of distribution to other parts of the colony. In 1872, and again in 1873, Amendment Acts were passed, and in the latter year the Schedule remained the same as before, except that *Ptarmigan* was added. Another Amendment was passed in 1875, and then in 1880 the preceding Acts were repealed, and a new Act was passed "to consolidate the Law for the Protection of Animals and for the encouragement of Acclimatisation Societies." Amendments to this Act were passed in 1881, 1884, 1886, 1889 and 1895. In the last of these was a clause which should have been enacted at a very much earlier date. Section 2 reads:

> From after the commencement of this Act no society, authority or person shall introduce or import into the colony, or turn at large, for the purposes of sport or acclimatisation, or as game, any animal or bird whatever without the consent in writing of the Minister for the time being in charge of the Department of Agriculture; nor shall any insect or reptile be introduced or imported into the colony without such consent as aforesaid.

Section 3 seems to be a piece of unworkable legislation, for it throws the onus of guarding against undesirable immigrants on persons who have very little means of stopping such. It states:

> It shall be the duty of the master, owner, charterer or agent of any vessel arriving at any port or place in New Zealand to effectually prevent any snake, scorpion, or *other noxious reptile* from being landed in New Zealand from such vessel, whether in the cargo or otherwise.

No one seems to have discovered the interesting zoological error of classing *scorpions as reptiles* till 1910, when in an Amending Act the section was made to read "*snake or other noxious reptile*, *scorpion or insect*." As a matter of fact an occasional snake has been found in foreign (chiefly Australian) cargoes, but has always been destroyed; but scorpions and large centipedes are frequently introduced along with fruit and hardwood timber from the South Sea Islands and Australia, and none of those responsible can possibly prevent these accidental importations.

Amendment Acts were passed in 1900 and 1903, and in the latter a clause was introduced (over-riding clauses in the Rabbit Act) giving the Governor in Council power to "declare that weasels, stoats, etc., declared to be a natural enemy of the Rabbit, and which have since proved to be the enemies of game and poultry may be killed" in any specified district.

In 1907 all the Acts passed between 1880 and 1903 were repealed and the "Animals Protection Act 1907" was passed, consolidating all previous legislation. The following is Section 17: "Every person who unlawfully takes or destroys, or wilfully destroys the eggs of any birds mentioned in the First, Second or Fifth Schedules hereto, is liable to a fine not exceeding Ten Pounds."

The First Schedule contains the following names: antelope, black game, deer, grouse, hares, imported wild duck of any species, moose, partridge, pheasant, plover, ptarmigan, quail, snipe, swans.

The Second Schedule contains the names of several native species, and the following introduced birds: black stilt plover (?), black swan, curlew and wild goose.

The Fifth Schedule deals with about thirty species of indigenous birds, and also includes the Tuatara lizard and opossums.

An Amending Act in 1914 gives the Governor in Council power to make sanctuaries for imported and native game, and the protection thus afforded has been freely availed of.

All the preceding legislation from 1861 onwards, dealing with the protection of certain animals, was really aimed at conserving imported animals for the purposes of sport. It was supplemented in 1862 by a "Birds Protection Act," which limited the period during which game could be shot. One of its provisions, one which certainly could not be passed to-day, was in Section 6, in which shooting on Sundays was prohibited.

The introduction of small birds into New Zealand was partly due to sentimental considerations, and partly to the necessity of checking the ravages of "army worms" and other caterpillars which threatened to arrest the cultivation of certain crops. However, within ten or fifteen years of the coming in of the birds, their numbers increased to such an extent, that the protection afforded them had to be taken away, and restrictive legislation imposed. In 1882 a "Small Birds Nuisance Act" was passed, "to authorize Local Governing Bodies to appropriate Funds and to levy Rates for the destruction of Sparrows or other birds injurious to crops." This Act was repealed in 1889, and replaced by a new Act (which was itself repealed and re-enacted in 1902), giving increased powers for the use of poisoned grain and other means of destruction. Finally in 1908 "The Injurious Birds Act" was passed, and under the provisions of this, wholesale destruction of birds and eggs is carried on every year, without, however, doing more than just keeping the danger within limits.

It will be seen from the history of the introduction of rabbits into New Zealand that it was not till about 1870 that these animals began to be abundant, but after that date their increase was so very great that they seriously affected the great pastoral industry of the country, and in some districts threatened to destroy it. Protection against them has led to the passing of much legislation, and of many regulations. This commenced in 1876 in the General Assembly, when the "Rabbit Nuisance Act" was passed. By this Act power was given to proclaim certain areas as districts in which the Act was to be enforced; trustees were to be elected annually, and were to have power to levy rates on the landowners not exceeding one halfpenny per acre. The occupiers or owners of affected lands were also to be called on to destroy all rabbits. In the following year an Amending Act was passed, and a bonus of one halfpenny per skin was to be offered for rabbit-skins. Any one introducing rabbits into any district without authority, rendered himself liable to be imprisoned for a period not exceeding six months, or to a fine up to fifty pounds. These Acts were repealed in 1880, and another was passed giving increased powers. Among other provisions was one giving the Trustees, in the event of any owner failing to control the pest, power to enter private lands and destroy the rabbits. Any owners failing to take efficient steps rendered themselves liable to a penalty of not less than one pound nor more than twenty pounds for each seven days of neglect. The sheep inspectors were also appointed rabbit inspectors. In 1881 the Act of the previous year was repealed, and another enacted. By this time ferrets, stoats and weasels had been introduced into the country, and they had to be protected. By Section 25 the Government had power to declare "any animals, the importation of which is not prohibited by any Act in force relating to the protection of animals, to be natural enemies of the rabbit," and such animals were to be protected. By Section 26, "Any person who shall be convicted of destroying or catching ferrets, weasels or such other animals as may from time to time be declared to be the natural enemies of the rabbit," rendered himself liable to a penalty not exceeding ten pounds or fourteen days' imprisonment.

I do not know that these sections were ever enforced to any extent, till recently (1921). In the rabbit-infested districts the enemies of the rabbit were protected, because the evil was ever present and the menace serious. In the outside districts to which ferrets, stoats and weasels quickly spread, these animals were destroyed to a considerable extent, and no one said anything about it.

By this same Act while rating powers within proclaimed districts were reduced to one farthing per acre, inspectors were authorised to enter on all unoccupied crown or native lands, and to destroy rabbits thereon.

In spite of all the legislation and regulations passed, the rabbit nuisance continued to get more severe, so in 1882 the previous Acts were repealed, and increased powers were given in a new Act to cope with the pest. In the case of an owner who would not pay for the cost of destruction of the rabbits, the Government could take his land and sell it. At the same time permission was given to private individuals to keep rabbits in cages or suitable enclosures, this being to provide material for biological classes in schools and university colleges. In 1885 a small Amendment Act was passed, and in the following year a fuller amending Act. In this the liberating of live rabbits was absolutely prohibited, and the keeping of them in captivity was prohibited to all except teachers of biology. The maximum penalty for any breaches of the Act was further increased to one hundred pounds. In addition, power was given to the Board to erect fences to check the spread of the pest, and under this provision many hundreds of miles of rabbit-proof fences were erected in Canterbury and other districts to arrest the onward march of the invading "bunny." In 1890 a further Rabbit Nuisance Act was passed; and an Amending Act in the following year. In 1898 "The Rabbit-proof Wire-netting Act," containing no fewer than 35 sections, was enacted, and this dealt with one phase of the question, as under it fencing districts were formed, Boards of Trustees set up, and rating powers were granted. While all the preceding legislation referred to areas which were rabbit-infested, this one enabled districts which were threatened to keep out the pest. In 1901 an Amendment Act to the Rabbit Nuisance Act was passed, and in 1902 a Consolidating Act of 116 sections dealt with the whole subject, repealing all past legislation, and re-enacting it in full detail. This is now the law of the whole important question, except that in 1910 a small Amendment

Act was passed, referring to a matter arising out of the main Act. Section 7 reads as follows:

> Notwithstanding anything in any Act it shall be lawful for any person to destroy hares, weasels or stoats: Provided that the Governor in Council may, on the recommendation of the Minister of Agriculture, suspend within any specified area the operation with respect to hares, weasels and stoats.

An interesting piece of legislation called "The Chatham Islands Animals Act" became law in 1884, and was re-enacted in 1908. Section 2 says: "No person shall introduce or allow to go at large any rabbit or hare in that part of the Colony called the Chatham Islands."

The legal position regarding opossums in New Zealand is rather interesting. By a Gazette Notice of 19th August, 1912, it was declared that opossums ceased to be imported game. Just a year later, owing to representations from acclimatisation societies, it was decided that the animals should be absolutely protected in certain districts; and a warrant to that effect was published in the Gazette of 7th August, 1913. In 1916 a further warrant was issued specifically protecting them in the Wellington Acclimatisation district.

They have now ceased to be imported game and they have been absolutely protected in certain areas. The position is now, in June, 1919, that:

> there is no existing law in force giving power to declare an open season for the taking or killing of these animals unless they were again declared to be either imported or native game, and this is not practicable as they would then automatically be protected in parts of the Dominion where protection is not desired; there being no existing power to enable them to be declared imported game in part only of the Dominion.

Probably an Amendment of the Law relating to these animals will shortly be enacted. (See Appendix A, p. 556.)

Insect pests due to imported species have received a good deal of attention from the legislature. In 1854 the Nelson Provincial Council passed an "Ordinance to prevent the increase of American Blight," which imposed a penalty of forty shillings on any one not clearing blight off any infected tree after due notice had been given, with five pounds for a repetition of the offence, and ten pounds for any one selling infected trees.

Thirty years later the New Zealand Parliament passed "the Codlin Moth Act 1884,"—"An Act to provide for the Destruction of the Insect known as *Carpocapsa pomonella*, or the 'Codlin Moth,'" under which power was given to proclaim certain districts in which it was to take effect, to levy contributions from all the orchards within these districts (such contributions not to exceed one halfpenny per tree), and to make regulations for the destruction and eradication of the pest.

Both the American blight and the codlin moth are still with us, but both are kept well under in all the fruit-growing districts. It is in small holdings where a few odd trees are grown and are not registered, or get overlooked, that all the dangerous fruit pests are found. Within recent years much more comprehensive legislation than these isolated Acts covered was passed. Thus in 1903 was enacted "The Orchard and Gardens Pest Act"—"An Act to prevent the Introduction into New Zealand of Diseases affecting Orchards and Gardens, and to provide for the eradication of such Diseases, and to prevent the spread thereof." Section 4 gave power to the Governor in Council to prohibit absolutely the introduction of any plant, fruit, fungus, parasite, insect or any other thing likely to introduce any disease into New Zealand. The First Schedule detailed the following insects which were to be kept out by all possible means:

Mediterranean or West Australian fruit fly (*Halterophora capitata*),
San José scale (*Aspidiotus perniciosus*),
Queensland fruit fly (*Tephrites tryoni*),
Vine louse or Phylloxera (*Phylloxera vastatrix*).

The Second Schedule contained the names of pests already in the country, whose eradication was left to district authorities to deal with:

American blight (*Schizoneura lanigera*),
Codlin moth (*Carpocapsa pomonella*),
Mussel- or oyster-scale (*Mytilaspis pomorum*).
Red mite (*Bryobia pratensis*).

This Act was re-enacted in 1908 in the Orchards and Gardens Diseases Act, which consolidated all previous legislation. This Act is still in force, but its operations have been greatly extended by numerous Orders in Council, specifying certain insects, fungi, etc., to be rigidly kept out[1].

[1] The following is a list of these prohibited insects, etc., the fungoid and bacterial pests being left out, as I have not attempted to deal with them in this work. Those marked † have never been introduced into New Zealand as yet. If fruit, timber, or other foreign material have shown any signs of these pests, they have been seized at the port of entry and destroyed.

1908, Sept. 7th:

Tomato fruit fly (*Lonchæa splendida*)

1911, Aug. 21st:

† *Dacus facialis*
† *Dacus virgatus*
† *Dacus melanotus*
† *Dacus cucumis*
† *Dacus tongiæ*
† *Dacus kirki*
† *Dacus raratongiensis*
† *Dacus passifloræ*
Currant-borer (*Ægeria tipuliformis*)
† Gypsy moth (*Ocneria dispar*)

Fishery legislation in New Zealand deals largely with marine fisheries. That dealing with imported fishes begins in 1867, with a "Salmon and Trout Act,"—"An Act to make provision for the preservation and propagation of Salmon and Trout in this Colony," and the preamble states: "Whereas it is contemplated to introduce Salmon and Trout into this Colony from abroad, etc." This Act gives the Governor power to make regulations for the protection of fish which had not then been introduced, "for the preservation and propagation of young salmon, salmon fry and spawn and young trout, trout fry and spawn upon its importation into the colony"; it placed restrictions on fishing in those streams into which such young fish or spawn were deposited; regulated the times and seasons for fishing; prohibited "the use of nets or other engines or devices for taking fish in any stream" so utilised; and prohibited the use of lime or other deleterious materials for destroying the fish. It is a curious little piece of legislation in advance of its object.

In 1877 "The Fish Protection Act" was passed, giving more

1915, Aug. 23rd:

† Apple aphis (*Aphis mali*)
Apple-blossom weevil *Anthonomus pomorum*)
† Apple-bud moth (*Hedya ocellana*)
† Apple-pith moths (*Blastoderma hillerella* and *B. vinolentella*)
† Apple root-borer (*Leptops hopei*)
† Apple saw-fly (*Hoplocampa testudinea*)
† Apple-sucker (*Psylla mali*)
† Apple-tree borers (*Rhizopertha collaris, Chrysobothris femorata* and *C. mali*)
† Bulb-mite (*Rhizoglyphus echinopus*)
† Colorado beetle (*Doryphora decemlineata*)
† Currant aphis (*Rhopalosiphon ribis*)
Currant clearwing (*Sesia tipuliformis*)
Currant gall-mite or big bud (*Eriophyces ribis*)
Currant-shoot borer (*Incurvaria capitella*)
† Fruit-bark beetle (*Scolytys rugulosus*)
† Gooseberry saw-fly (*Nematus ribesii*)
† Hop-aphis (*Phorodon humuli*)
† Lesser narcissus-fly (*Eumerus strigatus*)
† Light brown apple-moth (*Cacœcia postvittana*)
† Onion fly (*Hylemyia antigua*)
† Raspberry beetle (*Byturus tomentosus*)
Raspberry moth (*Lampronia rubiella*)
† Raspberry weevil (*Otiorhynchus picipes*)
† Round-headed tree-borer (*Saperda candida*)
† Rutherglen bug (*Nysius vinitor*)
† Woolly currant-scale (*Pulvinaria vitis*)

1917, 16th April:

† Banana scale (*Aspidiotus destructor transparens*)

1918, 23rd April:

Leaf-roller (*Cacœcia excessana*)

The last named is a native of New Zealand.

No doubt the object of including in these lists so many species which apparently have never yet been recorded as met with in the country, is to enable inspectors at the various ports of entry to at once hold up a consignment of fruit, etc., where any of these may be found. They occur in localities from whence they might be exported to New Zealand, and it is evident that up to the present time, if any of them have been met with, they have been destroyed at once, so that they have not got free into the country.

definite power to make regulations. (By some oversight this Act seems to have been overlooked when the Consolidating Act of 1908 was passed.) In the following year "The Fisheries Dynamite Act 1878" was passed to prohibit the use of dynamite or other explosives in public fisheries, whether marine or inland.

In 1884 an Amendment Act to that of 1867 was passed; and in the same year "The Fisheries Conservation Act," which gave power to make regulations for the protection of marine and fresh-water fisheries. In 1892 an Amendment Act to this was passed, giving, *inter alia*, power to acquire land for fish hatcheries. Further Amendment Acts were passed in 1903, 1906 and 1907. Then in 1908 "The Fisheries Act" consolidated all previous enactments. Among other provisions it prohibited the casting of sawdust or any saw-mill refuse into any waters, which has always been a sore point with saw-millers. (It apparently did not make any reference to refuse from flax-mills, which, in quantity, is a very deleterious substance.) It also fixed definitely close seasons for trout fishing, viz., from 1st day of May in each year to the following 30th day of September, and fixed the annual licence fee at £1 for men and 5*s*. for women and for boys under sixteen years of age. Licences to fish for perch were also provided for. This comprehensive Act of 99 sections is practically the complete law on the subject. An Amendment Act of 1912 deals chiefly with the encouragement of the whale fishery; but Section 9 authorises the construction of ponds for breeding and rearing trout for sale, a branch of industry which has not been really started to any extent in this country. A further small Amendment Act in 1914 completes the legislation on this subject.

I have already referred to early legislation directed towards the suppression or mitigation of the trouble caused by the vigorous growth and wide extension of thistles, gorse and broom. These, however, were only a few of the plants which soon got out of hand, and became a serious problem to the country. Spasmodic efforts to deal with the difficulty culminated in the passing of "The Noxious Weeds Act" in 1900. This was "An Act to prevent the spread of Noxious Weeds, and to enforce the Trimming of Hedges." This not only required owners to keep their land free of certain weeds, but also to clear the road-lines within their properties, and half the width of the boundary lines; with power to inspectors to enter upon the land, and, if necessary, to do the work at the owner's expense. The law has been more honoured in the breach than in the observance, and the Government has been itself one of the greatest offenders, some areas of Crown lands being perfect nurseries of weeds, which have sowed the country all round. Maori lands, too, have been an almost insuperable obstacle to the carrying out of the conditions of the Act, chiefly on account of the difficulties of joint ownership. There is a continual outcry about the necessity of eradicating noxious weeds, but the fault does not lie in the law, but in the difficulty of putting it into effect. Section 7 prohibits the sale of:

(*a*) Any noxious seeds, except in the case of gorse-seed to be sown for forage or fodder by permission in writing of the local authority, or for the planting of hedges or live fences; or

(*b*) Any grass-seed, or other seed or grain, which has not been thoroughly dressed by means of a seed-cleaning machine or other sufficient process for the purpose of removing all noxious seeds.

The Schedules of this Act are interesting, as giving some idea of the relative importance or seriousness of the weeds specified. Schedule I includes three plants which are always, and everywhere without exception, to be treated as enemies to be destroyed:

Blackberry (*Rubus fruticosus*),
Canadian or Californian thistle (*Cnicus arvensis*),
Sweet-briar (*Rosa rubiginosa*).

Schedule II specifies plants which are to be declared noxious weeds in certain restricted areas defined by the local authority:

Bathurst burr (*Xanthium spinosum*),
Broom (*Cytisus scoparius*),
Giant burdock (*Arctium majus*),
Gorse (*Ulex europæus*),
Hakea (*Hakea acicularis*),
Ragwort or ragweed (*Senecio jacobæa*).

Schedule III specifies noxious seeds, viz., those referred to in Section 7,—but the Governor may from time to time extend these Schedules II and III, by including other plants than those mentioned, on the recommendation of the Parliamentary Committee known as the Joint Agricultural, Pastoral and Stock Committee:

Bathburst burr (*Xanthium spinosum*),
Blackberry (*Rubus fruticosus*),
Broom (*Cytisus scoparius*),
Burdock (*Arctium*, any species),
Burr clovers (*Medicago denticulata* and *M. maculata*),
Clover dodder (*Cuscuta trifolii*),
Dock (*Rumex*, any species),
Fat-hen or white goose-foot (*Chenopodium album*),
Ox-eye daisy (*Chrysanthemum leucanthemum*),
Sweet-briar (*Rosa rubiginosa*),

Thistles (any species of *Carlina*, stemless thistle), *Carduus* (common plume or Scotch thistle), *Cnicus* (Californian thistle and woolly-headed thistle), *Centaurea* (star thistle), *Silybum* (milk thistle),

Wild turnip (*Brassica campestris*),

Ragwort or ragweed (*Senecio jacobæa*).

In 1908 "The Noxious Weeds Act" consolidated the preceding enactments, and, as far as machinery was concerned, was practically the same. Schedule III (the old Schedule II) specifies the same weeds with the addition of any species of *Arctium*, or burdock, and any species of *Rumex* or dock, elderberry (*Sambucus niger*), fennel (*Fœniculum vulgare*), foxglove (*Digitalis purpurea*), hemlock (*Conium maculatum*), kangaroo acacia (*Acacia armata*), lupin (*Lupinus luteus*), ox-eye daisy, pennyroyal (*Mentha pulegium*), periwinkle (*Vinca major*), St John's wort (*Hypericum perforatum* or *H. humifusum*), Tauhinu or New Zealand cotton-wood (*Cassinia leptophylla*), any species of thistle as specified above, tutsan (*Hypericum Androsæmum*), viper's bugloss (*Echium vulgare*), and wild borage (*Echium violaceum*).

The Noxious Seeds Schedule is similar to No. III of the previous Act with the addition of fennel, hemlock, pennyroyal, St John's wort and tutsan.

The record of the legislation passed by various parliaments in New Zealand is historically of interest, and of value from the point of view of the naturalist, as showing how various animals and plants developing under new conditions in a new country, "run away," as it were, and become so aggressive and so numerous in individuals as to constitute a serious menace to the well-being of the community. At the same time it is rather a curious record when looked at as a whole, for hardly had any Bill passed into law when amendments were found to be necessary, and these succeeded each other year after year with monotonous regularity. The point reached to-day, however, is one that can be contemplated with a certain amount of satisfaction. As far as new introductions of animals and plants are concerned, there is pretty close inspection at all ports of entry of seeds, fruits, etc., and few deleterious things pass the inspectors. Where objectionable introductions have got in during recent years, they have in several cases been followed up to the localities to which they were distributed, and have been eradicated. As to those animals and plants which are already in the country and which it is desirable to eradicate or keep in check, the conditions are more difficult, but in many cases the laws which have been passed have enabled them to be coped with. The hope for the future lies in two directions, viz., closer settlement of the land coupled with more intensive cultivation; and better education of all those concerned in the primitive industries of the country, which are mainly agricultural and pastoral, as to the economic waste which ensues whenever undesirable animals and plants are allowed to thrive. There is a growing desire for such education, and it is becoming more fully recognised that it is one of the important factors in the future success of the country.

APPENDIX A

OPOSSUMS IN NEW ZEALAND

In March, 1920, Professor H. B. Kirk, in response to a request from the New Zealand Government, made a report on the Australian Opossums in the country in answer to the following questions:

(1) Whether the damage to forests is likely to outweigh advantages to settlers in being able to earn a revenue by trapping or taking opossums in new country?

(2) On what areas these animals could be liberated with reasonable security against their overrunning and damaging State forests?

He found that these animals often do considerable damage in orchards by eating the leaves and young shoots of apple trees, of lemons, of peaches, and all other stone fruits; they bite fruit of all kinds, sometimes consuming the whole fruit, sometimes leaving it damaged on the trees, or causing it to fall. During the winter they do little damage.

They also bite off buds and shoots of roses and other garden shrubs, eat peas when the pods are filling, and occasionally eat other vegetables.

Though reported as very destructive to pine forests (*Pinus halepensis* and *P. maritima*) in South Australia, he found very little damage to pine plantations in New Zealand. North of Auckland they eat the young male cones of *P. radiata* (*P. insignis*).

In the native bush "the opossums eat leaves and young shoots of makomaka (*Aristotelia racemosa*), of karaka (*Corynocarpus lævigata*), of houhou (*Schefflera digitata*), of mahoe (*Melicytus ramiflorus*), of broadleaf (*Griselinia lucida*), of konini (*Fuchsia excorticata*), of matipo (*Pittosporum eugenioides* and other species), of kohekohe (*Dysoxylum spectabile*), the soft parts of miro-fruit and of the nikau-fruit (*Rhopalostylis sapida*), the fruit of the konini, and many others. By his weight he breaks young shoots, causing them to wither. I have examined the upper branches of many favourite food trees, but have never found that greater damage has been done than I have described, and the trees branch freely below the wound. I have found no native tree that has, in my opinion, been killed by an opossum. The favourite plants of the opossum are damaged by constant climbing and playing, but this generally happens near houses or at the edge of a clearing, but I have never seen serious damage of this kind in the forest." His general conclusions are that opossums do very little real damage in the bush.

He recommends that an open season be declared during which opossums may be lawfully taken, and that for the present this open season be the months of May, June and July. Also that a licence fee of £1 be charged to all trappers, and that for each skin taken a royalty of 1*s*. should be paid to the Crown. In fruit districts he recommends that any resident may kill opossums at any time without penalty, but he must report the same to the local postmaster. He further approves of the suggestion made by the Otago Acclimatisation Society to stock the forests on both sides of the great Alpine Range with opossums, and estimates that the fur trade would soon reach a value of £200,000 a year.

As an outcome of this report Regulations for the taking and killing of these animals appeared in the Supplement to the New Zealand *Gazette* of May 5th, 1921. A licence fee of £2. 10*s*. entitles the licencee to take or kill opossums; a royalty of 1*s*. per skin shall be paid to licensed brokers, who shall pay a licence fee of 21*s*., and shall receive 5 per cent. commission on the royalties collected. Every skin shall be stamped on payment of the royalty fee. Exportation of skins must be with the consent of the Under Secretary, Department of Internal Affairs.

Following on this *Gazette* notice the month of June, 1921, was declared an open season, several hundred licences were issued, and many thousands of skins were obtained, the exact number not being obtainable at the date of publication of this work.

APPENDIX B

LATER RECORDS

During the progress of this book notices of many new species, especially among the Insecta, havé been received, together with later accounts of those already referred to. As it was impossible to incorporate these into the text without disturbing the paging of the book, they have been collected together into an appendix, and each entry gives the reference to the page where it should be interpolated.

p. 241 Quinnat Salmon

Mr Ayson, Inspector of New Zealand Fisheries, informs me that while the run of this salmon in 1918 and 1919 in the Waitaki River was very small, so much so indeed that some people thought they had disappeared altogether, that of 1920 was quite large. Salmon are now running in several rivers to the north as far as the Waiau, and quite recently they have been reported from the Wairaripa in the North Island. These fish must have spawned also in the Hokitika River on the West Coast, as young salmon, only a few inches long, have been taken there.

During the present spawning season (1921) about 1,133,000 ova have been collected at the Hakataramea Hatchery, and fish from 4 ft. to 4 ft. 6 in. in length have been seen right up to the foothills of the Southern Alps. The run of fish this year in the Waitaki is quite phenomenal, and they are also showing up in the Clutha River in the south. The Quinnat Salmon is evidently firmly established in the South Island.

p. 258 MOLLUSCA

Family MURICIDÆ

Murex ramosus, Linn.

Within the five years—1903–08—two specimens of this introduced mollusc were found at Tauranga: one of these, 8½ in. long, was living when collected.

They must have been brought from one of the South Sea Islands, either purposely, or perhaps attached either to a ship's bottom, or to her anchors.

p. 265 *MYRIAPODA*

Order CHILOPODA

Family SCOLOPENDRIDÆ

Scolopendra subspinipes, Leach

This centipede is frequently introduced into New Zealand in fruit cases, usually from the South Sea Islands. The species is very widely spread in most tropical and sub-tropical countries. It does not seem to have established itself, unless perhaps from Whangarei northwards.

Ethmostigmus platycephalus, Newport

Recorded in the *Index Faunæ Novæ-Zealandiæ* (p. 235) as an indigenous species—*Heterostoma platycephala*—but without locality.

Ethmostigmus rubripes, Brandt

A specimen, now in the Christchurch Museum, was found in 1901 among timber imported from Australia.

INSECTA

p. 270 Sub-order PSEUDO-NEUROPTERA

Family ATROPIDÆ

Troctes divinatorius, Müller. Cabinet mite; Book-louse; Book-tick

This species appears to be equally common with *Atropos pulsatoria*.

p. 272 Order HYMENOPTERA

Family ICHNEUMONIDÆ

Pleurotropus (*Entodon*) *epigonus*, Walker, was recorded by the Agricultural Department on its introduction as *Semiotellus nigripes*. But this latter species is a parasite of *Oscinella* in Britain, not of the Hessian fly, and Mr D. Miller informs me that it is doubtful whether it occurs in New Zealand at all.

p. 273 Family CHALCIDIDÆ

Aphelinus mali, Haldane

This little wasp was introduced from America early in 1921 by Dr R. J. Tillyard, Chief of the Biological Department of the Cawthron Institute,

Nelson, and was recently stated by him to be "becoming established in the Institute's grounds." It is parasitic on the Woolly Aphis (*Schizoneura lanigera*) and the Mealy Bug (*Pseudococcus longispinus*).

In introducing these wasps Dr Tillyard found that other species of insects were unwittingly introduced along with them. There were two or three Chalcids, and also a small moth *Nepticula pomivorella*, Packard, whose caterpillar is a leaf-miner on apple trees.

p. 273 *Bruchophagus funebris*, Howard

This parasitic wasp was first identified and recorded by the Agricultural Department in 1921.

p. 282 Family Apidæ

Hive-bees (*Apis mellifica*, Linn., and *A. ligustica*, Spin.)

On 7th October, 1920, regulations under the Apiaries Amendment Act, 1913, were gazetted, defining the conditions under which bees, honey, and apiary appliances may be introduced into New Zealand.

Bees and honey can only be introduced at Auckland, Wellington, Lyttelton, Dunedin or Bluff. If they come by parcel post, they must be examined before being forwarded. Bees may be imported from any province of Italy, the United States of America, or the Commonwealth of Australia, but they must be accompanied by a certificate from the shipper of their freedom from disease. They have all to be examined by an Inspector on arrival.

p. 282 Family Vespidæ

Vespa germanica, Fabr.

Specimens of this wasp were taken in the Wairarapa district recently, and were identified by the officials of the Agricultural Department. It has not previously been recorded from New Zealand.

Order COLEOPTERA

p. 291 Family Bostrichidæ

Lyctus linearis, Goeze

This beetle, Mr D. Miller informs me, is frequently introduced into New Zealand in packing-cases and wooden furniture into which it bores. It does not seem to have established itself permanently.

It is noteworthy, however, that there is no restriction to the importation of insect-infected timbers, whether they are intended for building, packing-cases or furniture.

p. 299 Family Curculionidæ

Cyclas formicarius, Fabr.

This weevil probably spread originally from Cochin China, but is now distributed very widely over the tropics. It is frequently found in shipments of kumaras from the South Sea Islands, but it does not seem to have become established in New Zealand. Regulations to prevent its introduction have been gazetted.

Order LEPIDOPTERA

p. 300 Family BOMBYCIDÆ

Bombyx mori, Linn. Silkworm

As already stated at p. 300, Mr T. C. Batchelor introduced silkworms into Nelson in 1863. Seven years later he urged on the Government the importance of fostering the silk industry in New Zealand, stating that he had imported Tuscan mulberries, and had about 1800 trees ready to plant out. In the following year Mr Batchelor had two varieties of silkworms, viz. Lombardy Buffs and Japanese, and from these he reared two broods of cocoons, the latest forming in November. At the same time, and as an outcome of his action, the Government offered a bonus for the encouragement of the industry. Apparently a certain number, both of colonists and natives, commenced the cultivation of silkworms, but no one secured the bonus, and it was allowed to lapse.

In 1879 a case of crude silk from Auckland was shown at the Sydney Exhibition, but I have not been able to find out by whom it was grown.

In 1886–87 another effort to start sericulture was made by Mr G. A. Schoch of Auckland. He found that between 19,000 and 20,000 white mulberry trees suitable for silkworm-food were in the country. He ordered fresh seed from Italy, and this was distributed gratis. The Auckland Domain Board planted (in 1886) 1100 trees in the domain for distribution to silk-raisers. On December 9th, 1886, a shipment of silkworm eggs arrived from Italy, and about 600,000 were distributed gratis. Towards the end of January a further batch of eggs arrived from Japan—rather late in the season—and about two-thirds of these were distributed. Samples of Italian and Japanese cocoons, and skeins of both kinds of silk were later on forwarded to the Manchester Jubilee Exhibition. Mr Schoch was confident of success in his efforts, but the matter was not taken up seriously by others, and ultimately nothing came of it.

In times of agricultural depression and low prices, schemes for establishing sericulture as one of the industries of the Dominion are brought forward and command some attention, but whenever other primary industries are making good headway, the interest dies out.

Silkworms are still commonly reared about Auckland and other centres on a small scale, but not as commercial ventures.

Mr G. Howes informs me that in the neighbourhood of Sydney, New South Wales, silkworm escapes have established themselves in a wild state.

p. 327 Order HEMIPTERA

The insects of this order have not been at all commonly collected in New Zealand. As increasing attention is given to them, very many more species than are at present known will be recognised, and this applies to both native and introduced species. Since this work went to press, nine of the latter have been reported to me.

Sub-order HETEROPTERA

p. 328 Family CAPSIDÆ

Oncognathus binotatus, Fabr.

Mr D. Miller states that this European species, which feeds chiefly on grasses, is found at Wellington and at Wanganin, where it is very common. It is probably widely distributed.

Sub-order HOMOPTERA

p. 328 Family FULGORIDÆ

Siphanta acuta, Walk.

First recorded by Kirkaldy in 1909. This Australian species has been collected at Nelson, and in several localities in the North Island.

Saphena cinerea, Kirkaldy

Another Australian species, first recognised in 1921, and wrongly identified as *Siphanta granulata*, Kirk. It occurs generally throughout the Auckland provincial district.

p. 328 Family CICADELLIDÆ

Empoasca australis, Froggatt (?). Apple leaf-hopper

This insect has been introduced from Australia, but I have no information as to its range within New Zealand. Dr R. J. Tillyard informs me that the species has so far not been determined with certainty.

p. 331 Family APHIDIDÆ

Schizoneura ulmi, Linn. Leaf-crumpling Aphis of Elm

This aphis has been recently (1920–21) noticed for the first time in the Auckland district.

Mysus persicae, Sulzer. Green Peach Aphis

Mr D. Miller informs me that this species is widespread in New Zealand.

Aphis nerii, Fonsc.

First recorded in 1921 as occurring on Oleander.

Aphis bakeri, Lowen

Observed and recorded by Mr J. Meyrs on clover in 1921.

Pemphigus populi-transversus, Riley. The Poplar gall Aphis

This species was first observed on Poplar trees in Central Otago in 1920; it has since been reported from Nelson.

Myzaphis abietina, Walk. Spruce Aphis

This aphis, first recorded in 1919, is found generally throughout the North Island.

CRUSTACEA

p. 340 Division *SYNCARIDA*

Order ANASPIDACEÆ

Anaspides tasmaniæ, G. M. Thomson

In February, 1898, I received from Mr Leonard Rodway, the eminent Tasmanian botanist, a jar containing about twenty specimens of this most interesting shrimp. He had experienced great difficulty in keeping them alive in Hobart on account of the heat of the weather, and he also found that they were terrible cannibals. But this is probably true of all crustaceans when confined in a limited space without food. The remainder carried over to Dunedin remarkably well. But the mortality continued, though they were placed in a tank continuously supplied with fresh running water. All were dead within 16 days after arrival, the last eight of them being badly infested with fungus (*Saprolegnia* or *Achlya*).

In a communication to me in June, 1920, Mr Cheeseman adds the following notes of introduced species:

SAPINDACEÆ

p. 390 *Acer pseudo-platanus*, Linn.

Spreading in many localities in the neighbourhood of plantations.

p. 408 Hawthorn (*Cratægus oxyacantha*)

In the latter part of 1919 it was found that certain orchards in the Auckland district were infected with the bacterial disease known as Fire-blight (*Bacillus amylivorus*, Trev.). It is supposed that the disease was introduced from the Western States of America, and distributed by means of nursery stock from Auckland. An examination in 1920 showed that the infected area ranged from Warkworth in the north to Kihikihi in the south. The Waikato was the most widely infected district; there was a considerable spread of the trouble in Tauranga, but the orchards in the Thames district were free from it. Drastic restrictive measures were adopted by the Agricultural Department, and apparently in some areas, i.e. Warkworth and Silverdale, the disease was stamped out. As tested in the departmental laboratories the plants affected in New Zealand belong to the genera *Pyrus* (pears, apples, and quinces) and *Cratægus* (Hawthorn). Pear trees have suffered most, but the Department has come to the conclusion that the disease cannot be coped with in any district as long as the common Hawthorn (*C. oxyacantha*) is allowed to remain in evidence. Therefore at the end of last session of Parliament (1920) an Amendment of the Noxious Weeds Act 1908 was passed declaring Hawthorn a noxious weed. Clause 3 of this Act states that a special order may be made by any local authority limiting the operation of the Act to a part only of any district. Clause 4 reads as follows: "The planting of hawthorn is hereby prohibited, and every person commits an offence against the principal Act who propagates hawthorn in any manner, or who does any act with intent to propagate hawthorn, or who sells any seeds, plants, or cuttings of hawthorn."

p. 409 ROSACEÆ

Spiræa Lindleyana

Found near Arrowtown, where it was first observed by the Pastoral Runs Commission in May, 1920.

p. 409 SAXIFRAGEÆ

Escallonia macrantha

Common on banks in and near Wellington.

p. 410 ONAGRARIEÆ

Fuchsia Riccartoni

Found on old mining tailings near Ross, South Westland.

LABIATÆ

p. 460 *Thymus serpyllum*, Linn.

Abundantly naturalised near Clyde, in Central Otago.

p. 486 CYPERACEÆ

Kyllingia brevifolia, Rottb.

This species, doubtfully placed as indigenous in the *Manual N.Z. Flora* (p. 764), is now considered by Cheeseman as an introduction.

Cyperus tenellus, Linn.

Cyperus vegetus, Willd.

Both species are treated as doubtfully indigenous in the *Manual* (pp. 765 and 766), but are now considered to be introductions.

p. 496

In regard to *Festuca ovina* and *F. rubra* Mr Cheeseman has sent me the following note (July, 1921):

"For the common fescue-tussock I follow J. B. Armstrong in calling it *Festuca novæ-zealandiæ*. The introduced fescues are very different. They are common only in a comparatively few places where they have been sown purposely. The most widely spread is known as 'Chewing's fescue.' Red fescue (*F. rubra*) is taller and more tussocky, but is easily distinguished from *F. novæ-zealandiæ*. There are also several other indigenous, but non-tussocky species of *Festuca*, not properly classified as yet. Petrie has named a distinct one *F. multinodis*."

p. 501 LYCOPODIACEÆ

Selaginella Kraussiana

Common in damaged forest in the vicinity of Wellington.

APPENDIX C

THE TUTIRA DISTRICT

SINCE this work was in print, a very fine book entitled *Tutira, the Story of a New Zealand Sheep Station*, by Mr H. Guthrie-Smith, has been published by Messrs William Blackwood and Sons. It is a most interesting record of fine observational work, by a skilled naturalist, dealing with the physical and biological features and history of a comparatively small area of land in the Hawke's Bay district. No fewer than seven chapters deal with the naturalised plants, and seven more with naturalised animals. Much of the book consequently covers some of the ground gone over by me, but Mr Guthrie-Smith has given much greater detail regarding the introduction and spread of the species referred to than I have been able to do. Very many of his conclusions are identical with my own, and the book is a most valuable contribution to the biological literature of New Zealand. Unfortunately it has no index, and the table of contents only very partially supplies this want.

Without in any way attempting to summarise the information contained in this book, I here record a few of the facts, taking the various animals and plants in the order referred to in my own work.

Much valuable information is given as to the remarkable tracks made over the unbroken country by sheep (p. 181), while an interesting instance of melanism in sheep is recorded at p. 355, *et seq*. Each patch of bush and scrub, chiefly on Eastern Tutira, maintained its little herd of wild sheep. About 1892 a small flock of four or five black sheep was noticed on a block in the north-west of the Tutira district. In about ten or twelve years the flock had greatly increased, and when the land on which they had been running came to be cleared and fenced, some 220 wild sheep were rounded up and yarded. Of these over 90 were black, the larger number with white tips to their tails; about five or six were piebalds, and the rest white. All of these were pure merino, the rams carrying magnificent heads. Mr Guthrie-Smith's account of this flock and the explanation are worth careful study.

At pp. 308–9, four cases are given of the capture of rats, which were claimed to be of the species which occurred in New Zealand prior to Captain Cook's first visit, viz., the Kiore maori, or *Mus exulans*. Three of these captures, dating between 1879 and 1906, cannot be verified; it is possible that the animals caught belonged to this species. But the fourth specimen, captured by Captain Donne "in the forest path between Waikaremoana and Waikareiti not many years ago," seems from its photograph to be an example of the Maori rat. Though nearly extinct in New Zealand, it would appear that a very few specimens still survive in wild rough country.

In 1886 and succeeding years, stoats, ferrets and weasels were liberated in numbers in the Wairarapa district, and by 1901 they had reached the neighbourhood of Tutira. "Between 1902 and 1904 they had overrun the country between Tutira and the southern edge of the Poverty Bay Flat. Everywhere I heard of them. On every road and new-cut bridle-track during these two seasons I met or overtook weasels hurrying northwards, travelling as if life and death were in the matter. Three or four times also I came on weasels dead on the tracks. These weasels, alive or dead, were or had been travelling singly....For a short period weasels overran like fire the east coast between Tutira and Poverty Bay, and then like fire died out....Nowadays on Tutira I do not hear from shepherds or fencers of the weasel once in six years. I have not seen one for twenty years. There is something ridiculous in the fact that the weasel should have arrived on the station before the rabbit, and that later, when rabbits had become numerous, weasels should have practically passed out of the district."

Mr Guthrie-Smith says that the Australian Quail (*Synœcus australis*) was privately introduced in the sixties by Colonel Whitmore on Rissington, Hawke's Bay, but did not reach Tutira for more than thirty years. But it seems more likely to me that this bird came to Hawke's Bay from the Wellington Society's importations in 1875–76, for their game farm was in the Wairarapa.

Again it is said that the Californian Quail was introduced by the Hawke's Bay Provincial Council, and again at a later date by the Hawke's Bay Acclimatisation Society. Unfortunately I have been unable to obtain information from either of those sources. Mr Guthrie-Smith bears out my contention, originally suggested by Mr Cheeseman (see p. 116), that the failure of game birds in New Zealand is due to the food supply being destroyed by small birds. He says: "Californian Quail reached Tutira in the middle nineties, and although there was at first an increase in their numbers, it was a limited increase and soon ceased. Their advent as game-birds had come, in fact, too late to admit of any great success. The competition of innumerable goldfinches, yellow-hammers, larks, sparrows, and native species, several of which had also increased with the enlarged area of open country consequent on the destruction of bracken, had already affected the insect food-supply; the Californian quail is now disappearing from the run."

In connection with the naturalisation of the thrush and the blackbird, the spread of these birds in the North Island was almost certainly from Auckland, and Mr Guthrie-Smith gives reasons why their course was right round the coast of the Gulf of Thames, the Coromandel Peninsula, down the Bay of Plenty, round the East Cape, and so to Hawke's Bay, and he illustrates their probable route by an outline map. Mr Philpott's observations in Southland show that the thrush shuns the dense bush, hence the coastal route may have been followed in the case of those which penetrated to Tutira. But he found blackbirds in the heart of the forest country, and I see no reason why both species should not have spread over from the Thames valley direct to the east coast, the intervening strip of bush land being comparatively narrow.

The sparrow, on the other hand, is stated to have spread from Auckland up the Waikato as far as Taupo, and then across by the line of the Taupo-Napier road. This is quite probable, for the sparrow never seems to stray far from the haunts of men.

Minahs (*Acridotheres tristis*) were liberated by the Hawke's Bay Society in 1877, that is, after the Wellington Society's introductions of 1875–76, but they did not increase on Tutira till about 1890, though they became very common about Napier. This bird, like the sparrow, follows man in his migrations, and Mr Guthrie-Smith describes how "upon the approach of autumn, minahs largely use the roads, closing in on homesteads for scraps of fowl-feed and leavings of the gallows and kennels. The species has also of late developed a vulture-like habit of congregating near any sheep dead on the hills; in the vicinity of the carcase, awaiting the process of skinning, the expectant birds gather for their ghoulish meal." It is a most remarkable fact that starlings do not seem to have reached Tutira, though they have been extraordinarily abundant at Napier, only distant about a score of miles.

At p. 318, Mr Guthrie-Smith describes the mason fly (*Pison pruinosus*, Cameron) as one of the most remarkable of the alien insects on Tutira. It is really a wasp, and belongs to the Family Sphegidæ. He says it is believed to have reached New Zealand in chinks and knots of Australian lumber, and that it was noticed by the late Mr J. N. Williams in the late sixties. It is recorded in the *Index Faunæ Novæ-Zealandiæ* (p. 98) as an indigenous species, but Mr Guthrie-Smith says: "Unlike the black cricket, it seems never to have received a Maori name—a fact in itself pointing to a comparatively late naturalisation." Mr Howes is inclined to think it is an introduction, as it is slowly spreading south. He met with it at Waipori three years ago, and saw it for the first time in Dunedin later.

Chapters XXIV to XXX deal at length with the naturalised alien flora of Tutira, and give the date of introduction of many of the species recorded on the run, and frequently also the mode of their carriage and spread. A list of plants naturalised on Tutira prior to 1882 is followed by three lists bringing down the catalogue to 1920, and the names are in approximate order of their arrival. These lists, and the subsequent notes on many of the introduced plants, are extraordinarily valuable from a naturalist's point of view, for they represent careful and close observational work.

I cannot follow this record in detail, but will only select a few facts to supplement my own statements. It is stated at p. 252, that "owing to the great extent of second- and third-class country sown, also to the parlous state of the finances of the run in early days, cheap seeds were largely purchased; hundreds of bags of 'seconds,' of Yorkshire fog and warehouse sweepings, have been at various times scattered broadcast on its pumiceous areas."

On p. 256 a list of plants whose seeds were brought in or attached to sacks is given, and I quote the following passage, not only because it is a true picture, but also because it illustrates remarkably well the author's interesting and graphic style:

"The average life of a sack is, I daresay, about five years, each sack in its time playing many parts. Starting at the Bluff, the southernmost port of the South Island, a sack may only become finally useless in the far north of the North Island, having spread blights and noxious weeds from one end of the colony to the other. It may commence its career with all sorts of high ideals, with the determination to carry only Timaru wheat, Hawke's Bay ryegrass, and Akaroa cock's-foot, but has in later life to abate the lofty pretensions of youth and ultimately to submit to the carriage of ordinary grain, ordinary ryegrass, and ordinary cock's-foot. Later, still whole and presentable, our bag will be considered fit for tailings and oaten chaff. It will now perhaps cross Cook Strait and be passed about a farming district bearing perhaps in one short jolt apples, in another onions, becoming at each trip more stained with rain and marked with mud. It is now filled with potatoes—another downward stage—and forwarded to Auckland. By this time ragged, rent, disreputable, with senses blunted in regard to weed-carriage, it may reach some struggling settler's little home in the roadless north; there, with no pride left, it will cover a bee-hive, roof a leaky hen-coop, or in a buggy act as mat for dirty boots. Lastly, the poor creature takes to drink, and hangs in a besotted state about a native settlement. There, utterly degraded, it may serve as a saddle-cloth to some galled Maori hack, and ultimately dropped, hatch out some long-secreted weed, that like a wicked action comes to light at last. It is not very often that a stowaway is thus caught red-handed emerging from his hiding-place; yet white goose-foot (*Chenopodium album*) was seized by me in the very act, a magnificent specimen, his great roots embedded in a rotten sack, one of many strewn about the site of a Maori drainer's camp."

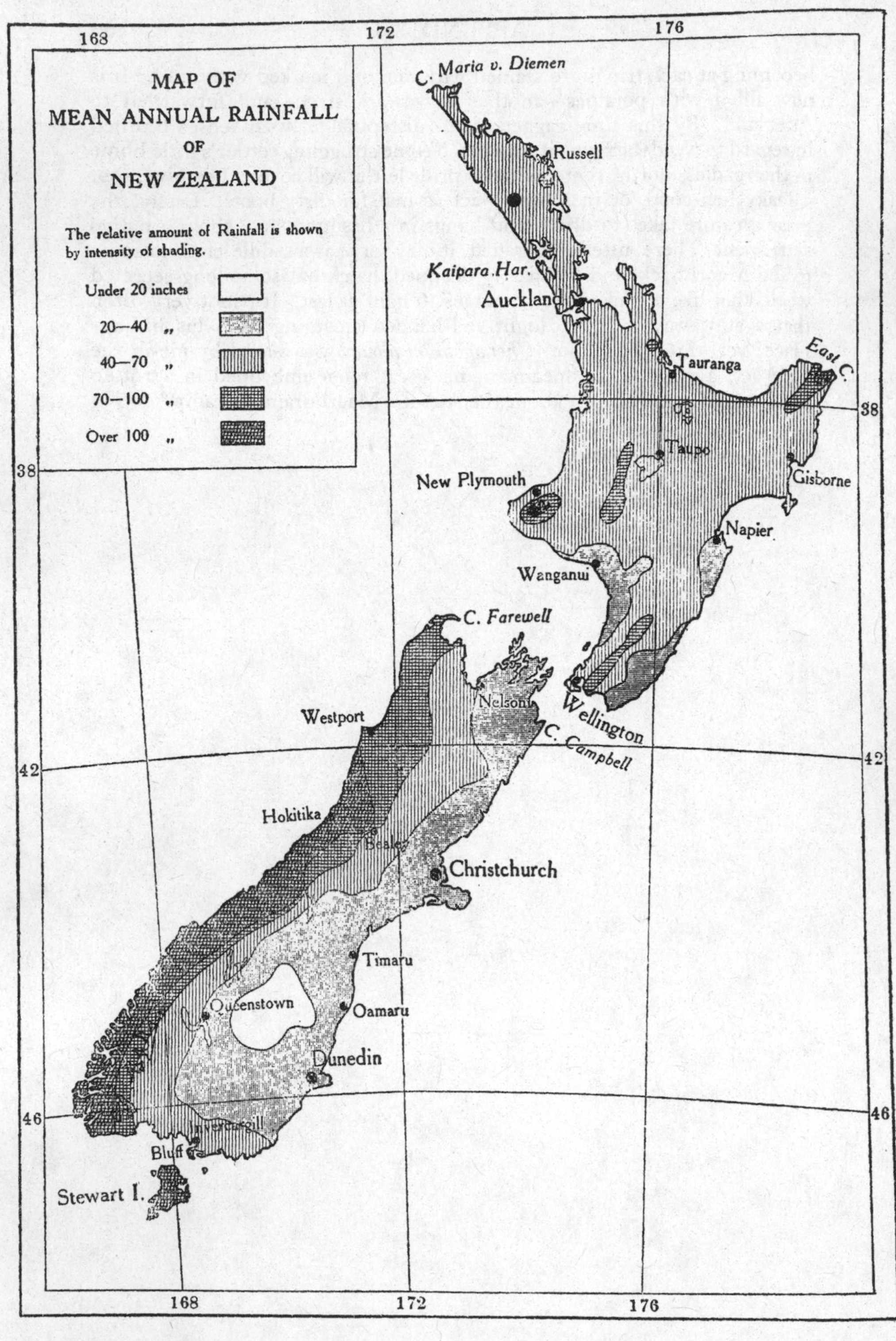

MAP OF
MEAN ANNUAL RAINFALL
OF
NEW ZEALAND
The relative amount of Rainfall is shown by intensity of shading.
Under 20 inches
20—40 „
40—70 „
70—100 „
Over 100 „
168
172
176
38
42
46
C. Maria v. Diemen
Russell
Kaipara Har.
Auckland
Tauranga
East C.
Taupo
Gisborne
New Plymouth
Napier
Wanganui
Wellington
C. Farewell
Nelson
C. Campbell
Westport
Hokitika
Bealey
Christchurch
Timaru
Queenstown
Oamaru
Dunedin
Invercargill
Bluff
Stewart I.

BIBLIOGRAPHY

ADAMS, T. W. "The species of the genus *Pinus* now growing in New Zealand, with some notes on their introduction and growth." Trans. N.Z.I. vol. XLVIII, pp. 216–233. 1907–1915.

Agricultural Department of New Zealand—Reports of the. Wellington. 1893–1920.

Agriculture, Journal of (published by the Agricultural Department of New Zealand). Wellington. 1909–1921.

ANGAS. "Savage Life and Scenes in Australia and New Zealand." 2 vols. London. 1847.

ANONYMOUS. "A journal of a voyage round the world in H.M.S. 'Endeavour' in the years 1768, 1769, 1770 and 1771." Publ. by T. Beckett. London. 1771.

—— "A Journal of a Voyage round the world in His Majesty's Ship 'Endeavour' in the years 1768, 1769, 1770 and 1771; undertaken in pursuit of Natural Knowledge, at the Desire of the Royal Society: containing all the various Occurrences of the Voyage, etc." London. 1771. Printed for T. Beckett and P. A. De Hondt.

—— "New Zealand in 1829." "From the Journal of an officer of the brig 'Hawes,' describing the capture of that vessel by the natives, and the cruelties exercised towards her crew; with some account of the country." United Service Journal, No. 18, June, 1830. London.

ARMSTRONG, J. B. "A short sketch of the Flora of the Province of Canterbury, with catalogue of Species." Trans. N.Z.I. vol. XII, pp. 317–323. 1879.

—— "Fertilisation of Red Clover in New Zealand." N.Z. Journal of Science, vol. I, pp. 500–504. 1883.

ARMSTRONG, JOHN F. "On the naturalized plants of the Province of Canterbury." Trans. N.Z.I. vol. IV, pp. 284–290. 1872.

ARTHUR, W. "On the Brown Trout introduced into Otago." Trans. N.Z.I. vol. XI, pp. 271–290. 1878.

—— "Notes on some specimens of migratory Salmonidæ." Trans. N.Z.I. vol. XIII, pp. 175–193. 1880.

—— "History of Fish Culture in New Zealand." Trans. N.Z.I. vol. XIV, pp. 180–210. 1881.

—— "On diseased Trout in Lake Wakatipu." Trans. N.Z.I. vol. XV, pp. 198–203. 1882.

—— "Notes on the Salmon disease in the Tweed and other rivers, and its remedy." N.Z. Journ. of Science, vol. I, pp. 347–353. 1883.

—— "On the Brown Trout introduced into Otago." Part II. Trans. N.Z.I. vol. XVI, pp. 467–512. 1883.

ASTON, B. C. "Some effects of imported animals on the indigenous vegetation." Trans. N.Z.I. vol. XLIV. Proceedings, pp. 19–24. 1911.

—— "The Vegetation of the Tarawera Mountains, New Zealand." Trans. N.Z.I. vol. XLVIII, pp. 304–314. 1915.

Auckland Acclimatisation Society. Annual Reports from the Commencement in 1867 to 1887; and from 1906–1921.

AYSON, L. F. "Introduction of American Fishes into New Zealand."

Bulletin of the Bureau of Fisheries, U.S.A. vol. XXVIII, part II, pp. 969–975. Washington. 1908.

BAKER, CARL F. "A Revision of American Siphonaptera or Fleas." Proc. U.S. National Museum, vol. XXVII, pp. 365–469. 1904.

BALLENY, JOHN. "Discoveries in the Antarctic Ocean in February, 1839." Journal Roy. Geog. Soc. London, 1839, pp. 517–526. (Reprinted in Antarctic Manual, 1901, pp. 336–347.) 1839–1901.

BANKS, Hon. Sir JOSEPH. Journal (1768–1771). Edited by Sir J. D. Hooker. London. 1896.

BARKER, Lady. "Station Life in New Zealand." London. 1870.

BATCHELOR, T. C. "On the varieties of the Mulberry Tree as food for the Silkworm." Trans. N.Z.I. vol. IV, pp. 424–426. 1871.

BATES, H. W. "On the Longicorn Coleoptera of New Zealand." Ann. and Mag. Nat. History; July and August. London. 1874. Reprinted in Trans. N.Z.I. vol. VII, pp. 315–332. 1874.

BATHGATE, ALEXANDER. "On the Lepidoptera of Otago." Trans. N.Z.I. vol. III, pp. 137–141. 1870.

—— "Notes on acclimatisation in New Zealand." Trans. N.Z.I. vol. XXX, pp. 266–279. 1897.

—— "The Sparrow plague and its remedy." Trans. N.Z.I. vol. XXXVI, pp. 67–79. 1903.

BEAN, TARLETON H. "Report on the Salmon and Salmon Rivers of Alaska." (Bulletin U.S. Fish Commission, vol. IX.) Washington. 1891.

BENHAM, W. BLAXLAND. "A re-examination of Hutton's types of New Zealand Earthworms." Trans. N.Z.I. vol. XXXI, pp. 156–163. 1898.

—— "On the flesh-eating propensity of the Kea." Trans. N.Z.I. vol. XXXIX, pp. 71–89. 1906.

BEST, ELSDON. "Food products of Tuhoeland: being notes on the food-supplies of non-agricultural tribes of the Natives of New Zealand." Trans. N.Z.I. vol. XXXV, pp. 45–111. 1902.

—— "Notes on ancient Polynesian migrants or voyagers to New Zealand, and voyage of the 'Aratawhao' canoe to Hawaiki." Trans. N.Z.I. vol. XXXVII, pp. 121–138. 1904.

BIDWILL, JOHN CARNE. "Rambles in New Zealand." London. 1841.

Biology and Horticulture, Divisions of. Bulletins 1–23. Issued by Agricultural Department of N.Z. Wellington. 1904–1907.

BRITTIN, G. "New Coccidae." Trans. N.Z.I. vol. XLVII, pp. 156–160. 1914.

BROAD, LOWTHER, His Honour. "The jubilee History of Nelson, from 1842 to 1892." Nelson. 1892.

BROUN, Captain T. "Notes on the Coleoptera of Auckland, New Zealand." Trans. N.Z.I. vol. VIII, pp. 262–275. 1875.

—— "Notes on Fruit-flies, with a description of a new species (*Dacus xanthodes*)." Trans. N.Z.I. vol. XXXVII, pp. 325–328. 1904.

—— "Notes on the destruction of Kumaras from the Friendly Islands (Tonga), caused by an imported Weevil, etc." Trans. N.Z.I. vol. XL, pp. 262–265. 1907.

BRUNNER, THOS. "Journal of an Expedition to Explore the Middle Island of New Zealand." 1846–1848.

BRYAN, WILLIAM ALANSON. "Natural History of Hawaii." Honolulu. 1915.

BUCHANAN, JOHN. "List of plants found on Miramar Peninsula, Wellington Harbour." Trans. N.Z.I. vol. V, pp. 349–352. 1872.

—— "On the Botany of Kawau Island; Physical features and causes influencing distribution of species." Trans. N.Z.I. vol. IX, pp. 503–527. 1876.

BUCHANAN, JOHN. "On some plants new to New Zealand, and description of a new species." Trans. N.Z.I. vol. XIV, pp. 356–357. 1881.

BUCKLAND, FRANK. "Notes and jottings from Animal Life." (Contains accounts of Salmon-egg collecting for Australia and New Zealand.) London. 1899.

BUICK, T. LINDSAY. "Old Marlborough, or the story of a Province." Palmerston North. 1900.

BULLER, A. P. "Notes on the occurrence of some rare species of Lepidoptera." Trans. N.Z.I. vol. XXXVII, pp. 331–333. 1904.

BULLER, WALTER LOWRY. "On the New Zealand Rat." (With illustrations.) Trans. N.Z.I. vol. III, pp. 1–4. 1870.

—— "A History of the Birds of New Zealand." London. 1873.

—— "A History of the Birds of New Zealand." 2nd edition. 2 vols. London. 1888.

—— "Note on *Mus maorium* (Hutton), with exhibition of specimen." Trans. N.Z.I. vol. XXV, pp. 49, 50. 1892.

—— "Supplement to the 'Birds of New Zealand.'" 2 vols. London. 1905.

BUTLER, A. G. "On two collections of Heterocerous Lepidoptera from New Zealand." Proc. Zool. Soc. London, pp. 379–386. 1877.

BUTLER, EDWARD A. "Our Household Insects." London. 1893.

CAMERON, P. "Notes on a collection of Hymenoptera from Greymouth, New Zealand, with descriptions of new species." Trans. Entom. Soc. of London, p. 202. 1883.

—— "Description of a new species of *Halictus* (Andrenidae) from Christchurch, New Zealand." Trans. N.Z.I. vol. XXXII, pp. 17–19. 1899.

—— "On a collection of Hymenoptera made in the neighbourhood of Wellington by Mr G. V. Hudson, with descriptions of new genera and species." Trans. N.Z.I. vol. XXXIII, pp. 104–120. 1900.

—— "A list of the Hymenoptera of New Zealand." Trans. N.Z.I. vol. XXXV, pp. 290–299. 1902.

Canterbury Philosophical Institute. "Report of a Committee on Native and introduced grasses"; with Appendices. (Trans. N.Z.I. vol. IV, pp. 292–310.) 1872.

—— "The Sub-Antarctic Islands of New Zealand." Edited by Chas. Chilton. Wellington. 1909.

CARRICK, R. (compiler and editor). "Historical Records of New Zealand South, prior to 1840." Dunedin. 1903.

CARSE, H. "On the Flora of the Mongonui County." Trans. N.Z.I. vol. XLIII, pp. 194–224. (List of naturalised plants at pp. 223, 224.) 1910.

—— "Some further additions to the Flora of the Mongonui County." Trans. N.Z.I. vol. XLVIII, pp. 237–243. 1915.

CHAPMAN, F. R. "A Handbook of Laws relating to Acclimatization, Fish, Fisheries, and the Protection of Animals and Birds." (Issued by the Otago Acclimatisation Society.) Dunedin. 1892.

CHEESEMAN, T. F. "On the Botany of the Titirangi District of the Province of Auckland." Trans. N.Z.I. vol. IV, pp. 270–284. 1872.

—— "On the Botany of the Pirongia Mountain." Trans. N.Z. Inst. vol. XII, pp. 317–323. 1879.

CHEESEMAN, T. F. "The naturalized plants of the Auckland Provincial District." Trans. N.Z.I. vol. XV, pp. 268–298. 1882.

—— "On the Flora of the North Cape District." Trans. N.Z.I. vol. XXIX, pp. 334–385. 1896.

—— "Notice of the establishment of *Vallisneria spiralis* in Lake Takapuna, together with some remarks on its life-history." Trans. N.Z.I. vol. XXIX, pp. 386–390. 1896.

—— "On the occurrence of *Ottelia* in New Zealand." Trans. N.Z.I. vol. XXXI, pp. 350–351. 1898.

—— "Notes on the cultivated food-plants of the Polynesians, with special reference to the Ti Pore (*Cordyline terminalis*)." Trans. N.Z.I. vol. XXXIII, pp. 306–311. 1900.

—— "Manual of the New Zealand Flora." Wellington. 1906.

—— "Contributions to a fuller knowledge of the Flora of New Zealand." No. 1. Trans. N.Z.I. vol. XXXIX, pp. 439–450. 1906.

—— "Contributions to a fuller knowledge of the Flora of New Zealand." No. 2. Trans. N.Z.I. vol. XL, pp. 270–285. 1907.

—— "Contributions to a fuller knowledge of the Flora of New Zealand." No. 4. Trans. N.Z.I. vol. XLIII, pp. 178–186. 1910.

—— "Contributions to a fuller knowledge of the Flora of New Zealand." No. 5. Trans. N.Z.I. vol. XLVI, pp. 1–9. 1913.

—— "Contributions to a fuller knowledge of the Flora of New Zealand." No. 6. Trans. N.Z.I. vol. LI, pp. 85–92. 1919.

CHILTON, CHAS. "Notes on the Distribution of some species of Terrestrial Isopoda introduced into Australasia." Ann. and Mag. of Natural History, series 7, vol. XVI, pp. 428–432. 1883.

CLARKE, F. E. "Notice of a tadpole found in a drain in Hokitika." Trans. N.Z.I. vol. XI, p. 573. 1878.

CLARKE, WILLIAM EAGLE. "Studies in Bird Migration." 2 vols. London. 1912.

CLELAND, J. BURTON. "The Food of Australian Birds." (Dept. of Agriculture, N.S.W.; Science Bulletin, No. 15.) Sydney. 1918.

—— "Presidential Address to the Roy. Soc. of N.S. Wales." May, 1918. Sydney. 1918.

COCHRAN, WILLIAM. "Tea and Silk Farming in New Zealand." (From Transactions of the Highland and Agricultural Society.) Edinburgh. 1882.

COCKAYNE, A. H. "Forest Diseases and their relation to afforestation." Report on State afforestation in New Zealand for 1910–1911, pp. 25–29. 1910–1911.

COCKAYNE, LEONARD. "A short account of the plant-covering of Chatham Island." Trans. N.Z.I. vol. XXXIV, pp. 243–325. 1901.

—— "A botanical excursion during midwinter to the Southern Islands of New Zealand." Trans. N.Z.I. vol. XXXVI, pp. 225–333. 1903.

—— "Some observations on the coastal vegetation of the South Island of New Zealand." Part I. Trans. N.Z.I. vol. XXXIX, pp. 312–359. 1906.

—— "Report on a botanical survey of Kapiti Island." Wellington. 1907.

—— "Some hitherto unrecorded plant-habitats." Part IV. Trans. N.Z.I. vol. XLI, pp. 399–403. 1908.

—— "Report on a botanical survey of the Tongariro National Park." Wellington. 1908.

—— "Report on a botanical survey of the Waipoua Kauri Forest." Wellington. 1908.

COCKAYNE, LEONARD. "Report on a botanical survey of Stewart Island." Wellington. 1909.

—— "Some hitherto-unrecorded plant habitats." Part v. Trans. N.Z.I. vol. XLII, pp. 311–319. 1909.

—— "New Zealand Plants and their Story." Wellington. 1910.

—— "Observations concerning Evolution, derived from ecological studies in New Zealand." Trans. N.Z.I. vol. XLIV, pp. 1–50. 1911.

—— "Report on the Dune-Areas of New Zealand." Wellington (Department of Lands), 1911. (Contains a list of naturalised plants at p. 42.)

COCKAYNE, L. and FOWERAKER, C. E. "Plant Associations at Cass." Trans. N.Z.I. vol. XLVIII, pp. 166–186. 1915.

COCKERELL, T. D. A. "On the Geographical Distribution of Slugs." Proc. Zool. Society of London, April 7th, 1891, pp. 214–226.

COLENSO, W. "Notes on the metamorphosis and development of one of our large Butterflies (*Danais berenice*), or a closely-allied species." Trans. N.Z.I. vol. x, pp. 276–280. 1877.

—— "On some new and undescribed New Zealand Ferns." Trans. N.Z.I. vol. XIII, pp. 376–384. 1880.

—— "An account of visits to, and crossings over, the Ruahine Mountain Range, performed in 1845–47." Napier. 1884.

—— "A brief list of some British plants (weeds) lately noticed, apparently of recent introduction into this part of the Colony; with a few notes thereon." Trans. N.Z.I. vol. XVIII, p. 288. 1885.

—— "A description of a species of *Orobanche* (supposed to be new) parasitical on a plant of Hydrocotyle." Trans. N.Z.I. vol. XXI, pp. 41–43. 1888.

—— "Observations on Mr T. White's paper 'On the Native Dog of New Zealand: Trans. N.Z.I. vol. XXIV, Art. 57.'" Trans. N.Z.I. vol. XXV, pp. 495–503. 1892.

—— "Notes, remarks and reminiscences of two peculiar introduced and naturalised South American plants." Trans. N.Z.I. vol. XXVI, pp. 323–332. 1893.

—— "An account of the finding of two Australian plants, hitherto unnoticed, here in New Zealand." Trans. N.Z.I. vol. XXVII, pp. 401, 402. 1894.

COOK, Captain JAMES. "A voyage towards the South Pole and round the world, performed in H.M. Ships the 'Resolution' and 'Adventure,' in the years 1772, 1773, 1774 and 1775." Written by James Cook, Commander of the 'Resolution'—in which is included Captain Furneaux's narrative of his proceedings in the 'Adventure' during the separation of the ships. (2nd Voyage.) 2 vols. London. W. Strahan. 1777.

COOK, Captain JAMES and KING, Captain JAMES. "A voyage to the Pacific Ocean undertaken by the command of His Majesty, for making Discoveries in the Northern Hemisphere...performed under the direction of Captains Cook, Clerke and Gore, in H.M. Ships the 'Resolution' and 'Discovery,' in the years 1776, 1777, 1778, 1779 and 1780. Vols. I and II written by Captain James Cook, F.R.S. Vol. III written by Captain James King, LL.D. and F.R.S. London. W. & A. Strahan. 1784.

CRAWFORD, JAMES COUTTS. "Recollections of Travel in New Zealand and Australia." London. 1880.

CROZET. "Voyage to Tasmania, New Zealand, the Ladrone Islands, and the Philippines in the years 1771–1772." Translated by H. Ling Koth. London. 1891.

CRUISE, RICHARD A., Major in the 84th Regt. Foot. "Journal of a ten months' residence in New Zealand." London. 1823.

DALMAS, le Comte de. "Araignées de Nouvelle-Zélande." Annales de la Société Entomologique de France, vol. LXXXVI, pp. 317–430. 1917.

DARWIN, CHARLES. "Journal of Researches into the natural history and geology of the countries visited during the voyage of H.M.S. 'Beagle' round the world." (2nd edition.) London. John Murray. 1845.

—— "On the agency of Bees in the Fertilisation of Papilionaceous Flowers, and on the Crossing of Kidney Beans." Ann. and Mag. Nat. Hist. Ser. III, vol. II. London. December, 1858.

—— "The variation of Animals and Plants under domestication." 2 vols. London. John Murray. 1868.

—— "The Origin of Species." 6th edition. London. 1875.

"Darwin, Charles, Life and Letters of." Edited by Francis Darwin. 3 vols. London. 1888.

DENDY, ARTHUR. "Notes on Land Planarians: Part I." Trans. N.Z.I. vol. XXVII, pp. 177–189. 1894.

DIEFFENBACH, ERNEST, M.D., late Naturalist to the New Zealand Company. "Travels in New Zealand." 2 vols. London. John Murray. 1843.

DRUMMOND, JAMES. "Dates on which introduced birds have been liberated, or have appeared, in different districts of New Zealand." Trans. N.Z.I. vol. XXXIX, pp. 503–508. 1906.

—— "On introduced birds." Trans. N.Z.I. vol. XXXIX, pp. 227–252. 1906. (A most valuable contribution to the subject.)

—— "Birds on Kapiti Island." Trans. N.Z.I. vol. XLI, pp. 30–32. 1908.

DRUMMOND, JAMES and HUTTON, Captain F. W. "The animals of New Zealand." Christchurch. 1904.

EARLE, AUGUSTUS. "A narrative of a nine months' residence in New Zealand in 1827." London. 1832. (Earle was draughtsman to H.M. Surveying Ship 'Beagle.')

FANTHAM, H. B., STEPHENS, J. W. W. and THEOBALD, F. V. "The Animal Parasites of Man." London. 1916.

Farmers, Leaflets for. Issued by Agricultural Department of New Zealand. Nos. 1–80. Wellington. 1893–1909.

FEREDAY, R. W. "On the direct injuries to vegetation in New Zealand by various insects, especially with reference to larvae of moths and butterflies feeding upon the Field Crops; and the expediency of introducing Insectivorous Birds as a remedy." Trans. N.Z.I. vol. V, p. 289. 1872.

—— "Observations on the occurrence of a Butterfly, new to New Zealand, of the genus *Danais*." Trans. N.Z.I. vol. VI, pp. 183–186. 1874.

—— "Description of a species of *Catocala*, new to science." Trans. N.Z.I. vol. IX, pp. 457–459. 1876.

FERGUSON, ALEXANDER. "On the cause of the disappearance of young trout from our streams." Trans. N.Z.I. vol. XXI, pp. 235–237. 1888. (Mr Ferguson finds Minnows (*Galaxias attenuatus*) are fond of very young trout fry.)

FIELD, H. C. "The Ferns of New Zealand" (illustrated). Wanganui. 1890.

Forestry, Report of Royal Commission on, pp. i–lxxviii and 1–87. Wellington. 1913.

FORSTER, GEORGE, F.R.S. "A voyage round the world in H.B.M.S.

'Resolution' during the years 1772, 1773, 1774 and 1775." 2 vols. London. 1777.

Gardeners and Fruit-growers, Leaflets for. (Mostly by T. W. Kirk.) Issued by Agricultural Department of New Zealand. Nos. 1–53. Wellington. 1894–1909.

GILLIES, ROBERT. "Notes on some changes in the Fauna of Otago." Trans. N.Z.I. vol. x, pp. 306–324. 1877.

GILRUTH, J. A. "Bubonic Plague: Investigations in New Zealand (April–July, 1900)." Wellington. 1902.

—— "The Toxic effects of the ingestion of *Senecio Jacobæa*." Aust. Assoc. Adv. of Science, 1904. Dunedin meeting, pp. 499–510. 1904.

GOVETT, R. H. "A bird-killing tree." Trans. N.Z.I. vol. XVI, pp. 364–366. 1883.

GUPPY, H. B. "Observations of a Naturalist in the Pacific between 1896 and 1899." 2 vols. London. 1906.

GUTHRIE-SMITH, H. "The Grasses of Tutira." Trans. N.Z.I. vol. XL, pp. 506–519. 1907.

—— "Birds of the Water, Wood and Waste." Christchurch. 1910.

—— "Mutton Birds and other birds." Christchurch. 1914.

HAMILTON, AUGUSTUS. "On an account of a massacre at the entrance of Dunedin Harbour in the Year 1817." Trans. N.Z.I. vol. XXVIII, pp. 141–147. 1895.

HARDCASTLE, E. "The Deer of New Zealand." (Issued by New Zealand Tourist Department.) Wellington. No date, but later than 1905.

"Hawke's Bay Acclimatisation Society." Minute Books from 1901–1915 (incl.) and Annual Reports to 1919.

HAY, JAMES. "Reminiscences of Earliest Canterbury and its settlers." Christchurch. 1915.

HEATH, NEIL. "Effects of Cold on Fishes." Trans. N.Z.I. vol. XVI, pp. 275–278. 1883. (Describes an experiment in which a gold-fish was frozen into a block of ice, and revived completely on thawing.)

HECTOR, JAMES. "On the Remains of a Dog found by Captain Rowan near White Cliffs, Taranaki." Trans. N.Z.I. vol. IX, pp. 243–248. 1876.

—— "On the occurrence of Salmon Trout in Nelson Harbour." Trans. N.Z.I. vol. XIV, pp. 211–213. 1881.

HELMS, R. "A Maori rat at Greymouth." N.Z. Journ. of Science, vol. I, p. 466. Dunedin. 1883.

HENRY, R. "Red Cats and Disease." Trans. N.Z.I. vol. XXXI, pp. 680–683. 1899.

HENSLOW, Rev. GEORGE. "The Origin of Floral Structures, through Insect and other agencies." London. 1888.

HILGENDORF, F. W. "Life History of *Plutella cruciferarum*, Zeller." Trans. N.Z.I. vol. XXXIII, pp. 145–146. 1900.

—— "Notes from the Canterbury College Mountain Biological Station, Cass. No. 6. Insect Life." Trans. N.Z.I. vol. L, pp. 133–144. 1918.

HOCHSTETTER, FERDINAND VON. "New Zealand. Its physical geography, geology and natural history." Translated from the German original (published in 1863) by Edward Sauter, A.M. Stuttgart. 1867.

HOCKEN, THOMAS MORLAND. "Contributions to the Early History of New Zealand." London. 1898. (Contains the diaries of Messrs Tuckett and D. Monro, made in 1844.)

—— "The Early History of New Zealand." Wellington. 1914.

HOOKER, JOSEPH DALTON. "The Botany of the Antarctic Voyage of

H.M. Discovery Ships 'Erebus' and 'Terror' in the years 1839–43. Flora Novæ-Zelandiæ. Part I. Flowering Plants." London. Lovell Reeve. 1853. (Contains no reference to introduced plants.)

HOOKER, JOSEPH DALTON. "Handbook of the New Zealand Flora." London. 1867.

—— "The Student's Flora of the British Islands." 2nd edition. London. Macmillan & Co. 1878.

HOPKINS, ISAAC. "Forty-two years of Bee-keeping in New Zealand: 1874–1916. Auckland. 1916.

HOWARD, L. O. "The House Fly: Disease Carrier." London. John Murray. 1912.

HOWES, GEORGE. "Fruit Destruction by Small Birds in Central Otago." Trans. N.Z.I. vol. XXXVIII, p. 304. 1905.

—— "Note on the occurrence of two rare and two introduced moths." Trans. N.Z.I. vol. XXXVIII, p. 509. 1905.

HUDSON, G. V. "*Eristalis tenax* and *Musca vomitoria* in New Zealand." Trans. N.Z.I. vol. XXII, pp. 187, 188. 1889.

—— "A few words on the Codlin-moths *Carpocapsa pomonella*, L., and *Cacœcia excessana*, Walk." Trans. N.Z.I. vol. XXIII, pp. 56–58. 1890. (The New Zealand Tortrix (*Cacœcia excessana*) attacks apples in the same way as the introduced *Carpocapsa pomonella*, and suggestions are given for dealing with the pests.)

—— "An elementary Manual of New Zealand Entomology." London. 1892.

—— "New Zealand Moths and Butterflies." London. 1898.

—— "New Zealand Neuroptera." London. 1904.

—— "Notes on Insect Swarms on mountain top in New Zealand." Trans. N.Z.I. vol. XXXVIII, pp. 335–337. Wellington. 1905.

—— "Recent observations on New Zealand Macrolepidoptera, including descriptions of new species." Trans. N.Z.I. vol. XL, pp. 104–107. 1907.

HUDSON, JAMES. "Notes on Blights." Trans. N.Z.I. vol. XXIII, p. 111. 1890.

HUTCHINSON, F., junior. "Scinde Island, from a Naturalist's point of view." Trans. N.Z.I. vol. XXXIII, pp. 213–221. 1900.

HUTTON, F. W. "On the Introduction of the Pheasant into the province of Auckland." Trans. N.Z.I. vol. II, p. 80. 1869.

—— "Catalogue of the Birds of New Zealand." Pp. iv–x, 1–85. Wellington. 1871.

—— "Description of a specimen of *Mus rattus*, L. in the Colonial Museum." Trans. N.Z.I. vol. IV, pp. 183, 184. 1871.

—— "Note on the Maori Rat." Trans. N.Z.I. vol. IX, p. 348. 1876. (Gives description of skulls and bones found at Shag Point, Otago.)

—— "Second Note on the Maori Rat." Trans. N.Z.I. vol. X, p. 288. 1877.

—— "Note on the Black Rat (*Mus rattus*, L.)" accompanying paper by Taylor White. Trans. N.Z.I. vol. XI, p. 344. 1878.

—— "Additions to the list of New Zealand Worms." Trans. N.Z.I. vol. XII, pp. 277, 278. 1879.

—— "Catalogues of the New Zealand Diptera, Orthoptera and Hymenoptera." Wellington. 1881.

—— "Note on the Rat that invaded Picton in March, 1884." Trans. N.Z.I. vol. XX, p. 43. 1887.

—— "The Grasshoppers and Locusts of New Zealand and the Kermadec Islands." Trans. N.Z.I. vol. XXX, pp. 135–150. 1897.

HUTTON, F. W. "Note on the ancient Maori Dog." Trans. N.Z.I. vol. XXX, pp. 151–155. 1897. (Gives measurement of skulls.)

—— "Synopsis of the Hemiptera of New Zealand, which have been described previous to 1896." Trans. N.Z.I. vol. XXX, pp. 167–187. 1897.

—— "Synopsis of the Diptera bracycera of New Zealand." Trans. N.Z.I. vol. XXXIII, pp. 1–95. 1900.

—— "Additions to the Diptera Fauna of New Zealand." Trans. N.Z.I. vol. XXXIV, pp. 179–196. 1901.

HUTTON, Captain F. W., F.R.S. and DRUMMOND, JAMES. "The Animals of New Zealand." Christchurch. 1904.

HUTTON, F. W. (Editor). "Index Faunæ Novæ Zealandiæ." London. Dulau & Co. 1904.

JAMESON, R. G. "New Zealand, South Australia and New South Wales; a record of recent travel in these colonies (1838–40)." London. 1842.

KERNER VON MARILAUN, ANTON. "Flowers and their unbidden guests." (Translated by Dr W. Ogle.) London. 1878.

—— "The Natural History of Plants." Translated by F. W. Oliver. 2 vols. London. 1902.

KEW, HARRY WALLIS. "The Dispersal of Shells." London. 1893.

KINGSLEY, R. I. "On the occurrence of the Black Vine-Weevil (*Otiorhynchus sulcatus*) in Nelson." Trans. N.Z.I. vol. XXII, pp. 338–340. 1889.

—— "On the occurrence of *Danais plexippus* and *Sphinx convolvuli* (?) in Nelson." Trans. N.Z.I. vol. XXIII, pp. 192–194. 1890.

—— "Zoological Notes: (1) Arboreal Nests of Bush rats (*Mus maorium*)." Trans. N.Z.I. vol. XXVII, pp. 238, 239. 1894.

—— "On the presence of some noxious weeds in Nelson district." Trans. N.Z.I. vol. XXVII, p. 407. 1894. (The plants are *Carduus arvensis* and *Melilotus arvensis*.)

KIRK, H. B. "On the occurrence of Starch and Glucose in Timber." Trans. N.Z.I. vol. XXXVII, pp. 379, 380. 1904.

KIRK, H. B. and BENDALL, W. E. "Report on Kapiti Island as a Plant and Animal Sanctuary." Trans. N.Z.I. vol. LI, pp. 468, 469. 1919.

KIRK, THOMAS, F.L.S. "On the Botany of the Great Barrier Island." Trans. N.Z.I. vol. I, pp. 144–157. 1868. (Contains lists of naturalised plants at the Bay of Islands, and on the Great and Little Barrier Islands.)

—— "On the occurrence of *Orobanche*, a genus new to the Flora of New Zealand." Trans. N.Z.I. vol. II, p. 106. 1869.

—— "Notes on the Botany of certain places in the Waikato district, April and May, 1870." Trans. N.Z.I. vol. III, pp. 142–147. 1870.

—— "On the Flora of the Isthmus of Auckland, and the Takapuna District." Trans. N.Z.I. vol. III, pp. 148–161. 1870.

—— "Notes on introduced Grasses in the Province of Auckland." Trans. N.Z.I. vol. IV, pp. 295–299. 1871.

—— "Notes on the Naturalized Plants of the Chatham Islands." Trans. N.Z.I. vol. V, pp. 320–322. 1872.

—— "Notes on the Flora of the Lake District of the North Island." Trans. N.Z.I. vol. V, pp. 322–345. 1872.

—— "Contributions to the Botany of Otago." Trans. N.Z.I. vol. X, pp. 406–417. 1877. (List of naturalized plants at pp. 414–417.)

—— "On the naturalized plants of Port Nicholson and the adjacent district." Trans. N.Z.I. vol. X, pp. 362–378. 1877.

KIRK, THOMAS, F.L.S. "Notice of the occurrence of *Lagenophora emphysopus* and other recorded plants in New Zealand." Trans. N.Z.I. vol. XII, pp. 397–399. 1879.

—— "New introduced plants." N.Z. Journ. of Science, vol. II, pp. 29–30. 1884.

—— "On the naturalized Dodders and Broom-rapes of New Zealand." Trans. N.Z.I. vol. XX, pp. 182–185. 1887.

—— "The Forest Flora of New Zealand." Wellington. 1889.

—— "On New Zealand Sowthistles." Trans. N.Z.I. vol. XXVI, pp. 263–266. 1893.

—— "Displacement of Species in New Zealand." Trans. N.Z.I. vol. XXVIII, pp. 1–27. 1895.

—— "Notice of the occurrence of an undescribed Palm Lily on the Auckland Peninsula." Trans. N.Z.I. vol. XXVIII, pp. 508, 509. 1895.

—— "The Students' Flora of New Zealand." Wellington. 1899. (Describes all the naturalised plants from Ranunculaceæ to Compositæ, inclusive.)

KIRK, T. W. "Note on the occurrence of English butterflies in New Zealand." N.Z. Journ. of Science, vol. II, p. 169. 1884.

—— "On the occurrence of the English Scaly Lizard (*Zootoca vivipara*) in New Zealand." Trans. N.Z.I. vol. XIX, p. 67. 1886.

—— "The Mole-cricket (*Gryllotalpa vulgaris*) in New Zealand." Trans. N.Z.I. vol. XXI, pp. 233–235. 1888.

—— "Note on the breeding habits of the European Sparrow (*Passer domesticus*) in New Zealand." Trans. N.Z.I. vol. XXIII, pp. 108–110. 1890. (1891. Abstract of, and discussion on, N.Z. Journ. of Science, 1891, p. 9.)

—— "On the occurrence of *Xanthium strumarium*, Linn. in New Zealand." Trans. N.Z.I. vol. XXVI, pp. 310–313. 1893.

KIRKCALDY, G. W. "Notes on the Hemiptera of the Index Faunae Novae-Zealandiae." Trans. N.Z.I. vol. XXXVIII, pp. 61, 62. 1905.

—— "A list of the Hemiptera of the Maorian Sub-region." Trans. N.Z.I. vol. XLI, pp. 22–29. 1908.

KNOX, F. J. "Observations on the Kiore or indigenous rat of New Zealand." Trans. N.Z.I. vol. IV, p. 362. 1872.

KNUTH, DR PAUL. "Handbook of Flower Pollination." (Translated by J. R. Ainsworth Davies.) 3 vols. Oxford. 1906.

LAING, R. M. and BLACKWELL, E. W. "Plants of New Zealand." Christchurch. 1906.

LIFFITON, EDW. N. "Notes on the decrease of pheasants on the West Coast of the North Island." Trans. N.Z.I. vol. XXI, pp. 225, 226. 1888.

LINDSAY, W. LAUDER. "Contributions to New Zealand Botany." Edinburgh. Williams & Norgate. 1868. (Contains many references to introduced plants and weeds, but no definite lists or names of species.)

LONG, HAROLD C. "Common weeds of the farm and garden." London. 1910.

LONGMAN, HEBER A. "List of Australasian and Austro-Pacific Muridæ." Brisbane. 1916.

—— "Notes on classification of Common Rodents with list of Australian species." Melbourne. 1916.

MACKAY, T. "On the American Blight on Apple Trees." (Trans. N.Z.I. vol. IV, pp. 429–432.) Wellington. 1871.

MCNAB, ROBERT. "Murihiku—a history of the South Island of New

Zealand and the islands adjacent and lying to the South, from 1642 to 1835." Wellington. Whitcombe & Tombs. 1909.

McNab, Robert. "Murihiku, some old-time events." Gore. 1904.

—— "Murihiku, some old-time events: being a series of 25 articles on the early history of the extreme southern portion of New Zealand." Gore. 1905. (The first 44 pages is a reprint of the 1904 edition.)

—— "Murihiku and the Southern Islands: a History of the West Coast Sounds, Foveaux Strait, Stewart Island, the Snares, Bounty, Antipodes, Campbell and Macquarie Islands from 1770 to 1829." Invercargill. 1907. (Contains *inter alia* "Journal of Lieut. Menzies, the Botanist of Vancouver's Expedition, while at Dusky in 1791.")

—— "Murihiku, a history of the South Island of New Zealand, and the isles adjacent and lying to the South, from 1642 to 1835." Wellington. 1909.

—— "The Old Whaling Days, a history of Southern New Zealand from 1830 to 1840." Christchurch. 1913.

—— "From Tasman to Marsden." Dunedin. 1914.

Marriner, George R. "On the presence of another Australian Frog in New Zealand." Trans. N.Z.I. vol. xxxix, pp. 144–149. 1906.

—— "Notes on the natural history of the Kea, with special reference to its reputed sheep-eating propensities." Trans. N.Z.I. vol. xxxix, pp. 271–305. 1906.

Marsden, Rev. S. "MS. Copy of a letter from the Rev. Samuel Marsden to the Governor of New South Wales, dated Paramatta, 30th May, 1815, being Report of his first visit to New Zealand." 1815.

Marshall, P. "New Zealand Diptera." Trans. N.Z.I. vol. xxviii, pp. 216–311. 1895.

Martin, H. B. "Objections to the introduction of Beasts of Prey to destroy the Rabbit." Trans. N.Z.I. vol. xvii, pp. 179–182. 1884.

Maskell, W. M. "An Aphidian insect infesting Pine Trees." N.Z. Journ. of Science, vol. ii, pp. 291, 292, also Trans. N.Z.I. vol. xvii, pp. 13–19. 1884.

—— "The Scale Insects (Coccididæ) of New Zealand." Wellington. 1887.

—— "Further Coccid Notes: with descriptions of new species from New Zealand, Australia and Fiji." Trans. N.Z.I. vol. xxiii, pp. 1–36. 1890.

—— "Further Coccid Notes: with descriptions of new species, and remarks on Coccids from New Zealand, Australia and elsewhere." Trans. N.Z.I. vol. xxiv, pp. 1–64. 1891.

—— "Further Coccid Notes: with descriptions of several new species, and discussion of various points of interest." Trans. N.Z.I. vol. xxvi, pp. 65–105. 1893.

—— "Further Coccid Notes: with descriptions of new species from New Zealand, Australia, Sandwich Islands, and elsewhere, and remarks upon many species already reported." Trans. N.Z.I. vol. xxvii, pp. 36–75. 1894.

—— "Synaptical list of Coccidae reported from Australasia and the Pacific Islands up to December, 1894." Trans. N.Z.I. vol. xxvii, pp. 1–35. 1894.

—— "Further Coccid Notes: with descriptions of new species, and discussion of questions of interest." Trans. N.Z.I. vol. xxviii, pp. 380–411. 1895.

MASKELL, W. M. "Further Coccid Notes: with descriptions of new species, and discussion of questions of interest." Trans. N.Z.I. vol. XXIX, pp. 293–331. 1896.

—— "Further Coccid Notes: with description of new species, and discussion of points of interest." Trans. N.Z.I. vol. XXX, pp. 219–252. 1897.

MEADE, The Hon. HERBERT, R.N. "A ride through the disturbed districts of New Zealand, &c." London. 1871.

MEESON, JOHN. "The plague of Rats in Nelson and Marlborough." Trans. N.Z.I. vol. XVII, p. 199. 1884.

MESTAYER, Miss M. K. "The occasional occurrence of Australian and South Sea Island Molluscs in New Zealand." N.Z. Journal of Science and Technology, vol. I, pp. 102–104. 1918.

MEYRICK, E. "On New Zealand Micro-Lepidoptera." Trans. N.Z.I. vol. XVI, pp. 2–49. 1883.

—— "Descriptions of New Zealand Micro-Lepidoptera." Trans. N.Z.I. vol. XVIII, pp. 162–183. 1885.

—— "Notes on New Zealand Pyralidina." Trans. N.Z.I. vol. XX, pp. 62–73. 1887.

—— "Revision of New Zealand Tineina." Trans. N.Z.I. vol. XLVII, pp. 205–244. 1914.

—— "Descriptions of New Zealand Micro-Lepidoptera." Trans. N.Z.I. vol. XXI, pp. 154–188. 1888.

MIERS, EDWARD J. "Catalogue of the Stalk- and Sessile-eyed Crustacea of New Zealand." London. 1876.

MILLER, DAVID. "Bionomic observations on certain New Zealand Diptera." Trans. N.Z.I. vol. XLII, pp. 226–235. 1910.

MILLER, DAVID and MORRIS N. WATT. "Contributions to the study of New Zealand Entomology, from an economical and biological standpoint." Trans. N.Z.I. vol. XLVII, pp. 274–284. 1914.

MONIEZ, R. "Sur quelques Arthropodes trouvés dans des fourmilières." La Revue Biologique du Nord de la France, vol. VI, pp. 201–215. 1894.

MOTTRAM, J. C. "Some observations on the Feeding-habits of Fish and Birds, with special reference to Warning, Coloration and Mimicry." Journal Linn. Soc. vol. XXXIV, pp. 47–60. 1918.

MÜLLER, HERMANN. "The Fertilisation of Flowers." (Translated by D'Arcy W. Thompson.) London. 1883.

MURISON, W. D. "Appendix to paper 'on the New Zealand Wild Dog' by R. Gillies." Trans. N.Z.I. vol. X, pp. 306–324. 1877.

MUSGRAVE, Captain JOHN. "Castaway on the Auckland Islands." Narrative of the wreck of the schooner 'Grafton'; from the Journals of." Edited by John L. Shillinglaw. London. 1866.

MUSSON, CHARLES T. "On the naturalised forms of land and fresh-water mollusca in Australasia." Proc. Linn. Soc. N.S. Wales, vol. V (series 2nd), pp. 883–896. 1890.

Nelson Acclimatisation Society. Annual Reports from 1911 to 1918.

NEUMANN, L. G. "A treatise on the parasites and parasitic diseases of the domesticated animals." London. 1905.

NICHOLAS, JOHN LIDDIARD. "Narrative of a voyage to New Zealand, performed in the years 1814 and 1815, in company with the Rev. Samuel Marsden." London. 1817.

NICOLS, ARTHUR. "The Acclimatisation of the Salmonidæ at the Antipodes, its History and Results." London. 1882.

North Canterbury Acclimatisation Society. Annual Reports from the commencement in 1864 to 1919 (inclusive).

Otago Acclimatisation Society. Annual Reports from the commencement in 1865 to 1919 (inclusive).

Pascoe, Francis P. "Descriptions of new genera and species of New Zealand Coleoptera." (Ann. and Mag. Nat. Hist. Sept. 1875 and Jan. 1876.) Trans. N.Z.I. vol. IX, pp. 402–427. 1876.

Petit-Thouars, Abel du. "Voyage autour du Monde sur la Frégate 'La Venus' pendant les Années 1836–1839, &c." Relation, Tome III. Paris. 1841.

Petrie, D. "Some effects of the Rabbit Pest." N Z. Journ. of Science, vol. I, pp. 412–414. 1883.

—— "On the naturalisation of *Calluna vulgaris*, Salisb., in the Taupo district." Trans. N.Z.I. vol. XLII, p. 199. 1909.

—— "Report on the Grass-denuded lands of Central Otago." N.Z. Dept. of Agriculture, Bulletin No. 23. 1912.

Phillips, Ernest. "Trout in Lakes and Reservoirs." London. 1914.

Phillips, Coleman. "Rabbit disease in the Wairarapa." Trans. N.Z.I. vol. XXI, pp. 429–438. 1888.

—— "Rabbit-disease in the South Wairarapa." Trans. N.Z.I. vol. XXII, pp. 308–325. 1889.

Philpott, Alfred. "Notes on certain introduced birds in Southland." The N.Z. Journal of Science and Technology, vol. I, pp. 328–330. 1918.

Polack, J. S. "New Zealand: being a narrative of travels and adventures during a residence in that country between the years 1831 and 1837." New edition. 2 vols. London, Richard Bentley. 1839.

Poppelwell, D. L. "Plant Covering; Codfish Island and Ragged Island." Trans. N.Z.I. vol. XLIV, pp. 76–85. 1911.

—— "Notes on the plant-covering of the Breaksea Islands, Stewart Island." Trans. N.Z.I. vol. XLVIII, pp. 246–252. 1915.

Potter, A. T. "On the habits of *Dermestes vulpinus*." Trans. N.Z.I. vol. XXXI, pp. 104–105. 1895.

Potts, T. H. "On recent changes in the Fauna of New Zealand." Letter in the "Field," Dec. 12th, 1872. (Reprinted in Christchurch, 1874.)

—— "On the Birds of New Zealand." Trans. N.Z.I. vol. VI, pp. 139–153. 1873.

—— "Out in the Open, a budget of scraps of natural history gathered in New Zealand." Christchurch. 1882.

—— "Some introduced birds in New Zealand." The Zoologist, vol. VIII, pp. 448–450. 1884.

Purdie, Alexander. "*Dermestes* introduced." N.Z. Journ. of Science, vol. II, pp. 166, 167. 1884.

Reischek, A. "Notes on the habits of the Polecat, Ferret, Mongoose, Stoat and Weasel." Trans. N.Z.I. vol. XVIII, p. 110. 1885. (Refers only to their habits in Europe.)

Rutherford, A. J. "Notes on Salmonidæ and their new home in the South Pacific." Trans. N.Z.I. vol. XXXIII, pp. 240–249. 1901.

Rutland, Joshua. "On the habits of the New Zealand Bush-rat (*Mus Maorium*)." Trans. N.Z.I. vol. XXII, pp. 300–307. 1889.

Sandager, F. "Observations on the Mokohinou Islands and the birds which visit them." Trans. N.Z.I. vol. XXII, pp. 286–294. 1889. (List of introduced visitants at p. 294.)

SANDAGER, F. "Note on some sea-trout (Salmon or Salmon-trout?)." Trans. N.Z.I. vol. XXV, p. 254. 1892.

SEEBOHM, HENRY. "Siberia in Europe." London. 1880.

—— "Siberia in Asia." London. 1882.

SHORTLAND, EDWARD. "The Southern Districts of New Zealand." London. 1851.

SLADEN, F. W. L. "The Humble Bee." London. 1912.

SMITH, W. W. "Notes on New Zealand Earthworms." Trans. N.Z.I. vol. XIX, p. 123. 1886.

—— "Notes on *Eristalis tenax* in New Zealand. Entom. Monthly Mag. vol. XXVI, pp. 240–242. 1890.

—— "Further notes on New Zealand Earthworms, with observations on the known aquatic species." Trans. N.Z.I. vol. XXV, pp. 111–146. 1892. (Contains a list of introduced species at p. 117.)

—— "Notes on certain species of New Zealand birds." Ibis, ser. VI, vol. V, pp. 509–521. 1893.

—— "Further notes on New Zealand Earthworms." Trans. N.Z.I. vol. XXVI, pp. 155–175. 1893.

—— "*Musca* (*Calliphora*) *vomitoria* in New Zealand." Entom. Monthly Mag. vol. XXX, pp. 54–57. 1894.

—— "Enemies of Humble Bees in New Zealand." The Entomologist, vol. XXIX, pp. 210, 211. 1896.

—— "Notes on *Phytomyza nigricornis*." The Entomologist, vol. XXX, p. 317. 1897.

—— "On the habits of *Liothula omnivora*, Fereday." Entom. Monthly Mag. vol. XXXIV, pp. 125–127. 1898.

—— "On mites attacking beetles and moths." Trans. N.Z.I. vol. XXXIV, pp. 199–201. 1901.

—— "Plants naturalised in the County of Ashburton." Trans. N.Z.I. vol. XXXVI, pp. 203–225. 1903.

Southland Acclimatisation Society, Annual Reports from 1867–1919.

STURM, F. W. C. "Further notes on *Danais berenice*." Trans. N.Z.I. vol. XI, p. 305. 1878.

SUTER, H. "List of the introduced land and fresh-water Mollusca of New Zealand." Trans. N.Z.I. vol. XXIV, pp. 279–281. 1891.

—— "Manual of the New Zealand Mollusca." Wellington. 1913.

TANCRED, Sir THOMAS. "Notes on the Natural History of the Province of Canterbury, in the Middle Island of New Zealand." Edinburgh New Philosophical Journal, No. 5, vol. III, Jan. 18th. 1856.

Taranaki Acclimatisation Society. Annual Reports (MS.) from the commencement in 1874 to 1915. (Some of the earlier reports were not well kept, and contain little or no information.)

TAYLOR, Rev. RICHARD. "Te Ika A Maui or New Zealand and its inhabitants." London. 1855. (2nd edition, 1870.)

—— "The Past and Present of New Zealand." London. 1868.

TERRY, CHARLES, F.R.S. "New Zealand; its advantages and prospects as a British Colony." London. 1842.

THOMSON, ARTHUR S. "The Story of New Zealand." 2 vols. London. 1859.

THOMSON, G. M. "On some naturalized plants of Otago." Trans. N.Z.I. vol. VI, pp. 446, 447. 1873.

—— "On some of the naturalized plants of Otago." Trans. N.Z.I. vol. VII, pp. 370–375. 1874.

THOMSON, G. M. "The Rabbit Pest." N.Z. Journ. of Science, vol. II, pp. 79, 80. 1884.

—— "Introduced moths in New Zealand." N.Z. Journ. of Science, vol. II, pp. 229, 230. 1884.

—— "Introduced Plants of Otago." N.Z. Journ. of Science, vol. II, pp. 573, 574. 1885.

—— "On some aspects of acclimatisation in New Zealand." Australasian Association for the Advancement of Science, vol. III, p. 194. 1891.

—— "Plant-acclimatisation in New Zealand." Trans. N.Z.I. vol. XXXIII, pp. 306–311. 1900.

—— "A New Zealand Naturalist's Calendar, and Notes by the Wayside." Dunedin. 1909.

TILLYARD, R. J. "Neuropteroid Insects of the Hot Springs Region, New Zealand, in relation to the problem of Trout-food." N.Z. Journal of Science and Technology, vol. III, pp. 271–279. 1921.

TOWNSON, W. "On the vegetation of the Westport District." Trans. N.Z.I. vol. XXXIX, pp. 380–433. 1906.

TRAVERS, W. T. L. "On the changes effected in the natural features of a new country by the introduction of civilized races." Trans. N.Z.I. vol. II, pp. 299–330. 1869.

—— "Notes on the Chatham Islands, extracted from letters from H. H. Travers." Trans. N.Z.I. vol. IV, pp. 63–66. 1871.

—— Presidential Address to the Wellington Philosophical Society, 1st July, 1871. Trans. N.Z.I. vol. IV, pp. 356–362. 1871.

—— "The Bird as the labourer of Man." Trans. N.Z.I. vol. XXXV, pp. 1–11. 1902.

URQUHART, A. T. "Notes on *Epacris microphylla* in New Zealand." Trans. N.Z.I. vol. XIV, pp. 364, 365. 1881.

—— "On the natural spread of the Eucalyptus in the Karaka district." Trans. N.Z.I. vol. XVI, p. 383. 1883.

—— "On the habits of Earthworms in New Zealand." Trans. N.Z.I. vol. XVI, pp. 266–275. 1883.

—— "On new species of Araneidæ." Trans. N.Z.I. vol. XIX, pp. 72–118. 1886.

—— "On the work of Earthworms in New Zealand." Trans. N.Z.I. vol. XIX, p. 119. 1886.

—— "Catalogue of the described species of New Zealand Araneidæ" Trans. N.Z.I. vol. XXIV, pp. 220–230. 1891.

Veterinary Science, Division of. Bulletins 1–15. Issued by Agricultural Department of N.Z. Wellington. 1903–1909.

Waitaki. Waimate Acclimatisation Society: Annual Reports and Minute Books from the commencement in 1877 to the end of 1890.

WAITE, EDGAR R. "On the habits of the Sydney Bush Rat (*Mus arboricola*)." Proc. Zool. Soc. 1897, p. 857.

WAKEFIELD, C. M. "Remarks on the Coleoptera of Canterbury, New Zealand." Trans. N.Z.I. vol. V, pp. 294–304. 1872.

—— "Note on the occurrence of *Dermestes lardarius* and *Phoracantha recurva* in Canterbury, New Zealand." Trans. N.Z.I. vol. VI, p. 153. 1873.

—— "Notes upon certain recently-described new genera and species of Coleoptera from Canterbury, New Zealand." Trans. N.Z.I. vol. VI, pp. 155–157. 1873.

WAKEFIELD, E. JERNINGHAM. "Adventure in New Zealand from 1839 to 1844." Christchurch. 1908.

WALLACE, ALFRED RUSSELL. "Island Life." 3rd edition. London. Macmillan & Co. 1902.
—— "Darwinism." 3rd edition. London. 1905.
WALSH, Rev. P. "The effect of Deer on the New Zealand Bush; a plea for the protection of our forest reserves." Trans. N.Z.I. vol. XXV, pp. 435–438. 1892.
—— "On the occurrence of *Cordyline terminalis* in New Zealand." Trans. N.Z.I. vol. XXXIII, pp. 301–306. 1900.
—— "The cultivation and treatment of the Kumara by the primitive Maories." Trans. N.Z.I. vol. XXXV, pp. 12–24. 1902.
Wellington Acclimatisation Society. Annual Reports, from commencement in 1884 to 1915.
WELLS, B. "The History of Taranaki." New Plymouth. 1878.
Westland Acclimatisation Society. Annual Reports and Minute Books from the commencement in 1887 to 1914 (inclusive).
WHITE, JOHN. "The Ancient History of the Maori." 6 vols. Wellington. 1888.
WHITE, TAYLOR. "Note accompanying specimens of the Black Rat (*Mus rattus*, L.)." Trans. N.Z.I. vol. XI, pp. 343, 344. 1878.
—— "Description of hybrid Ducks, bred from Common Duck (*A. boschus*) ♀ and Grey Duck (*A. superciliosus*) ♂." Trans. N.Z.I. vol. XVIII, p. 184. 1885.
—— "Notes on coloured sheep." Trans. N.Z.I. vol. XXI, pp. 402–406. 1888.
—— "On the Wild Dogs of New Zealand." Trans. N.Z.I. vol. XXII, pp. 327–330. 1889.
—— "Further notes on coloured sheep." Trans. N.Z.I. vol. XXIII, pp. 207–216. 1890.
—— "On Rats and Mice." Trans. N.Z.I. vol. XXIII, pp. 194–201. 1890.
—— "On Rabbits, Weasels and Sparrows." Trans. N.Z.I. vol. XXIII, pp. 201–207. 1890.
—— "On the native dog of New Zealand." Trans. N.Z.I. vol. XXIV, pp. 540–557. 1891. (Contains many letters and memoranda *re* wild dogs.)
—— "Te Kuri Maori (the Dog of New Zealand); a reply to the Rev. W. Colenso." Trans. N.Z.I. vol. XXVI, pp. 585–600. 1893.
—— "Remarks on the rats of New Zealand." Trans. N.Z.I. vol. XXVII, pp. 240–261. 1894.
—— "On rats, and their nesting in small branches of trees." Trans. N.Z.I. vol. XXX, pp. 303–309. 1897.
—— "Breeding Black Sheep; a study in colour." Trans. N.Z.I. vol. XXXIII, pp. 191–199. 1900.
—— "On Hybridism." Trans. N.Z.I. vol. XXXIII, pp. 199–206. 1900.
WOHLERS, J. F. H. "Memories of the life of; Missionary at Ruapuke. (Translated from the German by John Houghton.) Dunedin. 1895.
YATE, Rev. WILLIAM. "An account of New Zealand." London. 1835.

INDEX OF AUTHORITIES

Records and notes of animals and plants by Messrs Armstrong, Cheeseman, Hudson, T. Kirk, Miller, W. W. Smith and others, occur throughout Chapters VII to XII inclusive; they are too numerous to index.

INDEX OF ANIMALS AND PLANTS

The names of species of animals are indicated by †, those of species of plants by ‡

PRINTED IN ENGLAND BY J. B. PEACE, M.A.
AT THE CAMBRIDGE UNIVERSITY PRESS